Graduate Texts in Physics

Graduate Texts in Physics

Graduate Texts in Physics publishes core learning/teaching material for graduate- and advanced-level undergraduate courses on topics of current and emerging fields within physics, both pure and applied. These textbooks serve students at the MS- or PhD-level and their instructors as comprehensive sources of principles, definitions, derivations, experiments and applications (as relevant) for their mastery and teaching, respectively. International in scope and relevance, the textbooks correspond to course syllabi sufficiently to serve as required reading. Their didactic style, comprehensiveness and coverage of fundamental material also make them suitable as introductions or references for scientists entering, or requiring timely knowledge of, a research field.

More information about this series at http://www.springer.com/series/8431

Mildred Dresselhaus · Gene Dresselhaus
Stephen B. Cronin · Antonio Gomes Souza Filho

Solid State Properties

From Bulk to Nano

Mildred Dresselhaus
Department of Electrical Engineering and Computer Science and Department of Physics
Massachusetts Institute of Technology
Cambridge, MA
USA

Gene Dresselhaus
Francis Bitter Magnet Laboratory
Massachusetts Institute of Technology
Cambridge, CA
USA

Stephen B. Cronin
University Park Campus
University of Southern California
Los Angeles, CA
USA

Antonio Gomes Souza Filho
Departamento de Física
Universidade Federal do Ceará
Fortaleza, Ceará
Brazil

ISSN 1868-4513 ISSN 1868-4521 (electronic)
Graduate Texts in Physics
ISBN 978-3-662-57255-9 ISBN 978-3-662-55922-2 (eBook)
https://doi.org/10.1007/978-3-662-55922-2

Printed on acid-free paper

This Springer imprint is published by Springer Nature
The registered company is Springer-Verlag GmbH, DE
The registered company address is: Heidelberger Platz 3, 14197 Berlin, Germany

Foreword

What is solid state physics about? Take all the fundamentals of physics, including classical and quantum mechanics, electromagnetism, thermodynamics and statistical physics, and put them all together to study a piece of matter. Most physicists worldwide work in this field, largely interdisciplinary, overlapping with chemistry, engineering, biology and medicine. Sensors, solar panels, batteries, light emitting diodes, flat displays, touch screens, computer devices, optical fibers, field emitters, high performance coolers; these are all examples of technologies resulting from solid state physics applications. Cyber-physical systems in the 4.0 industry, internet-of-things (IoT), lab-on-a-chip; all rely on the applications of solid state systems.

Merging together all the fundamentals of physics into a piece of matter sounds beautiful, but it also sounds very complicated! The concepts and mathematical machinery that are needed to understand the main electrical, magnetic, thermal and optical properties of materials have to be in place. It is necessary to draw connections from the quantum mechanics to the physical properties of matter; from continuity relations to transport phenomena; from the Maxwell equations to optical observables. To achieve such an endeavor, a researcher needs a textbook that was tailored towards the best learning experience. And this is exactly what has been delivered by this book.

Solid State Properties—from bulk to nano was built based on the class notes of the MIT Professors Mildred Spiewak Dresselhaus and Gene Dresselhaus, with the contribution of many of their students and post-doctoral fellows, over decades. Professors Mildred and Gene Dresselhaus had the unique view of those who pioneered many discoveries on materials science, and followed the processes from the discovery of the basic concepts up to applications. One example was the study and development of graphite intercalated compounds in the 1970s, a material that is today the basis of chargeable cell phone and electric car batteries, and it keeps improving grateful to nanotechnology. Professors Stephen B. Cronin and Antonio Gomes Souza Filho were former students of Profs. Dresselhaus, and today they hold worldwide recognition in the fields of transport and optics of nanostructures. Because I had the experience of working with Profs. Dresselhaus on a similar project, the publication of their group theory class notes, I understand the important role played by Profs. Cronin and Souza Filho on transforming class notes into a self-contained textbook, working under the high standards of Professors Mildred and Gene Dresselhaus.

When travelling *from bulk to nano*, the book provides a sharp cut on the modern view of solid state science. It shows how the basic concepts that were introduced in the physics of bulk materials over the last century are developing while nanoscience and nanotechnology are taking place. This overview makes the reader capable not only of understanding the present materials science, but most importantly, the reader is able to evaluate the limitations of the modern concepts, to go beyond the well-established knowledge. *Solid State Properties—from bulk to nano* is the textbook for those who want to understand modern scientific papers related to novel materials properties, and for those who want to work and have an impact in the field.

Belo Horizonte
April 2017

Ado Jorio

Preface

We had the great pleasure and honor of having Mildred (Millie) Dresselhaus serve as our Ph.D. advisor. She taught us more than we could ever express in words, and she was patient as we worked to learn the material now contained in this book. Several years ago, when we approached Millie about publishing her class notes, we thought this would be an easy task since her husband Gene Dresselhaus had already formatted those notes beautifully in LaTex. But Millie, as in all her work, had very high standards and insisted that we update it to include new low-dimensional materials that would be of particular interest to present-day students. As a result, revisions went on for two and a half years. We were just completing the final checking of the textbook when Millie passed away on February 20, 2017. Despite her age, this news came as a shock because she had been working actively up until two weeks before her death. In fact, we were actually having trouble keeping up with her. We were deeply saddened to hear the news of her sudden passing. We had lost a great scientist, mentor, advisor, and friend.

This book is based on an introductory solid state physics class (MIT course number 6.732) that Millie started teaching in the mid-1970s and continued teaching and revising until 2005. Continuing in Millie's footsteps, we have taught a version of the course at the University of Southern California since 2007 as have several of her other former students and postdocs, including Ado Jorio at the Universidade Federal de Minas Gerais (Brazil) and Antonio Gomes Souza Filho at the Universidade Federal do Ceará (Brazil). For us, as for many of Millie's former students, this course was the single most important class that shaped our academic career.

Solid State Properties: From Bulk to Nano fills a gap between many of the basic solid state physics and materials science books that are currently available. It is written for a mixed audience of electrical engineering and applied physics students who have some knowledge of elementary undergraduate quantum mechanics and statistical mechanics. This book is organized into three parts: (I) Electronic

Structure, (II) Transport Properties, and (III) Optical Properties. Each topic is explained in the context of bulk materials and then extended to low-dimensional materials where applicable. The chapters end with homework problems to provide students an opportunity to engage with the material more intimately.

As we were finishing the book, Millie expressed her appreciation for the many former students who contributed to its development. Millie wrote: "For me, it was former students who took an old version of my introductory solid state course starting in the 1970s and continuing through decades of different students coming from different countries around the world with different interests and needs. Over my (45 years) active classroom teaching career (from the 1960s to 2005), I enjoyed being in the classroom, learning with the many students I had the opportunity to work with." This was a genuine trait of Millie while at the top of her field to think of herself as learning along with her students.

We are extremely grateful to all graduate students in Prof. Cronin's group at USC who from 2011 to 2017 worked diligently creating figures, implementing changes in LaTex, and checking sources in the literature. Without their massive effort, the publication of this book would not have been possible.

It has been a tremendous honor to have played a role in publishing this book. We hope that, because of Millie's diligence and expertise, it will continue to teach and inspire another generation of scientists and engineers.

Los Angeles, CA, USA
March 2017

Stephen B. Cronin
Professor of Electrical Engineering
Physics, and Chemistry
University of Southern California

Fortaleza, Brazil
March 2017

Antonio Gomes Souza Filho
Professor of Physics, Universidade
Federal do Ceará

In Memory Of

Photograph courtesy of Micheline Pelletier taken for Millie's L'Oreal/UNESCO Women in Science Prize 2007

Mildred Dresselhaus (1930–2017), also known as Queen of Carbon, had an illustrious career that spanned six decades. She was the first female Professor to receive full tenure at the Massachusetts Institute of Technology in 1968. She published more than 1700 scientific papers, co-wrote eight books and received various awards and accolades for her contributions to science and technology during the course of her life. She was awarded the National Medal of Science in 1990, the 11th Annual Heinz Award in 2005, the Oersted Medal in 2008, and the Kavli Prize in 2012. She was also co-recipient of the Enrico Fermi Award in 2012. She received the Presidential Medal of Freedom from President Obama in 2012. Mildred Dresselhaus served as the director of the Office of Science at the US Department of

Energy from 2000 to 2001 and as a chair of the governing board of the American Institute of Physics from 2003 to 2008. She was also the president of American Physical Society, the first female president of the American Association for the Advancement of Science, and the treasurer of the National Academy of Sciences.

Every morning, Millie would leave her house by 5:30 AM, and her car was always the first car in the parking lot at MIT. As a Ph.D. student, she studied under Enrico Fermi at the University of Chicago. He too was an early riser, and, since he lived nearby, they had plenty of time to discuss science as they walked to school together. Millie often spoke of those conversations and their influence on her studies in that challenging academic program. It was at Chicago that she met her future husband Gene Dresselhaus, and in 1960, they were both hired by MIT Lincoln Laboratory. One of the main reasons Millie decided to study carbon was that it was relatively unpopular. She wanted to work on a project that most people thought was hard and not that interesting, so that it would be okay if she had to stay home with a sick child. As an icon for women in science and a strong advocate for women in STEM, she worked tirelessly to expand opportunities for women in science.

Contents

Part I
Electronic Structure

Chapter 1
Crystal Lattices in Real and Reciprocal Space

1.1 Introduction

In describing the physical properties of solids, it is important to discuss periodicity of the lattice structures. Many of the properties we commonly discuss for solids are obtained by using the model of a perfect periodic lattice. Some examples of periodic lattices occur in one dimension (1D), two dimensions (2D) and in three dimensions (3D), as discussed in the next sections. In practice, it is very hard to find a solid which is perfectly crystalline, that is, without any defects or impurities. Most of the crystalline solids in reality are composed of small crystals, and grain boundaries also break the periodicity of the lattice. Such defects are very important for nanocrystals, because such systems are very small and are on the nanometer (nm) scale. The degree of disorder and density of defects are of great interest because defects provide interesting materials properties. The role of point defects, edges and grain boundaries will be treated when discussing electronic, optical and other material properties. There are some solids whose atomic structure is highly disordered and they are called amorphous solids, which are characterized by the absence of any long range order. However, some short range order between the atoms may still exist. Quasi-periodicity also exists in solid materials, being a special kind of symmetry exhibited by the so called quasi-crystals, but we do not discuss such materials in this book.

In this chapter, we focus on introducing the fundamental concepts of periodic solids, discussing periodic lattices in real and reciprocal space. We start the discussion by considering the structural arrangement of the atoms and their ordering in a three dimensional (3D) periodic structure. Afterwards, we describe the periodic lattices in reciprocal space. In many cases it is also of interest and convenience to introduce reciprocal space (or momentum space) where the length scale is measured in wave vectors discussed in Sect. 1.3. It is typically more convenient mathematically to use reciprocal space (k_x, k_y, k_z) rather than real space (x, y, z) for calculating the physical properties of crystalline solids. We also introduce the Brillouin zone in this chapter and give some examples of the Brillouin zone for cubic and two dimensional (2D) hexagonal lattices, including atom counting in unit cells in both real space and reciprocal space.

M. Dresselhaus et al., *Solid State Properties*, Graduate Texts in Physics,
https://doi.org/10.1007/978-3-662-55922-2_1

1.2 Crystalline Lattices: Real Space

1.2.1 Bravais Lattices

In crystalline solids, atoms (or groups of atoms) are arranged in uniform periodic arrays called lattices. We start by describing these lattices by their abstract geometric considerations, not specific to any material. A three-dimensional Bravais lattice is defined as the set of all points **R** given by the vector

$$\mathbf{R} = n_1\mathbf{a_1} + n_2\mathbf{a_2} + n_3\mathbf{a_3} \tag{1.1}$$

where $\mathbf{a_1}$, $\mathbf{a_2}$, and $\mathbf{a_3}$ are any three vectors not all in the same plane, and n_1, n_2, and n_3 are integers of all $\pm$ values including 0. The three vectors $\mathbf{a_i}$ are called primitive basis vectors that generate the entire lattice, which extends to infinity in all three directions. Figures 1.1a and 1.2b show examples of 2D and 3D Bravais lattices, which satisfy the definition given by (1.1). It is important to note that the primitive basis vectors for a given Bravais lattice are not unique. This is illustrated in Fig. 1.1c, where several possible pairs of primitive basis vectors are drawn for a 2D Bravais lattice.

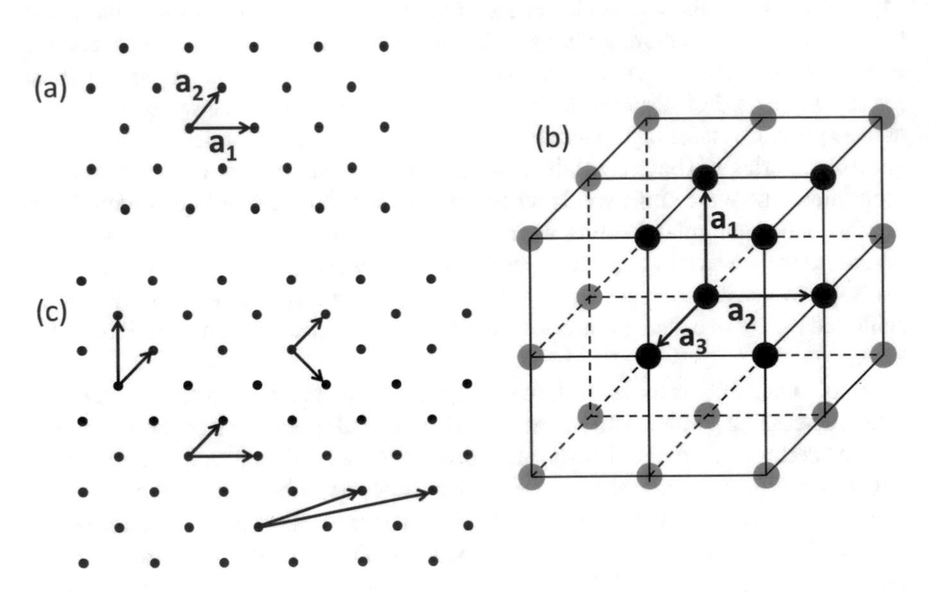

Fig. 1.1 **a** and **c** Examples of several 2D (two dimensional) lattices indicating several possible sets of primitive basis vectors. **b** A 3D cubic Bravais lattices

1.2.2 Unit Cells

A unit cell is a volume of space that fills all (3D) space when translated by the vectors of the periodic Bravais lattice vector **R**. The *primitive* unit cell has the minimum volume that, when translated by all the vectors in the Bravais lattice, fills all of space without unit cell overlapping with another, and without leaving any voids. A primitive unit cell must contain one lattice point. However, in a crystalline solid, there can be one or more atoms per lattice site and one or more atoms per unit cell. As with primitive basis vectors, primitive unit cells are not unique. By far, the most common primitive unit cell is the *Wigner–Seitz cell*, which is constructed by first drawing lines connecting a lattice point to all neighboring points in the lattice and then bisecting each line with a plane normal to this line, as illustrated in Fig. 1.2a. The smallest polyhedron bound by these planes is the Wigner–Seitz unit cell. Figure 1.2b shows the Wigner–Seitz unit cell for a body centered cubic Bravais lattice, where square and hexagonal faces are seen.

Two-dimensional Bravais lattices are classified into 5 lattice types, based on their symmetry considerations, as listed in Fig. 1.3. The most general lowest symmetry lattice is the oblique lattice, which is invariant only under rotation of π. There are four special lattice types that are invariant under various other rotations and reflection symmetry operators. The restrictions on the unit cell axes and angles for these special lattice types are listed in Fig. 1.3. Monolayered materials, such as graphene and boron nitride, would be described by such structures. Such monolayered structures can be large and have very many unit cells (quasi-infinite) or they can be of finite size, or they can be ribbons with a nanoscale width and a long length.

In three dimensions, there are 14 Bravais lattices, again classified by their symmetry considerations, as listed in Table 1.1. Here, the triclinic lattice is the most general, and there are 13 more special lattice types. In Table 1.1, these are grouped into seven types of unit cell symmetry categories according to the restrictions on the unit cell axes and angles for each of these 3D Bravais lattice types. For example, there are 3 cubic lattices, which include simple cubic (SC), body centered cubic (BCC), and face centered cubic (FCC).

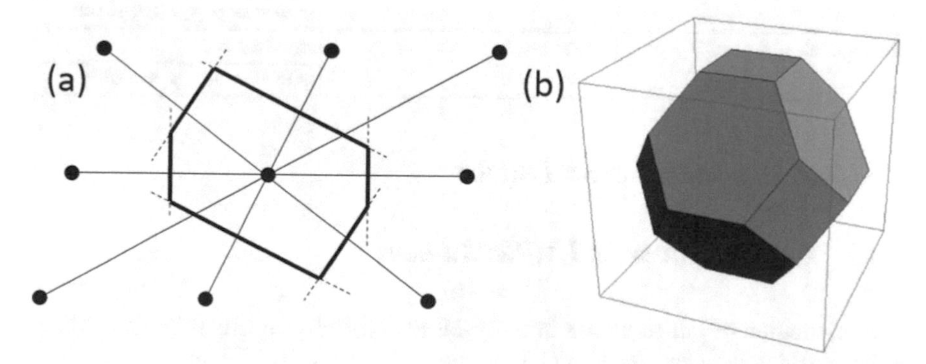

Fig. 1.2 The Wigner–Seitz unit cell for **a** a 2D and for **b** a 3D Bravais lattice

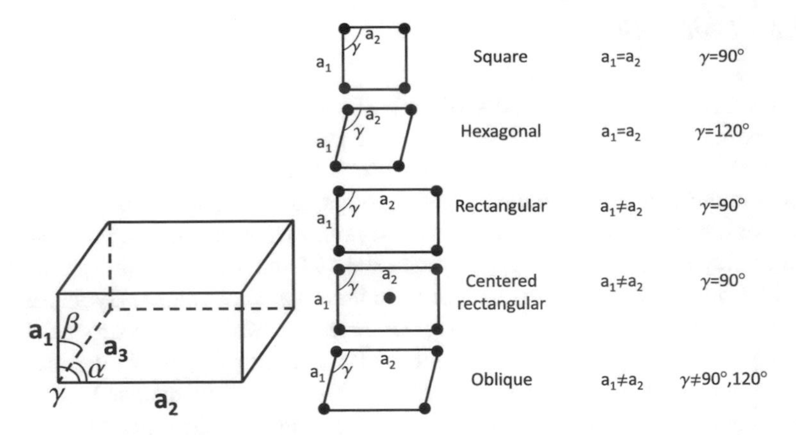

Fig. 1.3 General 3-dimensional unit cell and 5 types of two-dimensional Bravais Lattices in two dimensions

Table 1.1 Restrictions on cell axes and angles for the 14 types of Bravais Lattices in three dimensions. These 14 lattice types are distributed among the 7 crystal systems, and each lattice type is indicated by a symbol, which denotes the following 7 crystal systems P = primitive, C = centered, I = inversion symmetry, F = face centered, and R = rhombohedral

Crystal system	Number of lattices (symbol)	Restrictions on cell axes and angles
Triclinic	1 (P)	$a_1 \neq a_2 \neq a_3$
		$\alpha \neq \beta \neq \gamma$
Monoclinic	2 (P, C)	$a_1 \neq a_2 \neq a_3$
		$\alpha \neq \gamma, \beta = 90°$
Orthorhombic	4 (P, C, I, F)	$a_1 \neq a_2 \neq a_3$
		$\alpha = \beta = \gamma = 90°$
Tetragonal	2 (P, I)	$a_1 = a_2 \neq a_3$
		$\alpha = \beta = \gamma = 90°$
Cubic	3 (P, I, F)	$a_1 = a_2 = a_3$
		$\alpha = \beta = \gamma = 90°$
Trigonal	1 (R)	$a_1 = a_2 = a_3$
		$\alpha = \beta = \gamma \leq 120°, \neq 90°$
Hexagonal	1 (P)	$a_1 = a_2 \neq a_3$
		$\alpha = \beta = 90°, \gamma = 120°$

1.3 Lattices in Reciprocal Space

1.3.1 Crystal Planes and Miller Indices

The orientation of lattice planes in a crystal are typically specified by their *Miller* indices, which are determined by taking the reciprocal of the intercept of the three

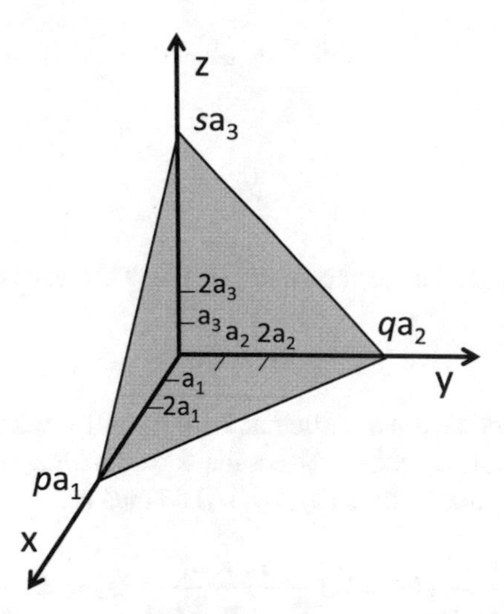

Fig. 1.4 General lattice plane intercepting the x, y, and z axes at pa_1, qa_2, and sa_3, respectively. The Miller indices (hkl) are obtained by multiplying the reciprocal of these intercepts ($1/p$, $1/q$, $1/s$) by their lowest common denominator as shown in the figure

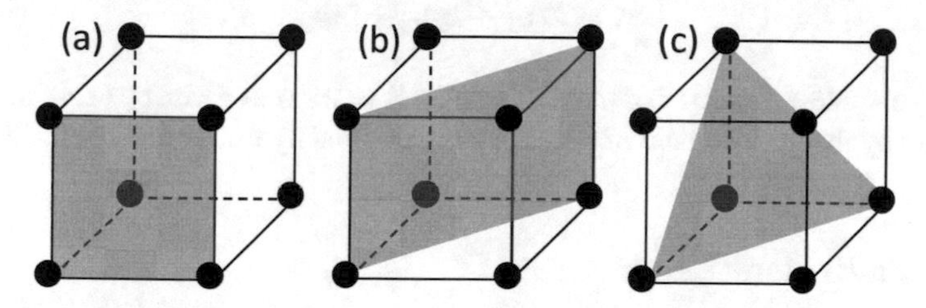

Fig. 1.5 Three lattice planes in a simple cubic lattice: **a** (100) plane, **b** (110) plane, and **c** (111) plane

vectors x, y, z with the planes shown in Fig. 1.4. This figure illustrates this general scheme in three dimensions with intercepts at integers p, q, and s. To describe the plane, we write the reciprocal of these intercepts as, ($1/p$, $1/q$, $1/s$), and we then multiply these numbers by their lowest common denominators to obtain a set of integers (hkl). We refer to these three numbers h, k, and l as the Miller indices. Figure 1.5 shows the lattice planes in a simple cubic lattice for the high symmetry (100), (110), and (111) planes, which are also labelled by their (hkl) indices.

1.3.2 Reciprocal Lattice Vectors

The reciprocal space lattice is defined by the set of all wavevectors **K** that yield plane waves ($e^{i\mathbf{k}\cdot\mathbf{r}}$) with the periodicity of the real space lattice, denoted by the real space lattice vectors **R**. Mathematically, this can be written for any **r** as

$$e^{i\mathbf{K}\cdot(\mathbf{R}+\mathbf{r})} = e^{i\mathbf{K}\cdot\mathbf{r}} \tag{1.2}$$

or

$$e^{i\mathbf{K}\cdot\mathbf{R}} = 1 \tag{1.3}$$

for all **R** in the real space lattice shown in Fig. 1.5. This leads to the general relation

$$\mathbf{K}\cdot\mathbf{R} = 2\pi N, \tag{1.4}$$

where N is any integer. Given a particular Bravais lattice with primitive vectors $\mathbf{a_1}$, $\mathbf{a_2}$, and $\mathbf{a_3}$, as in Fig. 1.4, we can construct the corresponding reciprocal space lattice using the reciprocal space vectors ($\mathbf{b_1}, \mathbf{b_2}, \mathbf{b_3}$) defined as.

$$\mathbf{b_1} = 2\pi\frac{\mathbf{a_2}\times\mathbf{a_3}}{\mathbf{a_1}\cdot(\mathbf{a_2}\times\mathbf{a_3})},\ \mathbf{b_2} = 2\pi\frac{\mathbf{a_3}\times\mathbf{a_1}}{\mathbf{a_2}\cdot(\mathbf{a_3}\times\mathbf{a_1})},\ \mathbf{b_3} = 2\pi\frac{\mathbf{a_1}\times\mathbf{a_2}}{\mathbf{a_3}\cdot(\mathbf{a_1}\times\mathbf{a_2})}. \tag{1.5}$$

These three vectors define a reciprocal lattice of the general form

$$\mathbf{K} = k_1\mathbf{b_1} + k_2\mathbf{b_2} + k_3\mathbf{b_3} \tag{1.6}$$

where k_i are integers, such that the product $\mathbf{K}\cdot\mathbf{R} = 2\pi N$ as given by (1.4), and (1.6), as shown by the lattice plane in Fig. 1.5 and specified by miller indices in Fig. 1.4.

1.4 The Brillouin Zone

The Brillouin zone is the Wigner–Seitz primitive cell of the reciprocal lattice. While the terms, the Brillouin zone and Wigner–Seitz unit cell both refer to the same geometrical construction, the Wigner–Seitz cell is typically reserved for describing real space lattices and the Brillouin zone is reserved for reciprocal space lattices. The Brillouin zone is particularly important in calculating and discussing the band theory of solids, which will be discussed in the following chapters. We now give examples of unit cells in real space and reciprocal space of materials that we use throughout the book for explaining concepts and for practical device applications in this book and generally.

1.4.1 Graphene and Boron Nitride

Figure 1.6 shows the real space and reciprocal space lattices for the 2D hexagonal lattice of graphene and Boron Nitride (BN). The vectors $\mathbf{a_i}$ and $\mathbf{b_i}$ are the real space and corresponding reciprocal space unit vectors, respectively. In the x, y coordinate system, these basis vectors can be written as

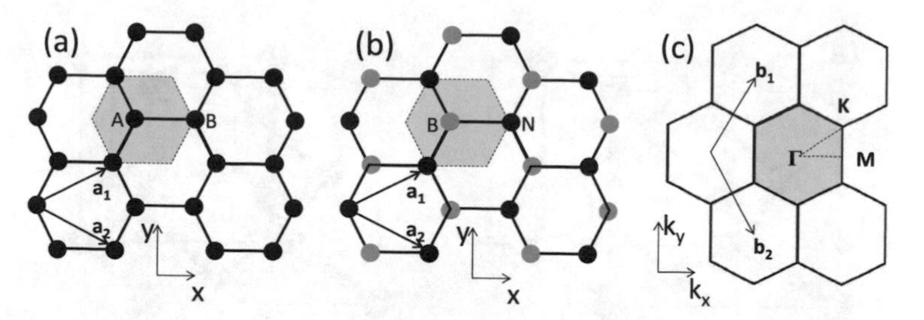

Fig. 1.6 Real space lattice for a hexagonal monolayer of **a** graphene and **b** boron nitride. **c** The reciprocal space lattice for a 2D hexagonal lattices generally. The shaded hexagon in **a** is the Wigner–Seitz unit cell in real space and in **c** is the corresponding Brillouin zone in reciprocal space

$$\mathbf{a}_1 = \left(\frac{\sqrt{3}}{2}a, \frac{a}{2}\right), \mathbf{a}_2 = \left(\frac{\sqrt{3}}{2}a, -\frac{a}{2}\right), \tag{1.7}$$

where $a = 1.42$ Å, which is the nearest neighbor distance in graphene, for which the graphene unit cell has two carbon atoms. Closely related to this nearest neighbor distance are the magnitude of the lattice vectors $|\mathbf{a}_1| = |\mathbf{a}_2| = 1.42 \times \sqrt{3}$ Å. The corresponding reciprocal lattice vectors are given by

$$\mathbf{b}_1 = \left(\frac{2\pi}{\sqrt{3}a}, \frac{2\pi}{a}\right), \mathbf{b}_2 = \left(\frac{2\pi}{\sqrt{3}a}, -\frac{2\pi}{a}\right), \tag{1.8}$$

and have for graphene a magnitude of $2\pi/1.42\sqrt{3}\,\text{Å}^{-1}$, as shown in Fig. 2.6.

The shaded hexagons in Fig. 1.6a, b indicate the graphene (Fig. 1.6a) and boron nitride (Fig. 1.6b) Wigner–Seitz unit cell, while the shaded hexagon in Fig. 1.6c depicts the Brillouin zone for these hexagonal lattices. These 2D crystal have two crystallographically equivalent atoms per unit cell labeled A and B, which are the same chemical element (Carbon) for graphene, but are different for BN, where the atomic species are Boron and Nitrogen. This difference in the chemical species has a profound consequences on the electronic properties leading graphene to be a semimetal and BN to be an insulator. Notice that the reciprocal space Brillouin zone is rotated by 30° or equivalently by 90° from the real space Wigner–Seitz unit cell. We will use graphene as a model 2D material throughout this book to demonstrate its unique physical and chemical properties.

1.4.2 Diamond and Zinc Blende Lattices

The diamond crystal structure (or related zinc blende structure) are two crystal structures found in many semiconducting and insulating materials. For example, silicon

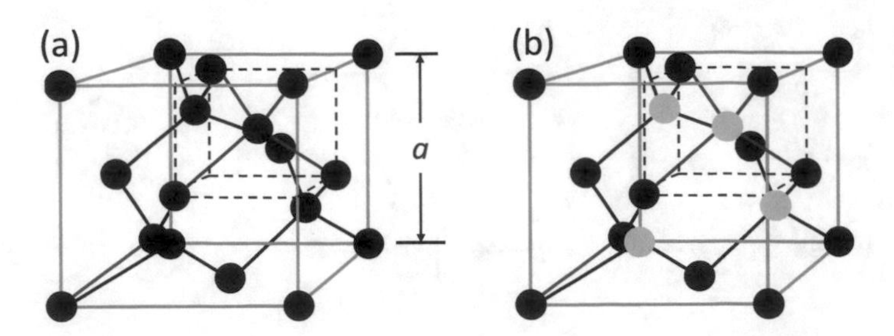

Fig. 1.7 Conventional cubic cell of **a** a diamond lattice and **b** a zinc blende lattice, highlighting that all atoms are the same in **a** but different in **b**

and germanium crystallize in the diamond structure (see Fig. 1.7a), while many III–V compound semiconductors, such as GaAs and InSb, crystallize in the zinc blende structure (see Fig. 1.7b). Both of these lattice structures consist of two interpenetrating FCC lattices displaced from one another along the body diagonal by one quarter of the diagonal length $(a/4, a/4, a/4)$, as shown in Fig. 1.7a, b. That is, the underlying lattice structure is FCC with two atoms per lattice unit cell, one at (0, 0, 0) and one at (1/4, 1/4, 1/4). For the diamond lattice, the two atoms are both of the same atomic species (see Fig. 1.7a), while in the zincblende structure (Fig. 1.7b), every other atomic site is occupied by a different kind of ion (e.g., Ga and As). The four nearest neighbor bonds have been drawn in the figure and they form a tetrahedral structure, whereby each black atom has a nearest neighbor gray atom and vice-versa.

The primitive translation vectors of the real space FCC lattice can be written as

$$\mathbf{a_1} = \frac{1}{2}a(\hat{y} + \hat{z}); \ \mathbf{a_2} = \frac{1}{2}a(\hat{z} + \hat{x}); \ \mathbf{a_3} = \frac{1}{2}a(\hat{x} + \hat{y}). \tag{1.9}$$

It follows that the corresponding primitive translation vectors of the reciprocal lattice are given by

$$\mathbf{b_1} = \frac{2\pi}{a}(-\hat{x} + \hat{y} + \hat{z}); \ \mathbf{b_2} = \frac{2\pi}{a}(\hat{x} - \hat{y} + \hat{z}); \ \mathbf{b_3} = \frac{2\pi}{a}(\hat{x} + \hat{y} - \hat{z}). \tag{1.10}$$

The shortest eight translation vectors in the reciprocal lattice are given by

$$\mathbf{K} = \frac{2\pi}{a}(\pm\hat{x} \pm \hat{y} \pm \hat{z}). \tag{1.11}$$

From this, we can show that the reciprocal space lattice, thereby, forms a BCC lattice. That is, the BCC lattice is the reciprocal to the FCC lattice, and vice versa. The Brillouin zone for the FCC lattice results in a truncated octahedron, as shown in Fig. 1.8. In this structure, the hexagonal faces bisect the lines joining the central point to the points on the vertices. The square faces bisect lines joining the central point to the central points in neighboring cubic cells.

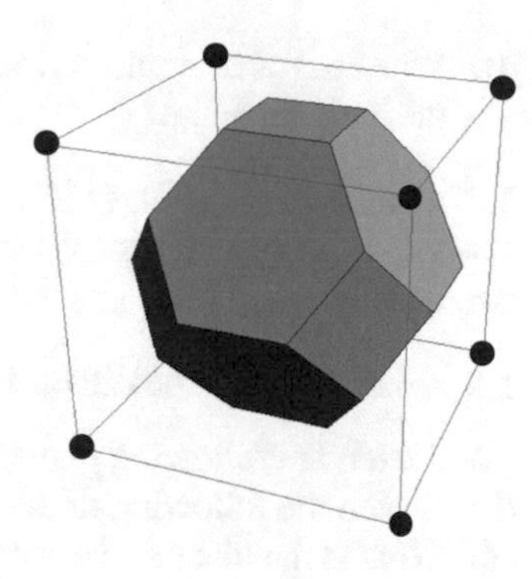

Fig. 1.8 Brillouin zone of a face centered cubic lattice (FCC)

Problems

1.1 Draw the Wigner–Seitz cell for:

(a) a 2D rectangular lattice.
(b) a 2D hexagonal lattice.

1.2 Consider a 2D rectangular lattice with $\mathbf{a}_1 > \mathbf{a}_2$.

(a) Write an expression for the set of real space lattice vectors **R**.
(b) Draw the real space lattice with the Wigner–Seitz unit cell.
(c) Write an expression for the set of reciprocal space lattice vectors **G**, in terms of the magnitudes of the real space lattice vectors $|\mathbf{a}_1|$ and $|\mathbf{a}_2|$.
(d) Draw the reciprocal space lattice with the Wigner–Seitz unit cell.

1.3 Repeat Problem 1.2 for a 2D hexagonal lattice.

1.4 GaAs has the zinc blende crystal structure. It has a lattice constant of 0.565 nm. Determine the number of Ga and As atoms per cm^3 in the material.

1.5 Ge has the diamond crystal structure and has the same lattice constant as GaAs.

(a) Determine the number of Ge atoms per cm^3.
(b) What is the distance (center to center) between nearest Ge atoms?
(c) What is the number density of Ge atoms?
(d) What is the mass density (grams/cm^3) of crystalline Ge?

1.6 A material consists of two atoms, A and B, that have effective radii, 0.21 nm and 0.18 nm, respectively. The lattice of the material has the BCC structure with atom A at the corners and atom B in the center of the cube. Assuming that the atoms are hard spheres and that the crystal structure is formed when the atoms that can touch one another are indeed touching, what is the lattice constant and the volume densities of atoms A and B.

(a) If the crystal is formed with atoms B at the corners and atom A at the center, what is the lattice constant and what are the atom densities of atoms A and B?

(b) When the smaller atoms touch, what is the size of the largest ball that fits along the body diagonal?

1.7 Consider the (100), (110), and (111) planes of Si. Which plane has the highest surface density of atoms? Which has the smallest surface density? What are those two densities?

1.8 Consider the three dimensional simple cubic lattice with a lattice constant a_0.

(a) Sketch the following planes: (100), (130), (230).
(b) Sketch the following directions in the lattice: [110], [311], [123].
(c) What is the distance between nearest (111) planes?

1.9 Determine the angle between tetrahedral bonds in Si. These angles are the same as the body diagonals of a cube. (Hint: use the law of cosines to find the angle.)

1.10 Show that the face centered cubic lattice is the reciprocal to the body centered cubic lattice.

1.11 Find the reciprocal lattice to an orthorhombic lattice with $\mathbf{a_1} > \mathbf{a_2} > \mathbf{a_3}$ and $\alpha = \beta = \gamma = 90°$.

1.12 Consider the two-dimensional simple triangular lattice for a free electron metal with two electrons/atom.

(a) Assuming a lattice constant of a, find the areas of the electron and hole pockets that are formed in the second and first Brillouin zones, respectively.
(b) Find the shapes of the electron and hole Fermi surfaces in the reduced Brillouin zone, obtained through translation by a reciprocal lattice vector.

Chapter 2
Electronic Properties of Solids

2.1 Introduction

In this chapter we present some methods that are employed in performing electronic structure calculations. We start by presenting a general quantum mechanical framework to describe a molecule or a solid. We then introduce the Born–Oppenheimer approximation (also called the adiabatic approximation), which allows us to reduce the problem to its corresponding electronic part. In the following, we introduce the independent-electron approximation and the main methods commonly used to solve the electronic problem.

2.2 Hamiltonian of the System

The system Hamiltonian is the starting point for calculating the electronic structure of molecules and solids. The Hamiltonian can be generally written as: Martin 2004

$$\hat{H} = -\frac{\hbar^2}{2m_e}\sum_i \nabla_i^2 - \frac{1}{4\pi\varepsilon_0}\sum_i\sum_I \frac{Z_I e^2}{|\mathbf{r}_i - \mathbf{R}_I|} + \frac{1}{8\pi\varepsilon_0}\sum_i\sum_{j\neq i}\frac{e^2}{|\mathbf{r}_i - \mathbf{r}_j|}$$
$$\sum_I \frac{\hbar^2}{2M_I}\nabla_I^2 + \frac{1}{8\pi\varepsilon_0}\sum_I\sum_{J\neq I}\frac{Z_I Z_J e^2}{|\mathbf{R}_I - \mathbf{R}_J|} \quad (2.1)$$

where i, j label the electrons and I, J label the related atomic nuclei. The quantities M_I, Z_I and $\mathbf{R}_I$ are, respectively, the mass, the atomic number, and the position of nucleus I. The electron charge and mass are, respectively, written as $-e$ and m_e, and $\mathbf{r}_i$ represents the position of the electron i. The first three terms in (2.1) account, respectively, for the kinetic energy of the electrons, the electron-nucleon Coulomb interaction, and the electron-electron Coulomb interaction. The fourth term describes the kinetic energy of the nuclei and the last term is the nucleon-nucleon Coulomb interaction.

An important aspect of this problem is the fact that the nuclei are much more massive than the electrons. This makes the kinetic energy of the nuclei small compared

M. Dresselhaus et al., *Solid State Properties*, Graduate Texts in Physics,
https://doi.org/10.1007/978-3-662-55922-2_2

to the other contributions to the Hamiltonian in (2.1). This allows us to decouple the nuclear and electronic contributions to (2.1) and to work on them separately. This constitutes a practical and useful simplification which makes both the structural optimization and electronic structure calculations easier to perform, but still provides sufficient accuracy for many physical problems. This is the so-called Born–Oppenheimer (BO) Martin 2004 or adiabatic approximation, which is a useful approximation for many purposes, such as for the calculation of the vibrational modes in solids. We should point out that the limitation of the BO approximation has been recently discussed for graphene and carbon nanotubes. A widely used approach is to consider the energy from the remaining electronic problem as an extra term added to the ion-ion interaction and to perform a subsequent geometrical optimization based on this effective interatomic potential. By considering the BO approximation, the problem is reduced to the electronic Hamiltonian $\hat{H}_e$ given by:

$$\hat{H}_e = -\frac{\hbar^2}{2m_e}\sum_i \nabla_i^2 - \frac{1}{4\pi\varepsilon_0}\sum_i\sum_I \frac{Z_I e^2}{|\mathbf{r}_i - \mathbf{R}_I|} + \frac{1}{8\pi\varepsilon_0}\sum_i\sum_{j\neq i}\frac{e^2}{|\mathbf{r}_i - \mathbf{r}_j|} \tag{2.2}$$

where the atomic positions entered as parameters.

2.3 The Electronic Problem

By considering the BO approximation, the electronic problem is simplified compared to the initial problem. However, it is still not easy to solve this problem within the BO approximation. A set of widely used strategies employs the independent-electron approximation. This mean-field approximation consists of defining one-electron wavefunctions that can be obtained from a one-electron Schrödinger equation. This is a significant simplification but gives very satisfactory results for many interesting physical systems and is used in most theoretical calculations of the electronic structure of molecules and solid state materials.

2.3.1 The Hartree Method

Hartree was a pioneer in developing the first quantitative electronic calculations for multi-electron systems (Hartree 1928). The Hartree method starts with the one electron equation which is written as:

$$\hat{H}_{HT}\psi_i^\sigma(\mathbf{r}) = -\frac{\hbar^2}{2m_e}\nabla^2\psi_i^\sigma(\mathbf{r}) + V_{HT}\psi_i^\sigma(\mathbf{r}) = \varepsilon_i^\sigma\psi_i^\sigma(\mathbf{r}) \tag{2.3}$$

where σ represents the spin and V_{HT} is an effective Hartree potential for each electron in the presence of the others. A different potential is defined for each electron in order to avoid self-interaction of the electron with itself. In order to obtain the ground state of the system, one fills the electronic states starting from the lowest energy levels, though always obeying the Pauli exclusion principle. (see Slater 1930)

2.3.2 Hartree–Fock (HF) Method

In 1930, Fock expanded Hartree's method by using an anti-symmetric wavefunction in terms of a Slater determinant written using one-electron Schrödinger wavefunctions. (see Fock 1930) The one-electron equations are then obtained by finding the corresponding wavefunctions that minimize the total energy obtained as the expectation energy ε_i^σ for the full Hamiltonian. This process yields the following equations:

$$\begin{aligned} \hat{H}_{HF}\psi_i^\sigma(\mathbf{r}) &= \Big[-\frac{\hbar^2}{2m_e}\nabla^2 + V_{ext}(\mathbf{r}) + V_{HT}(\mathbf{r}) + V_{xc}(\mathbf{r})\Big]\psi_i^\sigma(\mathbf{r}) \\ &= \varepsilon_i^\sigma\psi_i^\sigma(\mathbf{r}) \end{aligned} \tag{2.4}$$

in which V_{ext} is the external potential, V_{HT} is the Hartree potential and V_{xc} is the exchange potential. V_{HT} and V_{xc} are written as:

$$V_{HT}(\mathbf{r}) = \frac{e^2}{4\pi\varepsilon_0}\sum_{j,\sigma_j}\int d\mathbf{r}'\frac{\psi_j^{\sigma_j *}(\mathbf{r}')\psi_j^{\sigma_j}(\mathbf{r}')}{|\mathbf{r}-\mathbf{r}'|} \tag{2.5}$$

$$V_{xc}(\mathbf{r}) = -\frac{e^2}{4\pi\varepsilon_0}\sum_{j}\int d\mathbf{r}'\frac{\psi_j^{\sigma *}(\mathbf{r}')\psi_i^{\sigma}(\mathbf{r}')}{|\mathbf{r}-\mathbf{r}'|}\frac{\psi_j^\sigma(\mathbf{r})}{\psi_i^\sigma(\mathbf{r})}. \tag{2.6}$$

Note that, unlike the original Hartree approach, the mean Coulomb interaction in the Hartree–Fock approach V_{HT} includes a self-interaction contribution. The additional exchange term V_{xc}, which does not have a classical analogue, also contains such a self-interaction energy, but with an opposite sign so that the final result does not depend on self-interactions. The presence of the exchange potential V_{xc} is the main difference between the HF and the Hartree approaches. (see Fock 1930)

The meaning of the exchange term $V_{xc}(\mathbf{r})$ is not easy to understand at first because there is no classical analog. The physics behind it relies on the foundations of the independent electron approximation. Without considering any approximation, the solution for the electronic problem consists of a multi-electron wavefunction Ψ, which is a function of the coordinates for all the electrons $\mathbf{r}_i$, $i = 1, 2, 3, \ldots, N$. Since the electrons are Fermions, they should obey Pauli's exclusion principle which is reflected by the fact that the electronic wavefunction has to be anti-symmetric with respect to any permutation involving the positions of any two electrons i and j, such that:

$$\Psi(\mathbf{r}_1, \mathbf{r}_2, \ldots, \mathbf{r}_i, \ldots, \mathbf{r}_j, \ldots, \mathbf{r}_{N-1}, \mathbf{r}_N) = -\Psi(\mathbf{r}_1, \mathbf{r}_2, \ldots, \mathbf{r}_j, \ldots, \mathbf{r}_i, \ldots, \mathbf{r}_{N-1}, \mathbf{r}_N). \quad (2.7)$$

In addition, we expect the wavefunction Ψ dependence on the different $\mathbf{r}_i$ coordinates to be correlated in a more general way, so that the behavior of Ψ relative to a given $\mathbf{r}_i$ depends on the values of the $\mathbf{r}_j$ with $j \neq i$. However, when we use the independent electron approximation, we are restricting Ψ to have the form given by a Slater determinant. (see Slater 1929) By doing that, we are intrinsically losing information which can be directly associated with the electronic correlation. While the Hartree potential, in both the Hartree and HF approaches, represents the interaction of any electron with the system's electronic cloud, the correlation between electrons is related to the specific interaction of a given electron with any single electron in the system. This is not a simple problem to solve and accounting for this correlation is a central problem in the electronic structure research field. The HF method, however, is a first step in this direction because the exchange potential V_{xc} in (2.6) represents two aspects of such a correlation:

1. Self-interaction contributions are removed;
2. Short range interactions related to the Pauli's exclusion principle are accounted for.

As can be seen from (2.6), V_{xc} lowers the energy and V_{xc} can be interpreted as the interaction of the electron with an agent usually referred to as an "exchange hole". According to the expression for the exchange potential, this positive "exchange charge density" is determined by the electronic density (which is a sum over the j states) surrounding the electron i, and V_{ext} favors a ferromagnetic ordering of the electronic spins, since this interaction involves only electrons with the same spin. This is a consequence of Hund's rule which states that *as the number of electrons start to fill a set of degenerate atomic states, the electrons will evenly fill the available states so as to maximize the total spin as much as possible, only starting to occupy orbitals with opposite spin when there are no available empty states in the first spin state*. Note also that there is no energy lowering associated with two electrons with the same spin occupying the same electron orbital, since the $j = i$ contribution in the sum in (2.6) cancels with the corresponding self-interacting term in the sum from (2.5), constituting a clear manifestation of Pauli's exclusion principle. (see Pauli 1925)

2.3.3 Density Functional Theory

Not all correlation effects are accounted for by the exchange energy term V_{xc} which is added in the HF approach discussed in Sect. 2.3.2. In this regard, the introduction of Density Functional Theory (DFT) represents an important advancement in the field of electronic band structure calculations. (see Hohenberg and Kohn 1964) DFT has, in fact, become a standard tool for condensed matter physics and is widely accessible for many applications.

As we will see, DFT allows us to recast the many-electron problem into a set of one-electron Schrödinger-like equations. However, in contrast to the Hartree and HF approaches, the DFT approach is a more complete theory but its practical implementation demands other approximations to be made. The electronic density $n(\mathbf{r})$ is the main parameter in DFT and its key role can be understood in terms of the two Hohenberg–Kohn theorems given below, which constitute the basis of DFT: (see Hohenberg and Kohn 1964)

1st *Theorem*: If a system of interacting electrons is immersed in an external potential V_{ext}, this potential is uniquely determined (except by a constant) by the electronic density n_0 of the ground state (GS).
2nd *Theorem*: Let $E[n]$ be the functional for the energy relative to the electronic density $n(\mathbf{r})$ for a given $V_{ext}(\mathbf{r})$. Then this functional has its global minimum (GS energy) for the exact electronic density $n_0(\mathbf{r})$ corresponding to the ground state.

The first theorem states that all the system properties are determined by the electronic density for the ground state since n_0 determines V_{ext}, which determines the Hamiltonian, which in turn defines the ground state and all the excited states. Also, we can use the energy functional $E[n]$ (see (2.8) where this functional is written as $E_{el}[n]$) to determine the exact ground state energy and density. It is important to note, however, that DFT is not only a ground state theory, but instead gives the system's Hamiltonian which is, in principle, all we need to obtain the ground state and all the excited states. However, the ground state can be obtained in a systematic way within DFT. The energy functional $E_{HK}[n]$ in the Hohenberg–Kohn approach is written as:

$$E_{HK}[n] = T[n] + E_{el}[n] + \int d\mathbf{r} V_{ext}(\mathbf{r}) n(\mathbf{r}) \tag{2.8}$$

where $T[n]$ is the kinetic energy functional, V_{ext} is the external potential felt by the electrons (including the contribution from the nuclei) and the electronic energy $E_{el}[n]$ accounts for all the electron-electron interactions.

Despite having the correct tool to obtain the electronic ground state (i.e., the minimization of $E[n]$ relative to n), it is still not clear how to proceed in using this tool. The necessary recipe is given by the Kohn–Sham ansatz. (see Kohn and Sham 1965) According to this ansatz, the ground state electronic density of our system can be written as the ground state of an auxiliary system of non-interacting electrons. The one-electron wavefunctions ψ_i^σ for this auxiliary system are determined by Schrödinger-like equations of the form:

$$\hat{H}^{\sigma}_{aux}\psi^{\sigma}_i(\mathbf{r}) = -\frac{\hbar^2}{2m_e}\nabla^2\psi^{\sigma}_i(\mathbf{r}) + V^{\sigma}\psi^{\sigma}_i(\mathbf{r}) = \varepsilon^{\sigma}_i\psi^{\sigma}_i(\mathbf{r}) \tag{2.9}$$

where σ labels the electron spin. The electronic density $n(\mathbf{r})$ is written as:

$$n(\mathbf{r}) = \sum_{\sigma}\sum_{i=1}^{N^{\sigma}}|\psi^{\sigma}_i(\mathbf{r})|^2 \tag{2.10}$$

where the N^{σ} is the total number of electrons in each spin state σ. The corresponding auxiliary kinetic energy is:

$$T_{aux} = -\frac{\hbar^2}{2m_e}\sum_{\sigma}\sum_{i=1}^{N^{\sigma}}\langle\psi^{\sigma}_i|\nabla^2|\psi^{\sigma}_i\rangle = \frac{\hbar^2}{2m_e}\sum_{\sigma}\sum_{i=1}^{N^{\sigma}}\int d\mathbf{r}|\nabla\psi^{\sigma}_i(\mathbf{r})|^2. \tag{2.11}$$

The classical Coulomb interaction for the electron-electron repulsion E_{CI} is given by

$$E_{CI}[n] = \frac{1}{8\pi\varepsilon_0}\int\int d\mathbf{r}d\mathbf{r}'\frac{n(\mathbf{r})n(\mathbf{r}')}{|\mathbf{r}-\mathbf{r}'|}, \tag{2.12}$$

where the ε_0 is the vacuum dielectric permittivity. By summing up these terms, the expression for the Kohn–Sham energy functional reads:

$$E_{KS}[n] = T_{aux}[n] + \int d\mathbf{r}V_{ext}(\mathbf{r})n(\mathbf{r}) + E_{CI}[n] + E_{xc}[n] \tag{2.13}$$

where V_{ext} is the external potential (including the contribution from the nuclei) and E_{xc} is the energy functional which accounts for the exchange and all the correlation effects. If we consider the Hohenberg–Kohn energy $E_{HK} = E_{KS}$, we have:

$$E_{xc}[n] = T[n] - T_{aux}[n] + E_{el}[n] - E_{CI}[n] \tag{2.14}$$

which indicates that E_{xc} contains the exchange contribution and all the other correlation effects related to both the kinetic energy and the electron-electron interactions. Here lies the main problem of DFT: we do not know the exact form of E_{xc}. Even though DFT yields the exact solution for the electronic problem, its practical implementation requires an approximation regarding the form of the exchange and correlation energy terms. The usual approach is to write this energy as:

$$E_{xc}[n] = \int d\mathbf{r}n(\mathbf{r})\varepsilon_{xc}([n],\mathbf{r}) \tag{2.15}$$

where $\varepsilon_{xc}([n],\mathbf{r})$ is the exchange-correlation energy per electron at the position $\mathbf{r}$ for a given density $n(\mathbf{r})$. The minimization of the energy functional is obtained by varying the one-electron wavefunctions ψ_i^σ and using the Lagrange multipliers ε_i^σ corresponding to the normalization constraint $\langle \psi_i^\sigma | \psi_i^\sigma \rangle = 1$:

$$\frac{\delta}{\delta \psi_i^{\sigma *}} \Big(E_{KS}[n] - \sum_\sigma \sum_{j=1}^{N_\sigma} \varepsilon_j^\sigma \big(\int d\mathbf{r} \psi_j^{\sigma *} \psi_j^\sigma - 1 \big) \Big) =$$

$$= \frac{\delta T_{aux}[n]}{\delta \psi_i^{\sigma *}} + \frac{\delta E_{ext}[n]}{\delta \psi_i^{\sigma *}} + \frac{\delta E_{HT}[n]}{\delta \psi_i^{\sigma *}} + \frac{\delta E_{xc}[n]}{\delta \psi_i^{\sigma *}} - \varepsilon_i^\sigma \psi_i^\sigma$$

$$= -\frac{\hbar^2}{2m_e} \nabla^2 \psi_i^\sigma(\mathbf{r}) + \left(\frac{\delta E_{ext}[n]}{\delta n(\mathbf{r})} + \frac{\delta E_{HT}[n]}{\delta n(\mathbf{r})} + \frac{\delta E_{xc}[n]}{\delta n(\mathbf{r})} \right) \frac{\delta n(\mathbf{r})}{\delta \psi_i^{\sigma *}} - \varepsilon_i^\sigma \psi_i^\sigma = 0 \quad (2.16)$$

This minimization of the energy functional is carried out so that the electrons obey:

$$-\frac{\hbar^2}{2m_e} \nabla^2 \psi_i^\sigma(\mathbf{r}) + \Bigg(V_{ext}(\mathbf{r}) + V_{HT}(\mathbf{r}) + \varepsilon_{xc}(\mathbf{r}) \Bigg) \psi_i^\sigma = \varepsilon_i^\sigma \psi_i^\sigma . \quad (2.17)$$

Here the Hartree potential $V_{HT}(\mathbf{r})$ which represents the interaction of any electron with its surrounding electronic cloud, is given by:

$$V_{HT}[n] = \frac{1}{8\pi \varepsilon_0} \int d\mathbf{r}' \frac{n(\mathbf{r})}{|\mathbf{r} - \mathbf{r}'|} \quad (2.18)$$

in which V_{ext} and ε_{ext} are the external potential and exchange-correlation energy per electron, respectively. Equation (2.17) is the well-known Kohn–Sham equation for the auxiliary problem. This is the basis for many theoretical calculations of the electronic structure that have been performed on molecules and solids.

2.4 Plane Wave and Localized Basis Sets

In order to solve Schrödinger's equation for a molecule or solid, one first has to choose a basis-set to use for the electronic wavefunctions. In order to obtain precise results, the first property we expect from a basis set is completeness:

$$\sum_i |\phi_i\rangle \langle \phi_i| = 1. \quad (2.19)$$

It turns out that, in practice, it is never possible to use such a complete set. Plane waves, for instance, constitute a basis set which is naturally complete, but only as long as an infinite number of plane waves are explicitly included. However, in a numerical implementation, one always has to use a finite discretized set of states which is a subset of the total plane wave set. In this case, the systematic way to improve the accuracy of the calculation is to increase the number of functions in the basis set. Such an improvement is not boundless, since computational resources have a finite processing capability.

One alternative to plane waves is to use the Linear Combination of Atomic Orbitals (LCAO) method. Here, the basis consists of functions corresponding to the electronic states from the isolated atoms. Despite its simplicity, this method yields quite accurate results, and it constitutes the foundation of several computational packages and studies in the literature. The main advantages of this method are the reduced computational cost and the easy association of the molecular levels with the atomic orbitals. One major drawback of this approach is the difficulty in assessing its validity, given the impossibility to systematically improve the basis set, and to calculate the remaining error.

Let us examine the use of the LCAO by expanding the electronic wavefunctions from a crystal in terms of a local orbital basis. In such a basis, each orbital basis function is associated with an atom in the structure. One appropriate choice for these orbitals are functions centered on the atomic sites. These functions can be written as:

$$\phi_j(\mathbf{r}-\mathbf{R}) = \phi_\alpha(\mathbf{r}-\mathbf{r}_P-\mathbf{R}) = \phi^{\mathbf{R}}_{n_j l_j m_j}(\boldsymbol{\rho}) = \phi^{\mathbf{R}}_{n_j l_j}(\rho) Y_{l_j m_j}(\hat{\rho}) \tag{2.20}$$

where the coordinate ρ is

$$\boldsymbol{\rho} = \mathbf{r} - \mathbf{r}_P - \mathbf{R} \tag{2.21}$$

and $\mathbf{r}_P$ is the position of the Pth atom in the crystal unit cell (relative to the origin of the unit cell), α enumerates the atomic orbitals centered at P, and $\mathbf{R}$ is a lattice vector from the Bravais lattice (which localizes the origin of its corresponding cell). In this terminology, we define j to represent the (P, α) pair. The l_j, m_j and n_j in (2.20) represent the angular momentum, its projection on a given axis, and the number of different functions with the same angular momentum, respectively. Also, the $Y_{l_j m_j}(\hat{\rho})$ functions denote the spherical harmonics, which provide basis functions for the (l_j, m_j) states. We list the spherical harmonics for $l_j = 0, 1, 2$ below:

$$Y_{0,0}(\theta,\phi) = \frac{1}{2}\sqrt{\frac{1}{\pi}} \tag{2.22}$$

$$Y_{1,0}(\theta,\phi) = \frac{1}{2}\sqrt{\frac{3}{2\pi}}\cos\theta \tag{2.23}$$

$$Y_{1,\pm1}(\theta,\phi) = \mp\frac{1}{2}\sqrt{\frac{3}{2\pi}}e^{\pm i\phi}\sin\theta \tag{2.24}$$

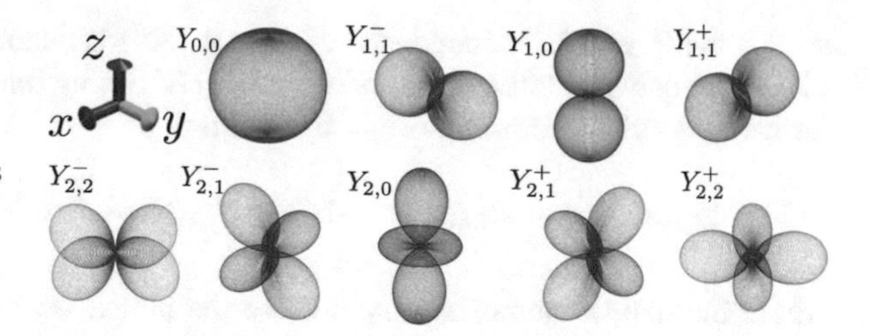

Fig. 2.1 Spherical harmonics in the $Y_{l,|m|}^{\pm}$ form. These harmonics are used to describe s- (l=0,m=0); p- (l=1, m=-1,0,+1) and d-states (l=2, m=-2,-1,0,+1,+2)

$$Y_{2,0}(\theta,\phi) = \frac{1}{4}\sqrt{\frac{5}{\pi}}(3\cos^2\theta - 1) \tag{2.25}$$

$$Y_{2,\pm1}(\theta,\phi) = \mp\frac{1}{2}\sqrt{\frac{15}{2\pi}}e^{\pm i\phi}\sin\theta\cos\theta \tag{2.26}$$

$$Y_{2,\pm2}(\theta,\phi) = \frac{1}{4}\sqrt{\frac{15}{2\pi}}e^{\pm 2i\phi}\sin^2\theta. \tag{2.27}$$

Orbitals with $m = 0$ are real, while real orbitals for the $m \neq 0$ cases can be obtained by the following transformation:

$$Y_{l,|m|}^{\pm} = \frac{1}{2}(Y_{l,m} \pm Y_{l,-m}). \tag{2.28}$$

Plots for the individual $Y_{l,|m|}^{\pm}$ are shown in Fig. 2.1.

It is often more convenient to use the complex $e^{\pm im\phi}$ expressions in actual calculations since this functional form allows us to simplify the calculation of the two-centered integrals contributing to the Hamiltonian matrix elements as discussed in the next section.

2.5 Hamiltonian Matrix Elements

The Hamiltonian matrix elements are written in terms of the localized basis functions used in the form of angular momentum spherical harmonics in Sect. 2.4:

$$H_{j,l}(\mathbf{R}',\mathbf{R}'') = \int dr^3\phi_j^*(\mathbf{r}-\mathbf{R}')\hat{H}\phi_l(\mathbf{r}-\mathbf{R}''). \tag{2.29}$$

In addition, the translational crystal symmetry allows us to write:

$$H_{j,l}(\mathbf{R}'-\mathbf{R}''',\mathbf{R}''-\mathbf{R}''') = H_{j,l}(\mathbf{R}',\mathbf{R}'') \tag{2.30}$$

so that the $\mathbf{R}'$ and $\mathbf{R}''$ dependence of $H_{j,l}(\mathbf{R}', \mathbf{R}'')$ is determined exclusively by the difference between lattice vectors $\mathbf{R} = \mathbf{R}'' - \mathbf{R}'$. Using this fact, we can refer all the lattice vectors to a common origin by writing

$$H_{j,l}(\mathbf{R}', \mathbf{R}'') = H_{j,l}(\mathbf{0}, \mathbf{R}'' - \mathbf{R}') = H_{j,l}(\mathbf{R}) = \int dr^3 \phi_j^*(\mathbf{r}) \hat{H} \phi_l(\mathbf{r} - \mathbf{R}), \quad (2.31)$$

where the matrix elements only involve the lattice vector to the common origin. Similarly, the wavefunction overlap terms are given by:

$$S_{j,l}(\mathbf{R}) = \int dr^3 \phi_j^*(\mathbf{r}) \phi_l(\mathbf{r} - \mathbf{R}). \quad (2.32)$$

The one electron Hamiltonian operator then has the form:

$$\hat{H} = \hat{T} + \sum_{p,\mathbf{R}} V(|\mathbf{r} - \mathbf{r}_p - \mathbf{R}|) \quad (2.33)$$

where $\hat{T}$ is the one-electron kinetic energy operator and $V(|\mathbf{r} - \mathbf{r}_p - \mathbf{R}|)$ is the potential energy decomposed into a sum of spherically symmetric terms centered at the atoms located at positions $\mathbf{r}_p$ relative to the unit cell located at $\mathbf{R}$. The kinetic energy contribution to the Hamiltonian matrix elements can be composed of one- or two-centered integrals depending on whether or not the orbitals i and j are centered at the same atom. Since the potential can be viewed as a sum of spherically symmetric terms, the contributions of the potential to the Hamiltonian matrix element can also have three-center integrals as well as one- and two-center integrals. We can readily notice four different types of potential energy contributions:

- **One-center:** when both orbitals and the potential are centered on the same atom;
- **Two-center 1:** when the orbitals are centered on different atoms and the potential is on one of these atoms;
- **Two-center 2:** when both orbitals are centered on the same atom and the potential is on another atom;
- **Three-center:** when both orbitals and the potential are all centered on different atoms.

The overlap terms are always composed of one- or two-center integrals. The important aspects of the integration can be easily addressed for the two-center integrals. Let $M_{lm,l'm'}$ be a two-center integral, between two orbitals from different atoms, corresponding to the kinetic or potential energy contributions to a Hamiltonian matrix element or to an overlap matrix element. For simplicity, let us suppose that the line joining the two centers corresponds to the z-axis. We can then write:

$$M_{lm,l'm'} = \int f_1(\rho_1) f_2(\rho_2) Y_{l,m}^*(\hat{\rho}_1) Y_{l',m'}(\hat{\rho}_2) d^3\mathbf{r} \quad (2.34)$$

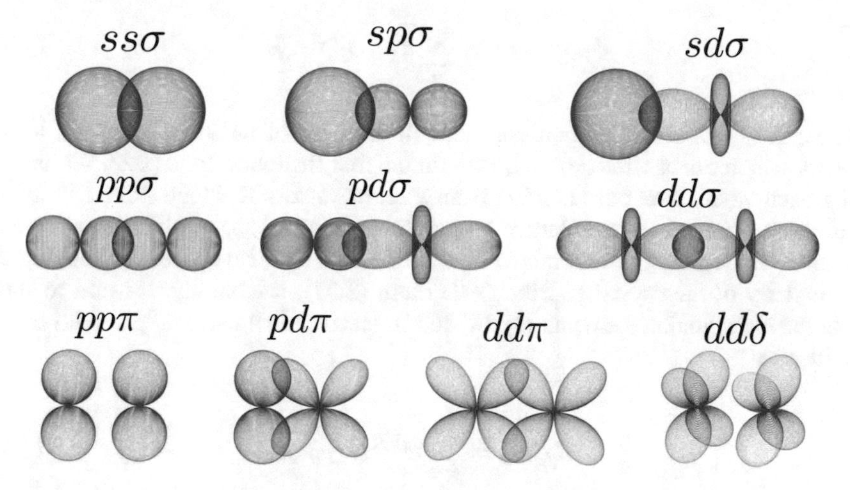

Fig. 2.2 Different two-centered integral schemes for the Hamiltonian matrix elements using a localized basis, and using the following notation: $l = 0, 1, 2, 3, \ldots$ denotes respectively $s, p, d, f, \ldots$ and $m = 0, \pm1, \pm2, \ldots$ denotes respectively σ, π, δ angular momentum states

with

$$\hat{\rho}_i = \frac{\mathbf{r} - \mathbf{r}_i}{|\mathbf{r} - \mathbf{r}_i|} \quad i = 1, 2. \tag{2.35}$$

The ϕ dependence from this integral can be isolated so that:

$$M_{lm,l'm'} = \frac{M_{ll'm}}{2\pi} \int_0^{2\pi} e^{-im\phi} e^{im'\phi} d\phi = M_{ll'm} \delta_{mm'}. \tag{2.36}$$

Equation 2.36 represents a significant simplification for the calculation of these angular momentum terms. The usual nomenclature for such quantities is to denote orbital quantum number $l = 0, 1, 2, 3, \ldots$ by $s, p, d, f, \ldots$ and $m = 0, \pm1, \pm2, \ldots$ by $\sigma, \pi, \delta, \ldots$. In Fig. 2.2 we illustrate the different integral schemes for $l = 0, 1, 2$.

2.6 Bloch Functions

Even with the simplifications introduced in the last section, it is impractical to work with the Hamiltonian in the simple atomic orbitals representation for a periodic solid. Instead we take the advantage of the periodicity V(r) of a crystalline lattice. Bloch's theorem (see Bloch 1928) indicates that we can write the eigenfunctions $\phi_{j\mathbf{k}}(\mathbf{r})$ for electrons in a periodic potential and in a single unit cell as:

$$\phi_{j\mathbf{k}}(\mathbf{r}) = N_{j\mathbf{k}} \sum_{\mathbf{R}} e^{i\mathbf{k}\cdot\mathbf{R}} \phi_j(\mathbf{r} - \mathbf{R}) \tag{2.37}$$

where $N_{j\mathbf{k}}$ is a normalization constant. In the case of an infinite crystal, $\mathbf{k}$ is a vector which can assume any value within the first Brillouin Zone (BZ). Within this approach we redirect our attention from a set of vectors $\mathbf{R}$ which extend along the infinite real space to a set of lattice vectors $\mathbf{k}$ in reciprocal space which are contained within a finite portion (determined by the BZ) of reciprocal space. By utilizing the periodicity of the crystal in Bloch's theorem (2.37), it is straightforward to show that the Hamiltonian elements $H_{j,l}(\mathbf{k})$ for an electron in a periodic potential can be written as

$$H_{j,l}(\mathbf{k}) = \sum_{\mathbf{R}} e^{i\mathbf{k}\cdot\mathbf{R}} H_{j,l}(\mathbf{R}). \tag{2.38}$$

Analogously, for the overlap matrix elements $S_{i,j}$, we can also utilize Bloch's theorem to write

$$S_{j,l}(\mathbf{k}) = \sum_{\mathbf{R}} e^{i\mathbf{k}\cdot\mathbf{R}} S_{j,l}(\mathbf{R}). \tag{2.39}$$

Now, we can then expand the electronic eigenfunctions ψ in terms of the eigenfunctions $\phi_{j\mathbf{k}}(\mathbf{r})$ as:

$$\Psi_\alpha(\mathbf{r}) = \sum_j c_{j\alpha} \phi_{j\mathbf{k}}(\mathbf{r}). \tag{2.40}$$

Hence, the Schrödinger equation now reads:

$$\hat{H} \Psi_\alpha(\mathbf{r}) = E_\alpha \Psi_\alpha(\mathbf{r}) \tag{2.41}$$

or alternatively, when using the LCAO basis we can write:

$$\sum_j c_{j\alpha} \hat{H} \phi_{j\mathbf{k}}(\mathbf{r}) = E_\alpha \sum_j c_{j\alpha} \phi_{j\mathbf{k}}(\mathbf{r}). \tag{2.42}$$

If we multiply on the left by $\phi^*_{l\mathbf{k}}(\mathbf{r})$ and integrate over space, we end up with:

$$\sum_j c_{j\alpha} \int \phi^*_{l\mathbf{k}}(\mathbf{r}) \hat{H} \phi_{j\mathbf{k}}(\mathbf{r}) d^3r = E_\alpha \sum_j c_{j\alpha} \int \phi^*_{l\mathbf{k}}(\mathbf{r}) \phi_{j\mathbf{k}}(\mathbf{r}) d^3r \tag{2.43}$$

$$\sum_j c_{j\alpha} H_{l,j}(\mathbf{k}) = \sum_j c_{j\alpha} E_\alpha S_{l,j}(\mathbf{k}) \tag{2.44}$$

$$\mathbf{H}\mathbf{c}_\alpha = E_\alpha \mathbf{S}\mathbf{c}_\alpha, \tag{2.45}$$

where $\mathbf{H}$ and $\mathbf{S}$ are square matrices with elements $H_{l,j}(\mathbf{k})$ and $S_{l,j}(\mathbf{k})$, respectively, and $\mathbf{c}_\alpha$ is a column vector with matrix elements $c_{j\alpha}$. The energy eigenvalues are then obtained algebraically by the secular equation:

$$|\mathbf{H} - E_\alpha \mathbf{S}| = 0 \tag{2.46}$$

where | | denotes the determinant commonly used to solve the eigenvalue problem explicitly.

A computational procedure is then used to calculate the electronic structure $E(\mathbf{k})$, where we use a discrete set of vectors $\mathbf{k}_i$, $i = 1, 2, 3, \ldots, N$. These $\mathbf{k}$ vectors represent how the electronic states behave as a function of $\mathbf{k}$. Since the Hamiltonian matrix elements coupling different $\mathbf{k}$ vectors are zero, we can write the secular (2.46) in block diagonal form in terms of the energy levels E_i for the various i eigenvalues each of which can have degenerate energy levels depending on the symmetry of the crystal structures. More explicitly

$$\begin{vmatrix} \mathbf{H}(\mathbf{k}_1) - E_\alpha \mathbf{S}(\mathbf{k}_1) & 0 & 0 & \ldots & 0 \\ 0 & \mathbf{H}(\mathbf{k}_2) - E_\alpha \mathbf{S}(\mathbf{k}_2) & 0 & \ldots & 0 \\ 0 & 0 & \mathbf{H}(\mathbf{k}_3) - E_\alpha \mathbf{S}(\mathbf{k}_3) & \ldots & 0 \\ \vdots & \vdots & \vdots & \ddots & \vdots \\ 0 & 0 & 0 & \ldots & \mathbf{H}(\mathbf{k}_N) - E_\alpha \mathbf{S}(\mathbf{k}_N) \end{vmatrix} = 0 \tag{2.47}$$

and such a matrix diagonalization can be broken into smaller sub-blocks:

$$|\mathbf{H}(\mathbf{k}_i) - E_\alpha \mathbf{S}(\mathbf{k}_i)| = 0, \qquad i = 1, 2, 3, \ldots, N \tag{2.48}$$

for each $\mathbf{k}$-point $\mathbf{k}_i$, and at high symmetry points the various blocks will show the appropriate degeneracies satisfying the symmetry requirements of the potential. For these k-points where degeneracies in energy occur, we select appropriate linear combinations of the wave functions which are each orthogonal to one another.

2.7 The Slater–Koster Approach

Felix Bloch introduced the concept of an electronic energy band structure $E(\mathbf{k})$ and his famous "Bloch's Theorem" to handle the symmetry of a periodic lattice. (see Bloch 1928) Later, Jones and co-workers were the first to expand the original s-symmetry-only approach to take into account a basis of different orbitals. (see Jones et al. 1934) However, the Tight-Binding (TB) model in the form it is widely used today was presented by Slater and Koster. (see Slater and Koster 1954) This is the simplest model to solve the electronic problem of periodic systems and, despite its simplicity, it gives excellent results and deep insight into the solid state lattice periodicity and surface phenomena. In this TB approach, one uses a basis of highly localized atomic orbitals and considers the Hamiltonian matrix elements of the system using empirical parameters that work well for rapid calculations of real materials. (see M. Martin 1970)

The TB parameters are further simplified by discarding three-center-integral contributions to the Hamiltonian martix elements. (see Slater and Koster 1954) We are then restricted to considering only the one-center and two-center contributions. The two-center integrals are then simplified using (2.36). However when applying (2.36), one can argue that the choice for the axis will in general not coincide with the line joining the atoms. However it is always possible to write the spherical harmonics relative to the bond line as a linear combinations of the spherical harmonics relative to the z-axis. Using these transformations we can write the two-center-integral contributions to the Hamiltonian matrix elements as a linear combination of the $M_{ll'm}$ terms using (2.36). Slater and Koster came up with expressions for the elements involving the s, p and d orbitals. (see Slater and Koster 1954) Below we reproduce these relations for the case of s and p orbitals, using the same notation used in (2.36) and are described in the caption to Fig. 2.2.

$$M_{s,s} = M_{s,s,\sigma} \tag{2.49}$$

$$M_{s,p_z} = z^2 M_{s,p,\sigma} \tag{2.50}$$

$$M_{p_x,p_x} = x^2 M_{p,p,\sigma} + (1 - x^2) M_{p,p,\pi} \tag{2.51}$$

$$M_{p_x,p_y} = xy(M_{p,p,\sigma} - M_{p,p,\pi}). \tag{2.52}$$

The TB parameters are fitted to reproduce the crystal properties (such as electronic energy bands or lattice parameters) of a given model system. In addition, one also has to define a cutoff radius for the distance between the atoms so that the Hamiltonian matrix elements for the atomic orbitals are zero when the atoms are separated by a distance larger than the specified cutoff.

References

M. Born, J.R. Oppenheimer, Zur quantentheorie der molekeln. Annalen der Physik **84**, 457 (1927)

R.M. Martin, *Electronic structure*, Basic Theory and Practical Methods (Cambridge University Press, Cambridge, 2004)

D. R. Hartree, The wave mechanics of an atom with non-coulombic central fields: parts i, ii, iii. Mathematical Proceedings of The Cambridge Philosophical Society 24, 89, 111, 426 (1928)

J.C. Slater, Note on Hartree's method. Phys. Rev. **35**, 210 (1930)

V. Fock, Nherungsmethode zur Lsung des quantenmechanischen Mehrkrperproblems. Z. Phys. **61**, 126 (1930)

J.C. Slater, The theory of complex spectra. Phys. Rev. **34**, 1293 (1929)

W. Pauli, ber den Zusammenhang des Abschlusses der Elektronengruppen im Atom mit der Komplexstruktur der Spektren. Z. Phys. **31**, 765 (1925)
P. Hohenberg, W. Kohn, Inhomogeneous electron gas. Phys. Rev. **136**, B864 (1964)
W. Kohn, J. Sham, Self-consistent equations including exchange and correlation effects. Phys. Rev. **140**, A1133 (1965)
F. Bloch, Uber die Quantenmechanik der Elektronen in Kristallgittern. Z. Phyzik **52**, 555 (1928)
H. Jones, N.F. Mott, H.W.B. Skinner, A theory of the form of the X-Ray emission bands of metals. Phys. Rev. **45**, 379 (1934)
J.C. Slater, G.F. Koster, Simplified LCAO method for the periodic potential problem. Phys. Rev. **94**, 1498 (1954)
M. Martin, Elastic properties of ZnS structure semiconductors. Phys. Rev. B **1**, 4005 (1970)

Chapter 3
Weak and Tight Binding Approximations for Simple Solid State Models

3.1 Introduction

Many of the physical properties of solids are closely related to the electronic energy dispersion relations $E(\mathbf{k})$ in these materials, and in particular to the behavior of $E(\mathbf{k})$ within a few electron volts (eV) in energy from the Fermi level E_f. Conversely, the analysis of transport (discussed in Part I) and other physical measurements (discussed in other parts of this book) provides a great deal of information about $E(\mathbf{k})$. Although transport measurements do not generally provide the most sensitive tool for studying $E(\mathbf{k})$, measurements of the electronic conductivity and Hall effect are fundamental to solid state physics because they can be carried out on nearly all materials and therefore provide a valuable tool for characterizing the carrier density, carrier type and carrier mobility of materials. To provide the necessary background for the discussion of transport properties, we give here a brief review of the Energy dispersion relations $E(\mathbf{k})$ in solids. In this connection, we consider in Chap. 3 the two limiting cases of weak and tight binding, which are useful and simple approximations for giving physical insights. In Chap. 4 we discuss $E(\mathbf{k})$ for real solids including prototype metals, semiconductors, semimetals and insulators.

3.2 One Electron $E(\mathrm{K})$ in Solids

3.2.1 *Weak Binding or the Nearly Free Electron Approximation*

The simplest model for discussing the electronic behavior of electrons in solids is the so called weak binding or nearly free electron approximation. This model is based on the following four assumptions:

M. Dresselhaus et al., *Solid State Properties*, Graduate Texts in Physics,
https://doi.org/10.1007/978-3-662-55922-2_3

1. The periodic potential $V(\mathbf{r}) = V(\mathbf{r} + \mathbf{R_n})$ is sufficiently weak so that the electrons behave if they were almost free.
2. The effect of a periodic potential on the nearly free electron is treated within the framework of perturbation theory.
3. The potential $V(\mathbf{r})$ can be an arbitrary periodic potential.
4. This model is appropriate for describing valence electrons in simple metals.

The weak binding approximation has achieved some success in describing the electronic properties of valence electrons in many simple metals. For the core electrons closest to the nucleus, however, the potential energy is comparable to the kinetic energy so that core electrons are tightly bound and the weak binding approximation is not appropriate. If the electrons are, on the other hand, strongly bound to the atomic nucleus, the tight binding approximation, discussed in Sect. 3.2.2 is appropriate.

In the weak binding approximation, we solve the Schrödinger equation in the limit of a very weak periodic potential $V(\mathbf{r})$

$$\mathscr{H}\psi = \mathscr{H}_0 + V(\mathbf{r}) = E\psi. \tag{3.1}$$

Using time–independent perturbation theory (see Appendix A), we write the energy $E(\mathbf{k})$ as

$$E(\mathbf{k}) = E^{(0)}(\mathbf{k}) + E^{(1)}(\mathbf{k}) + E^{(2)}(\mathbf{k}) + ... \tag{3.2}$$

and take the unperturbed (free electron solution) $E^{(0)}(\mathbf{k})$ to correspond to $V(\mathbf{r}) = 0$ so that $E^{(0)}(\mathbf{k})$ is the plane wave, free electron, solution

$$E^{(0)}(\mathbf{k}) = \frac{\hbar^2 k^2}{2m}, \tag{3.3}$$

in which m is the free electron mass. The corresponding normalized eigenfunctions are the plane wave states

$$\psi_{\mathbf{k}}^{(0)}(\mathbf{r}) = \frac{e^{i\mathbf{k}\cdot\mathbf{r}}}{\Omega^{1/2}} \tag{3.4}$$

in which Ω is the volume of the crystal.

The first order correction to the energy $E^{(1)}(\mathbf{k})$ is the diagonal matrix element of the perturbation potential taken between the unperturbed states:

$$\begin{aligned} E^{(1)}(\mathbf{k})\ \langle\psi_{\mathbf{k}}^{(0)} \mid V(\mathbf{r}) \mid \psi_{\mathbf{k}}^{(0)}\rangle &= \tfrac{1}{\Omega}\textstyle\int_{\Omega} e^{-i\mathbf{k}\cdot\mathbf{r}}V(\mathbf{r})e^{i\mathbf{k}\cdot\mathbf{r}}d^3r \\ \tfrac{1}{\Omega_0}\textstyle\int_{\Omega_0} V(\mathbf{r})d^3r &= \overline{V(\mathbf{r})} \end{aligned} \tag{3.5}$$

where $\overline{V(\mathbf{r})}$, denotes the average over the unit cell of the crystal, and the value of $\overline{V(\mathbf{r})}$ is independent of $\mathbf{k}$, where Ω_0 is the volume of the unit cell and is independent of the choice of the origin of the coordinate system. Thus, in first order perturbation theory, we merely add a constant energy $\overline{V(\mathbf{r})}$ to the energy of the free electron particle energy, and that constant term is exactly the mean potential energy seen by the electron, averaged over the unit cell. The terms of particular interest arise in second order perturbation theory and are

$$E^{(2)}(\mathbf{k}) = \sum_{\mathbf{k}'}{}' \frac{|\langle \mathbf{k}'|V(\mathbf{r})|\mathbf{k}\rangle|^2}{E^{(0)}(\mathbf{k}) - E^{(0)}(\mathbf{k}')} \tag{3.6}$$

where the prime over the summation indicates that $\mathbf{k}' \neq \mathbf{k}$ or more explicitly that the coupling of wave vectors is second order couple different wave vector. We next compute the off-diagonal matrix element $\langle \mathbf{k}'|V(\mathbf{r})|\mathbf{k}\rangle$ in (3.6) as follows:

$$\begin{aligned} \langle \mathbf{k}'|V(\mathbf{r})|\mathbf{k}\rangle &= \int_\Omega \psi_{\mathbf{k}'}^{(0)*} V(\mathbf{r}) \psi_{\mathbf{k}}^{(0)} d^3r \\ &= \tfrac{1}{\Omega} \int_\Omega e^{-i(\mathbf{k}'-\mathbf{k})\cdot\mathbf{r}}\, V(\mathbf{r}) d^3r \\ &= \tfrac{1}{\Omega} \int_\Omega e^{i\mathbf{q}\cdot\mathbf{r}} V(\mathbf{r}) d^3r \end{aligned} \tag{3.7}$$

where $\mathbf{q}$ is the difference wave vector $\mathbf{q} = \mathbf{k} - \mathbf{k}'$ and the integration is over the whole crystal. We now exploit the periodicity of the potential $V(\mathbf{r})$. Let $\mathbf{r} = \mathbf{r}' + \mathbf{R}_n$ where $\mathbf{r}'$ is an arbitrary vector in a unit cell and $\mathbf{R}_n$ is a periodic lattice vector. Then because of the periodicity $V(\mathbf{r}) = V(\mathbf{r}')$

$$\langle \mathbf{k}'|V(\mathbf{r})|\mathbf{k}\rangle = \frac{1}{\Omega} \sum_n \int_{\Omega_0} e^{i\mathbf{q}\cdot(\mathbf{r}'+\mathbf{R}_n)} V(\mathbf{r}') d^3r' \tag{3.8}$$

where the sum is over all unit cells and the integration is over the volume Ω_0 of one unit cell. Then

$$\langle \mathbf{k}'|V(\mathbf{r})|\mathbf{k}\rangle = \frac{1}{\Omega} \sum_n e^{i\mathbf{q}\cdot\mathbf{R}_n} \int_{\Omega_0} e^{i\mathbf{q}\cdot\mathbf{r}'} V(\mathbf{r}') d^3r'. \tag{3.9}$$

Writing the following expressions for the lattice vectors $\mathbf{R}_n$ and for the difference wave vectors $\mathbf{q}$

$$\begin{aligned} \mathbf{R}_n &= \textstyle\sum_{j=1}^3 n_j \mathbf{a}_j \\ \mathbf{q} &= \textstyle\sum_{j=1}^3 \alpha_j \mathbf{b}_j \end{aligned} \tag{3.10}$$

where n_j is an integer, and $\mathbf{a}_j$ and $\mathbf{b}_j$ are unit vectors in real and reciprocal space, respectively. The lattice sum $\sum_n e^{i\mathbf{q}\cdot\mathbf{R}_n}$ can then be carried out exactly to yield

$$\sum_n e^{i\mathbf{q}\cdot\mathbf{R}_n} = \left[\prod_{j=1}^{3} \frac{1 - e^{2\pi i N_j \alpha_j}}{1 - e^{2\pi i \alpha_j}}\right] \tag{3.11}$$

where $N = N_1 N_2 N_3$ is the total number of unit cells in the crystal and α_j is a real number. This sum generally fluctuates wildly as $\mathbf{q}$ varies but averages to zero. The total sum is appreciable only if

$$\mathbf{q} = \sum_{j=1}^{3} m_j \mathbf{b}_j \tag{3.12}$$

where m_j is an integer and $\mathbf{b}_j$ is a primitive vector in reciprocal space, so that $\mathbf{q}$ must be a reciprocal lattice vector $\mathbf{G}$. Hence, since $\mathbf{b_j} \cdot \mathbf{R_n} = 2\pi l_{jn}$ where l_{jn} is an integer, we can write

$$\sum_n e^{i\mathbf{q}\cdot\mathbf{R}_n} = N\delta_{\mathbf{q},\mathbf{G}} \tag{3.13}$$

where δ is the Kronecker delta function which vanishes unless $\mathbf{q} = \mathbf{G}$.

This discussion shows that the matrix element $\langle \mathbf{k}'|V(\mathbf{r})|\mathbf{k}\rangle$ is only important when $\mathbf{q} = \mathbf{G}$ is a reciprocal lattice vector $= \mathbf{k} - \mathbf{k}'$ from which we conclude that the periodic potential $V(\mathbf{r})$ only connects wave vectors $\mathbf{k}$ and $\mathbf{k}'$ separated by a reciprocal lattice vector. We note that this is the same relation that determines the Brillouin zone boundary. The matrix element $\langle \mathbf{k}'|V(\mathbf{r})|\mathbf{k}\rangle$ is then

$$\langle \mathbf{k}'|V(\mathbf{r})|\mathbf{k}\rangle = \frac{N}{\Omega}\int_{\Omega_0} e^{i\mathbf{G}\cdot\mathbf{r}'} V(\mathbf{r}')d^3r'\delta_{\mathbf{k}'-\mathbf{k},\mathbf{G}} \tag{3.14}$$

where

$$\frac{N}{\Omega} = \frac{1}{\Omega_0} \tag{3.15}$$

and the integration in (3.14) is over the unit cell with a volume Ω_0. We introduce $V_{\mathbf{G}}$, which is the Fourier coefficient of the periodic potential $V(\mathbf{r})$ where

$$V_{\mathbf{G}} = \frac{1}{\Omega_0}\int_{\Omega_0} e^{i\mathbf{G}\cdot\mathbf{r}'} V(\mathbf{r}')d^3r' \tag{3.16}$$

so that

$$\langle \mathbf{k}'|V(\mathbf{r})|\mathbf{k}\rangle = \delta_{\mathbf{k}-\mathbf{k}',\mathbf{G}}\ V_{\mathbf{G}}. \tag{3.17}$$

We can now use this matrix element to calculate the 2nd order change in the energy based on perturbation theory (see Appendix A)

$$E^{(2)}(\mathbf{k}) = \sum_{\mathbf{G}} \frac{|V_{\mathbf{G}}|^2}{k^2 - (k')^2}\left(\frac{2m}{\hbar^2}\right) = \frac{2m}{\hbar^2}\sum_{\mathbf{G}} \frac{|V_{\mathbf{G}}|^2}{k^2 - |\mathbf{G}+\mathbf{k}|^2}. \tag{3.18}$$

We observe that when $k^2 = |\mathbf{G}+\mathbf{k}|^2$, the denominator in (3.18) vanishes and $E^{(2)}(\mathbf{k})$ can become very large. This condition is identical with the Laue X-ray diffraction condition. Thus, at a Brillouin zone boundary, the weak perturbing potential can have a very large effect and therefore non–degenerate perturbation theory will not work in this case.

For **k** values near a Brillouin zone boundary, we must then use degenerate perturbation theory (see Appendix B). Since the matrix elements coupling the plane wave states **k** and **k + G** do not vanish, *first-order degenerate* perturbation theory is sufficient and leads to the determinantal equation coupling two states separated by a reciprocal lattice vector **G**.

$$\begin{vmatrix} E^{(0)}(\mathbf{k}) + E^{(1)}(\mathbf{k}) - E & \langle \mathbf{k}+\mathbf{G}|V(\mathbf{r})|\mathbf{k}\rangle \\ \langle \mathbf{k}|V(\mathbf{r})|\mathbf{k}+\mathbf{G}\rangle & E^{(0)}(\mathbf{k}+\mathbf{G}) + E^{(1)}(\mathbf{k}+\mathbf{G}) - E \end{vmatrix} = 0 \tag{3.19}$$

in which

$$\begin{array}{c} E^{(0)}(\mathbf{k})\ \hbar^2 k^2/2m \\ E^{(0)}(\mathbf{k}+\mathbf{G})\ \left[\hbar^2|\mathbf{k}+\mathbf{G}|^2\right]/2m \end{array} \tag{3.20}$$

and both

$$E^{(1)}(\mathbf{k}) = \langle \mathbf{k}|V(\mathbf{r})|\mathbf{k}\rangle = \overline{V(\mathbf{r})} = V_0 \tag{3.21}$$

and

$$E^{(1)}(\mathbf{k}+\mathbf{G}) = \langle \mathbf{k}+\mathbf{G}|V(\mathbf{r})|\mathbf{k}+\mathbf{G}\rangle = V_0. \tag{3.22}$$

are equal to the same constant value V_0. Solution of the determinantal equation (3.19) yields:

$$[E - V_0 - E^{(0)}(\mathbf{k})][E - V_0 - E^{(0)}(\mathbf{k}+\mathbf{G})] - |V_{\mathbf{G}}|^2 = 0, \tag{3.23}$$

or equivalently we can write this in one expanded algebraic form

$$E^2 - E[2V_0 + E^{(0)}(\mathbf{k}) + E^{(0)}(\mathbf{k}+\mathbf{G})] + [V_0 + E^{(0)}(\mathbf{k})][V_0 + E^{(0)}(\mathbf{k}+\mathbf{G})] - |V_{\mathbf{G}}|^2 = 0. \tag{3.24}$$

Solution of the quadratic equation (3.24) yields

$$E^{\pm} = V_0 + \frac{1}{2}[E^{(0)}(\mathbf{k}) + E^{(0)}(\mathbf{k}+\mathbf{G})] \pm \sqrt{\frac{1}{4}[E^{(0)}(\mathbf{k}) - E^{(0)}(\mathbf{k}+\mathbf{G})]^2 + |V_{\mathbf{G}}|^2} \tag{3.25}$$

and we come out with two solutions for the two strongly coupled states. It is of interest to look at these two solutions in two limiting cases:

Case (i)

$$|V_{\mathbf{G}}| \ll \frac{1}{2}|[E^{(0)}(\mathbf{k}) - E^{(0)}(\mathbf{k}+\mathbf{G})]|$$

In the case of small $|V_{\mathbf{G}}|$, we can expand the square root expression in (3.25) for small $|V_{\mathbf{G}}|$ to obtain:

$$\begin{aligned} E(\mathbf{k}) &= V_0 + \tfrac{1}{2}[E^{(0)}(\mathbf{k}) + E^{(0)}(\mathbf{k}+\mathbf{G})] \\ &\pm \tfrac{1}{2}[E^{(0)}(\mathbf{k}) - E^{(0)}(\mathbf{k}+\mathbf{G})] \cdot [1 + \tfrac{2|V_{\mathbf{G}}|^2}{[E^{(0)}(\mathbf{k}) - E^{(0)}(\mathbf{k}+\mathbf{G})]^2} + \ldots] \end{aligned} \tag{3.26}$$

which simplifies to the two solutions:

$$E^-(\mathbf{k}) = V_0 + E^{(0)}(\mathbf{k}) + \frac{|V_{\mathbf{G}}|^2}{E^{(0)}(\mathbf{k}) - E^{(0)}(\mathbf{k}+\mathbf{G})} \tag{3.27}$$

$$E^+(\mathbf{k}) = V_0 + E^{(0)}(\mathbf{k}+\mathbf{G}) + \frac{|V_{\mathbf{G}}|^2}{E^{(0)}(\mathbf{k}+\mathbf{G}) - E^{(0)}(\mathbf{k})} \tag{3.28}$$

and we recover the result in (3.18) obtained before using non–degenerate perturbation theory. This result in (3.18) is valid far from the Brillouin zone boundary, but near the zone boundary the more complete expression of (3.25) must be used.

Case (ii)

$$|V_{\mathbf{G}}| \gg \frac{1}{2}|[E^{(0)}(\mathbf{k}) - E^{(0)}(\mathbf{k}+\mathbf{G})]|$$

Sufficiently close to the Brillouin zone boundary we have large $|V_{\mathbf{G}}|$ yielding the relation:

$$|E^{(0)}(\mathbf{k}) - E^{(0)}(\mathbf{k}+\mathbf{G})| \ll |V_{\mathbf{G}}| \tag{3.29}$$

so we can expand $E(\mathbf{k})$ as given by (3.25) to obtain

$$E^{\pm}(\mathbf{k}) = \frac{1}{2}[E^{(0)}(\mathbf{k}) + E^{(0)}(\mathbf{k}+\mathbf{G})] + V_0 \pm \left[|V_{\mathbf{G}}| + \frac{1}{8}\frac{[E^{(0)}(\mathbf{k}) - E^{(0)}(\mathbf{k}+\mathbf{G})]^2}{|V_{\mathbf{G}}|} + \ldots \right] \tag{3.30}$$

$$\cong \frac{1}{2}[E^{(0)}(\mathbf{k}) + E^{(0)}(\mathbf{k}+\mathbf{G})] + V_0 \pm |V_{\mathbf{G}}|, \tag{3.31}$$

so that at the Brillouin zone boundary $E^+(\mathbf{k})$ is elevated by $|V_{\mathbf{G}}|$, while $E^-(\mathbf{k})$ is depressed by $|V_{\mathbf{G}}|$ and the band gap that is formed is $2|V_{\mathbf{G}}|$, where $\mathbf{G}$ is the reciprocal lattice vector for which $E(\mathbf{k}_{B.Z.}) = E(\mathbf{k}_{B.Z.} + \mathbf{G})$ in which the subscript B.Z. denotes

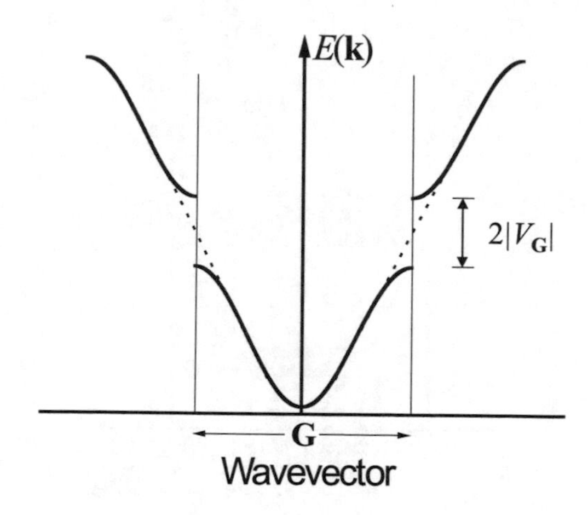

Fig. 3.1 One dimensional electronic energy bands for the nearly free electron model shown in the extended Brillouin zone scheme. The dashed curve corresponds to the case of free electrons and the solid curves to the case where a weak periodic potential is present. The band gaps at the zone boundaries have a magnitude of $2|V_{\mathbf{G}}|$

the Brillouin zone boundary and $V_{\mathbf{G}}$ is the Fourier transform of the periodic potential in real space given by

$$V_{\mathbf{G}} = \frac{1}{\Omega_0} \int_{\Omega_0} e^{i\mathbf{G}\cdot\mathbf{r}} V(\mathbf{r}) d^3r. \tag{3.32}$$

From this discussion it is clear that every Fourier component of the periodic potential gives rise to a specific band gap in reciprocal space. We see further that the *band gap* represents a range of energy values for which there is no solution to the eigenvalue problem of (3.19) for real k (see Fig. 3.1). In the band gap we assign an imaginary value to the wave vector which can be interpreted as a highly damped and non–propagating plane wave with wavevector $\mathbf{G}$.

We note that for such a plane wave the larger the value of $\mathbf{G}$, the smaller the value of $V_{\mathbf{G}}$, so that higher Fourier components give rise to smaller band gaps. Near these energy discontinuities at the Brillouin zone boundary, thc wave functions become linear combinations of the unperturbed states

$$\begin{aligned} \psi_{\mathbf{k}} &= \alpha_1 \psi_{\mathbf{k}}^{(0)} + \beta_1 \psi_{\mathbf{k}+\mathbf{G}}^{(0)} \\ \psi_{\mathbf{k}+\mathbf{G}} &= \alpha_2 \psi_{\mathbf{k}}^{(0)} + \beta_2 \psi_{\mathbf{k}+\mathbf{G}}^{(0)} \end{aligned} \tag{3.33}$$

and at the zone boundary itself, instead of traveling waves $e^{i\mathbf{k}\cdot\mathbf{r}}$, the wave functions become standing waves $\cos(\mathbf{k}\cdot\mathbf{r})$ and $\sin(\mathbf{k}\cdot\mathbf{r})$. We note that the $\cos(\mathbf{k}\cdot\mathbf{r})$ solution corresponds to a maximum in the charge density at the lattice sites and therefore corresponds to an energy minimum (for which the lower level in some cases would be identified with a valence band extrema). Likewise, the $\sin(\mathbf{k}\cdot\mathbf{r})$ solution corresponds to a minimum in the charge density and therefore corresponds to a maximum in the energy, thus forming the upper level which in some cases would be identified with an unoccupied conduction band state.

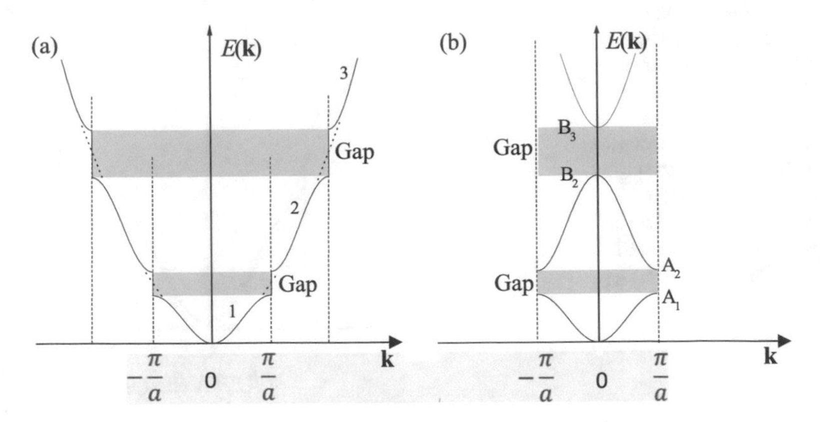

Fig. 3.2 **a** One dimensional electronic energy bands for the nearly free electron model shown in the reduced Brillouin zone scheme for the three bands of lowest energy. **b** The same $E(\mathbf{k})$ as in **(a)** but now shown in the extended zone scheme. The shaded areas denote the band gaps between bands n and $n+1$ and the white areas denote the band states

In constructing $E(\mathbf{k})$ for the reduced Brillouin zone scheme, we make use of the periodicity of $E(\mathbf{k})$ in reciprocal space

$$E(\mathbf{k}+\mathbf{G}) = E(\mathbf{k}), \tag{3.34}$$

and consider only one unit cell in reciprocal space. The reduced zone scheme more clearly illustrates the formation of energy bands (labeled (1) and (2) in Fig. 3.2), band gaps E_g and band widths (defined in Fig. 3.2 as the range of energy between the minimum (E_{min}) and the maximum (E_{max}) energy for a given band).

We now discuss the connection between the $E(\mathbf{k})$ relations shown above and the transport properties of solids, which can be illustrated by considering the case of a semiconductor. An intrinsic semiconductor at temperature $T = 0$ has no carriers so that the Fermi level runs right through the band gap. On the diagram of Fig. 3.2, this would mean that the Fermi level might run between bands (1) and (2), so that band (1) is completely occupied and band (2) is completely empty. One further property of the semiconductor is that the band gap E_g be small enough so that at some temperature (e.g., room temperature) there will be a reasonable number of thermally excited carriers, perhaps $10^{15}/\text{cm}^3$. The doping with donor (electron donating) impurities will raise the Fermi level above the conduction band edge and doping with acceptor (electron extracting) impurities will lower the Fermi level below the valence band edge. Neglecting for the moment the effect of impurities on the $E(\mathbf{k})$ relations for the perfectly periodic crystal, let us consider what happens when we raise the Fermi level into the bands. If we know the shape of the $E(\mathbf{k})$ curve, we can estimate the velocity of the electrons and also the so–called m^* of the electrons. From the diagram in Fig. 3.2 we see that the conduction bands tend to fill up electron states starting at their energy band extrema.

Since the energy bands have zero slope about their extrema, we can write $E(\mathbf{k})$ as a quadratic form in $\mathbf{k}$. It is convenient to write the proportionality in terms of the quantity called the effective mass m^* defined as

$$E(\mathbf{k}) = E(0) + \frac{\hbar^2 k^2}{2m^*} \tag{3.35}$$

so that m^* is defined by

$$\frac{1}{m^*} \equiv \frac{\partial^2 E(\mathbf{k})}{\hbar^2 \partial k^2} \tag{3.36}$$

and we can say in some approximate way that an electron in a solid moves as if it were a free electron but with an effective mass m^* rather than a free electron mass m. The larger the band curvature, the smaller the effective mass. The mean velocity of the electron $\mathbf{v_k}$ is also found from $E(\mathbf{k})$, according to the relation

$$\mathbf{v}_k = \frac{1}{\hbar}\frac{\partial E(\mathbf{k})}{\partial \mathbf{k}}. \tag{3.37}$$

For this reason the energy dispersion relations $E(\mathbf{k})$ are very important in the determination of the transport properties for electrons and hole carriers in solids, where a hole is defined as a positively charged carrier in the valence band state from which an electron has been excited to the conduction band.

3.2.2 Tight Binding Approximation

In the tight binding approximation a number of assumptions are made and these are different from the assumptions that are made for the weak binding approximation. The assumptions for the tight binding approximation are:

1. The energy eigenvalues and eigenfunctions are known for an electron in an isolated atom.
2. When the atoms are brought together to form a solid, the atoms remain sufficiently far apart from each other, so that each electron can be assigned to a particular atomic site. This assumption is not valid for valence electrons in metals which are not localized, and for this reason, the valence electrons are best treated by the weak binding approximation.
3. The periodic potential of the lattice V(r) is approximated by a superposition of atomic potentials.

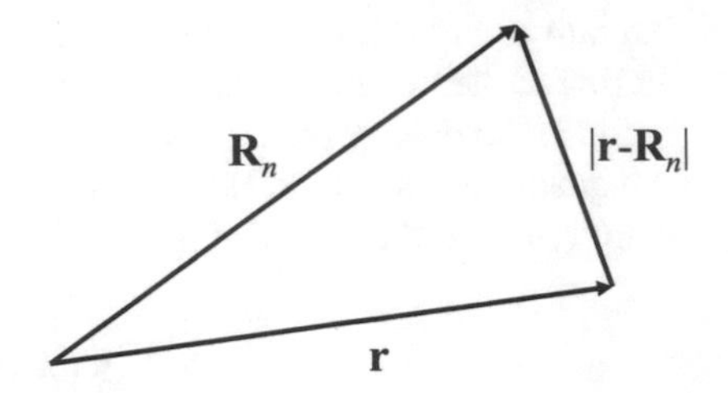

Fig. 3.3 Definition of the vectors used in the tight binding approximation

4. Perturbation theory can be used to treat the difference between the actual potential and the atomic potential.

Thus, both the weak and tight binding approximations are based on perturbation theory. For the weak binding approximation the unperturbed state is the free electron plane–wave state, while for the tight binding approximation, the unperturbed state is the atomic state. In the case of the weak binding approximation, the perturbation Hamiltonian is the weak periodic potential itself, while for the tight binding case, the perturbation is the *difference* between the periodic potential and the atomic potential around which the electron is localized.

We review here the major features of the tight binding approximation. Let $\phi(\mathbf{r} - \mathbf{R}_n)$ denote the atomic wave function for an atom where $\mathbf{r}$ and $\mathbf{R}_n$ are each measured with respect to the atom, and the vector $\mathbf{r} - \mathbf{R}_n$ is the atomic position measured relative to the lattice vector $\mathbf{R}_n$ as shown in Fig. 3.3. The Schrödinger equation for an electron in an isolated atom is then:

$$\left[-\frac{\hbar^2}{2m}\nabla^2 + U(\mathbf{r} - \mathbf{R}_n) - E^{(0)}\right]\phi(\mathbf{r} - \mathbf{R}_n) = 0 \tag{3.38}$$

where $U(\mathbf{r} - \mathbf{R}_n)$ is the atomic potential and $E^{(0)}$ is the atomic eigenvalue (see Fig. 3.3).

We now assume that the atoms are brought together to form the crystal for which $V(\mathbf{r})$ is the periodic potential, and $\psi(\mathbf{r})$ and $E(\mathbf{k})$ are, respectively, the wave function and energy eigenvalue for the electron in the crystal:

$$\left[-\frac{\hbar^2}{2m}\nabla^2 + V(\mathbf{r}) - E\right]\psi(\mathbf{r}) = 0. \tag{3.39}$$

In the tight binding approximation, we write $V(\mathbf{r})$ as a sum of atomic potentials:

$$V(\mathbf{r}) \simeq \sum_n U(\mathbf{r} - \mathbf{R}_n). \tag{3.40}$$

If the interaction between neighboring atoms is ignored, then each state has a degeneracy of $N =$ number of atoms in the crystal. However, the interaction between the atoms in the crystal lifts this degeneracy (see Fig. 3.4).

The energy eigenvalues $E(\mathbf{k})$ in the tight binding approximation for a non–degenerate s–state is simply given by

$$E(\mathbf{k}) = \frac{\langle \mathbf{k}|\mathcal{H}|\mathbf{k}\rangle}{\langle \mathbf{k}|\mathbf{k}\rangle} \tag{3.41}$$

because s states are non-degenerate if spin is not considered. The normalization factor in the denominator $\langle \mathbf{k}|\mathbf{k}\rangle$ is inserted because the wave functions $\psi_{\mathbf{k}}(\mathbf{r})$ in the tight binding approximation are usually not normalized. The Hamiltonian in the tight binding approximation is written as

$$\mathcal{H} = -\frac{\hbar^2}{2m}\nabla^2 + V(\mathbf{r}) = \left\{-\frac{\hbar^2}{2m}\nabla^2 + [V(\mathbf{r}) - U(\mathbf{r} - \mathbf{R}_n)] + U(\mathbf{r} - \mathbf{R}_n)\right\} \tag{3.42}$$

$$\mathcal{H} = \mathcal{H}_0 + \mathcal{H}' \tag{3.43}$$

in which $\mathcal{H}_0$ is the atomic Hamiltonian at site n

$$\mathcal{H}_0 = -\frac{\hbar^2}{2m}\nabla^2 + U(\mathbf{r} - \mathbf{R}_n) \tag{3.44}$$

and the perturbation $\mathcal{H}'$ is the difference between the actual periodic potential and the atomic potential at lattice site n

$$\mathcal{H}' = V(\mathbf{r}) - U(\mathbf{r} - \mathbf{R}_n). \tag{3.45}$$

We construct the wave functions for the unperturbed problem as a linear combination of atomic functions $\phi_j(\mathbf{r} - \mathbf{R}_n)$ labeled by quantum number j

$$\psi_j(\mathbf{r}) = \sum_{n=1}^{N} C_{j,n}\phi_j(\mathbf{r} - \mathbf{R}_n) \tag{3.46}$$

so that $\psi_j(\mathbf{r})$ is an eigenstate of a Hamiltonian satisfying the periodic potential of the lattice. In this treatment we assume that the tight binding wave–functions ψ_j can be identified with a *single* atomic state ϕ_j; this approximation must be relaxed in dealing with degenerate levels. According to Bloch's theorem, $\psi_j(\mathbf{r})$ in the solid must satisfy the relation:

$$\psi_j(\mathbf{r} + \mathbf{R}_m) = e^{i\mathbf{k}\cdot\mathbf{R}_m}\psi_j(\mathbf{r}) \tag{3.47}$$

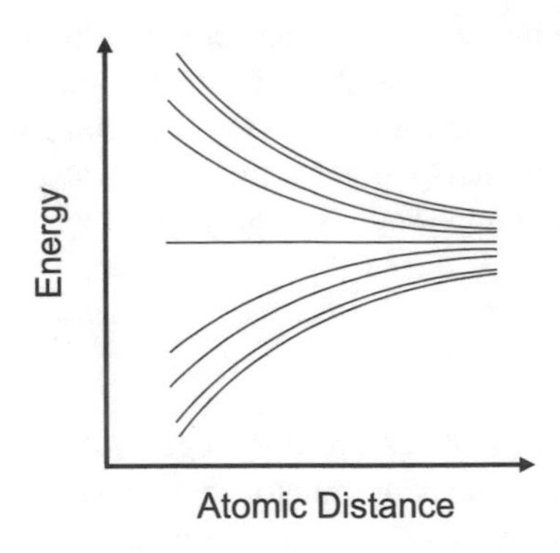

Fig. 3.4 The relation between atomic states on the right and the broadening due to the presence of neighboring atoms seen on the left. As the interatomic distance decreases (going to the left in the diagram), the level broadening increases so that a band of levels occurs at atomic separations characteristic of solids

where $\mathbf{R}_m$ is an arbitrary lattice vector. This restriction imposes a special form on the coefficients $C_{j,n}$ as described below.

Substitution of the expansion in the atomic functions $\psi_j(\mathbf{r})$ from (3.46) into the left side of (3.47) yields:

$$\begin{aligned}\psi_j(\mathbf{r}+\mathbf{R}_m) &= \textstyle\sum_n C_{j,n}\, \phi_j(\mathbf{r}-\mathbf{R}_n+\mathbf{R}_m)\\ &= \textstyle\sum_Q C_{j,Q+m}\, \phi_j(\mathbf{r}-\mathbf{R}_Q)\\ &= \textstyle\sum_n C_{j,n+m}\, \phi(\mathbf{r}-\mathbf{R}_n)\end{aligned} \tag{3.48}$$

where we have utilized the substitution $\mathbf{R}_Q = \mathbf{R}_n - \mathbf{R}_m$ and the fact that Q is a dummy index. Now for the right side of the Bloch theorem (3.47) we have

$$e^{i\mathbf{k}\cdot\mathbf{R}_m}\psi_j(\mathbf{r}) = \sum_n C_{j,n} e^{i\mathbf{k}\cdot\mathbf{R}_m}\phi_j(\mathbf{r}-\mathbf{R}_n). \tag{3.49}$$

The coefficients $C_{j,n}$ which relate the actual wave function $\psi_j(\mathbf{r})$ to the atomic functions $\phi_j(\mathbf{r}-\mathbf{R}_n)$ are therefore not arbitrary but must thus satisfy both the periodicity of the potential in the variable $\mathbf{r}$ as well as Bloch's theorem:

$$C_{j,n+m} = e^{i\mathbf{k}\cdot\mathbf{R}_m} C_{j,n} \tag{3.50}$$

which can be accomplished by setting:

$$C_{j,n} = \xi_j e^{i\mathbf{k}\cdot\mathbf{R}_n} \tag{3.51}$$

where the new coefficient ξ_j is independent of the band index n. We therefore obtain:

$$\psi_{j,\mathbf{k}}(\mathbf{r}) = \xi_j \sum_n e^{i\mathbf{k}\cdot\mathbf{R}_n}\phi_j(\mathbf{r}-\mathbf{R}_n) \tag{3.52}$$

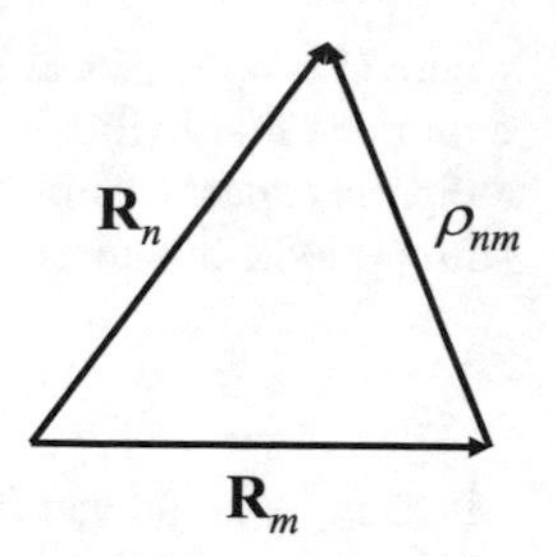

Fig. 3.5 Definition of $\boldsymbol{\rho}_{nm}$, with a magnitude denoting the distance between atoms at $\mathbf{R}_m$ and $\mathbf{R}_n$ in the crystal lattice

where j is an index labeling the particular atomic state of degeneracy N, and $\mathbf{k}$ is the quantum number for the translation operator that labels the Bloch state $\psi_{j,\mathbf{k}}(\mathbf{r})$.

For simplicity, we will limit the present discussion of the tight binding approximation to s–bands (non–degenerate atomic states) and therefore we can suppress the j index of the wave functions. (The treatment for p–bands is similar to what we will do here, but more complicated because of the degeneracy of the atomic p states and also higher momentum states, should they be important.) To find the matrix elements of the Hamiltonian we write

$$\langle \mathbf{k}'|\mathscr{H}|\mathbf{k}\rangle = |\xi|^2 \sum_{n,m} e^{i(\mathbf{k}\cdot\mathbf{R}_n - \mathbf{k}'\cdot\mathbf{R}_m)} \int_{\Omega} \phi^*(\mathbf{r}-\mathbf{R}_m)\mathscr{H}\phi(\mathbf{r}-\mathbf{R}_n)d^3r \tag{3.53}$$

in which the integration is carried out throughout the volume of the crystal. Since $\mathscr{H}$ is a function which is periodic in the lattice, and the only important distance (see Fig. 3.5) is

$$(\mathbf{R}_n - \mathbf{R}_m) = \boldsymbol{\rho}_{nm}. \tag{3.54}$$

We then write the integral in (3.53) as:

$$\langle \mathbf{k}'|\mathscr{H}|\mathbf{k}\rangle = |\xi|^2 \sum_{\mathbf{R}_m} e^{i(\mathbf{k}-\mathbf{k}')\cdot\mathbf{R}_m} \sum_{\boldsymbol{\rho}_{nm}} e^{i\mathbf{k}\cdot\boldsymbol{\rho}_{nm}} \mathscr{H}_{mn}(\boldsymbol{\rho}_{nm}) \tag{3.55}$$

where we have written the matrix element $\mathscr{H}_{mn}(\boldsymbol{\rho}_{nm})$ as

$$\mathscr{H}_{mn}(\boldsymbol{\rho}_{nm}) = \int_{\Omega} \phi^*(\mathbf{r}-\mathbf{R}_m)\mathscr{H}\phi(\mathbf{r}-\mathbf{R}_m-\boldsymbol{\rho}_{nm})d^3r = \int_{\Omega} \phi^*(\mathbf{r}')\mathscr{H}\phi(\mathbf{r}'-\boldsymbol{\rho}_{nm})d^3r'. \tag{3.56}$$

We note here that because of the lattice periodicity of the Hamiltonian $\mathscr{H}$ and the wave function ϕ, the integral in (3.56) depends only on $\boldsymbol{\rho}_{nm}$ and not on $\mathbf{R}_m$. According to (3.13), the first and largest term is the sum in (3.55) is

$$\sum_{\mathbf{R}_m} e^{i(\mathbf{k}-\mathbf{k}')\cdot\mathbf{R}_m} = \delta_{\mathbf{k}',\mathbf{k}+\mathbf{G}}N \tag{3.57}$$

where $\mathbf{G}$ is a reciprocal lattice vector. It is convenient to restrict the $\mathbf{k}$ vectors to lie within the first Brillouin zone (i.e., we limit ourselves to reduced wave vectors $\mathbf{k}$ in reciprocal space). This is consistent with the manner of counting states for a crystal with periodic boundary conditions of length d on a side

$$k_i d = 2\pi m_i \qquad \text{for each direction } i \tag{3.58}$$

where m_i is an integer in the range $1 \leq m_i < N_i$, and $N_i \approx N^{1/3}$ where N is the total number of unit cells in the crystal. From (3.58) we have

$$k_i = \frac{2\pi m_i}{d}. \tag{3.59}$$

The maximum value that a particular m_i can assume is N_i and the maximum value for k_i is $2\pi/a$ at the Brillouin zone boundary, since $N_i/d = 1/a$. With this restriction, $\mathbf{k}$ and $\mathbf{k}'$ must both lie within the 1st B.Z. and thus cannot differ by any reciprocal lattice vector other than $\mathbf{G} = 0$. We thus obtain the following form for the matrix element of $\mathscr{H}$ (and also for the corresponding forms for the matrix elements of $\mathscr{H}_0$ and $\mathscr{H}'$):

$$\langle \mathbf{k}' | \mathscr{H} | \mathbf{k} \rangle = |\xi|^2 N \delta_{\mathbf{k},\mathbf{k}'} \sum_{\boldsymbol{\rho}_{nm}} e^{i\mathbf{k}\cdot\boldsymbol{\rho}_{nm}} \mathscr{H}_{mn}(\boldsymbol{\rho}_{nm}) \tag{3.60}$$

yielding the following result for energy eigenvalues

$$E(\mathbf{k}) = \frac{\langle \mathbf{k} | \mathscr{H} | \mathbf{k} \rangle}{\langle \mathbf{k} | \mathbf{k} \rangle} = \frac{\sum_{\boldsymbol{\rho}_{nm}} e^{i\mathbf{k}\cdot\boldsymbol{\rho}_{nm}} \mathscr{H}_{mn}(\boldsymbol{\rho}_{nm})}{\sum_{\boldsymbol{\rho}_{nm}} e^{i\mathbf{k}\cdot\boldsymbol{\rho}_{nm}} \mathscr{S}_{mn}(\boldsymbol{\rho}_{nm})} \tag{3.61}$$

in which the orthonormalization is given by

$$\langle \mathbf{k}' | \mathbf{k} \rangle = |\xi|^2 \delta_{\mathbf{k},\mathbf{k}'} N \sum_{\boldsymbol{\rho}_{nm}} e^{i\mathbf{k}\cdot\boldsymbol{\rho}_{nm}} \mathscr{S}_{mn}(\boldsymbol{\rho}_{nm}) \tag{3.62}$$

where the matrix element $\mathscr{S}_{mn}(\boldsymbol{\rho}_{nm})$ measures the overlap of atomic functions on different sites

$$\mathscr{S}_{mn}(\boldsymbol{\rho}_{nm}) = \int_{\Omega} \phi^*(\mathbf{r}) \phi(\mathbf{r} - \boldsymbol{\rho}_{nm}) d^3 r. \tag{3.63}$$

The overlap integral $\mathscr{S}_{mn}(\boldsymbol{\rho}_{nm})$ will be nearly 1 when $\boldsymbol{\rho}_{nm} = 0$ and will fall off rapidly as $\boldsymbol{\rho}_{nm}$ increases, which exemplifies the spirit of the tight binding approximation. By selecting $\mathbf{k}$ vectors that lie within the first Brillouin zone, the orthogonality condition on the wave function $\psi_{\mathbf{k}}(\mathbf{r})$ is automatically satisfied. Writing $\mathscr{H} = \mathscr{H}_0 + \mathscr{H}'$ yields:

$$\begin{aligned}\mathscr{H}_{mn} &= \int_{\Omega} \phi^*(\mathbf{r}-\mathbf{R}_m)\left[-\frac{\hbar^2}{2m}\nabla^2 + U(\mathbf{r}-\mathbf{R}_n)\right]\phi(\mathbf{r}-\mathbf{R}_n)d^3r \\ &+ \int_{\Omega} \phi^*(\mathbf{r}-\mathbf{R}_m)[V(\mathbf{r}) - U(\mathbf{r}-\mathbf{R}_n)]\phi(\mathbf{r}-\mathbf{R}_n)d^3r\end{aligned} \tag{3.64}$$

or

$$\mathscr{H}_{mn} = E^{(0)}\mathscr{S}_{mn}(\boldsymbol{\rho}_{nm}) + \mathscr{H}'_{mn}(\boldsymbol{\rho}_{nm}) \tag{3.65}$$

in which the perturbation Hamiltonian $\mathscr{H}' = V(\mathbf{r}) - U(\mathbf{r}-\mathbf{R_n})$ is the difference between the periodic potential and the atomic potential in which $\mathscr{H}'$ gets large only close to an atomic site. The general expression for the tight binding approximation thus becomes:

$$E(\mathbf{k}) = E^{(0)} + \frac{\sum_{\boldsymbol{\rho}_{nm}} e^{i\mathbf{k}\cdot\boldsymbol{\rho}_{nm}}\mathscr{H}'_{mn}(\boldsymbol{\rho}_{nm})}{\sum_{\boldsymbol{\rho}_{nm}} e^{i\mathbf{k}\cdot\boldsymbol{\rho}_{nm}}\mathscr{S}_{mn}(\boldsymbol{\rho}_{nm})}. \tag{3.66}$$

In the spirit of the tight binding approximation, the second term in (3.66) is assumed to be small, which is a good approximation if the overlap of the atomic wave functions is small. We classify the sum over $\boldsymbol{\rho}_{nm}$ according to the distance between site m and site n: (i) zero distance, (ii) the nearest neighbor distance, (iii) the next nearest neighbor distance, etc.

$$\sum_{\boldsymbol{\rho}_{nm}} e^{i\mathbf{k}\cdot\boldsymbol{\rho}_{nm}}\mathscr{H}'_{mn}(\boldsymbol{\rho}_{nm}) = \mathscr{H}'_{nn}(0) + \sum_{\boldsymbol{\rho}_1} e^{i\mathbf{k}\cdot\boldsymbol{\rho}_{nm}}\mathscr{H}'_{mn}(\boldsymbol{\rho}_{nm}) + \ldots. \tag{3.67}$$

The zeroth neighbor term $\mathscr{H}'_{nn}(0)$ in (3.67) results in a constant additive energy, independent of $\mathbf{k}$ that comes directly from the atomic potential. The sum over nearest neighbor distances $\boldsymbol{\rho}_1$ gives rise to a $\mathbf{k}$–dependent perturbation, and hence is of particular interest in calculating the electronic band structure. The terms $\mathscr{H}'_{nn}(0)$ and the sum over the nearest neighbor terms in (3.67) are of comparable magnitude, as can be seen by the following argument. In the integral of the k-independent term

$$\mathscr{H}'_{nn}(0) = \int \phi^*(\mathbf{r}-\mathbf{R}_n)[V - U(\mathbf{r}-\mathbf{R}_n)]\phi(\mathbf{r}-\mathbf{R}_n)d^3r \tag{3.68}$$

we note that $|\phi(\mathbf{r}-\mathbf{R}_n)|^2$ has an appreciable amplitude only in the vicinity of the site $\mathbf{R}_n$. But at site $\mathbf{R}_n$, the potential energy term $[V - U(\mathbf{r}-\mathbf{R}_n)] = \mathscr{H}'$ is a small term, so that $\mathscr{H}'_{nn}(0)$ represents the product of a small term times a large term. On the other hand, the integral $\mathscr{H}'_{mn}(\boldsymbol{\rho}_{nm})$ taken over nearest neighbor distances has a factor $[V - U(\mathbf{r}-\mathbf{R}_n)]$ which is large near the mth site; however, in this case the wave functions $\phi^*(\mathbf{r}-\mathbf{R}_m)$ and $\phi(\mathbf{r}-\mathbf{R}_n)$ are on different atomic sites and have only a small overlap on nearest neighbor sites. Therefore $\mathscr{H}'_{mn}(\boldsymbol{\rho}_{nm})$ over nearest neighbor sites also results in the product of a large quantity times a small quantity.

In treating the denominator in the perturbation term of (3.66), we must carry out the summation:

$$\sum_{\rho_{nm}} e^{i\mathbf{k}\cdot\boldsymbol{\rho}_{nm}} \mathscr{S}_{mn}(\boldsymbol{\rho}_{nm}) = \mathscr{S}_{nn}(0) + \sum_{\rho_1} e^{i\mathbf{k}\cdot\boldsymbol{\rho}_{nm}} \mathscr{S}_{mn}(\boldsymbol{\rho}_{nm}) + \ldots. \tag{3.69}$$

In this case the leading term $\mathscr{S}_{nn}(0)$ is approximately unity and the overlap integral $\mathscr{S}_{mn}(\boldsymbol{\rho}_{nm})$ over nearest neighbor sites is small, and can be neglected to lowest order in comparison with unity. The nearest neighbor term in (3.69) is of comparable relative magnitude to the next nearest neighbor terms arising from $\mathscr{H}'_{mn}(\boldsymbol{\rho}_{nm})$ in (3.67).

We will next make *several explicit evaluations* of $E(\mathbf{k})$ in the tight–binding limit to show how this method directly incorporates the crystal symmetry. For illustrative purposes we will give results for the simple cubic lattice (SC), the body centered cubic lattice (BCC), and the face centered cubic lattice (FCC). We shall assume here that the overlap of atomic potentials on neighboring sites is sufficiently weak so that only nearest neighbor terms need be considered in the sum on $\mathscr{H}'_{mn}$ and only the leading term need be considered in the sum on $\mathscr{S}_{mn}$.

For the simple cubic structure there are 6 terms in the nearest neighbor sum on $\mathscr{H}'_{mn}$ in (3.66) with the $\boldsymbol{\rho}_1$ vectors given by:

$$\boldsymbol{\rho}_1 = a(\pm 1, 0, 0),\quad a(0, \pm 1, 0),\quad a(0, 0, \pm 1). \tag{3.70}$$

By symmetry, $\mathscr{H}'_{mn}(\boldsymbol{\rho}_1)$ is the same for all of the $\boldsymbol{\rho}_1$ vectors so that

$$E(\mathbf{k}) = E^{(0)} + \mathscr{H}'_{nn}(0) + 2\mathscr{H}'_{mn}(\boldsymbol{\rho}_1)[\cos k_x a + \cos k_y a + \cos k_z a] + \ldots \tag{3.71}$$

where $\boldsymbol{\rho}_1$ = the nearest neighbor separation and k_x, k_y, k_z are components of the wave vector $\mathbf{k}$ in the first Brillouin zone.

The dispersion relation $E(\mathbf{k})$ in (3.71) clearly satisfies three properties which characterize the energy eigenvalues in typical periodic structures:

1. Periodicity in $\mathbf{k}$ space under translation by a reciprocal lattice vector $\mathbf{k} \rightarrow \mathbf{k}+\mathbf{G}$,
2. $E(\mathbf{k})$ is an even function of $\mathbf{k}$ (i.e., $E(\mathbf{k}) = E(-\mathbf{k})$)
3. The derivative of $E(k)$ vanishes ($\partial E/\partial k = 0$) at the Brillouin zone boundary

In the above expression (3.71) for $E(\mathbf{k})$, the maximum value for the term in brackets is $\pm$ 3. Therefore for a simple cubic lattice in the tight binding approximation we obtain a bandwidth of 12 $\mathscr{H}'_{mn}(\rho_1)$ from nearest neighbor interactions as shown in Fig. 3.6.

Because of the different locations of the nearest neighbor atoms in the case of the BCC and FCC lattices, the expressions for $E(\mathbf{k})$ will be different for the various cubic

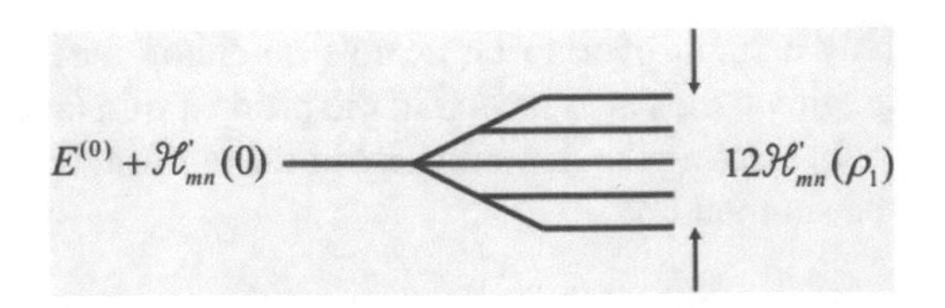

Fig. 3.6 The relation between the atomic levels and the broadened level in the tight binding approximation

lattices. Thus the form of the tight binding approximation explicitly takes account of the crystal structure. The results for the simple cubic, body centered cubic and face centered cubic lattices are summarized below.

Simple cubic

$$E(\mathbf{k}) = \text{const} + 2\mathscr{H}'_{mn}(\boldsymbol{\rho}_1)[\cos k_x a + \cos k_y a + \cos k_z a] + ... \tag{3.72}$$

Body centered cubic

The eight $\boldsymbol{\rho}_1$ vectors for the nearest neighbor distances in the BCC structure are $(\pm a/2, \pm a/2, \pm a/2)$ so that there are 8 exponential terms which combine in pairs such as:

$$\left[\exp\frac{ik_x a}{2}\exp\frac{ik_y a}{2}\exp\frac{ik_z a}{2} + \exp\frac{-ik_x a}{2}\exp\frac{ik_y a}{2}\exp\frac{ik_z a}{2}\right] \tag{3.73}$$

to yield for an intermediate sum

$$2\cos(\frac{k_x a}{2})\exp\frac{ik_y a}{2}\exp\frac{ik_z a}{2}. \tag{3.74}$$

Thus by carrying out the summation over x, y, and z for the BCC structure, we obtain:

$$E(\mathbf{k}) = \text{const} + 8\mathscr{H}'_{mn}(\boldsymbol{\rho}_1)\cos(\frac{k_x a}{2})\cos(\frac{k_y a}{2})\cos(\frac{k_z a}{2}) + ... \tag{3.75}$$

where $\mathscr{H}'_{mn}(\boldsymbol{\rho}_1)$ is the matrix element of the perturbation Hamiltonian taken between nearest neighbor atomic orbitals for the BCC lattice.

Face centered cubic

For the FCC structure there are 12 nearest neighbor distances $\boldsymbol{\rho}_1$: $(0, \pm\frac{a}{2}, \pm\frac{a}{2})$, $(\pm\frac{a}{2}, \pm\frac{a}{2}, 0)$, $(\pm\frac{a}{2}, 0, \pm\frac{a}{2})$, so that the twelve exponential terms combine in groups of 4 to yield:

$$\begin{aligned}\exp\tfrac{ik_x a}{2}\exp\tfrac{ik_y a}{2} + \exp\tfrac{ik_x a}{2}\exp\tfrac{-ik_y a}{2} + \exp\tfrac{-ik_x a}{2}\exp\tfrac{ik_y a}{2} +\\ \exp\tfrac{-ik_x a}{2}\exp\tfrac{-ik_y a}{2} = 4\cos(\tfrac{k_x a}{2})\cos(\tfrac{k_y a}{2}),\end{aligned} \tag{3.76}$$

thus resulting in the energy dispersion relation

$$\begin{aligned}E(\mathbf{k}) = \text{const} + 4\mathscr{H}'_{mn}(\boldsymbol{\rho}_1)\\ \left[\cos(\tfrac{k_y a}{2})\cos(\tfrac{k_z a}{2}) + \cos(\tfrac{k_x a}{2})\cos(\tfrac{k_z a}{2}) + \cos(\tfrac{k_x a}{2})\cos(\tfrac{k_y a}{2})\right] + ...\end{aligned} \tag{3.77}$$

We note that $E(\mathbf{k})$ for the FCC is different from that for the SC or BCC structures. We thus see that the tight-binding approximation has symmetry considerations built into its formulation by consideration of the symmetrical arrangement of the atoms in each crystal lattice. The situation is quite different in the weak binding approximation where symmetry enters into the form of $V(\mathbf{r})$ and determines which Fourier components $V_{\mathbf{G}}$ will be important in creating band gaps.

3.2.3 Comparison of Weak and Tight Binding Approximations

We will now make some general statements about bandwidths and forbidden band gaps which follow from either the tight binding or the weak binding (nearly free electron) approximations. With increasing energy, the bandwidth tends to increase. When using the tight–binding picture, the higher energy atomic states are less closely bound to the nucleus, and the resulting increased overlap of the wave functions results in a larger value for $\mathscr{H}'_{mn}(\boldsymbol{\rho}_1)$ in the case of the higher atomic states: that is, for silicon, which has 4 valence electrons in the $n = 3$ shell, the overlap integral $\mathscr{H}'_{mn}(\boldsymbol{\rho}_1)$ will be smaller than for germanium which is isoelectronic to silicon but has instead 4 valence electrons in the $n = 4$ atomic shell. On the weak–binding picture, the same result follows, since for higher energies, the electrons are more nearly free; therefore, there are more allowed energy ranges available, or equivalently, the energy range of the forbidden states is smaller. Also in the weak–binding approximation, the band gap of $2|V_{\mathbf{G}}|$ tends to decrease as $\mathbf{G}$ increases, because of the oscillatory character of $e^{-i\mathbf{G}\cdot\mathbf{r}}$ in

$$V_{\mathbf{G}} = \frac{1}{\Omega_0}\int_{\Omega_0} e^{-i\mathbf{G}\cdot\mathbf{r}} V(\mathbf{r}) d^3r. \tag{3.78}$$

From the point of view of the tight–binding approximation, the increasing bandwidth with increasing energy (see Fig. 3.7) is also equivalent to a decrease in the

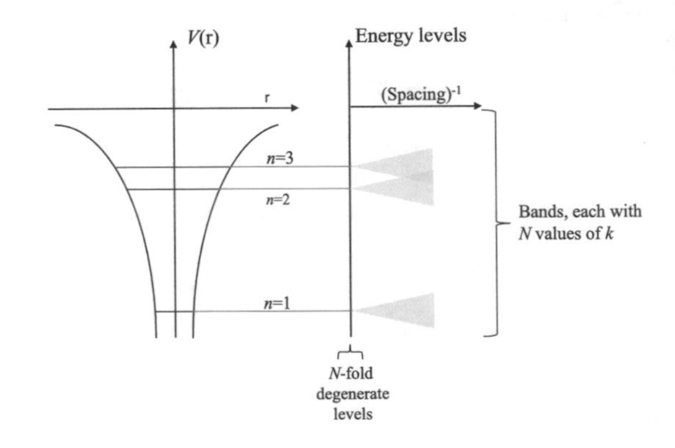

Fig. 3.7 Schematic diagram of the quantized energy levels $n = 1, 2, 3$ showing the increased bandwidth and decreased band gap in the tight binding approximation as n increases and the interatomic separation decreases

forbidden band gap. At the same time, the atomic states at higher energies become more closely spaced, so that the increased bandwidth eventually results in increased band overlaps.

When band overlaps occur, the tight-binding approximation, as given above, must be generalized to treat coupled or interacting bands using degenerate perturbation theory (see Appendix B).

3.2.4 *Tight Binding Approximation with 2 Atoms/Unit Cell*

1D Solid: Polyacetylene

We present here a simple example of the tight binding approximation for a simplified version of polyacetylene, which has two carbon atoms (with their appended hydrogens) per unit cell. In Fig. 3.8 we show, within the box defined by the dotted lines, the unit cell for *trans*-polyacetylene $(\mathrm{CH})_x$. This unit cell of an infinite one-dimensional chain contains two inequivalent carbon atoms, A and B. There is one π-electron per carbon atom, thus giving rise to two π-energy bands in the first Brillouin zone. These two bands are called bonding π-bands for the valence band, and anti-bonding π-bands for the conduction band.

The lattice unit vector and the reciprocal lattice unit vector of this one-dimensional polyacetylene chain are given by $\mathbf{a}_1 = (a, 0, 0)$ and $\mathbf{b}_1 = (2\pi/a, 0, 0)$, respectively. The Brillouin zone in 1D is the line segment $-\pi/a < k < \pi/a$ and the Brillouin zone boundary is at $k = \pm\pi/a$. The Bloch orbitals consisting of A and B atoms are given by

$$\psi_j(r) = \frac{1}{\sqrt{N}} \sum_{R_\alpha} e^{ikR_\alpha} \phi_j(r - R_\alpha), \quad (\alpha = \mathrm{A}, \mathrm{B}) \tag{3.79}$$

where the summation is taken over the atom site coordinate R_α for the A or the B carbon atoms in the solid.

To solve for the energy eigenvalues and wavefunctions, we need to solve the general equation:

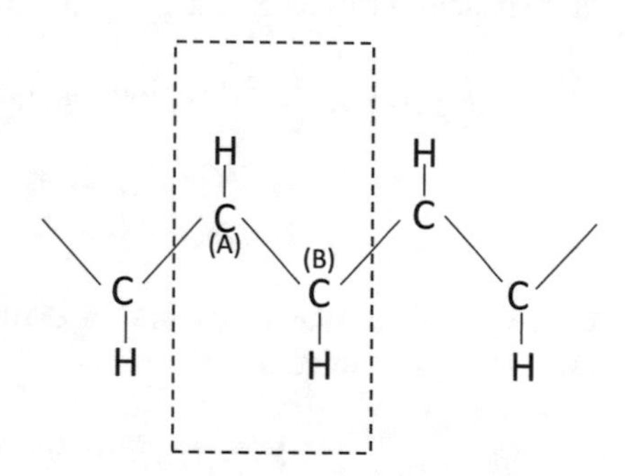

Fig. 3.8 The unit cell of *trans*-polyacetylene bounded by a box defined by the dashed lines, and showing two inequivalent carbon atoms, A and B bonded to hydrogen atoms, within the unit cell (See Saito et al. 1998)

$$\mathscr{H}\psi = E\mathscr{S}\psi \tag{3.80}$$

where $\mathscr{H}$ is the $n \times n$ tight binding matrix Hamiltonian for the n coupled bands ($n = 2$ in the case of polyacetylene) and $\mathscr{S}$ is the corresponding $n \times n$ overlap integral matrix. To obtain a solution to this matrix equation, we require that the determinant $|\mathscr{H} - E\mathscr{S}|$ vanish. This approach is easily generalized to periodic structures with more than 2 atoms per unit cell.

The (2×2) matrix Hamiltonian, $\mathscr{H}_{\alpha\beta}$, $(\alpha, \beta = A, B)$ is obtained by substituting (3.79) into (3.80) and normalizing the wave functions appropriately

$$\mathscr{H}_{jj'}(\mathbf{k}) = \langle\psi_j|\mathscr{H}|\psi_{j'}\rangle, \quad \mathscr{S}_{jj'}(\mathbf{k}) = \langle\psi_j|\psi_{j'}\rangle \qquad (j, j' = 1, 2), \tag{3.81}$$

where the integrals over the Bloch orbitals, $\mathscr{H}_{jj'}(\mathbf{k})$ and $\mathscr{S}_{jj'}(\mathbf{k})$, are called transfer integral matrices and the overlap integral matrices, respectively. When $\alpha = \beta = $ A, we obtain the diagonal matrix element

$$\begin{aligned}
\mathscr{H}_{AA}(r) &= \frac{1}{N}\sum_{R,R'} e^{ik(R-R')}\langle\phi_A(r-R')|\mathscr{H}|\phi_A(r-R)\rangle \\
&= \frac{1}{N}\sum_{R'=R} E_{2p} + \frac{1}{N}\sum_{R'=R\pm a} e^{\pm ika}\langle\phi_A(r-R')|\mathscr{H}|\phi_A(r-R)\rangle \\
&\quad + (\text{terms equal to or more distant than } R' = R \pm 2a) \\
&= E_{2p} + (\text{terms equal to or more distant than } R' = R \pm 2a).
\end{aligned} \tag{3.82}$$

In (3.82) the main contribution to the matrix element $\mathscr{H}_{AA}$ comes from $R' = R$, and this gives the orbital energy of the $2p$ level, E_{2p}. We note that E_{2p} is not simply the atomic energy value for the free atom, because the Hamiltonian $\mathscr{H}$ also includes a crystal potential contribution. The next order contribution to $\mathscr{H}_{AA}$ in (3.82) comes from terms in $R' = R \pm a$, which are here neglected for simplicity. Similarly, $\mathscr{H}_{BB}$ also gives E_{2p} to the same order of approximation.

Next let us consider the off-diagonal matrix element $\mathscr{H}_{AB}(r)$ which explicitly couples the A unit to the B unit. The largest contribution to $\mathscr{H}_{AB}(r)$ arises when atoms A and B are nearest neighbors. Thus, in the summation over R', we only consider the leading terms with $R' = R \pm a/2$ as a first approximation and neglect more distant terms to obtain

$$\begin{aligned}
\mathscr{H}_{AB}(r) &= \frac{1}{N}\sum_{R}\big\{e^{-ika/2}\langle\phi_A(r-R)|\mathscr{H}|\phi_B(r-R-a/2)\rangle \\
&\quad + e^{ika/2}\langle\phi_A(r-R)|\mathscr{H}|\phi_B(r-R+a/2)\rangle\big\} \\
&= 2\gamma_0\cos(ka/2)
\end{aligned} \tag{3.83}$$

where γ_0 is the transfer integral or carbon-carbon interaction energy appearing in (3.83) and is denoted by

$$\gamma_0 = \langle\phi_A(r-R)|\mathscr{H}|\phi_B(r-R\pm a/2)\rangle. \tag{3.84}$$

Here we have assumed that all the π bonding orbitals are of equal length (1.5Å bonds). In the real $(\mathrm{CH})_x$ compound, bond alternation occurs, in which the bonding between adjacent carbon atoms alternates between single bonds (1.7Å) and double bonds (1.3Å). With this bond alternation, the two matrix elements between atomic wavefunctions in (3.83) are no longer equal. Although the distortion of the lattice lowers the total energy, the electronic energy always decreases more than the lattice energy in a one-dimensional material. This distortion that makes the lattice by a process called the Peierls instability. This instability arises, for example, when a distortion is introduced into a system previously containing degenerate states with 2 equivalent atoms per unit cell. The distortion making the atoms inequivalent increases the unit cell by a factor of 2 and concurrently decreases the reciprocal lattice by a factor of 2. If the energy band was formally half filled, a band gap is introduced by the Peierls instability at the Fermi level, which lowers the total energy of the system. It is stressed here that γ_0 has a negative value which means that γ_0 is an attractive potential that bonds atoms together to form a condensed state of matter. The matrix element $\mathscr{H}_{BA}(r)$ is obtained from $\mathscr{H}_{AB}(r)$ through the Hermitian conjugation relation $\mathscr{H}_{BA} = \mathscr{H}^*_{AB}$, but since $\mathscr{H}_{AB}$ is real in this case, we obtain $\mathscr{H}_{BA} = \mathscr{H}_{AB}$.

The overlap matrix $\mathscr{S}_{ij}$ can be calculated by a similar method as was used for $\mathscr{H}_{ij}$, except that the intra-atomic integral $\mathscr{S}_{ij}$ yields a unit matrix in the limit of large interatomic distances, if we also assume that the atomic wavefunction is normalized so that $\mathscr{S}_{AA} = \mathscr{S}_{BB} = 1$. It is assumed that for polyacetylene, the $\mathscr{S}_{AA}$ and $\mathscr{S}_{BB}$ matrix elements are still approximately unity. For the off-diagonal matrix element for polyacetylene, we have $\mathscr{S}_{AB} = \mathscr{S}_{BA} = 2s\cos(ka/2)$, where s is an overlap integral between the nearest A and B atoms,

$$s = \langle \phi_A(r-R)|\phi_B(r-R\pm a/2)\rangle. \tag{3.85}$$

The secular equation for the $2p_z$ orbital of CH_x is obtained by setting the determinant of $|\mathscr{H} - E\mathscr{S}|$ to zero to obtain

$$\begin{aligned} &\begin{vmatrix} E_{2p}-E & 2(\gamma_0 - sE)\cos(ka/2) \\ 2(\gamma_0 - sE)\cos(ka/2) & E_{2p}-E \end{vmatrix} \\ &= (E_{2p}-E)^2 - 4(\gamma_0 - sE)^2\cos^2(ka/2) \\ &= 0 \end{aligned} \tag{3.86}$$

yielding the eigenvalues of the energy dispersion relations of (3.86)

$$E_{\pm}(\mathbf{k}) = \frac{E_{2p} \pm 2\gamma_0\cos(ka/2)}{1 \pm 2s\cos(ka/2)}, \quad (-\frac{\pi}{a} < k < \frac{\pi}{a}) \tag{3.87}$$

in which the $+$ sign is associated with the bonding π-band and the $-$ sign is associated with the antibonding π^*-band, as shown in Fig. 3.9. Here it is noted that by setting E_{2p} to zero (thereby defining the origin of the energy), the levels E_+ and E_- are degenerate at $ka = \pm\pi$. Figure 3.9 is constructed for $\gamma_0 < 0$ and $s > 0$.

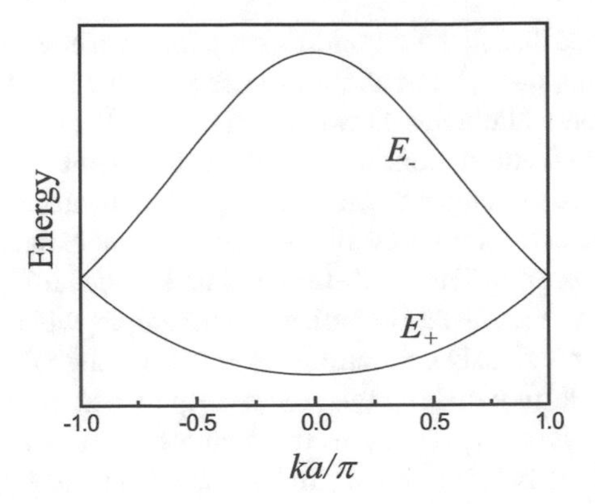

Fig. 3.9 The energy dispersion relation $E_{\pm}(\mathbf{k})$ for polyacetylene $[(CH)_x]$, given by (3.87) with values for the parameters $t\gamma_0 = -1$ and $s = 0.2$. Curves $E_+(\mathbf{k})$ and $E_-(\mathbf{k})$ are called bonding π and antibonding π^* energy bands, respectively, and the energy is plotted in units of γ_0 (See Saito et al. 1998)

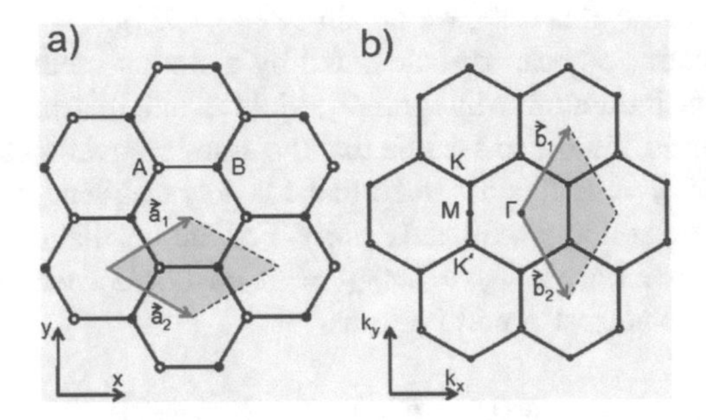

Fig. 3.10 Real and reciprocal space for graphene. **(a)** The unit cell (rhombus) with their basis vectors $\mathbf{a_1}$ and $\mathbf{a_2}$. A and B stand for the two non-equivalent carbon atoms in the unit cell. **(b)** Brillouin zone (enclosed in the rhombus) with reciprocal lattice vectros $\mathbf{b_1}$ and $\mathbf{b_2}$. The reciprocal space structure has two inequivalent Dirac points K and K'

Since there are two π electrons per unit cell, each with a different spin orientation, both electrons occupy the bonding π energy band. The effect of the inter-atomic bonding is to lower the total energy below E_{2p}.

2D Solid: Graphene

A simple and elegant example for ilustrating band structure calculations using tight binding is graphene. In this truly two-dimensional (2D) system, carbon atoms are arranged in a hexagonal lattice as schematically shown in Fig. 3.10(a). The graphene unit vectors $\mathbf{a_1}$ and $\mathbf{a_2}$ can be written in cartesian coordinates as $\mathbf{a_1} = a(\sqrt{3}/2\hat{\mathbf{i}}+1/2\hat{\mathbf{j}})$ and $\mathbf{a_2} = a(\sqrt{3}/2\hat{\mathbf{i}} - 1/2\hat{\mathbf{j}})$, where $a = |\mathbf{a_1}| = |\mathbf{a_2}|$, and $a = 2.46\text{Å} = 0.246\,nm$ while $\hat{\mathbf{i}}$ and $\hat{\mathbf{j}}$ are unit vectors along the x and y directions, respectively. Direct calculation show that these vectors do not form an orthogonal basis: $\mathbf{a_1} \cdot \mathbf{a_2} = a^2/2$ as shown in Fig. 3.10(a) the unit cell has two inequivalent atoms labeled A and B.

In Fig. 3.10(b) the unit vectors in reciprocal space are shown and the unit cell here is also hexagonal and has two inequivalent Dirac K and K' points in the reciprocal lattice which are called Dirac points for monolayer graphene. At these high symmetry points the electronic energy in the valence and conduction bands are degenerate, and linear $E(k)$ relations can be used closed to the K and K' points in the Brillouin zone.

The electronic energy dispersion relations for graphene are calculated by solving the eigenvalue problem for a Hamiltonian $\mathscr{H}$ (2x2) and an overlap matrix $\mathscr{S}$ (2x2), associated with the two non-equivalent carbon atoms in the honeycomb 2D lattice [see Fig. 3.10], within the tight-binding approximation,

$$\mathscr{H} = \begin{pmatrix} \varepsilon_{2p} & -\gamma_0 f(k) \\ -\gamma_0 f(k)^* & \varepsilon_{2p} \end{pmatrix} \text{ and } \quad \mathscr{S} = \begin{pmatrix} 1 & sf(k) \\ sf(k)^* & 1 \end{pmatrix}, \tag{3.88}$$

where ε_{2p} is the site energy of the $2p$ atomic orbital and

$$f(k) = e^{ik_x a/\sqrt{3}} + 2e^{-ik_x a/2\sqrt{3}} \cos \frac{k_y a}{2}, \tag{3.89}$$

where $a = |\mathbf{a}_1| = |\mathbf{a}_2| = \sqrt{3}a_{\mathrm{C-C}}$. Solution of the secular equation $\det(\mathscr{H} - E\mathscr{S}) = 0$ implied by (3.88) leads to the eigenvalues

$$E_{\pm}(\mathbf{k}) = \frac{\varepsilon_{2p} \pm \gamma_0 w(\mathbf{k})}{1 \mp s w(\mathbf{k})}, \tag{3.90}$$

for the C-C nearest neighbor overlap energy $\gamma_0 > 0$ (here we conventionally use γ_0 as a positive value) where s denotes the overlap of the electronic wavefunctions on adjacent sites, and E_+ and E_- correspond to the π^* conduction band and the π valence band, respectively. The function $w(\mathbf{k})$ in (3.90) is given by

$$w(\mathbf{k}) = \sqrt{|f(\mathbf{k})|^2} = \sqrt{1 + 4\cos\frac{\sqrt{3}k_x a}{2}\cos\frac{k_y a}{2} + 4\cos^2\frac{k_y a}{2}} \tag{3.91}$$

leading in the limit $s = 0$ and $\varepsilon_{2p} = 0$ to a symmetric form for the dispersion relations $E(k_x, k_y)$ for electrons in graphene

$$E_{\pm}(k_x, k_y) = \pm\gamma_o \left\{1 + 4\cos\frac{\sqrt{3}k_x a}{2}\cos\frac{k_y a}{2} + 4\cos^2\frac{k_y a}{2}\right\}, \tag{3.92}$$

where $a = 1.421 \times \sqrt{3}$Å is the lattice constant for a 2D graphene layer and γ_o is the nearest-neighbor C-C energy overlap integral.

In Fig. 3.11, a plot is shown of the electronic dispersion relations for a 2D graphene lattice as a function of (k_x, k_y) in the 2D hexagonal Brillouin zone [see Fig. 3.10] that is obtained by using (3.91) and adopting the parameters $\gamma_o = 3.013\,\mathrm{eV}$, $s = 0.129$,

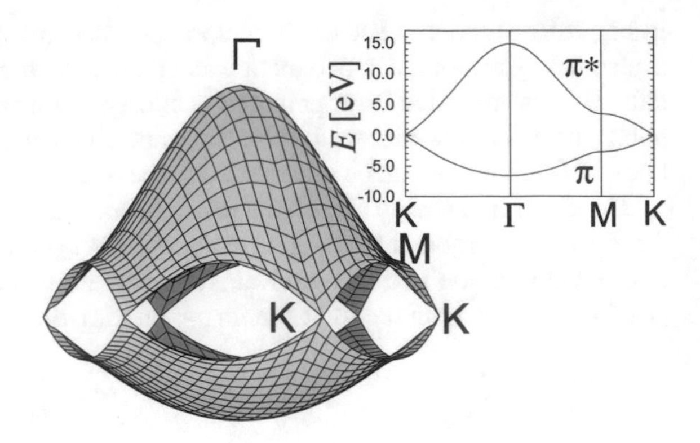

Fig. 3.11 Energy dispersion of the bonding π and anti-bonding π^* bands for the 2D graphene layer using $\gamma_0 = 3.03\,\text{eV}$ and $s = 0.129$ [see (3.90)]. The inset on the right shows the energy dispersion calculated for graphene along high symmetry directions in the Brillouin zone (See Saito et al. 1998)

and $\varepsilon_{2p} = 0$. In the limit $s = 0$ and $\varepsilon_{2p} = 0$, the valence π and conduction π^* bands become symmetric with respect to each other. Near the K point in the Brillouin zone, $E_{\pm}(\mathbf{k})$ has a linear dependence on $k = |\mathbf{k}|$ measured from the K point so that,

$$\omega(\mathbf{k}) = \frac{\sqrt{3}}{2} ka + \ldots \tag{3.93}$$

For $ka << 1$, (3.90) reduces to

$$E(k) = \pm\frac{\sqrt{3}}{2}\gamma_0 ka = \pm\frac{3}{2}\gamma_0 ka_{\text{C-C}}, \tag{3.94}$$

where $a_{\text{C-C}}$ is the nearest neighbor carbon-carbon distance. If the physical phenomena under consideration only involve small k vectors, it is convenient to use (3.94) to calculate the electronic transition energies. For larger k vectors a more detailed $E(k)$ is necessary to reflect the trigonal warping effect that occurs for this lattice. The linear dispersion is a special feature of graphene for low k values, thus leading value to interpretation that low energy electrons of graphene propagate as massless particles that mimic photon propagation in this limit.

Problems

3.1 This problem is a review of the nearly free electron approximation.

(a) Write a general expression for the $E(\mathbf{k})$ relations for the empty lattice (i.e., $V(\mathbf{r}) = 0$) for a two dimensional square lattice.

(b) Find $E(\mathbf{k})$ explicitly along $\Gamma - X$ and $X - L$ for the lowest 3 energy levels including the degeneracies of each level. Plot $E(\mathbf{k})$ for these levels. (Note: the Γ point is $(\pi/a)(0, 0)$; the X point is $(\pi/a)(1, 0)$; and the L point is $(\pi/a)(1, 1)$).

(c) Suppose that small carrier pockets are formed in the energy bands about points Γ, X and L of the square lattice Brillouin Zone. In each case, indicate the shape of this carrier pocket and the number of equivalent full carrier pockets that are formed. Be more quantitative here. Suppose that the carrier pocket allows 0.01% occupation of the zone with carriers.
(d) Find the wave functions corresponding to the three lowest X point energy levels in the empty lattice model.
(e) Using first order degenerate perturbation theory, find the effect of a small periodic potential $V(\mathbf{r})$ on producing band gaps for these X point energy levels according to the nearly free electron approximation. Which degeneracies in (d) are thus lifted?

3.2 This problem is to review the tight binding approximation as applied to graphene.

(a) Suppose that the overlap integral for the electrons on adjacent sites vanishes ($s = 0$). Sketch the effect of taking $s = 0$ on the dispersion relations (Fig. 3.11). This approximation is sometimes used for describing the electronic properties of carbon nanotubes.
(b) By considering k points close to the K or K' points in the Brillouin zone, find the effect of (3.94) and show that the resulting constant energy surface forms a cone. Due to the linear k dependence, this has been called the Dirac cone.
(c) Discuss the consequences of this linear electronic dispersion relation on the effective mass of the charge carrier as a function of k.

3.3 This problem is to review the tight binding approximation as applied to polyacetylene.

(a) To satisfy the bonding requirements of carbon, polyacetylene has alternating single and double bonds with bond lengths of 1.7Å and 1.3Å, respectively (see Sect. 3.2.4). Please clarify why single and double bonds are needed from a chemical point of view. What modification to the electronic dispersion relations does this bond alternation give rise to (see Fig. 3.9)? Sketch the effect on Fig. 3.8 of introducing these differences in the bond lengths.
(b) In this treatment, the effect of the hydrogen atoms has been ignored. What physical argument can you give to justify the approximation that has been made here?

3.4 Consider the nearly free electron picture of a crystalline solid in the limit that the periodic potential vanishes $V(n) = 0$.

(a) What is the difference in energy between a simple cubic lattice and a face centered cubic lattice for the lowest two energy levels at the Γ point ($k = 0$) and the X point at the Brillouin zone boundary in the (100) direction? To start this problem specify the lattice vectors that you are using to specify the simple cubic lattice and the face centered cubic lattice.
(b) What is the difference in the corresponding degeneracies for the levels in part (a)?

(c) Suppose that a stress is applied along the (001) direction to lower the symmetry of the lattice for the simple cubic lattice and for the face centered cubic lattice, giving each a tetragonal distortion. What splitting in the energy levels in parts (a) and (b) would you expect?

3.5 Imagine that we have a single (100) monolayer of sodium. Treating this monolayer as a 2D metal, find an expression for the Fermi energy in terms of the lattice constant a, assuming 1 free electron/atom and an effective mass equal to the free electron mass.

3.6 In many practical cases, the transport properties depend primarily on two energy bands that are strongly coupled. (This is called the simple two band model.)

(a) For the case of the two band model, write down the differential equation satisfied by the periodic part of the Bloch function $\Psi_{nk} = e^{ik \cdot r} u_{nk}(\mathbf{r})$, where the Bloch function is $u_{nk}(\mathbf{r})$ and $V(\mathbf{r})$ is the periodic potential.
(b) Using degenerate first order perturbation theory for two coupled bands, derive the $\mathbf{k}$ dependence of the energy $E(\mathbf{k})$ for bands with extrema at $\mathbf{k} = 0$. This solution for 2 coupled bands is exact.
(c) Show that the effective mass for each of the bands increases with increasing $|k|$. The results you have derived in this problem form the basis of $\mathbf{k} \cdot \mathbf{p}$ perturbation theory, an important topic in solid state physics.

3.7 Consider the two-dimensional simple equilateral triangular lattice for a free electron metal with two electrons/atom.

(a) Assuming a lattice constant of a, find the areas of the electron and hole pockets that are formed in the second and first Brillouin zones, respectively, for this 2D simple triangular lattice.
(b) Find the shapes of the electron and hole Fermi surfaces in the reduced Brillouin zone, obtained through translation by the appropriate reciprocal lattice vector, and write dow the appropriate reciprocal lattice vector.

Suggested Reading

R. Saito, G. Dresselhaus, M.S. Dresselhaus, Physical Properties of Carbon Nanotubes (Imperial College Press, 1998)

Chapter 4
Examples of Energy Bands in Solids

4.1 Introduction

In Fig. 4.1 we present some schematic examples of energy bands which are representative of metals, semiconductors and insulators, and we point out some of the characteristic features in each case. Figure 4.1 distinguishes in a schematic way between insulators (a), metals (b), semimetals (c), a thermally excited semiconductor (d) for which at temperature $T = 0$ all states in the valence band are occupied and all states in the conduction band are unoccupied, assuming no impurities or crystal defects. Finally in Fig. 4.1e, we see a p-doped semiconductor which is deficient in electrons, not having sufficient electrons to fill the valence band completely, as in the case of (d). The semiconductor (e) will have a non-zero carrier density at $T = 0$, while for a semiconductor (d), the carrier density will be zero at $T = 0$.

Figure 4.2 shows a schematic view of the electron dispersion relations for an insulator (a), while (c) shows dispersion relations for a metal. In the case of Fig. 4.2b, we have a semimetal with no dopants, so that at $T = 0$, we have a semimetal where the number of electrons equals the number of holes, but a metal with a low carrier density if the electron and hole densities are not equal to one another at $T = 0$.

In this chapter we examine a number of representative $E(\mathbf{k})$ diagrams for illustrative materials. For each of the $E(\mathbf{k})$ diagrams we consider the following questions:

1. Is the material a metal, a semiconductor (with a direct or indirect gap), a semimetal or an insulator?
2. To which atomic (molecular) levels do the bands in the band diagram correspond? Which bands are important in determining the electronic structure? What are the bandwidths, bandgaps?
3. What information does the $E(\mathbf{k})$ diagram provide concerning the following questions:

   a. Where are the carriers located in the Brillouin zone?
   b. Are the carriers electrons or holes?
   c. Are there many or few carriers?
   d. How many carrier pockets of each type are there in the Brillouin zone?
   e. What is the shape of the Fermi surface?
   f. Are the carrier velocities high or low?

M. Dresselhaus et al., *Solid State Properties*, Graduate Texts in Physics,
https://doi.org/10.1007/978-3-662-55922-2_4

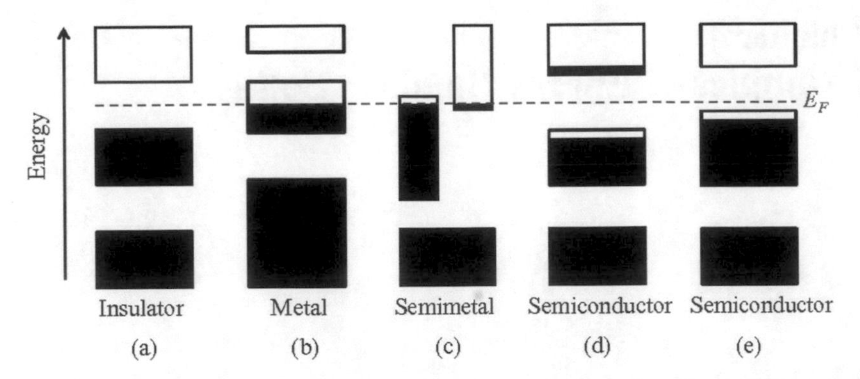

Fig. 4.1 Schematic electron occupancy of allowed energy bands for an insulator, metal, semimetal and semiconductor. The vertical extent of the boxes indicates the allowed energy regions: the shaded areas indicate the regions filled with electrons. In a semimetal (such as bismuth) one band is almost filled and another band is nearly empty at a temperature of absolute zero ($T = 0$) with the electron density equal to the hole density. A pure semiconductor (such as silicon) becomes an insulator at $T = 0$. Panel **d** shows an intrinsic semiconductor at a finite temperature, with carriers that are thermally excited. Panel **e** shows a p-doped semiconductor that is electron-deficient, as, for example, because of the introduction of acceptor impurities

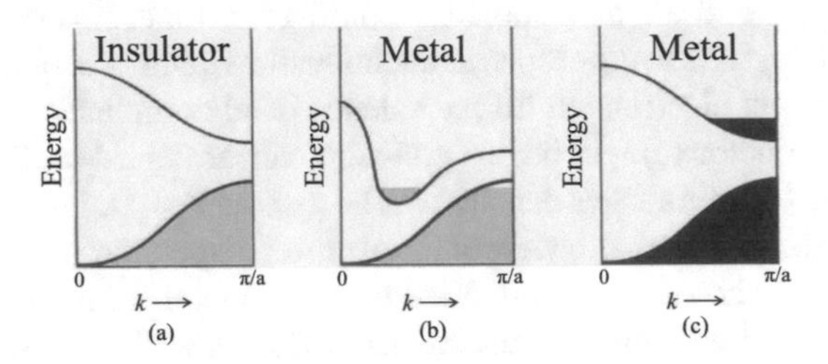

Fig. 4.2 Occupied states and band structures giving **a** an insulator, **b** a metal or a semimetal because of band overlap, and **c** a metal because of partial occupation of an electron band. In **b** the band overlap for a 3D solid need not occur at the same wave vector k in the Brillouin zone

g. Are the carrier mobilities for each carrier pocket high or low?

4. What information is provided concerning the optical properties?

   a. Where in the Brillouin Zone is the threshold for optical transitions?
   b. At what photon energy does the optical threshold occur?
   c. For semiconductors, does the threshold correspond to a direct energy gap or an indirect gap involving a phonon-assisted transition?

In order to illustrate how to answers these questions, we now consider the electronic band structure of some illustrative materials in some detail.

4.2 Metals

4.2.1 Alkali Metals–e.g., Sodium

For the alkali metals the valence electrons are nearly free and the weak binding approximation describes these electrons quite well. The Fermi surface is nearly spherical and the energy band gaps are small. The crystal structure for the alkali metals is body centered cubic (BCC) (the unit cell in reciprocal space is shown in Fig. 4.3b) and the $E(\mathbf{k})$ diagram (Fig. 4.3a) is drawn starting with the bottom of the half–filled conduction band. For example, the $E(\mathbf{k})$ diagram in Fig. 4.3 for sodium, representing the $3s$ conduction band, begins at ~ -0.6 Rydberg. The electron energy is given in Rydbergs, where 1 Rydberg $= 13.6$ eV, the ionization energy of a hydrogen atom. The filled valence bands, corresponding to the $1s$, $2s$ and $2p$ atomic levels, lie much lower in energy and are not shown in Fig. 4.3.

For the case of sodium, the $3s$ conduction band is very nearly free electron–like and the $E(\mathbf{k})$ relations are closely isotropic. Thus the $E(\mathbf{k})$ relations along the $\Delta(100)$, $\Sigma(110)$ and $\Lambda(111)$ directions [see Fig. 4.3b] are essentially coincident and can be so plotted, as shown in Fig. 4.3a. For these metals, the Fermi level is determined so that the $3s$ band is exactly half–occupied, since the Brillouin zone is large enough to accommodate 2 electrons per unit cell. Thus the radius of the Fermi surface k_F satisfies the relation

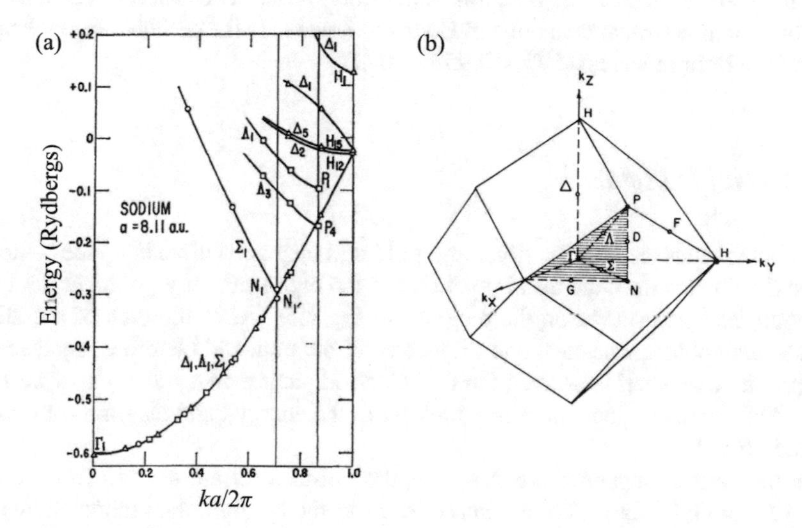

Fig. 4.3 **a** Energy dispersion relations $E(\mathbf{k})$ for the nearly free electron metal sodium which has an atomic configuration $1s^2 2s^2 2p^6 3s$. **b** The Brillouin zone for the BCC lattice showing the high symmetry points and axes. Sodium can be considered as a prototype alkali metal crystalline solid for discussing the dispersion relations for nearly free electron metals. Reprinted with permission from Physical Review, vol. 128 p. 82. Copyright (1962) and Physical Review B, vol. 7 p. 2416 Copyright (1973) American Physical Society

$$\frac{4}{3}\pi k_F^3 = \frac{1}{2}V_{\text{B.Z.}} = \frac{1}{2}(2)\left(\frac{2\pi}{a}\right)^3, \quad \text{or} \quad \frac{k_F\, a}{2\pi} \sim 0.63, \tag{4.1}$$

where $V_{\text{B.Z.}}$ and a are, respectively, the volume of the Brillouin zone and the lattice constant. For the alkali metals, the effective mass m^* is nearly equal to the free electron mass m and the Fermi surface is nearly spherical and never comes close to the Brillouin zone boundary. The zone boundary for the Σ, Λ and Δ directions are indicated in the $E(\mathbf{k})$ diagram of Fig. 4.3a by vertical lines. For the alkali metals, the band gaps are very small compared to the band widths and the $E(\mathbf{k})$ relations are parabolic ($E = \hbar^2k^2/2m^*$) almost up to the Brillouin zone boundaries. By comparing $E(\mathbf{k})$ for Na with the BCC empty lattice bands (see Fig. 4.4) for which the potential $V(r) = 0$, we can see the effect of the very weak periodic potential in partially lifting the band degeneracy at the various high symmetry points in the Brillouin zone. The threshold for optical transitions corresponds to photons having sufficient energy to take an electron from an occupied state at k_F to an unoccupied state at k_F, since the wave vector for photons is very small compared with the Fermi wave vector k_F and since wave vector conservation (also called crystal momentum conservation) is required for optical transitions. The threshold for optical transitions from the highest occupied state to an empty state with the same wave vector, as indicated by the $\hbar\omega$ value and a vertical arrow in Fig. 4.3a. Because of the low density of initial and final states for a given energy separation, we would expect optical interband transitions for alkali metals to be very weak and this is in agreement with experimental observations for all the alkali metals. The notation "a.u." in Fig. 4.3a stands for atomic units and expresses lattice constants in units of Bohr radii, which is the radius of the hydrogen atom (one Bohr radius is 0.529 Å(0.0529 nm)).

4.2.2 Noble Metals

The noble metals are copper, silver and gold and they crystallize in a face centered cubic (FCC) structure; the usual notation for the high symmetry points in the FCC Brillouin zone are shown on the diagram in Fig. 4.5a. As in the case of the alkali metals, the noble metals have one valence electron/atom and therefore one electron per primitive unit cell. However, the free electron picture does not work so well for the noble metals, as you can see by looking at the energy band diagram for copper given in Fig. 4.5b.

In the case of copper, the bands near the Fermi level are derived from the $4s$ and $3d$ atomic levels which are clearly seen in the empty lattice model shown in Fig. 4.5. The so-called $4s$ and $3d$ bands accommodate a total of 12 electrons, while the number of available electrons is 11. Therefore the Fermi level must cross these bands. Consequently copper is metallic. In Fig. 4.5b we see that the $3d$ valence bands are relatively flat and show little dependence on wave vector $\mathbf{k}$. We can trace the $3d$ bands by starting at $\mathbf{k} = 0$ with the $\Gamma_{25'}$ and Γ_{12} levels derived from the angular momentum $L = 2$ state. On the other hand, the $4s$ band has a strong k–dependence

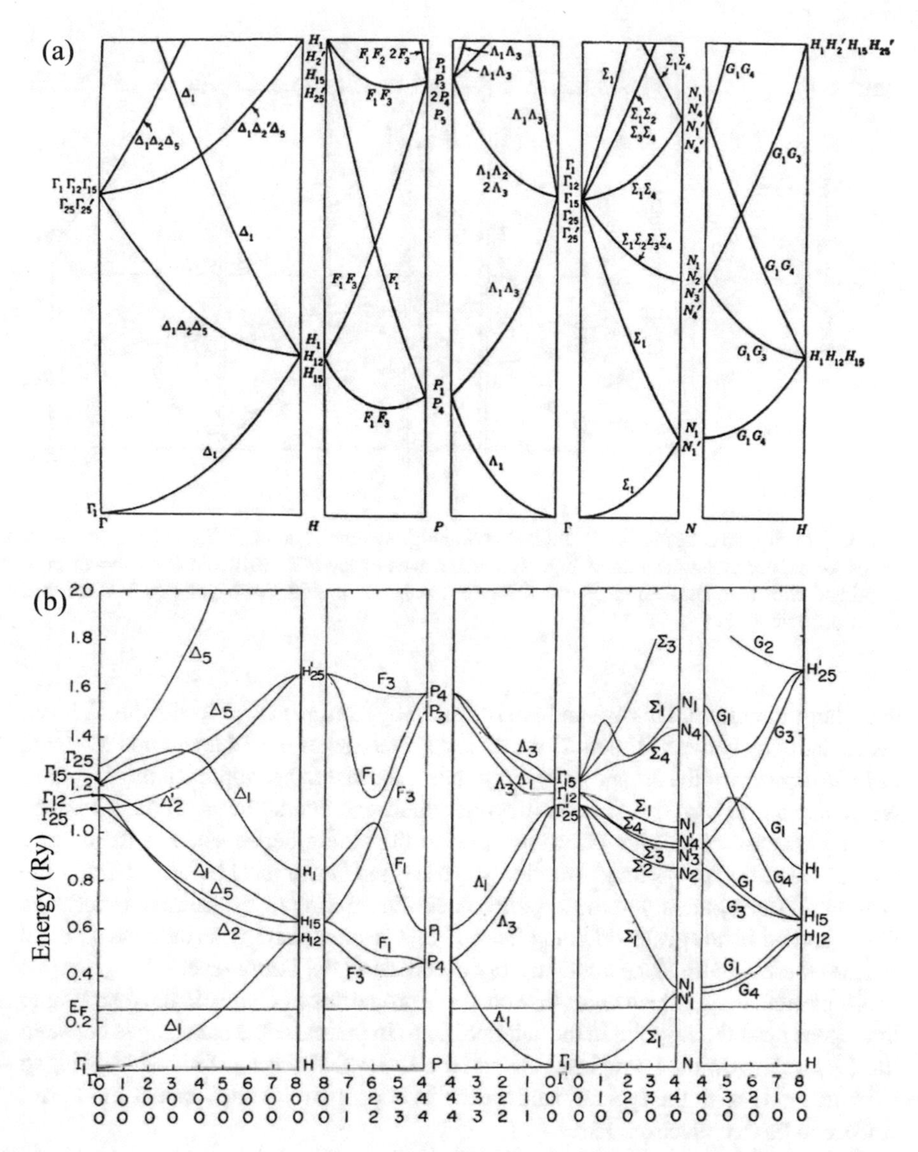

Fig. 4.4 **a** $E(\mathbf{k})$ for a BCC lattice in the empty lattice approximation, $V \equiv 0$. **b** $E(\mathbf{k})$ for sodium, showing the effect of a weak periodic potential $V(r)$ in lifting accidental band degeneracies at $k = 0$ and at the zone boundaries (high symmetry points) in the Brillouin zone, such as $\Gamma(\mathbf{k} = 0, 0, 0)$, $H(\mathbf{k} = 8, 0, 0)$, $P(\mathbf{k} = 4, 4, 4)$, $N(\mathbf{k} = 4, 4, 0)$. The Fermi level goes through the 3*s* band in the solid state. Note that the splittings of the energy bands are quite different for the various energy bands $E(\mathbf{k})$ and at different high symmetry points. Reprinted with permission from Physical Review B, vol. 7 p. 2416 Copyright (1973) American Physical Society

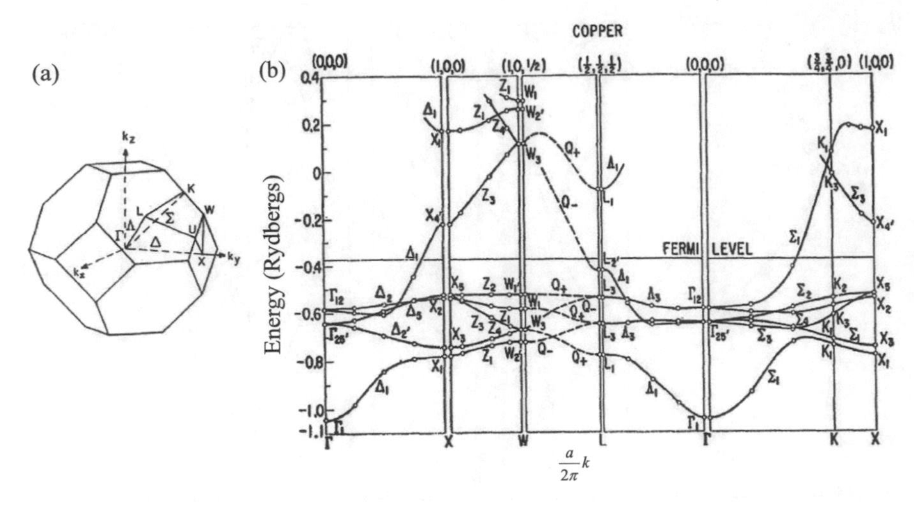

Fig. 4.5 **a** Brillouin zone for a FCC lattice showing high symmetry points. **b** The calculated energy bands for copper along the various high symmetry axes of the FCC Brillouin zone shown in **a**. Reprinted with permission from Physical Review, vol. 125 p. 109 Copyright (1962) American Physical Society

and a large curvature. This 4*s* band can be traced by starting at $\mathbf{k} = 0$ with the Γ_1 level. About halfway between Γ and X, the 4*s* level approaches the 3*d* levels and a *mixing* or *hybridization* of the 3*d* and 4*s* levels occurs. As we further approach the X–point, we can again pick up the 4*s* band (beyond where the interaction with the 3*d* bands occurs) because of its high curvature (due to the strong dependence of the energy on wavevector). This 4*s* band eventually crosses the Fermi level before reaching the Brillouin Zone boundary at the X point. A similar mixing or hybridization between the 4*s* and 3*d* bands occurs in going from Γ to L, except that in this case the 4*s* band reaches the Brillouin Zone boundary *before* crossing the Fermi level.

Of particular significance for the transport properties of copper is the energy gap that opens up at the L–point in the valence band. In this case, the band gap is between the $L_{2'}$ level below the Fermi level E_F and the L_1 level above E_F. Since this bandgap is comparable with the typical bandwidths in copper, we cannot expect the Fermi surface to be free electron–like.

By looking at the energy bands $E(\mathbf{k})$ along the major high symmetry directions, such as the (100), (110) and (111) directions, we can readily trace the origin of the copper Fermi surface [see Fig. 4.6]. Here we see basically a spherical Fermi surface with necks pulled out in the (111) directions and making contact with the Brillouin zone boundary through these necks, thereby linking the Fermi surface in one Brillouin zone to that in the next Brillouin zone in the extended zone scheme. In the (100) direction, the cross section of the Fermi surface is nearly circular, indicative of the nearly parabolic $E(\mathbf{k})$ relation of the 4*s* band at the Fermi level in going from Γ to X. In contrast, in going from Γ to L, the 4*s* band never crosses the Fermi level. Instead, the 4*s* level is depressed from the free electron parabolic curve as

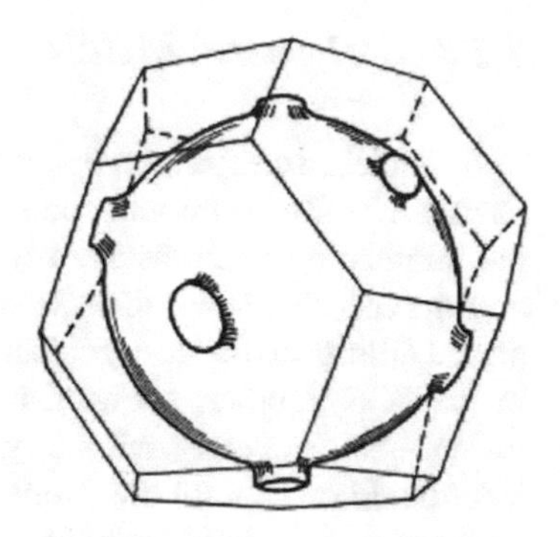

Fig. 4.6 A sketch of the Fermi surface of copper inscribed within the FCC Brillouin zone. Reprinted with permission from Physical Review, vol. 125 p. 109 Copyright (1962) American Physical Society

the Brillouin zone boundary is reached, thereby producing a high density of states near the Brillouin zone boundary. Thus, near the zone boundary, more electrons can be accommodated per unit energy range, or to say this another way, there will be increasingly more **k** vectors with approximately the same energy. This causes the constant energy surfaces to be pulled out in the direction of the Brillouin zone boundary [see Fig. 4.6]. This "pulling out" effect follows both from the weak binding and tight binding approximations and results in a high density of states near the zone boundary that was also seen in Fig. 4.5. The effect is more pronounced as the strength of the periodic potential ($V_{\mathbf{r}}$) increases.

If the periodic potential is sufficiently strong so that the resulting bandgap at the zone boundary straddles the Fermi level, as occurs at the L–point in copper, the Fermi surface makes contact with the Brillouin zone boundary. The resulting Fermi surfaces are called *open surfaces* because the Fermi surfaces between neighboring Brillouin zones are connected to each other. The electrons associated with the necks of the Fermi surface are contained in the electron pocket shown in the $E(\mathbf{k})$ diagram away from the L–point in the LW direction which is $\perp$ to the $\{111\}$ direction in reciprocal space. The copper Fermi surface shown in Fig. 4.6 bounds *electron* states. Hole pockets are formed in copper and constitute the unoccupied space between the electron surfaces in the extended zone scheme. Direct evidence for hole pockets is provided by Fermi surface measurements to be described later in this book.

From the $E(\mathbf{k})$ diagram for copper [Fig. 4.5b] we see that the threshold for optical interband transitions occurs for photon energies large enough to take an electron at constant **k**–vector from a filled $3d$ level to an unoccupied state above the Fermi level. Such interband transitions can be made near the L–point in the Brillouin zone [as shown in Fig. 4.5b]. Because of the high density of initial states in the d–band, these transitions will be quite intense. The occurrence of these interband transitions at $\sim$2 eV gives rise to a large absorption of electromagnetic energy in this photon energy region. The reddish color of copper metal is thus due to a higher reflectivity for photons in the red (below the threshold for interband transitions) than for photons in the blue (above this energy threshold).

4.2.3 Polyvalent Metals

The simplest example of a polyvalent metal is aluminum with 3 electrons/atom and having a $3s^2 3p$ electronic configuration for its three valence electrons. (As far as the number of electrons/atom is concerned, two electrons/atom completely fill a non-degenerate bands—one for spin up, the other for spin down.) Because of the partial filling of the $3s^2 3p^6$ bands, aluminum is a metal. Aluminum crystallizes in the FCC structure so we can use the same notation as for the Brillouin zone in Fig. 4.5a. The electronic energy bands for aluminum (see Fig. 4.7) are very free electron–like. This follows from the small magnitudes of the band gaps relative to the band widths on the energy band diagram shown in Fig. 4.7. The lowest valence band shown in Fig. 4.7 is the 3s band which can be traced by starting at zero energy at the Γ point ($\mathbf{k} = 0$) and going out to X_4 at the X–point, to W_3 at the W–point, to L'_2 at the L–point and back to Γ_1 at the Γ point ($\mathbf{k} = 0$). Since this 3s band always lies below the Fermi level, it is completely filled, containing 2 electrons. The third valence electron partially occupies the second and third p–bands (which are more accurately described as hybridized 3p–bands with some admixture of the 3s bands with which they hybridize). From Fig. 4.7 we can see that the second band is partly filled; the occupied states extend from the Brillouin zone boundary inward toward the center of the zone; this can be seen in going from the X point to Γ, on the curve labeled Δ_1. Since the second band starts near the center of the Brillouin zone remain unoccupied, the volume enclosed by the Fermi surface in the second band is a hole pocket. The aluminum Fermi surface showing the holes in Zone 2 is presented in

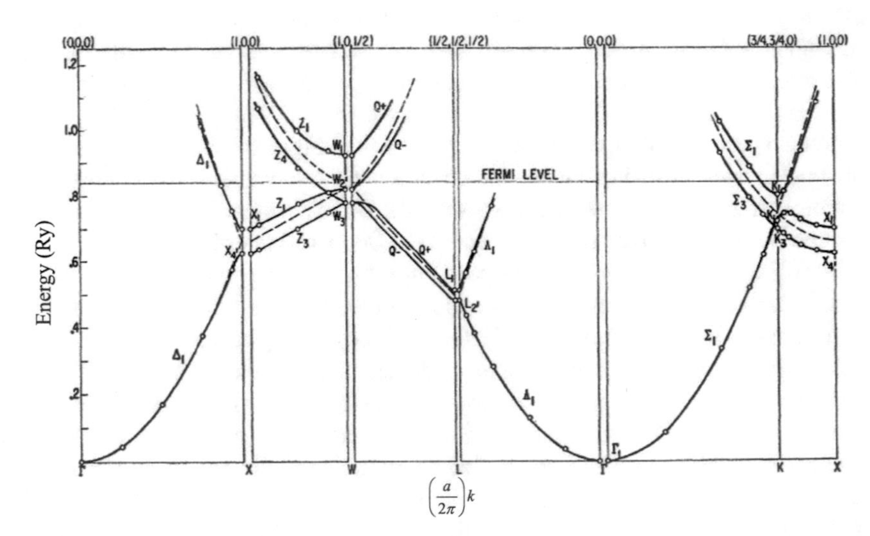

Fig. 4.7 Electronic energy band diagram for aluminum which is metallic and crystallizes in a FCC structure. The dashed lines correspond to the free electron model and the solid curves include the effect of the periodic potential $V(\mathbf{r})$. Reprinted with permission from Physical Review, vol. 125 p. 109 Copyright (1962) American Physical Society

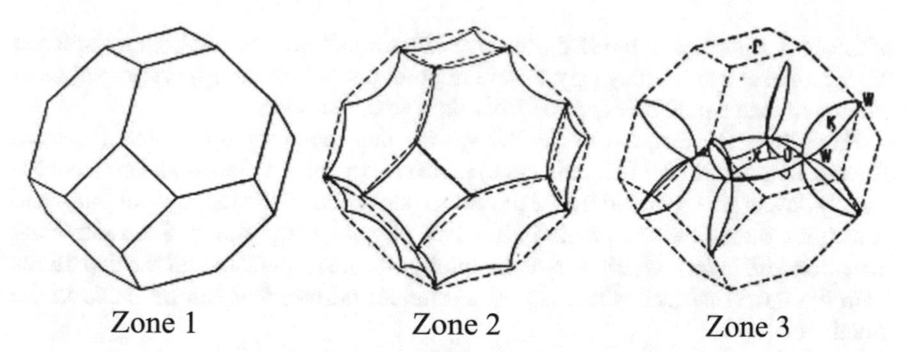

Fig. 4.8 The three valence electrons for aluminum occupy three Brillouin zones. Zone 1 is completely occupied. Zone 2 in nearly filled with electrons and is best described as a hole surface, where the holes occupy the interior portion of the second zone, shown in the figure. Zone 3 is a complex electron structure with occupied electron states near the Brillouin zone boundaries, and the occupied states are shown only in part for clarity. Reprinted with permission from Physical Review, vol. 118 p. 1190 Copyright (1960) American Physical Society

Fig. 4.8. Because $E(\mathbf{k})$ for the second band in the vicinity of E_F is free electron–like, the masses for the holes are approximately equal to the free electron mass.

The 3rd zone electron pockets are small and are found around the $K-$ and $W-$ points as can be seen in Fig. 4.7. These electron pockets are **k** space volumes that enclose electron states (see Fig. 4.8), and because of the large curvature of $E(\mathbf{k})$, these electrons have relatively small effective masses. Figure 4.7 gives no evidence for any 4th zone pieces of Fermi surface, and for this reason we can conclude that all the electrons are either in the second band or in the third zone pieces of Fermi surface. Therefore we conclude that the total electron concentration is sufficient to exactly fill a half of the volume of the Brillouin zone V_{BZ}:

$$V_{e,2} + V_{e,3} = \frac{V_{BZ}}{2}. \tag{4.2}$$

With regard to the second zone, it is partially filled with electrons and the rest of the zone is empty (since holes correspond to the unfilled states):

$$V_{h,2} + V_{e,2} = V_{BZ}, \tag{4.3}$$

so that the volume that is empty slightly exceeds the volume that is occupied. Therefore we focus attention on the more dominant second zone holes. Substitution of (4.2) into (4.3) then yields volumes for the second zone holes and the third zone electrons

$$V_{h,2} - V_{e,3} = \frac{V_{BZ}}{2} \tag{4.4}$$

where the subscripts e, h on the volumes in **k** space refer to electrons and holes. This notation also contains the Brillouin zone (B.Z.) index which is given for each of the carrier pockets in the B.Z. Because of the small masses and high mobility of the 3rd zone electrons, they play a more important role in the transport properties of aluminum than would be expected from their small numbers.

From the $E(\mathbf{k})$ diagram in Fig. 4.7 we see that at the same **k**–points (near the K– and W–points in the Brillouin zone) there are occupied $3s$ levels and are unoccupied $3p$ levels are separated by $\sim$1 eV. From this we conclude that optical interband transitions should be observable in the 1eV photon energy range. Such interband transitions are in fact observed experimentally and are responsible for the departures from the nearly perfect reflectivity of aluminum mirrors that can be made in the vicinity of 1 eV.

4.3 Semiconductors

Assume that we have a semiconductor at $T = 0$ K with no impurities. The Fermi level will then lie within a band gap. Under these conditions, there are no carriers, and there is no Fermi surface. We now illustrate the energy band structure for several representative semiconductors in the limit of $T = 0$ K and no impurities. Semiconductors having no impurities or defects and no carriers are called *intrinsic semiconductors*.

4.3.1 PbTe

In Fig. 4.9 we illustrate a simplified version of the electronic energy bands for PbTe. This direct gap semiconductor [see Fig. 4.10a] is chosen initially for illustrative purposes because the energy bands in the valence and conduction bands that are of most importance in determining the physical properties of PbTe when this semiconductor is non-degenerate. Therefore, the energy states in PbTe near E_F are simpler to understand than for the more common semiconductors silicon and germanium, and for many of the III–V and II–VI compound semiconductors, for which the valence band is degenerate.

In Fig. 4.9, we show the position of E_F for the idealized conditions of the intrinsic (no carriers at $T = 0$) semiconductor PbTe. From a diagram like this, we can obtain a great deal of information which could be useful for making semiconductor devices. For example, we can calculate the effective masses from the band curvatures, and also the electron velocities from the slopes of the $E(\mathbf{k})$ dispersion relations shown in Fig. 4.9. PbTe is also a commonly used thermoelectric material, and for this reason density functional theory based calculations of the band structure have been made (Hummer et al. 2007). Since 2015, this material has become interesting for its topological behaviors.

Suppose we add impurities (e.g., donor impurities) to PbTe. The donor impurities will raise the Fermi level and an electron pocket will eventually be formed in the

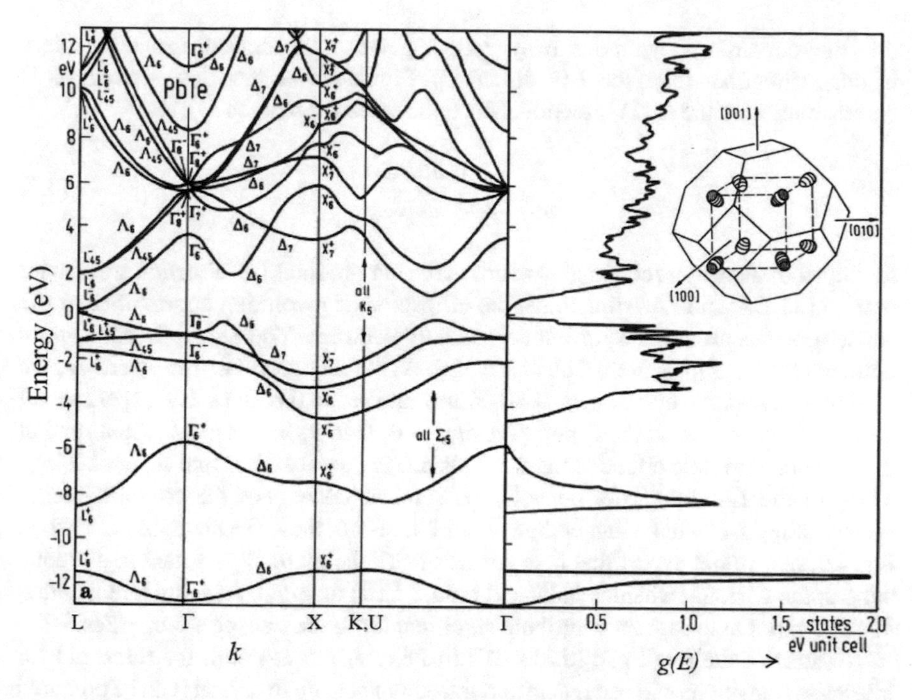

Fig. 4.9 **a** Energy band structure and density of states for PbTe obtained from an empirical pseudopotential calculation. The inset shows the localization of pockets at the Brillouin zone. Reprinted with permission from Physical Review B, vol. 11 p. 651 Copyright (1962) American Physical Society

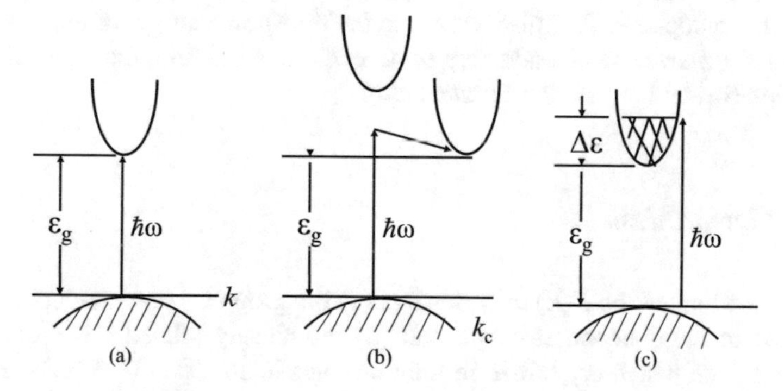

Fig. 4.10 Optical absorption processes for **a** a direct band gap semiconductor without doping, and for **b** an indirect band gap semiconductor where the optical transition is from one point in the BZ to another point and a phonon is necessary to conserve energy and momentum, and **c** a direct band gap semiconductor with the conduction band filled to the level shown

L_6^- conduction band about the L–point (see Fig. 4.9). This electron pocket has been described in terms of an ellipsoidal Fermi surface because the band curvature is different as we move away from the L point in the $L\Gamma$ direction as compared with the band curvature as we move away from L point on the Brillouin zone boundary in other directions (e.g., the LW direction). Figure 4.9 shows $E(\mathbf{k})$ from L to Γ corresponding to the (111) direction. Since the effective masses

$$\frac{1}{m_{ij}^*} = \frac{1}{\hbar^2}\frac{\partial^2 E(\mathbf{k})}{\partial k_i \partial k_j} \tag{4.5}$$

for both the valence and conduction bands in the longitudinal $L\Gamma$ direction are heavier than in the LK and LW directions, the ellipsoids of revolution approximating the carrier pockets are prolate for both holes and electrons. The L and Σ point room temperature band gaps are 0.311 and 0.360 eV, respectively. For the electrons, the effective mass parameters are $m_\perp = 0.053m_e$ and $m_\parallel = 0.620m_e$. The experimental hole effective masses at the L point are $m_\perp = 0.0246m_e$ and $m_\parallel = 0.236m_e$ and at the Σ point, the hole effective mass values are $m_\perp = 0.124m_e$ and $m_\parallel = 1.24m_e$. Thus for the L-point carrier pockets, the semi-major axis of the constant energy surface along $L\Gamma$ will be longer than along LK. From the $E(\mathbf{k})$ diagram for PbTe in Fig. 4.9, one would expect that hole carriers could be thermally excited to a second band at the Σ point, which is indicated on the $E(\mathbf{k})$ diagram. At room temperature, these Σ point hole carriers contribute significantly to the transport properties.

Because of the small gap (0.311 eV) in PbTe at the L–point, the threshold for interband transitions will occur at infrared frequencies. PbTe crystals can be prepared either p–type or n–type, but they are never perfectly stoichiometric. Therefore, at room temperature the Fermi level E_F often lies in either the valence or conduction band for actual PbTe crystals. Since optical transitions conserve wavevector, the interband transitions will occur at k_F [see Fig. 4.10c] and at a higher photon energy than for the undoped PbTe. This increase in the threshold energy for interband transitions in *degenerate* semiconductors (where E_F lies within either the valence or conduction bands) is called the *Burstein shift*.

4.3.2 Germanium

We will next look at the $E(\mathbf{k})$ relations for: (1) the group IV semiconductors which crystallize in the diamond structure and (2) the closely related III–V compound semiconductors which crystallize in the zinc blende structure (see Fig. 4.11 for a schematic diagram for common group IV and III–V semiconductors). These semiconductors have degenerate valence bands at $\mathbf{k} = 0$ [see Fig. 4.11d] and for this reason have more complicated $E(\mathbf{k})$ relations for hole carriers than is the case for the lead salts discussed in Sect. 4.3.1 for a simplified version of Ge that neglects the spin-orbit interaction. The $E(\mathbf{k})$ diagram for germanium is shown in Fig. 4.12. Ge is an indirect gap semiconductor with a bandgap occurring between the top of the valence band at $\Gamma_{25'}$, and the bottom of the lowest conduction band at L_1. Since the valence and conduction band extrema occur at *different* points in the Brillouin zone, Ge is

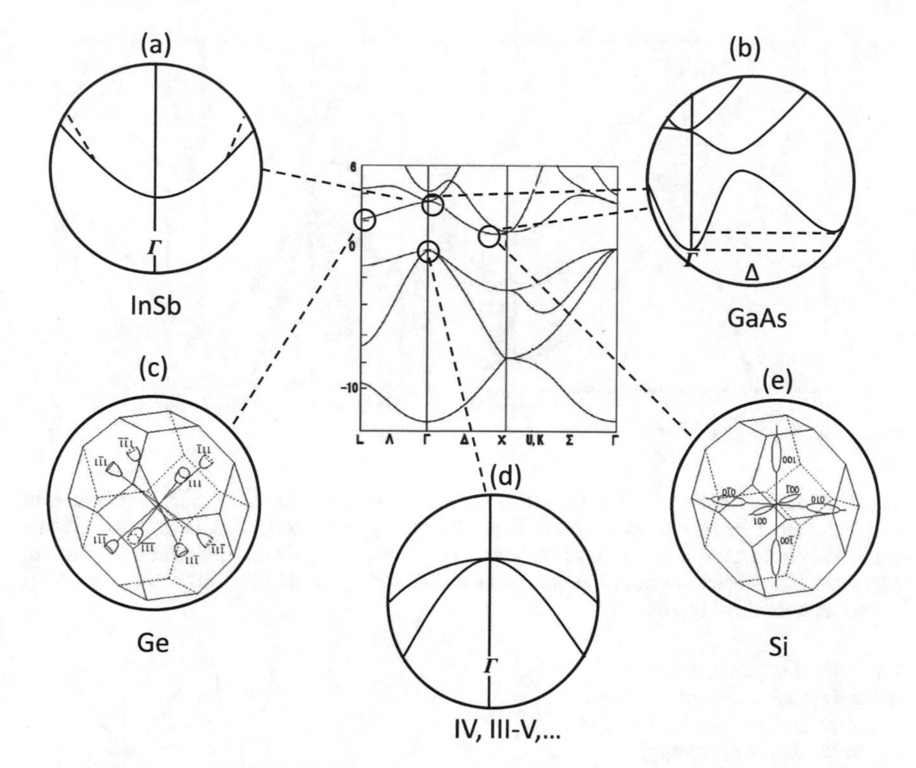

Fig. 4.11 Important details of the band structure of typical group IV semiconductors (like Si and Ge) and for various III–V semiconductors. Reprinted with permission from Physical Review B, vol. 10 p. 5095 Copyright (1974) American Physical Society

an *indirect gap* semiconductor [see Fig. 4.10b]. Using the same arguments as were given in Sect. 4.3.1 for the Fermi surface of PbTe, we see that the constant energy surfaces for electrons in germanium are ellipsoids of revolution [see Fig. 4.11c]. As for the case of PbTe, the ellipsoids of revolution are elongated along ΓL which is the heavy mass direction in this case. Since the multiplicity of L–points is 8, we have 8 half–ellipsoids of this kind within the first Brillouin zone, just as for the case of PbTe. By translation of these half–ellipsoids by a reciprocal lattice vector, we can form 4 full–ellipsoids. The $E(\mathbf{k})$ diagram for germanium (see Fig. 4.12) further shows that the next highest conduction band above the L point minimum is at the Γ–point ($\mathbf{k} = 0$) and after that along the ΓX axis at a point commonly labelled as a Δ–point. Because of the degeneracy of the highest valence band, the Fermi surface for holes in germanium is more complicated than for electrons. The lowest direct band gap in germanium is at $\mathbf{k} = 0$ between the $\Gamma_{25'}$ valence band and the $\Gamma_{2'}$ conduction band. From the $E(\mathbf{k})$ diagram we note that the electron effective mass for the $\Gamma_{2'}$ conduction band is very small because of the high curvature of the $\Gamma_{2'}$ band about $\mathbf{k} = 0$, and this effective mass is isotropic so that the constant energy surfaces are spheres.

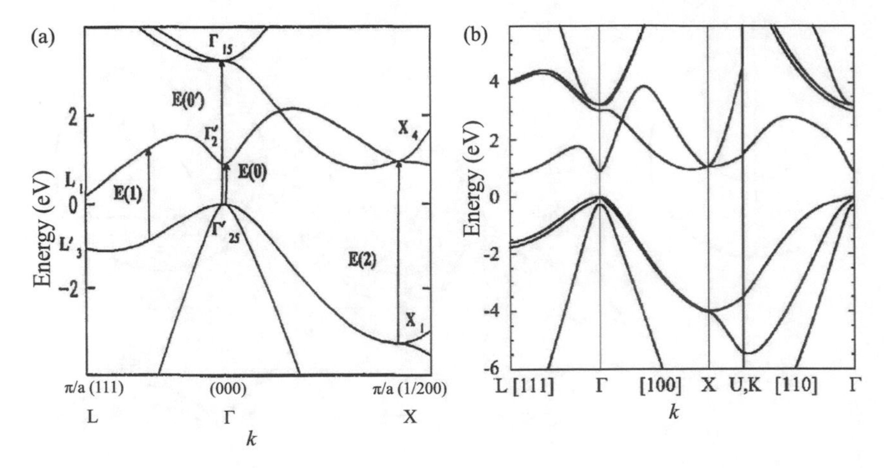

Fig. 4.12 **a** Electronic energy band structure of Ge without spin-orbit interaction. Reprinted with permission from Physical Review B, vol. 70, p. 235204 Copyright (2004) American Physical Society. **b** The electronic energy bands of Ge near $k = 0$ when the spin-orbit interaction is included. Reprinted with permission from Journal of Chemical Physics, vol. 101 p. 1607 Copyright (1994) American Institute of Physics

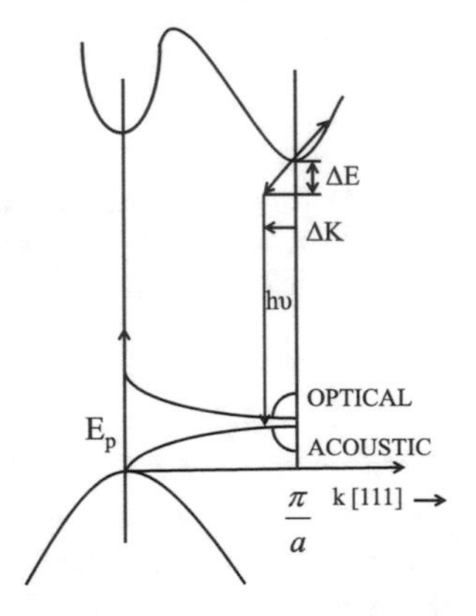

Fig. 4.13 Illustration of the indirect emission of light due to carriers and phonons in Ge. [$h\nu$ is the photon energy; ΔE is the energy delivered to an electron; E_p is the energy delivered to the lattice (phonon energy)]

The optical properties for germanium show a very weak optical absorption for photon energies corresponding to the indirect gap (see Fig. 4.13). Since the valence and conduction band extrema occur at a different **k**–point in the Brillouin zone, the indirect gap excitation requires a phonon to conserve crystal momentum. Hence the threshold for this indirect transition is

$$(\hbar\omega)_{\text{threshold}} = E_{L_1} - E_{\Gamma_{25'}} - E_{\text{phonon}}. \quad (4.6)$$

The optical absorption for germanium increases rapidly above the photon energy corresponding to the direct band gap $E_{\Gamma_{2'}} - E_{\Gamma_{25'}}$, because of the higher probability for the direct optical excitation process. However, the absorption here remains low

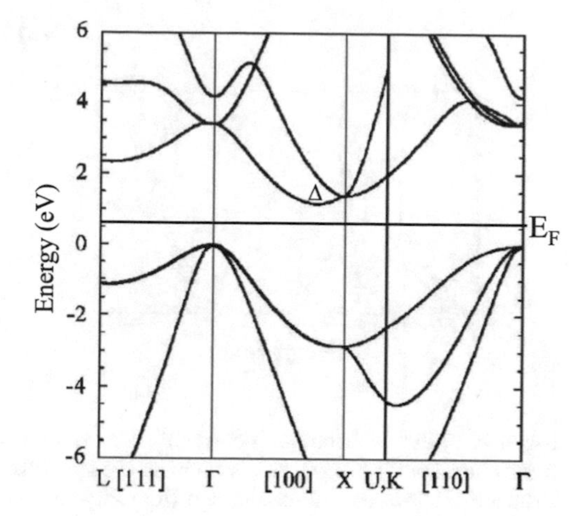

Fig. 4.14 Electronic energy band structure of Si. Reprinted with permission from Physical Review B, vol. 70 p. 235204 Copyright (2004) American Physical Society

compared with the absorption at yet higher photon energies because of the low density of states for the Γ–point transition, as seen from the $E(\mathbf{k})$ diagram. Very high optical absorption, however, occurs for photon energies corresponding to the energy separation between the $L_{3'}$ and L_1 bands. This energy band separation is approximately the same for a large range of **k** values, thereby giving rise to a very large joint density of states (the number of states with constant energy separation per unit energy range). A large *joint density of states* arising from the *tracking* of conduction and valence bands is found for germanium, silicon and the III–V compound semiconductors, and for this reason these materials tend to have high dielectric constants (to be discussed in Part III of this book which focuses on optical properties).

4.3.3 Silicon

From the energy band diagram for silicon shown in Fig. 4.14, we see that the energy bands of Si are quite similar to those for germanium. $E(\mathbf{k})$ for Si and Ge, however, differ in detail. For example, in the case of silicon, the electron pockets are formed around a Δ point located along the ΓX (100) direction. For silicon there are 6 electron pockets within the first Brillouin zone instead of the 8 half–pockets which occur in germanium. The constant energy surfaces are again ellipsoids of revolution with a heavy longitudinal mass and a light transverse effective mass [see Fig. 4.11e]. The second type of electron pocket that is energetically favored is about the L_1 point, but to fill electrons there, we would need to raise the Fermi energy by $\sim$1 eV for Si.

Silicon is of course the most important semiconductor for device applications and is at the heart of semiconductor technology for transistors, integrated circuits, and many electronic devices. The optical properties of silicon also have many similarities to those in germanium, but show differences in detail. For Si, the indirect gap [see Fig. 4.10b] occurs at $\sim$1 eV and is between the $\Gamma_{25'}$ valence band and the Δ conduction

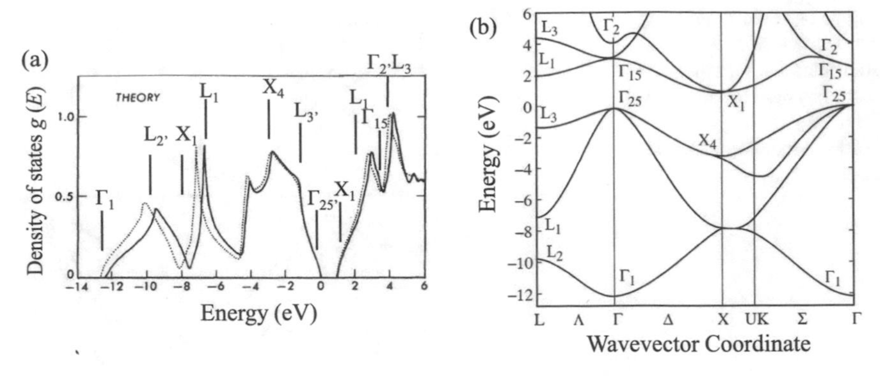

Fig. 4.15 **a** Plot of theoretical density of states in the valence and conduction bands of silicon, and **b** the corresponding $E(\mathbf{k})$ curves showing the symbols for the high symmetry points of the band structure. Reprinted with permission from Physical Review B, vol. 14 p. 556 Copyright (1976) American Physical Society

band extrema. Just as in the case for germanium, strong optical absorption occurs for large volumes of the Brillouin zone at energies comparable to the $L_{3'} \rightarrow L_1$ energy separation for germanium, because of the "tracking" of the valence and conduction bands in momentum space. The density of electron states for Si covering a wide energy range is shown in Fig. 4.15a, and the corresponding energy band diagram is shown in Fig. 4.15b. Most of the features in the density of states can be identified from the band model, as is shown in Fig. 4.15a for Si.

4.3.4 III–V Compound Semiconductors

Another important class of semiconductors is the III–V compound semiconductors which crystallize in the zinc blende structure; this structure is like the diamond structure except that the two atoms/unit cell are of a different chemical species. The III–V compounds also have many practical applications, such as semiconductor lasers for fast electronics and communications, GaAs in light emitting diodes, and InSb for infrared detectors. In Fig. 4.16 the $E(\mathbf{k})$ diagram for GaAs is shown and we see that the electronic levels are very similar to those of Si and Ge. One exception is that the lowest conduction band for GaAs is at $\mathbf{k} = 0$ so that both valence and conduction band extrema are at $\mathbf{k} = 0$. Thus GaAs is a *direct* gap semiconductor [see Fig. 4.10a], and for this reason, GaAs shows a stronger and more sharply defined optical absorption threshold than Si or Ge. Figure 4.11b shows a schematic of the conduction bands for GaAs. Here we see that the lowest conduction band for GaAs has high curvature and therefore a small effective mass that is noteworthy. This effective mass is isotropic so that the constant energy surface for electrons in GaAs is a sphere and there is just one such sphere in the Brillouin zone. The next lowest

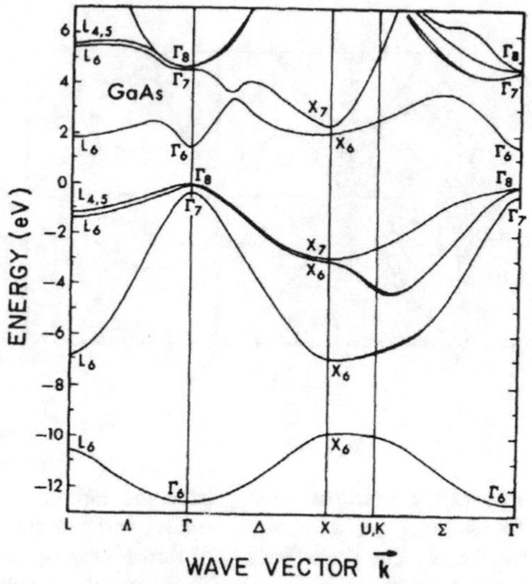

Fig. 4.16 Electronic energy band structure of the III–V compound GaAs. Reprinted with permission from Physical Review B, vol. 14 p. 556 Copyright (1976) American Physical Society

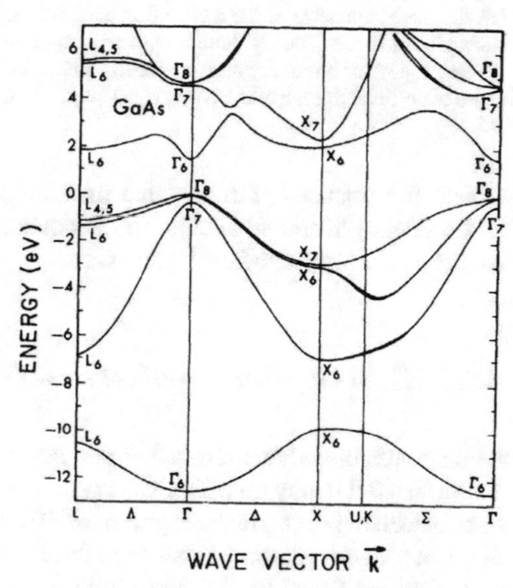

Fig. 4.17 Electronic energy band structure of the III–V compound InSb. Reprinted with permission from Physical Review B, vol. 14 p. 556 Copyright (1976) American Physical Society

conduction band is at a Δ point, and a significant carrier density can be excited into this Δ point pocket at high temperatures.

The constant energy surface for electrons in the direct gap semiconductor InSb shown in Fig. 4.17 is likewise a sphere, because InSb is also a direct gap semiconductor. InSb differs from GaAs insofar as InSb has a very small band gap (~0.2 eV), occurring in the infrared. Both direct and indirect band gap materials are found in the III–V compound semiconductor family. Except for optical phenomena close to the band gap, these compound semiconductors all exhibit very similar optical properties

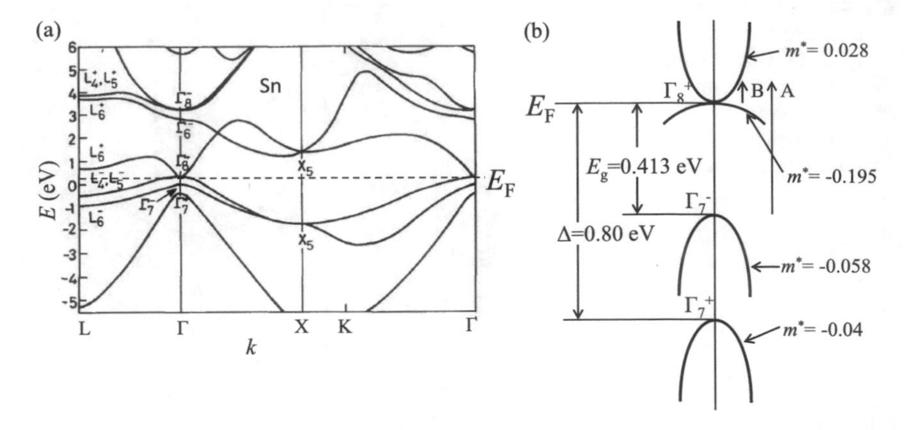

Fig. 4.18 **a** Electronic energy band structure of gray tin, including the spin-orbit interaction. **b** Detailed diagram of the energy bands of gray tin near $k = 0$, including details about the spin-orbit interaction. The Fermi level goes through the degeneracy point between the filled valence band and the empty conduction band in the idealized model for gray tin at $\Gamma = 0$. The Γ_7^- hole band has the same symmetry as the conduction band for Ge when spin–orbit interaction is included, as shown in Fig. 4.12b. The Γ_7^+ hole band has the same symmetry as the "split–off" valence band for Ge when spin–orbit interaction is included. Reprinted with permission from IEEE Proceedings-I Communications Speech and Vision, vol. 129 p. 189 Copyright (1982) IEEE

which are associated with the band-tracking phenomena, whereby an energy band in the valence band and another in the conduction band are separated by a similar energy over a large range of k values.

4.3.5 Zero Gap Semiconductors – Gray Tin

It is also possible to have a so-called *zero gap* semiconductor material. An example of such a material is gray tin which also crystallizes in the diamond structure. The energy band model for gray tin including spin–orbit interaction is shown in Fig. 4.18a. On this diagram the zero gap occurs between the Γ_{8^+} valence band and the Γ_{8^-} conduction band, and the Fermi level runs right through this degeneracy point between these bands. Spin–orbit interaction (to be discussed later in this book) is very important for gray tin in the region of the $\mathbf{k} = 0$ band degeneracy. A detailed diagram of the energy bands near $\mathbf{k} = 0$ and including the effect of spin–orbit interaction is shown in Fig. 4.18b. In gray tin the effective mass for the conduction band is much lighter than for the valence band, and this effect can be clearly seen by the band curvatures shown in Fig. 4.18b.

Optical transitions in Fig. 4.18b, labeled B, occur in the far infrared spectral region. These transitions occur from the upper valence band to the conduction band. In the near infrared, interband transitions labeled A are induced from the Γ_7^- valence band to the Γ_8^+ conduction band. We note that gray tin is classified as a zero gap semiconductor rather than a semimetal (see Sect. 4.4) because there are no band overlaps in a zero-gap semiconductor with interband transitions occurring anywhere in the Brillouin zone. Because of the zero band gap in grey tin, impurities play a major role in determining the position of the Fermi level. Gray tin is normally prepared n-type which means that there are some electrons present in the conduction band (for example, a typical electron concentration would be $10^{15}/\text{cm}^3$ which amounts to less than 1 carrier/10^7 atoms).

4.3.6 Transition Metal Dichalcogenides, Such as MoS_2 and WS_2

The absence of an intrinsic band gap in graphene presents a challenge for the use of graphene in electronic and energy conversion devices. As a result, other 2D layered materials with finite band gap energies have begun to receive increasing attention. Of particular interest are the layered transition metal dichalcogenides, such as MoS_2 and WSe_2. While bulk MoS_2 is an indirect band gap semiconductor with a band gap of 1.2 eV, quantum confinement increases the indirect band gap of bulk MoS_2 beyond the direct band gap at the K-point in the Brillouin zone, as shown in Fig. 4.19a. As a result, monolayer MoS_2 is a direct band gap semiconductor with a band gap of 1.9 eV and the luminescence quantum efficiency of atomically thin MoS_2 is four orders of magnitude higher than for bulk MoS_2. Monolayer crystals of other transition metal dichalcogenides, such as WS_2 and WSe_2, also shift to direct band gap semiconductors from their indirect bulk counterparts, as shown in Fig. 4.19b. These monolayer crystals have created a new class of direct band gap semiconductors with promising optoelectronic properties, and these have received a large amount of international attention.

4.3.7 Molecular Semiconductors – Fullerenes

Other examples of semiconductors are molecular solids such as C_{60} (see Fig. 4.20). For the case of solid C_{60}, we show in Fig. 4.20a a C_{60} molecule, which crystallizes in a FCC structure with four C_{60} molecules per conventional simple cubic unit cell. A small distortion of the bonds, lengthening the C–C bond lengths of the single bonds to 1.46Å and shortening the double bonds to 1.40Å, stabilizes a band gap of $\sim$1.5 eV [see Fig. 4.20b]. In this semiconductor, the energy bandwidths are very small compared with the band gaps, so that this material can be considered as an organic molecular semiconductor. The transport properties of C_{60} differ markedly from those for conventional group IV or III–V semiconductors, which have much wider band widths.

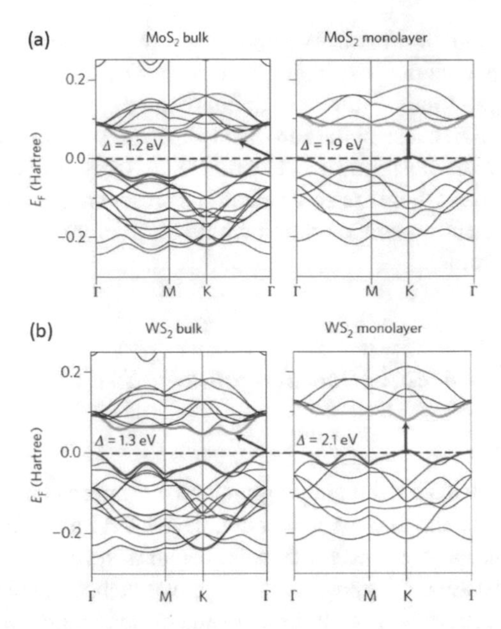

Fig. 4.19 Band structures calculated from first-principles density functional theory (DFT) for bulk and monolayer MoS_2 (**a**) and WS_2 (**b**). Reprinted with permission from Nature Nanotechnology, vol. 7, p. 699 (2012)

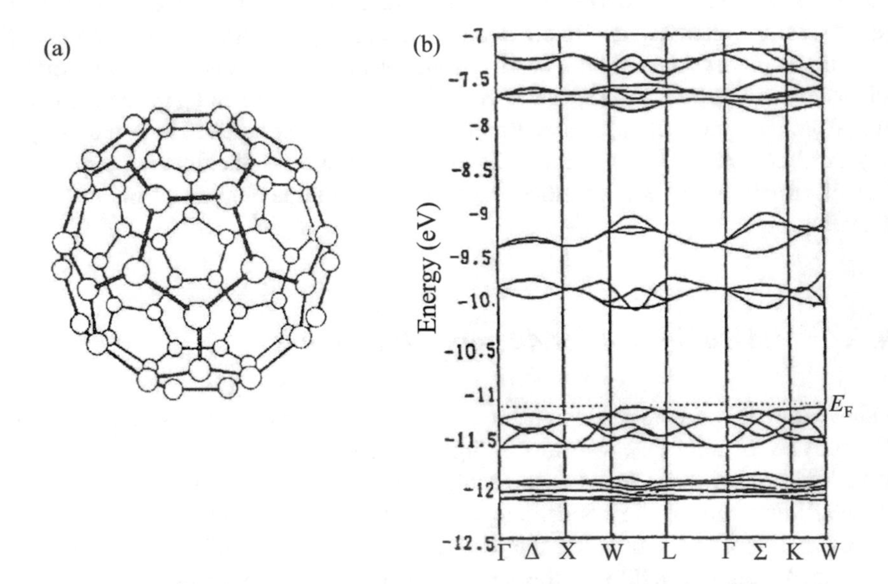

Fig. 4.20 **a** Structure of the icosahedral C_{60} molecule, and **b** the calculated one-electron electronic energy band structure of FCC solid C_{60}. The Fermi energy lies between the occupied valence levels and the empty conduction levels. Reprinted with permission from the International Journal of Quantum Chemistry, vol. 54 p. 265 Copyright (1995) Wiley

4.4 Semimetals

Another type of material that commonly occurs in nature is the *semimetal*. Semimetals have exactly the correct number of electrons to completely fill an integral number of Brillouin zones. Nevertheless, in a semimetal the highest occupied Brillouin zone is not filled up completely, since some of the electrons find lower energy states in "higher" zones (see Fig. 4.2). For semimetals the number of electrons that spill over into a higher Brillouin zone is exactly equal to the number of holes that are left behind. This is illustrated schematically in Fig. 4.21a where a two-dimensional Brillouin zone is shown and a circular Fermi surface of equal area is inscribed. Here we can easily see the electrons in the second zone at the zone edges and the holes at the zone corners that are left behind in the first zone. Translation by a reciprocal lattice vector brings two pieces of the electron surface together to form a surface in the shape of a lens, and the 4 pieces at the zone corners form a rosette shaped hole pocket. Typical examples of semimetals are bismuth and graphite. For these semimetals the carrier density is very low, on the order of one carrier/10^6 atoms.

The carrier density of a semimetal is thus not very different from that which occurs in doped semiconductors, but the behavior of the conductivity $\sigma(T)$ as a function of temperature is very different. For intrinsic semiconductors, the carriers which are excited thermally contribute significantly to conduction. Consequently, the conductivity tends to rise rapidly with increasing temperature. For a semimetal, the carrier concentration does not change strongly with temperature because the carrier density is determined by the band overlap. Since the electron scattering by lattice vibrations increases with increasing temperature, the conductivity of semimetals tends to fall as the temperature increases.

4.4.1 Graphene

Graphene is a special case of a semimetal. It consists of a single atomic layer and its electronic structure was theoretically investigated in 1947 in a pioneering work of Wallace. A breakthrough in carbon science was the isolation of graphene single layers by the micromechanical cleavage of graphite which allowed a series of pioneering experiments to be performed, thus revealing striking physical phenomena in graphene, such as ballistic transport, and quantum Hall effect at room temperature and typical relativistic phenomena, such as the Berry phase and the Klein paradox. In graphene, the electronic dispersion relation E(**k**) is linear in **k** and isotropic near the K point, thus forming the so called Dirac cone. In a neutral graphene system, the valence and conduction bands touch each other at the six K and K' points (see Fig. 3.11) in the Brillouin zone which defines the Dirac point or the Fermi surface.

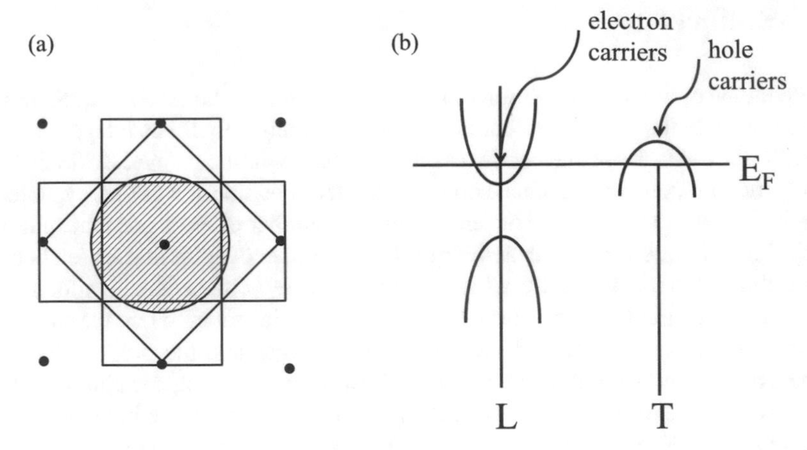

Fig. 4.21 **a** Schematic diagram of a semimetal in two dimensions. **b** Schematic diagram of the energy bands $E(\mathbf{k})$ of bismuth showing electron pockets at the L point and a hole pocket at the T point. The T point is the point at the Brillouin zone boundary in the $\{111\}$ direction along which a stretching distortion occurs in real space, and the L points refer to the 3 other equivalent $\{1\bar{1}\bar{1}\}$, $\{\bar{1}1\bar{1}\}$, and $\{\bar{1}\bar{1}1\}$ directions, before the distortions are introduced

4.4.2 Bismuth

A schematic diagram of the energy bands of the semimetal bismuth is shown in Fig. 4.21b. In the absence of external doping, electron and hole carriers exist in bismuth in equal numbers but at different locations in the Brillouin zone. For Bi, electrons are at the L–point, and holes are at the T–point [see Fig. 4.21b]. The crystal structure for Bi can be understood from the NaCl structure by considering a very small displacement of the Na FCC structure relative to the Cl FCC structure along one of the body 111 diagonals and an elongation of that body diagonal relative to the other 3 body diagonals. The special $\{111\}$ direction corresponds to $T - L$ in the Brillouin zone (see Fig. 4.21b), while the other three $\{111\}$ directions are labelled in this figure as $T - L$.

Instead of a band gap between the valence and conduction bands (as occurs for semiconductors), semimetals are characterized by a band overlap in the meV range. In bismuth, a small band gap also occurs at the L–point between the conduction band and a lower filled valence band. Because the coupling between these L-point valence and conduction bands is very strong, some of the effective mass components for the electrons in bismuth are anomalously small. As far as the optical properties of bismuth are concerned, bismuth behaves much like a metal with a high reflectivity at low frequencies due to the presence of intrinsic high mobility free carriers.

4.5 Insulators

The electronic structure of insulators is similar to that of semiconductors, in so far as both insulators and semiconductors have a band gap separating the valence and conduction bands. However, in the case of insulators, the band gap is so large that thermal energies are not sufficient to excite a significant number of carriers across the band gap.

Even in insulators there is often a measurable electrical conductivity. For these materials, the band electronic transport processes become less important relative to charge hopping from one atom to another by over-coming a potential barrier. Ionic conduction can also occur in insulating ionic crystals. From a practical point of view, one of the most important applications of insulators is for the control of electrical breakdown phenomena.

The principal experimental methods for studying the electronic energy bands depend on the nature of the solid. For insulators, the optical properties are the most important, while for semiconductors both optical and transport studies are important. For metals, optical properties are less important and Fermi surface studies become more important.

In the case of insulators, electrical conductivity can arise through the motion of lattice ions as they move from one lattice vacancy to another, or from one interstitial site to another. Ionic conduction therefore occurs through the presence of lattice defects, and is promoted in materials with open crystal structures. In ionic crystals, there are relatively few mobile electrons or holes even at high temperature, so that conduction in these materials is predominantly due to the motions of ions.

Ionic conductivity (σ_{ionic}) is proportional both to the density of lattice defects (vacancies and interstitials) and to the diffusion rate, so that we can write

$$\sigma_{\text{ionic}} \sim e^{-(E+E_0)/k_B T} \tag{4.7}$$

where E_0 here denotes the activation energy for ionic motion and E is the energy for the formation of a defect (a vacancy, a vacancy pair, or an interstitial). Being an activated process, ionic conduction is enhanced at elevated temperatures. Since defects in ionic crystals can be observed visibly as the migration of color through the crystal, ionic conductivity can be distinguished from electronic conductivity by comparing the transport of charge with the transport of mass, as can, for example, be measured by the amount of material that is plated out on electrodes in contact with the ionic crystal.

4.5.1 Rare Gas and Ionic Crystals

The simplest insulator is a solid formed of rare gas atoms. An example of a rare gas insulator is solid argon, which crystallizes in the FCC structure with one Ar

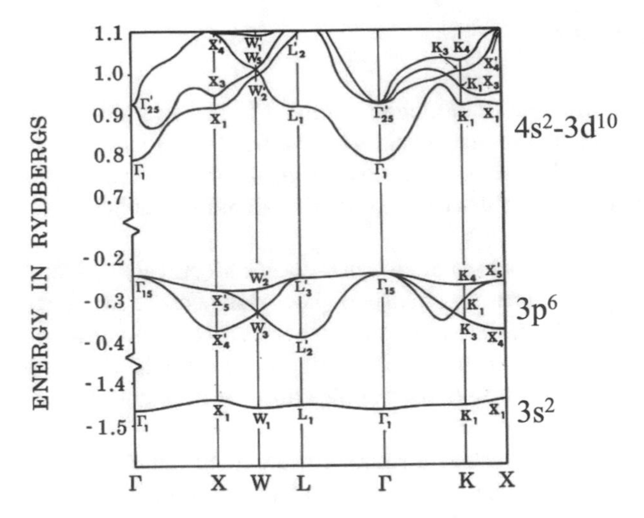

Fig. 4.22 Electronic energy band structure of Argon. Reprinted with permission from Physical Review B, Vol. 1 p. 3464 Copyright (1970) American Physical Society

atom/primitive unit cell. With an atomic configuration $3s^2 3p^6$, argon has filled $3s$ and $3p$ bands which are easily identified in the energy band diagram in Fig. 4.22. These occupied bands have very narrow band widths compared to their band gaps and are therefore well described by the tight binding approximation. Figure 4.22 shows that the higher energy states forming the conduction bands (the hybridized $4s$ and $3d$ bands) show more dispersion than the more tightly bound valence band states. The band diagram (Fig. 4.22) shows argon to have a direct band gap at the Γ point which is about 1 Rydberg or 13.6 eV. Although the $4s$ and $3d$ bands have similar energies, identification with the atomic levels can easily be made near $k = 0$ where the lower lying Γ_1 $4s$-band has considerably more band curvature than the $3d$ levels which are easily identified because of their degeneracies [the so called three-fold t_g ($\Gamma_{25'}$) and the two-fold e_g (Γ_{12}) crystal field levels for d-bands in a cubic crystal].

Another example of an insulator formed from a closed shell configuration is found in Fig. 4.23. Here the closed shell configuration results from charge transfer, as occurs in all ionic crystals. For example in the ionic crystal LiF (or in other alkali halide compounds which have a fcc structure), the valence band is identified with the filled anion orbitals (fluorine p–orbitals in this case) and at much higher energy, the empty cation conduction band levels will lie (lithium s–orbitals, in this case). Because of the wide band gap separation in the alkali halides between the valence and conduction bands, such materials are transparent at optical frequencies.

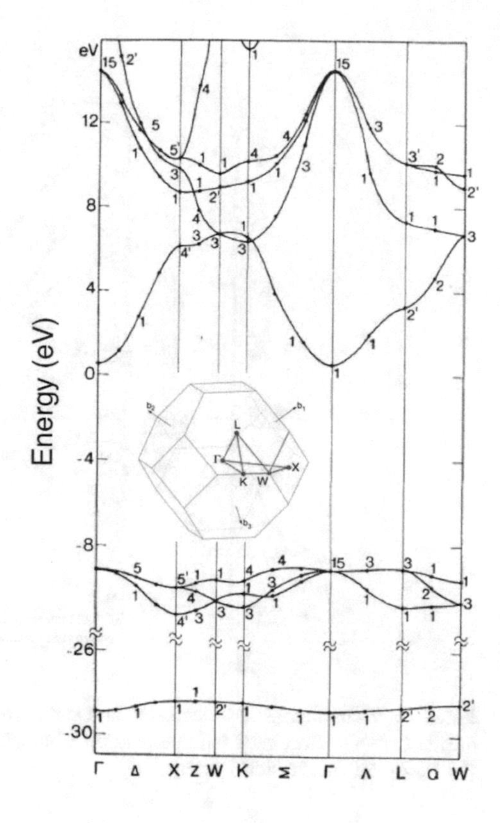

Fig. 4.23 Band structure of the alkali halide insulator LiF. This ionic crystal is used extensively for UV optical components because of its large band gap. The inset shows the Brillouin zone of LiF and the line symmetry used for plotting the energy dispersion relations. Reprinted with permission from Physical Review B, vol. 21 p. 799 Copyright (1980) American Physical Society

4.5.2 *Boron Nitride*

In Chap. 1, the real space and reciprocal space lattices were given for the 2D hexagonal lattice of boron nitride (BN) (see Fig. 1.6b). Two-dimensional hexagon BN is an insulator with a band gap of 5.8eV. The electronic band structure and the corresponding Brillouin zone of 2D hexagonal boron nitride is shown in Fig. 4.24. Here, a direct band gap can be seen at the P-point (or K-point) in the Brillouin zone. This corresponds to the same point at which the conduction and valence band of graphene touch each other, and in this case the band gap is opened because of the breaking symmetry of the A and B sublattices.

4.5.3 *Wide Bandgap Semiconductors*

Insulating behavior can also occur for wide bandgap semiconductors with covalent bonding, such as diamond, ZnS and GaP (see Fig. 4.25). The $E(\mathbf{k})$ diagrams for these materials are very similar to the dispersion relations for typical III–V semiconducting compounds and for the group IV semiconductors silicon and germanium; the main

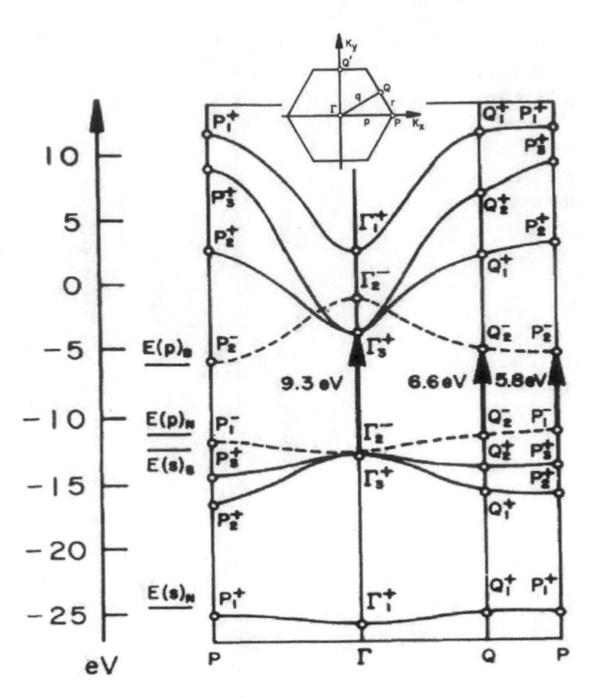

Fig. 4.24 Two–dimensional band structure and corresponding Brillouin zone of hexagonal boron nitride (h-BN). Reprinted with permission from Physical Review B, vol. 30 p. 6051 Copyright (1984) American Physical Society

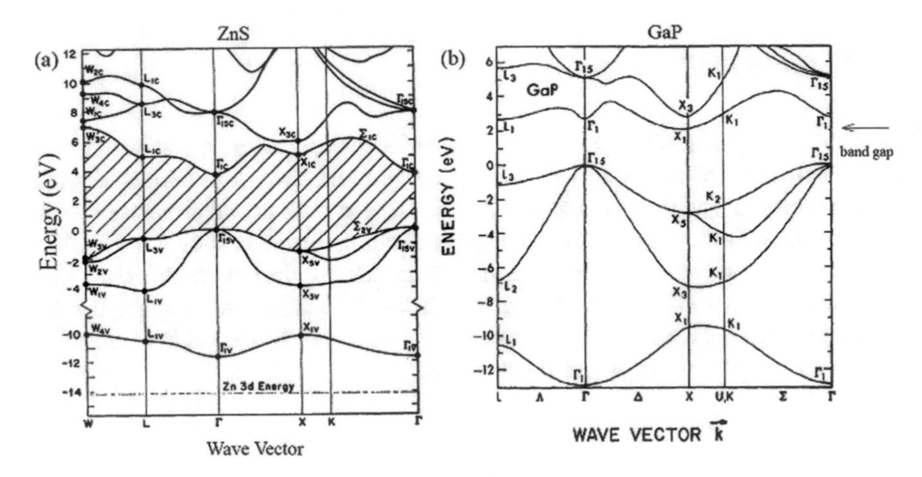

Fig. 4.25 Electronic energy band structure of **a** cubic ZnS, a direct gap semi–insulating II–VI semiconductor, where the lined region defines the bandgap occurring at different wavevectors, and in **b** the band structure for cubic GaP, an indirect gap semi–insulating III–V semiconductor is shown. These wide bandgap semiconductors are of interest for their optical properties. Reprinted with permission from Physical Review, vol. 179 p. 740 Copyright (1969) and Physical Review B, vol. 14 p. 556 Copyright (1976) American Physical Society

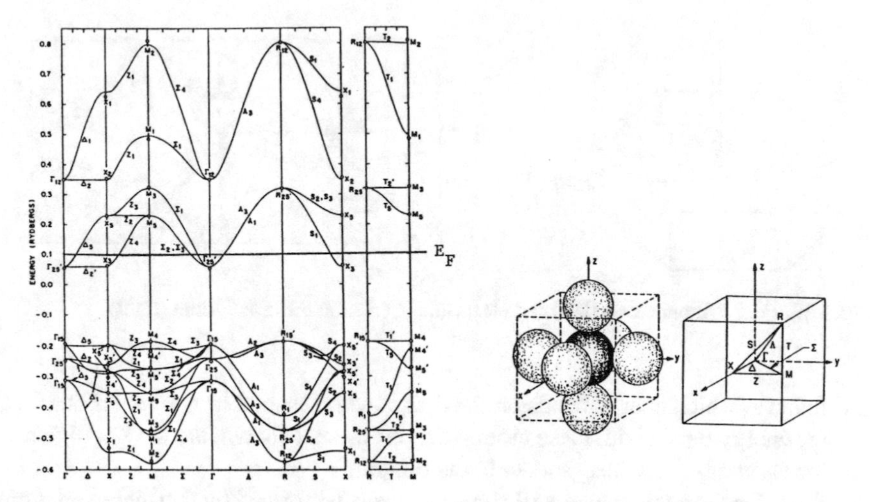

Fig. 4.26 Band structure and unit cell of ReO_3. (L.F. Mattheiss Phys. Rev. **181**, 987, (1969))

difference, however, is the large band gap separating valence and conduction bands in hexagonal crystals like h-BN.

Problems

4.1 Consider the diagram for the electronic dispersion relations $E(\boldsymbol{k})$ for ReO_3 shown below. The atomic configuration for Re is $4f^{14}5d^56s^2$ and for O is $2s^22p^4$, and the ReO_3 unit cell and Brillouin zone are shown in Fig. 4.26.

(a) How many atoms per unit cell are there (see diagram in Fig. 4.26))?
(b) Which bands are associated with the Re and which with the oxygen?
(c) Is ReO_3 a semiconductor or a metal?
(d) Where in the Brillouin zone are the carrier pockets located? Estimate the effective masses for the carriers qualitatively.
(e) How many electrons are contained in the carrier pockets?
(f) What is the shape of the Fermi surface?
(g) Where in the Brillouin zone does the lowest energy optical transition occur? Is this transition expected to be strong or weak? Why?

4.2 Consider the band structure diagram given below for tellurium which crystallizes in a hexagonal structure with three Te atoms/unit cell. The atomic configuration for tellurium is $5s^25p^4$. (Note: in the diagram the authors use Z rather than A in the hexagonal Brillouin zone.)

(a) Sketch the approximate position of the Fermi Level E_F on the band diagram and give your reasons for this placement of E_F.

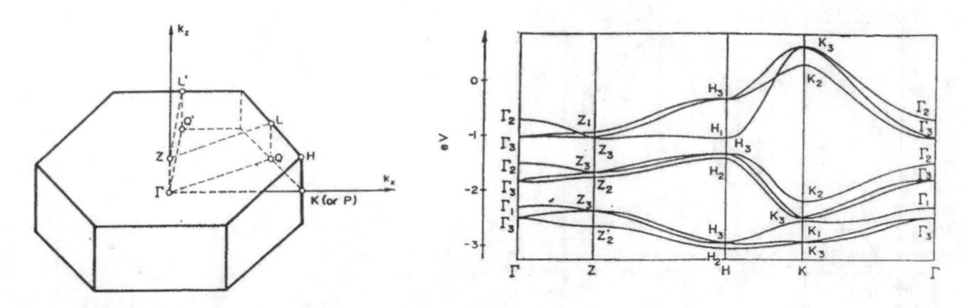

Fig. 4.27 First Brillouin zone (left) and electronic band structure of tellurium (right)

(b) Indicate which energy bands on the diagram correspond to s, p and d bands. If the energy bands with these atomic origins are not shown, are they at higher or lower energy than those shown in the diagram?
(c) From the diagram, where will carrier pockets be formed by thermal excitation? Identify the carrier pockets for electrons or holes. How many carrier pockets of each type are there?
(d) From the band diagram, what is the shape of the constant energy surfaces for electrons and for holes? Along which directions will electrons (holes) have light effective masses and along which directions will the masses be heavy?
(e) Is tellurium transparent to visible light ($\lambda = 500\,\text{nm}$)? Explain! Is the optical absorption strong or weak at the threshold for optical transitions (e.g., in comparison to GaAs)? Explain! (Fig. 4.27)

4.3 MoS_2 is a layered material, like graphite, that can be exfoliated to form monolayer MoS_2 using the Scotch tape? method. The energy band diagrams for bulk and monolayer MoS_2 are shown in Fig. 4.19.

(a) At what point in the Brillouin zone do the band gaps of bulk and monolayer MoS_2 occur?
(b) Do these materials (bulk and monolayer MoS_2) have direct or indirect band gaps? Explain from the $E(\mathbf{k})$ diagram.
(c) Where in the Brillouin zone does the lowest energy optical transition occur in bulk MoS_2 and in monolayer MoS_2?

4.4 Figure 4.28 shows the energy band diagrams for 3 different types of carbon nanotubes. Doubly degenerate bands are plotted with bold lines. For each of these nanotubes

(a) Indicate whether a particular nanotube is metallic or semiconducting.
(b) Where in the Brillouin zone is the band gap?
(c) How many atoms are there in the unit cell?
(d) Estimate the band gap of the semiconducting nanotube?

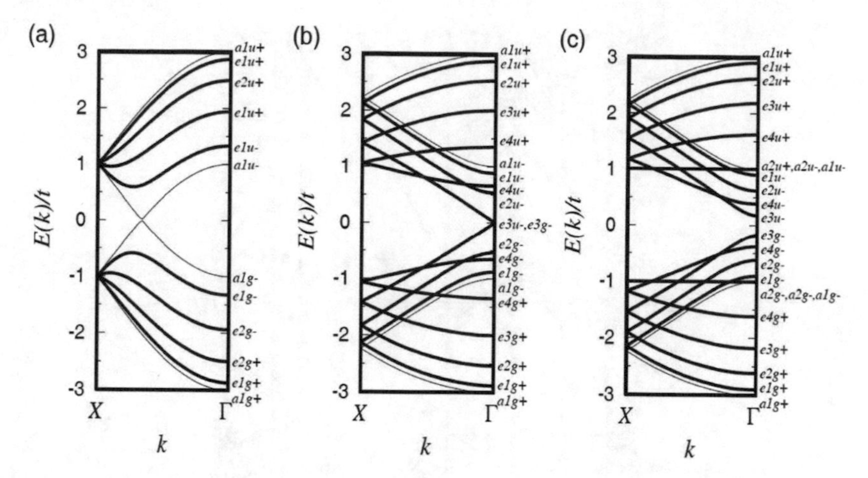

Fig. 4.28 One-dimensional energy dispersion relations for **a** (5,5), **b** (9,0), and **c** (10,0) carbon nanotubes. The energies are normalized by the parameter t = 2.9 eV (R. Saito, Physical Properties of Carbon Nanotubes, Imperial College Press, 1998)

4.5 $KMoO_3$ crystallizes in a cubic perovskite structure, where the atomic electronic configurations are K: $4s$, Mo: $4d^5 5s$, and O: $2s^2 2p^4$. The calculated electronic band structure is shown in Fig. 4.29. Note 1 Rydberg is 13.6 eV.

(a) Based on the one electron band diagram, is the material a metal, semiconductor or insulator?
(b) Where is the Fermi level in this material?
(c) Which energy bands at $k = 0$ (the Γ-point) are derived from oxygen atomic levels?
(d) Which energy bands at $k = 0$ are derived from K atomic levels? and which from Mo atomic levels? Which are hybridized?
(e) What is the shape of the Fermi surface (or surfaces) around $k = 0$? Are the carriers at the Fermi level electrons or holes?
(f) Suppose that we could prepare $KNbO_3$ and $KZrO_3$ in the same cubic crystal structure (Nb: $4d^4 5s$ and Zr: $4d^2 5s^2$). What would you expect the shape of the Fermi surfaces to be in these cases? What is the nature of the electronic transport for each of the 3 cases (Mo, Nb, & Zr)?

4.6 Consider the $E(\boldsymbol{k})$ diagram shown (Fig. 4.30) for a skutterudite $CoSb_3$ crystal and the diagram for the atoms in the unit cell.

(a) How many Co and Sb atoms are there per unit cell?
(b) In the tight binding limit, which bands of Co and which bands of Sb would you expect to be bonding states and which to be antibonding states? The atomic configurations for Co and Sb are: Co $3d^7 4s^2$ and Sb $5s^2 5p^3$. The band diagram shown in Fig. 4.30 may provide some clues, but you will need to use physical arguments beyond what you see in the diagram.

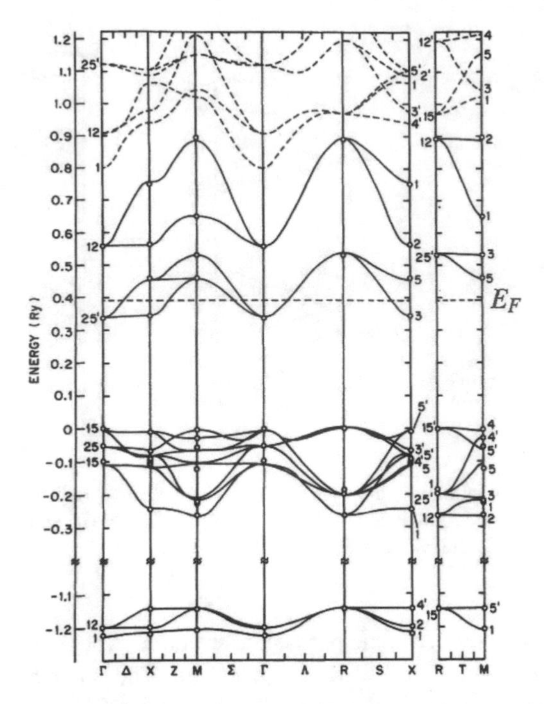

Fig. 4.29 The calculated electronic band structure of $KMoO_3$. From Mattheiss, L. F. "Energy Bands for $KNiF_3$, $SrTiO_3$, $KMoO_3$ and $KTaO_3$."*Physical Review B*, 6 (1972) 4718

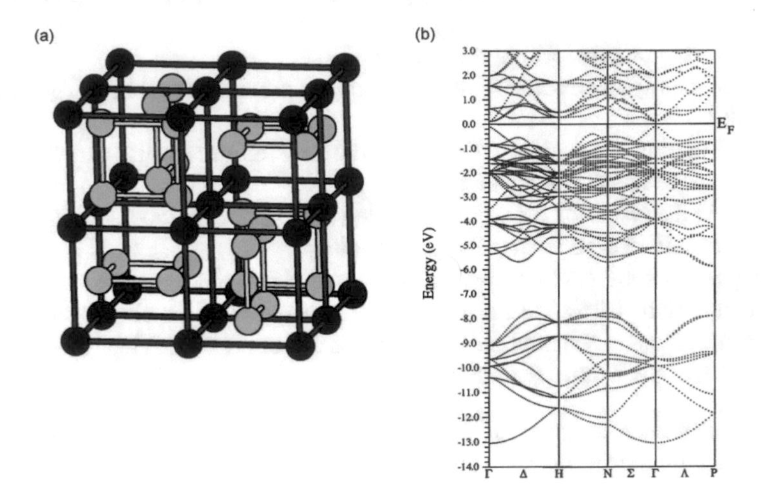

Fig. 4.30 **a** The skutterudite structure of $CoSb_3$. Black balls represent Co and gray balls represent Sb. The size of the atoms is arbitrary. **b** Electronic band structure of the skutterudite $CoSb_3$. From Sofo, J.O., and G.D. Mahan. "Electronic structure of $CoSb_3$: A narrow-band-gap semiconductor. "*Physical Review B* 58.23 (1998): 15620

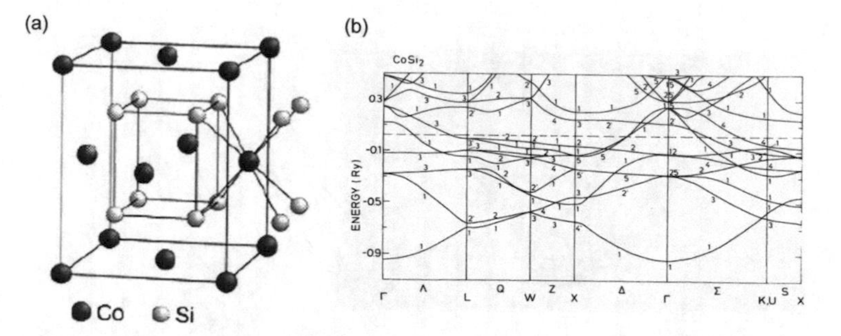

Fig. 4.31 **a** Structure of $CoSi_2$. Black balls represent Co and gray Si. The size of the atoms is arbitrary. **b** Electronic band structure of $CoSi_2$. From Lambrecht, Walter RL, Niels E. Christensen, and Peter Blöchl. "Electronic structure and properties of $NiSi_2$ and $CoSi_2$ in the fluorite and adamantane structures."*Physical Review B* 36.5 (1987): 2493

(c) Is $CoSb_3$ a semiconductor, semimetal or a metal?
(d) Where are the electron and hole carrier pockets expected to form? What is their multiplicity within the Brillouin zone?
(e) Do you expect the carriers to have high or low mobility relative to Cu or to Si? What is the basis for your conclusion?
(f) Upon doping, this material has a high potential for thermoelectric applications. Can you offer an explanation for why this might occur?

4.7 Find an expression for the Fermi energy E_F for the following cases:

(a) a 3D nearly free electron metal like Na which crystallizes in a BCC lattice, with lattice constant a.
(b) a 3D metal $Na_{0.5}K_{0.5}$ alloy with the BCC structure where the lattice constant for Na is a and that for K is b.
(c) a 2D monolayer of Na atoms.
(d) For which case would you expect the carrier mobility to be the greatest and why?

4.8 Consider a two-dimensional (2D) honeycomb lattice (see Fig. 3.10) with two atoms per unit cell.

(a) Find the smallest unit cell in real space. How many atoms are in a unit cell? Assume that a carbon atom sits at each corner of the hexagon and that the sheet is infinite in the plane but only one atomic layer thick.
(b) What is the corresponding unit cell in reciprocal space?
(c) Write an expression for the electronic dispersion relations $E(\mathbf{k})$ for a non-degenerate band in the 2D honeycomb lattice.
(d) Plot your results in (c) preferably using MATLAB, along the x and y directions.

4.9 Consider the picture for the cubic crystal $CoSi_2$ shown (Fig. 4.31). The atomic configuration for a free Co atom is $3d^7 4s^2$ and for a Si atom is $3s^2 3p^2$ (Fig. 4.31).

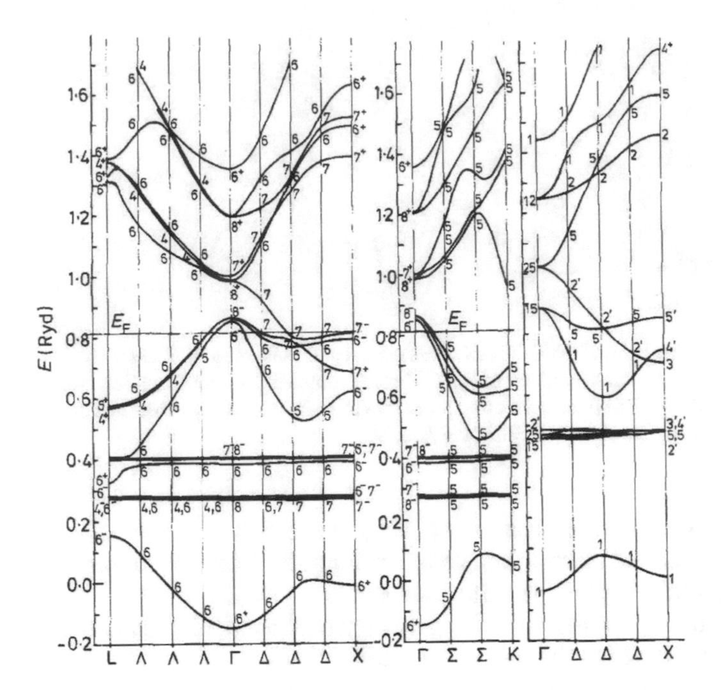

Fig. 4.32 Electronic band structure of HfC. From Weinberger, P., et al. "On the electronic structure of HfC, TaC and UC."Journal of Physics C: Solid State Physics 12.5 (1979): 801

(a) From the band diagram, is $CoSi_2$ a metal, semiconductor, semimetal, or insulator? Why?
(b) Find the crystal structure for $CoSi_2$ using your favorite reference, e.g., Wyckoff.
(c) Indicate at the zone center which bands are Co d-bands.
(d) Which are the Si derived s and p-bands at the center of the Brillouin zone?
(e) Where are the holes?
(f) Where are the electrons?
(g) Do you expect $CoSi_2$ to be a high mobility material? Why?
(h) What is the threshold for optical transitions? Will the intensity for these optical transitions be small or large, and why?

4.10 The diagram (Fig. 4.32) shows the electron energy bands of HfC which crystallizes in the NaCl structure (two interpenetrating fcc lattices). The free atom electronic configurations which constitute the valence and conduction bands in the solid are $5d^2 6s^2$ for hafnium and $2s^2 2p^2$ for carbon.

(a) Is hafnium carbide (HfC) a metal, semiconductor, or an insulator? Explain the reason for your answer.
(b) How many atoms are in its primitive unit cell?
(c) How many valence electrons are there per unit cell? Enough to fill how many energy bands?

(d) Identify in the diagram (e.g., at the Γ point) the carbon 2s and 2p bands; the hafnium 5d and 6s bands. Are there other bands of importance in the diagram? If so, what is their identification?
(e) Using the results in the last part of this problem, state which bands are fully occupied, totally empty, or partly filled. Where then in the Brillouin zone are the electrons and the holes located?
(f) How many carrier pockets are there about $\mathbf{k} = 0$ and what is the shape of their Fermi surfaces? Sketch the (100) cross-section of the Fermi surface.
(g) Indicate on the diagram the lowest energy for optical transitions near $\mathbf{k} = 0$.

Suggested Readings

F. Bassani, G. Pastori Paravicini, Electronic States and Optical Transitions in Solids. Chapter 4 (1993)
R.E. Peierls, Quantum Theory of Solids. Chapter 4
B. Segall, Energy Band of Aluminum. Phys. Rev. B **124**, 1797 (1961)
J.C. Slater, Quantum Theory of Atoms and Molecules. Chapter 10

Reference

K. Hummer, A. Gruneis, G. Kresse, Structural and electronic properties of lead chalcogenides from first principles. Phys. Rev. B **75**, 195211 (2007)

Chapter 5
Effective Mass Theory

5.1 Introduction

The effective mass model for a crystalline solid is simple but it is very instructive for use in discussing the electronic properties of solids because it allows a simple treatment of the electrons for certain energies and wave vectors (maxima and minima in $E(\boldsymbol{k})$ curves) as if the electrons were almost free. In this simple theory, the only effect of the crystal lattice is to change the mass of the electrons in the sense that, in the solid, the effective mass of the carrier of electrical charge can be larger or smaller than the mass of free electron. (In some special cases, such as graphene, this mass can be zero close to the Dirac points.) Furthermore, this mass also depends on the direction the electron is moving. Therefore, the effective mass can be used for describing many electronic properties of the solid state using a classical theory, just by considering the electrons as having a wavevector dependent effective mass, which as we will see, carries information about the electronic band structure and its anisotropy. The fact that, in the solid, electrons have an effective mass different from that of free electrons is a consequence of a more profound concept, in the sense that in the solid we should not treat the electron as a classical particle, but rather as a quantum particle interacting with a periodic potential provided by the lattice.

5.2 Wavepackets in Crystals and the Group Velocity of Electrons in Solids

In a crystal lattice, the electronic motion is induced by an applied field which is conveniently described by a wavepacket composed of eigenstates of the unperturbed crystal. These eigenstates are Bloch functions which reflect the lattice symmetry

$$\psi_{nk}(\mathbf{r}) = e^{i\mathbf{k}\cdot\mathbf{r}} u_{nk}(\mathbf{r}) \tag{5.1}$$

M. Dresselhaus et al., *Solid State Properties*, Graduate Texts in Physics,
https://doi.org/10.1007/978-3-662-55922-2_5

and are associated with an electronic band n, where n is a quantum number. These wavepackets are solutions of the time-dependent Schrödinger equation

$$\mathscr{H}_0 \psi_n(\mathbf{r}, t) = i\hbar \frac{\partial \psi_n(\mathbf{r}, t)}{\partial t} \tag{5.2}$$

where the time independent part of the Hamiltonian $\mathscr{H}_0$ can be written as

$$\mathscr{H}_0 = \frac{p^2}{2m} + V(\mathbf{r}), \tag{5.3}$$

where $V(\mathbf{r}) = V(\mathbf{r} + \mathbf{R}_n)$ is a periodic potential reflecting the crystal symmetry of the lattice. The wave packets $\psi_n(\mathbf{r}, t)$ can be written in terms of the Bloch states $\psi_{nk}(\mathbf{r})$ as

$$\psi_n(\mathbf{r}, t) = \sum_k A_{n,k}(t)\psi_{nk}(\mathbf{r}) = \int d^3k A_{n,k}(t)\psi_{nk}(\mathbf{r}) \tag{5.4}$$

where we have replaced the sum by an integral over the Brillouin zone, since the allowed $\mathbf{k}$ values for a macroscopic solid are *very* closely spaced, thus being appropriately treated as continuous in this approximation. If this Hamiltonian $\mathscr{H}_0$ is time-independent, as is often the case, we can write

$$A_{n,k}(t) = A_{n,k} e^{-i\omega_n(\mathbf{k})t} \tag{5.5}$$

where

$$\hbar\omega_n(\mathbf{k}) = E_n(\mathbf{k}) \tag{5.6}$$

and thereby obtain

$$\psi_n(\mathbf{r}, t) = \int d^3k A_{n,k} u_{nk}(\mathbf{r}) e^{i[\mathbf{k}\cdot\mathbf{r} - \omega_n(\mathbf{k})t]}. \tag{5.7}$$

We can localize the wavepacket in $\mathbf{k}$-space by requiring that the coefficients $A_{n,k}$ be large only in a confined region of $\mathbf{k}$-space, centered at $\mathbf{k} = \mathbf{k}_0$. If we now expand the band energy in a Taylor series around $\mathbf{k} = \mathbf{k}_0$, we obtain:

$$E_n(\mathbf{k}) = E_n(\mathbf{k}_0) + (\mathbf{k} - \mathbf{k}_0) \cdot \left.\frac{\partial E_n(\mathbf{k})}{\partial \mathbf{k}}\right|_{\mathbf{k}=\mathbf{k}_0} + \cdots, \tag{5.8}$$

where we have written $\mathbf{k}$ as

$$\mathbf{k} = \mathbf{k}_0 + (\mathbf{k} - \mathbf{k}_0). \tag{5.9}$$

Since $|\mathbf{k} - \mathbf{k}_0|$ is assumed to be small compared with Brillouin zone dimensions, we are justified in retaining only the first two terms of the Taylor expansion in (5.8) given above. Substitution into (5.4) for the wavepacket yields:

$$\psi_n(\mathbf{r}, t) \simeq e^{i(\mathbf{k}_0\cdot\mathbf{r}-\omega_n(\mathbf{k}_0)t)} \int d^3k A_{n,k}\, u_{nk}(\mathbf{r}) e^{i(\mathbf{k}-\mathbf{k}_0)\cdot\left[\mathbf{r}-\frac{\partial\omega_n(\mathbf{k})}{\partial\mathbf{k}}t\right]} \tag{5.10}$$

where

$$\hbar\omega_n(\mathbf{k}_0) = E_n(\mathbf{k}_0) \tag{5.11}$$

and

$$\hbar\frac{\partial\omega_n(\mathbf{k})}{\partial\mathbf{k}} = \frac{\partial E_n(\mathbf{k})}{\partial\mathbf{k}}. \tag{5.12}$$

The derivative $\partial\omega_n(\mathbf{k})/\partial\mathbf{k}$ which appears in the phase factor of (5.10) is evaluated at $\mathbf{k} = \mathbf{k}_0$. Except for the periodic function $u_{nk}(\mathbf{r})$, (5.10) is in the standard form for a wavepacket moving with *"group velocity"* $\mathbf{v}_g$

$$\mathbf{v}_g \equiv \frac{\partial\omega_n(\mathbf{k})}{\partial\mathbf{k}}, \tag{5.13}$$

so that

$$\mathbf{v}_g = \frac{1}{\hbar}\frac{\partial E_n(\mathbf{k})}{\partial\mathbf{k}}, \tag{5.14}$$

while the phase velocity

$$\mathbf{v}_p = \frac{\omega_n(\mathbf{k})}{\mathbf{k}} = \frac{\partial E_n(\mathbf{k})}{\hbar\partial\mathbf{k}}. \tag{5.15}$$

In the limit of free electrons the group velocity becomes

$$\mathbf{v}_g = \frac{\mathbf{p}}{m} = \frac{\hbar\mathbf{k}}{m} \tag{5.16}$$

and $\mathbf{v}_g = \mathbf{v}_p$ in the free electron limit. This result also follows from the above discussion using

$$E_n(\mathbf{k}) = \frac{\hbar^2k^2}{2m} \tag{5.17}$$

$$\frac{\partial E_n(\mathbf{k})}{\hbar\partial\mathbf{k}} = \frac{\hbar\mathbf{k}}{m}. \tag{5.18}$$

We shall show later that the electron wavepacket moves through the crystal very much like a free electron, provided that the wavepacket remains localized in k space during the time interval of interest, for the particular problem under consideration. Because of the uncertainty principle, the localization of a wavepacket in reciprocal space implies a delocalization of the wavepacket in real space.

We use wavepackets to describe electronic states in a solid when the crystal is perturbed in some way (e.g., by an applied electric or magnetic field). We make frequent applications of wavepackets to transport theory (e.g., electrical conductivity). In many practical applications of transport theory, use is made of the Effective-Mass Theorem, which is one of the most important results of transport theory.

We note that the above discussion for the wavepacket is given in terms of the **perfect crystal**. In our discussion of the Effective-Mass Theorem we will see that these wavepackets are also of use in describing situations where the Hamiltonian which enters Schrödinger's equation contains both the unperturbed Hamiltonian of the perfect crystal $\mathscr{H}_0$ and the perturbation Hamiltonian $\mathscr{H}'$ arising from an external perturbation. Common perturbations are applied electric or magnetic fields, or a lattice defect, or the presence of an impurity atom at a lattice site.

5.3 The Effective Mass Theorem

We shall now present the Effective Mass theorem, which is central to the consideration of the electrical and optical properties of solids. An elementary proof of the theorem will be given here for a simple but important case, namely a non-degenerate band which can be identified with the corresponding atomic state where the interaction with the lattice is negligible. The theorem can be also discussed from a more advanced point of view which considers also the case of degenerate bands but this is not considered here.

For many practical situations, we find a solid material to be present in the environment of some perturbing field (e.g., an externally applied electric field, or the perturbation created by an impurity atom or a crystal defect). The perturbation may be either time-dependent or time-independent and it can be treated in the effective mass approximation whereby the periodic potential is replaced by an effective Hamiltonian based on the $E(\mathbf{k})$ relations for the perfect crystal.

Let us start with the time-dependent Schrödinger's equation

$$(\mathscr{H}_0 + \mathscr{H}')\psi_n(\mathbf{r}, t) = i\hbar\frac{\partial\psi_n(\mathbf{r}, t)}{\partial t}. \tag{5.19}$$

We then substitute the expansion for the wave packet

$$\psi_n(\mathbf{r}, t) = \int d^3k A_{nk}(t)e^{i\mathbf{k}\cdot\mathbf{r}}u_{nk}(\mathbf{r}) \tag{5.20}$$

into Schrödinger's equation and make use of the Bloch solution

$$\mathscr{H}_0 e^{i\mathbf{k}\cdot\mathbf{r}}u_{nk}(\mathbf{r}) = E_n(\mathbf{k})e^{i\mathbf{k}\cdot\mathbf{r}}u_{nk}(\mathbf{r}) \tag{5.21}$$

to obtain:

$$\begin{aligned}(\mathscr{H}_0 + \mathscr{H}')\psi_n(\mathbf{r}, t) &= \textstyle\int d^3k[E_n(\mathbf{k}) + \mathscr{H}']A_{nk}(t)e^{i\mathbf{k}\cdot\mathbf{r}}u_{nk}(\mathbf{r}) = i\hbar(\partial\psi_n(\mathbf{r}, t)/\partial t)\\ &= i\hbar \textstyle\int d^3k\dot{A}_{nk}(t)e^{i\mathbf{k}\cdot\mathbf{r}}u_{nk}(\mathbf{r}).\end{aligned} \tag{5.22}$$

It then follows from Bloch's theorem that $E_n(\mathbf{k})$ is a periodic function in the reciprocal lattice. We can therefore expand $E_n(\mathbf{k})$ in a Fourier series in the direct lattice

$$E_n(\mathbf{k}) = \sum_{\mathbf{R}_\ell} E_{n\ell}\, e^{i\mathbf{k}\cdot\mathbf{R}_\ell} \tag{5.23}$$

where the $\mathbf{R}_\ell$ are lattice vectors. Now consider the differential operator $E_n(-i\nabla)$ formed by replacing $\mathbf{k}$ by $-i\nabla$

$$E_n(-i\nabla) = \sum_{\mathbf{R}_\ell} E_{n\ell}\, e^{\mathbf{R}_\ell\cdot\nabla}. \tag{5.24}$$

Consider the effect of $E_n(-i\nabla)$ on an arbitrary function $f(\mathbf{r})$. Since $e^{\mathbf{R}_\ell\cdot\nabla}$ can be expanded in a Taylor series, we obtain

$$\begin{aligned} e^{\mathbf{R}_\ell\cdot\nabla} f(\mathbf{r}) &= [1 + \mathbf{R}_\ell\cdot\nabla + \tfrac{1}{2}(\mathbf{R}_\ell\cdot\nabla)(\mathbf{R}_\ell\cdot\nabla) + \cdots] f(\mathbf{r}) \\ &= f(\mathbf{r}) + \mathbf{R}_\ell\cdot\nabla f(\mathbf{r}) + \tfrac{1}{2!} R_{\ell,\alpha} R_{\ell,\beta} \frac{\partial^2}{\partial r_\alpha \partial r_\beta} f(\mathbf{r}) + \cdots \\ &= f(\mathbf{r} + \mathbf{R}_\ell). \end{aligned} \tag{5.25}$$

Thus the effect of $E_n(-i\nabla)$ on a Bloch state is to produce $E_n(\mathbf{k})$ because

$$E_n(-i\nabla)\psi_{nk}(\mathbf{r}) = \sum_{\mathbf{R}_\ell} E_{n\ell}\psi_{nk}(\mathbf{r}+\mathbf{R}_\ell) = \sum_{\mathbf{R}_\ell} E_{n\ell} e^{i\mathbf{k}\cdot\mathbf{R}_\ell} e^{i\mathbf{k}\cdot\mathbf{r}} u_{nk}(\mathbf{r}) = E_n(\mathbf{k})\psi_{nk}(\mathbf{r}), \tag{5.26}$$

since from Bloch's theorem $\psi_{nk}(\mathbf{r} + \mathbf{R}_\ell)$ can be written as

$$\psi_{nk}(\mathbf{r} + \mathbf{R}_\ell) = e^{i\mathbf{k}\cdot\mathbf{R}_\ell} \left[e^{i\mathbf{k}\cdot\mathbf{r}} u_{nk}(\mathbf{r}) \right]. \tag{5.27}$$

Substitution of

$$E_n(-i\nabla)\psi_{nk}(\mathbf{r}) = E_n(\mathbf{k})\psi_{nk}(\mathbf{r}) \tag{5.28}$$

from (5.26) into Schrödinger's equation (5.22) yields:

$$\int d^3k \left[E_n(-i\nabla) + \mathscr{H}' \right] A_{nk}(t) e^{i\mathbf{k}\cdot\mathbf{r}} u_{nk}(\mathbf{r}) = \left[E_n(-i\nabla) + \mathscr{H}' \right] \int d^3k\, A_{nk}(t) e^{i\mathbf{k}\cdot\mathbf{r}} u_{nk}(\mathbf{r}) \tag{5.29}$$

so that

$$\left[E_n(-i\nabla) + \mathscr{H}' \right] \psi_n(\mathbf{r}, t) = i\hbar \frac{\partial \psi_n(\mathbf{r}, t)}{\partial t}. \tag{5.30}$$

The result of (5.30) is called the *effective mass theorem*. We observe that the original crystal Hamiltonian $p^2/2m + V(\mathbf{r})$ does not appear in this equation. It has instead been replaced by an effective Hamiltonian which is an operator formed from the solution $E(\mathbf{k})$ for the perfect crystal in which we replace $\mathbf{k}$ by $-i\nabla$. For example, for the free electron ($V(\mathbf{r}) \equiv 0$)

$$E_n(-i\nabla) \to -\frac{\hbar^2\nabla^2}{2m}. \tag{5.31}$$

In applying the effective mass theorem, we assume that $E(\mathbf{k})$ is known either from the results of a theoretical calculation or from the analysis of experimental results. What is important here is that once $E(\mathbf{k})$ is known, the effect of various perturbations on the ideal crystal can be treated in terms of the solution to the energy levels of the perfect crystal, without recourse to consideration of the full Hamiltonian. In practical cases, the solution to the effective mass equation is much easier to carry out than the solution to the original Schrödinger's equation.

According to the above discussion, we have assumed that $E(\mathbf{k})$ is specified throughout the Brillouin zone. For many practical applications, the region of $\mathbf{k}$-space which is of importance is confined to a small portion of the Brillouin zone. In such cases it is only necessary to specify $E(\mathbf{k})$ in a local region (or regions) and to localize our wavepacket solutions to these local regions of $\mathbf{k}$-space. Suppose that we localize the wavepacket around $\mathbf{k} = \mathbf{k}_0$, and we correspondingly expand our Bloch functions around $\mathbf{k}_0$,

$$\psi_{nk}(\mathbf{r}) = e^{i\mathbf{k}\cdot\mathbf{r}}u_{nk}(\mathbf{r}) \simeq e^{i\mathbf{k}\cdot\mathbf{r}}u_{nk_0}(\mathbf{r}) = e^{i(\mathbf{k}-\mathbf{k}_0)\cdot\mathbf{r}}\psi_{nk_0}(\mathbf{r}) \tag{5.32}$$

where we have noted that $u_{nk}(\mathbf{r}) \simeq u_{nk_0}(\mathbf{r})$ has only a weak dependence on $\mathbf{k}$. Then our wavepacket can be written as

$$\psi_n(\mathbf{r}, t) = \int d^3k\ A_{nk}(t)e^{i(\mathbf{k}-\mathbf{k}_0)\cdot\mathbf{r}}\psi_{nk_0}(\mathbf{r}) = F(\mathbf{r}, t)\psi_{nk_0}(\mathbf{r}) \tag{5.33}$$

where $F(\mathbf{r}, t)$ is called the *amplitude* or *envelope* function and is defined by

$$F(\mathbf{r}, t) = \int d^3k\ A_{nk}(t)e^{i(\mathbf{k}-\mathbf{k}_0)\cdot\mathbf{r}}. \tag{5.34}$$

Since the time dependent Fourier coefficients $A_{nk}(t)$ are assumed here to be large only near $\mathbf{k} = \mathbf{k}_0$, then $F(\mathbf{r}, t)$ will be a slowly varying function of $\mathbf{r}$, because in this case

$$e^{i(\mathbf{k}-\mathbf{k}_0)\mathbf{r}} \simeq 1 + i(\mathbf{k} - \mathbf{k}_0)\cdot\mathbf{r} + \cdots. \tag{5.35}$$

It can be shown that the envelope function also satisfies the effective mass equation

$$\left[E_n(-i\nabla) + \mathscr{H}'\right]F(\mathbf{r}, t) = i\hbar\frac{\partial F(\mathbf{r}, t)}{\partial t} \tag{5.36}$$

where we now replace $\mathbf{k} - \mathbf{k}_0$ in $E_n(\mathbf{k})$ by $-i\nabla$. This form of the effective mass equation is useful for treating the problem of donor and acceptor impurity states in semiconductors, where $\mathbf{k}_0$ is taken as the wavevector of where the energy band extremum occurs.

5.4 Application of the Effective Mass Theorem to Donor Impurity Levels in a Semiconductor

Suppose that we add an impurity from column V in the Periodic Table to a semiconductor such as silicon or germanium, which are both members of column IV of the periodic table. This impurity atom will have one more electron than is needed to satisfy the valency requirements for the tetrahedral bonds which the germanium or silicon atoms form with their 4 valence electrons (see Fig. 5.1).

This extra electron from the impurity atom will be free to wander through the lattice, subject, of course, to the coulomb attraction of this electron to the ion core which will have one unit of positive charge. We will consider here the case where we add just a small number of these impurity atoms so that we may focus our attention on a single, isolated substitutional impurity atom in an otherwise perfect lattice. In the course of this discussion we will define more carefully what the limits on the impurity concentration must be so that the treatment given here is applicable.

Let us also assume that the conduction band of the host semiconductor in the vicinity of the band "minimum" at $\mathbf{k}_0$ has the simple analytic form

$$E_c(\mathbf{k}) \simeq E_c(\mathbf{k}_0) + \frac{\hbar^2(k - k_0)^2}{2m^*}. \tag{5.37}$$

We can consider this expression for the conduction band level $E_c(\mathbf{k})$ as a special case of the Taylor expansion of $E(\mathbf{k})$ about an energy band minimum at $\mathbf{k} = \mathbf{k}_0$. For the present discussion, $E(\mathbf{k})$ is assumed to be isotropic in $\mathbf{k}$; this typically occurs in cubic semiconductors with band extrema at $\mathbf{k} = 0$. The quantity m^* in this equation is the *effective mass* for the electrons. We will see that the energy levels corresponding to the donor electron will lie in the band gap below the conduction band minimum as indicated in the diagram in Fig. 5.2. To solve for the impurity levels explicitly, we may use the time-independent form of the effective mass theorem derived from (5.36)

$$\left[E_n(-i\nabla) + \mathscr{H}'\right] F(\mathbf{r}) = (E - E_c)F(\mathbf{r}). \tag{5.38}$$

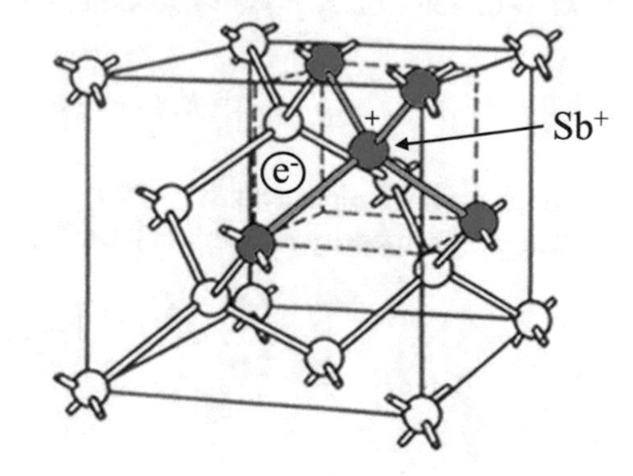

Fig. 5.1 Crystal structure of diamond, showing the tetrahedral bonding arrangement of the carbon atoms with an Sb^+ ion on one of the diamond lattice sites and a free donor electron available for conduction. Reprinted with permission from John Wiley & Sons Inc, Sze and Ng, Physics of Semiconductor Devices

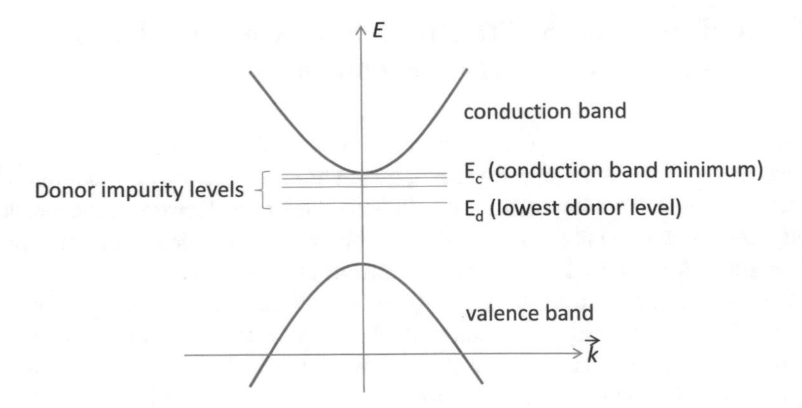

Fig. 5.2 Schematic band diagram showing donor levels in a semiconductor

Equation (5.38) is applicable to the impurity problem in a semiconductor provided that the amplitude function $F(\mathbf{r})$ is sufficiently slowly varying over a unit cell. In the course of this discussion, we will see that the donor electron in a column IV semiconductor (or III–V or II–VI compound semiconductor) will wander over many lattice sites before being scattered by a lattice defect and therefore this approximation on $F(\mathbf{r})$ will be justified.

For a singly ionized donor impurity (such as arsenic in germanium), the perturbing potential $\mathscr{H}'$ can be represented as a Coulomb potential

$$\mathscr{H}' = -\frac{e^2}{\varepsilon r} \tag{5.39}$$

where ε is an average dielectric constant of the crystal medium which the donor electron sees as it wanders through the crystal. Experimental data on donor impurity states indicate that ε is very closely equal to the low frequency limit of the electronic dielectric constant $\varepsilon_1(\omega)|_{\omega=0}$, which we will discuss extensively in treating the optical properties of solids (Part III of this book). The above discussion involving an isotropic $E(\mathbf{k})$ is also appropriate for some semiconductors with a conduction band minimum at $\mathbf{k}_0 = 0$. The Effective Mass equation for the unperturbed crystal is then

$$E_n(-i\nabla) = -\frac{\hbar^2}{2m^*}\nabla^2 \tag{5.40}$$

in which we have replaced $\mathbf{k}$ by $-i\nabla$.

The donor impurity problem in the effective mass approximation thus becomes

$$\left[-\frac{\hbar^2}{2m^*}\nabla^2 - \frac{e^2}{\varepsilon r}\right]F(\mathbf{r}) = (E - E_c)F(\mathbf{r}) \tag{5.41}$$

where all energies are measured with respect to the bottom of the conduction band E_c. If we replace m^* by m and e^2/ε by e^2, we immediately recognize this equation as Schrödinger's equation for a hydrogen atom under the identification of the energy eigenvalues with

$$E_n = \frac{e^2}{2n^2 a_0} = \frac{me^4}{2n^2\hbar^2} \tag{5.42}$$

where a_0 is the Bohr radius $a_0 = \hbar^2/me^2$. This identification immediately allows us to write E_ℓ for the donor energy levels as

$$E_\ell = E_c - \frac{m^* e^4}{2\varepsilon^2 \ell^2 \hbar^2} \tag{5.43}$$

where $\ell = 1, 2, 3, \ldots$ is an integer denoting the donor level quantum numbers and we identify the bottom of the conduction band E_c as the ionization energy for this effective hydrogenic problem. Physically, this means that the donor levels correspond to bound (localized) states, while the band states above E_c correspond to delocalized nearly-free electron-like states. The lowest or "ground-state" donor energy level is then written as

$$E_d = E_{\ell=1} = E_c - \frac{m^* e^4}{2\varepsilon^2\hbar^2}. \tag{5.44}$$

It is convenient to identify the "effective" first Bohr radius a_0^* for the donor level as

$$a_0^* = \frac{\varepsilon\hbar^2}{m^* e^2} \tag{5.45}$$

and to recognize that the wave function for the ground state donor level will be of the form

$$F(\mathbf{r}) = \mathscr{C} e^{-r/a_0^*} \tag{5.46}$$

where $\mathscr{C}$ is the normalization constant and a_0^* is an effective ground state donor level. Thus using the effective mass theorem the solutions to (5.41) for a semiconductor are hydrogenic energy levels with the substitutions $m \to m^*$, $e^2 \to (e^2/\varepsilon)$ and the ionization energy, usually taken as the zero of energy for the hydrogen atom, now becomes, the conduction band extremum E_c.

For a semiconductor like germanium, we have a very large dielectric constant, $\varepsilon \simeq 16$. The value for the effective mass is somewhat more difficult to specify in germanium since the constant energy surfaces for germanium are located about the L-points in the Brillouin zone (see Sect. 4.3.2) and are ellipsoids of revolution. Since the constant energy surfaces for such semiconductors are non-spherical, the effective mass tensor is anisotropic. However we will write down an average effective mass value $m^*/m \simeq 0.12$ so that we can estimate pertinent magnitudes for the donor levels in a typical semiconductor. With these values for ε and m^*, we obtain:

$$E_c - E_d \simeq 0.007\,\text{eV} \tag{5.47}$$

and the effective Bohr radius

$$a_0^* \simeq 70\,\text{Å}. \tag{5.48}$$

These values are to be compared with the ionization energy of 13.6 eV for the hydrogen atom and with the hydrogenic Bohr radius of $a_0 = \hbar^2/me^2 = 0.5\,\text{Å}$.

Thus, we see that a_0^* is indeed large enough to satisfy the requirement that $F(\mathbf{r})$ be slowly varying over a unit cell. On the other hand, if a_0^* were to be comparable to a lattice unit cell dimension, then $F(\mathbf{r})$ could not be considered as a slowly varying function of $\mathbf{r}$ and generalizations of the above treatment would have to be made. Such generalizations involve: (1) treating $E(\mathbf{k})$ over a wider region of $\mathbf{k}$-space, and (2) relaxing the condition that impurity levels are to be associated with a single band. From the uncertainty principle, the localization in momentum space for the impurity state requires a delocalization in real space; and likewise, the converse is true, that a localized impurity in real space corresponds to a delocalized description in $\mathbf{k}$-space. Thus "shallow" hydrogenic donor levels (close in energy to the band extremum) can be attributed to a specific band at a specific energy extremum at $\mathbf{k}_0$ in the Brillouin zone. On the other hand, "deep" donor levels (far in energy from the band extremum) are not hydrogenic and have a more complicated energy level $E(\mathbf{k})$ structure. Deep donor levels cannot be readily associated with a specific band or a specific $\mathbf{k}$ point in the Brillouin zone.

In dealing with this impurity problem, it is helpful to discuss the donor levels in silicon and germanium. For example in silicon where the conduction band extrema are at the Δ point (see Sect. 4.3.3), the effective mass theorem requires us to replace $E(-i\nabla)$ by

$$E_n(-i\nabla) \rightarrow -\frac{\hbar^2}{2m_\ell^*}\frac{\partial^2}{\partial x^2} - \frac{\hbar^2}{2m_t^*}\left(\frac{\partial^2}{\partial y^2} + \frac{\partial^2}{\partial z^2}\right) \tag{5.49}$$

where m_l^* and m_t^* denote the longitudinal and transverse effective mass components of the effective mass tensor, and the resulting Schrödinger's equation can no longer be solved analytically. Although this is a very interesting problem from a practical point of view, it is important to note that numerical solutions are needed in this case.

5.5 Quasi-classical Electron Dynamics

According to the "Correspondence Principle" of Quantum Mechanics, wavepacket solutions of Schrödinger's equation (see Sect. 5.2) follow the trajectories of classical particles and satisfy Newton's laws. One can also give a Correspondence Principle argument for the form which is assumed by the velocity and acceleration of a wavepacket. According to the Correspondence Principle, the connection between the classical Hamiltonian and the quantum mechanical Hamiltonian is made by the identification of $\mathbf{p} \rightarrow (\hbar/i)\nabla$. Thus, we write

$$E_n(-i\nabla) + \mathscr{H}'(\mathbf{r}) \leftrightarrow E_n(\mathbf{p}/\hbar) + \mathscr{H}'(\mathbf{r}) = \mathscr{H}_{\text{classical}}(\mathbf{p}, \mathbf{r}). \tag{5.50}$$

In classical mechanics, Hamilton's equations give the velocity according to :

$$\dot{\mathbf{r}} = \frac{\partial \mathscr{H}}{\partial \mathbf{p}} = \nabla_p \mathscr{H} = \frac{\partial E(\mathbf{k})}{\hbar \partial \mathbf{k}} \tag{5.51}$$

in agreement with the group velocity for a wavepacket given by (5.13). Hamilton's equation for the acceleration is given by:

$$\dot{\mathbf{p}} = -\frac{\partial \mathscr{H}}{\partial \mathbf{r}} = -\frac{\partial \mathscr{H}'(\mathbf{r})}{\partial \mathbf{r}}. \tag{5.52}$$

For example, in the case of an applied electric field $\mathbf{E}$, the perturbation Hamiltonian is

$$\mathscr{H}'(\mathbf{r}) = -e\mathbf{r} \cdot \mathbf{E} \tag{5.53}$$

so that

$$\dot{\mathbf{p}} = \hbar\dot{\mathbf{k}} = e\mathbf{E}. \tag{5.54}$$

In this equation $e\mathbf{E}$ is the classical Coulomb force on an electric charge due to an applied field $\mathbf{E}$. It can be shown that in the presence of a magnetic field $\mathbf{B}$, the acceleration theorem follows the Lorentz force equation

$$\dot{\mathbf{p}} = \hbar\dot{\mathbf{k}} = e[\mathbf{E} + (1/c)\mathbf{v} \times \mathbf{B}] \tag{5.55}$$

where

$$\mathbf{v} = \frac{\partial E(\mathbf{k})}{\hbar \partial \mathbf{k}}. \tag{5.56}$$

In the crystal, the crystal momentum $\hbar\mathbf{k}$ for the wavepacket plays the role of the momentum for a classical particle.

5.6 Quasi-classical Theory of Electrical Conductivity – Ohm's Law

We will now apply the idea of the quasi-classical electron dynamics in a solid to the problem of the electrical conductivity for a metal with an arbitrary Fermi surface and band structure. The electron is treated here as a wavepacket with momentum $\hbar\mathbf{k}$ moving in an external electric field $\mathbf{E}$ in compliance with Newton's laws. Because of the acceleration theorem, we can think of the electric field as creating a

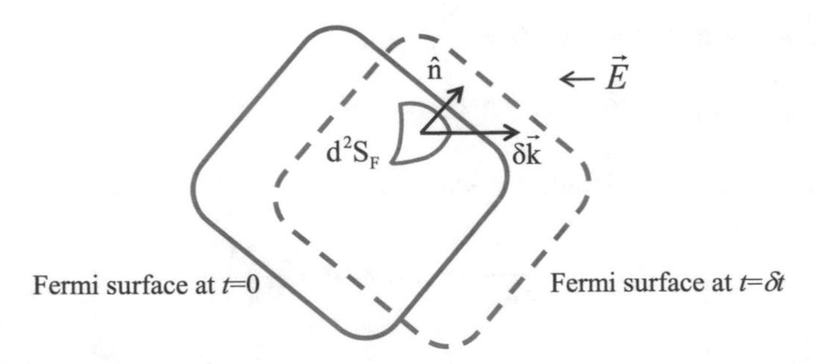

Fig. 5.3 Displaced Fermi surface at $t = \delta t$ under the action of an electric field **E**

"displacement" of the electron distribution in **k**-space. We remember that the Fermi surface encloses the region of occupied states within the Brillouin zone. The effect of the electric field is to change the wave vector **k** of an electron by

$$\delta \mathbf{k} = \frac{e}{\hbar}\mathbf{E}\delta t \tag{5.57}$$

(where we note that the charge on the electron e is a negative number). We picture the displacement $\delta\mathbf{k}$ of (5.57) by the displacement of the Fermi surface in a time δt shown in Fig. 5.3. From this diagram we see that the incremental volume of **k**-space $\delta^3 V_{\mathbf{k}}$ which is "swept out" in the time δt due to the presence of the field **E** is

$$\delta^3 V_{\mathbf{k}} = d^2 S_F \ \hat{n} \cdot \delta\mathbf{k} = d^2 S_F \ \hat{n} \cdot \left(\frac{e}{\hbar}\mathbf{E}\delta t\right) \tag{5.58}$$

and the electron density n is found from

$$n = \frac{2}{(2\pi)^3} \int_{E \le E_F} d^3k \tag{5.59}$$

where $d^2 S_F$ is the element of area on the Fermi surface and $\hat{n}$ in (5.58) is a unit vector normal to this element of area and $\delta^3 V_{\mathbf{k}} \to d^3k$ both denote elements of volume in **k**-space. The definition of the electrical current density is the current flowing through a unit area in real space and is given by the product of the [number of electrons per unit volume] with the [charge per electron] and with the [group velocity] so that the element of the current density $\delta\mathbf{j}$ given in (5.60) created by applying the electric field **E** for a time interval δt is given by

$$\delta\mathbf{j} = \int [2/(2\pi)^3] \ \cdot \ [\delta^3 V_{\mathbf{k}}] \cdot [e] \cdot [\mathbf{v}_g] \tag{5.60}$$

where $\mathbf{v}_g$ is the group velocity for an electron wavepacket and $2/(2\pi)^3$ is the density of electronic states in **k**-space (including the spin degeneracy of two) because we can put 2 electrons (with ↑ and ↓ spins) in each phase space state. Substitution for $\delta^3 V_{\mathbf{k}}$ in (5.58) yields the instantaneous rate of change of the current density averaged over the Fermi surface

$$\frac{\partial \mathbf{j}}{\partial t} = \frac{e^2}{4\pi^3\hbar} \oint\oint \mathbf{v}_g\, \hat{n} \cdot \mathbf{E}\, d^2 S_F = \frac{e^2}{4\pi^3\hbar} \oint\oint \mathbf{v}_g \left(\frac{\mathbf{v}_g \cdot \mathbf{E}}{|v_g|} \right) (d^2 S_F) \tag{5.61}$$

since the group velocity $\mathbf{v_g}$ given by (5.13) is directed normal to the Fermi surface. In a real solid, the electrons will not be accelerated indefinitely, but will eventually collide with an impurity, or a lattice defect or a lattice vibration (phonon).

These collisions will serve to maintain the displacement of the Fermi surface at some steady state value, depending on τ, the average time between collisions, which defines the relaxation time τ to return to equilibrium. This relaxation time τ can be introduced through the expression

$$n(t) = n(0)e^{-t/\tau} \tag{5.62}$$

where $n(t)$ is the number of electrons that have not yet made a collision at time t, assuming that the last collision had been made at time $t = 0$. The relaxation time τ is here given by the average time between collisions

$$\langle t \rangle = \frac{1}{\tau} \int_0^\infty t e^{-t/\tau} dt = \tau. \tag{5.63}$$

If in (5.57), we set $\langle \delta t \rangle = \tau$ and write the average current density as $\mathbf{j} = \langle \delta \mathbf{j} \rangle$, then we obtain from (5.64)

$$\mathbf{j} = \frac{e^2 \tau}{4\pi^3\hbar} \int \mathbf{v}_g \frac{\mathbf{v}_g \cdot \mathbf{E}}{|v_g|} (d^2 S_F). \tag{5.64}$$

We define the conductivity tensor $\overset{\leftrightarrow}{\sigma}$ as $\mathbf{j} = \overset{\leftrightarrow}{\sigma} \cdot \mathbf{E}$, so that (5.64) provides an explicit expression for the symmetric second rank tensor $\overset{\leftrightarrow}{\sigma}$.

$$\overset{\leftrightarrow}{\sigma} = \frac{e^2 \tau}{4\pi^3\hbar} \int \frac{\mathbf{v}_g \mathbf{v}_g}{|v_g|} (d^2 S_F). \tag{5.65}$$

In the free electron limit $\overset{\leftrightarrow}{\sigma}$ becomes a scalar (isotropic conduction) and is given by the Drude formula which we derive below from (5.65). Using the equations for the free electron limit

$$\begin{aligned} E &= \hbar^2 k^2/2m \\ E_F &= \hbar^2 k_F^2/2m \\ \mathbf{v}_g &= \hbar \mathbf{k}_F/m. \end{aligned} \tag{5.66}$$

We then obtain

$$\mathbf{v}_g \mathbf{v}_g \to v_x^2 = v_y^2 = v_z^2 = v^2/3 \tag{5.67}$$

$$\int d^2 S_F = 4\pi k_F^2, \tag{5.68}$$

so that the number of electrons/unit volume can be written as:

$$n = \frac{1}{4\pi^3}\frac{4\pi}{3}\, k_F^3. \tag{5.69}$$

Therefore

$$\mathbf{j} = \frac{e^2\tau}{4\pi^3\hbar}\left(\frac{\hbar k_F}{m}\right)\frac{1}{3}\,\mathbf{E}(4\pi k_F^2) = \frac{ne^2\tau}{m}\,\mathbf{E}. \tag{5.70}$$

Thus the free electron limit gives Ohm's law in the familiar form

$$\sigma = \frac{ne^2\tau}{m} = ne\mu, \tag{5.71}$$

showing that the electrical conductivity, in the diffusion regime where carrier scattering is important, depends on both the carrier density n and the carrier mobility μ. For low dimensional systems, that are important on the nano-scale, ballistic transport is dominant, and in this regime the carriers can go from the anode to the cathode without scattering. This regime will be considered in a later chapter in this book.

A slightly modified form of Ohm's law is also applicable to conduction in a material for which the energy dispersion relations have a simple parabolic form and m has been replaced by the effective mass m^*, $E(\mathbf{k}) = \hbar^2 k^2/2m^*$. In this case σ is given by

$$\sigma = ne^2\tau/m^* \tag{5.72}$$

where the effective mass is found from the band curvature $1/m^* = \partial^2 E/\hbar^2\partial k^2$. The generalization of Ohm's law can also be made to deal with solids for which the effective mass tensor is *anisotropic* and this will be discussed in Chap. 7.

Problems

5.1 Show that the spatial probability density for a free electron propagating in one dimension is constant.

5.2 Show that the envelope function $F(\mathbf{r}, t)$ in (5.34) also satisfies the effective mass equation

$$\left[E_n(-i\nabla) + \mathcal{H}'\right] F(\mathbf{r}, t) = i\hbar\frac{\partial F(\mathbf{r}, t)}{\partial t}$$

5.3 Consider a one dimensional linear chain containing N atoms and with lattice constant a, such that the chain length is $L = Na$. Show that the average group velocity for a filled band in this one dimensional system is zero.

5.4 Consider that an electron moving in a crystal can be described by a superposition of Bloch waves such that the Bloch wave functions are

$$\psi_k(x,t) = e^{i(kx-\omega(k)t}u_k(x)$$

and the wave packet is

$$\Psi(x,t) = \int_{\infty}^{+\infty} e^{(k-k_o)^2/\alpha^2} e^{i(kx-\omega(k)t} u_k(x) u_k(x) dk.$$

(a) Identify the envelope function in the wave packet $\Psi(x,t)$.

(b) If α is much smaller than the size of the Brillouin zone, the envelope function is peaked around k_0 $u_k(x) = u_{k0}(x)$ does not depend on k and the dispersion of electronic states $\omega(k) = E(k)/\hbar$ is linear, that is

$$\omega(k) = v_g(k - k_0) + \omega(k_0).$$

Show that the maximum of the probability distribution associated with this wave packet moves with group velocity v_g.

5.5 Consider a crystal with electrons having the dispersion relation

$$E_e = \frac{h^2}{2}\left(\frac{k_x^2}{m_{11}} + \frac{k_y^2}{m_{22}} + \frac{k_z^2}{m_{33}}\right) \tag{5.73}$$

with the values: $m_{11} = m_0$, $m_{22} = \frac{m_0}{3}$, $m_{33} = \frac{m_0}{9}$ and denote the Fermi energy by E_F.

(a) Find an expression for the length of the Fermi wave vector along the shortest distance in momentum space.

(b) What is the length of the Fermi wave vector along a (111) direction?

5.6 (a) Derive an expression for the temperature dependence of the Fermi energy E_F for an intrinsic semiconductor (e.g., GaAs). Assume the electrons ($m_e = 0.067m_0$) go into a spherical carrier pocket at $k = 0$ and that the holes are in a degenerate band at $k = 0$ with heavy holes ($m_{hh} = 0.62m_0$) and light holes ($m_{lh} = 0.074m_0$). Assume the split off band is fully occupied by electrons.

(b) Does E_F increase or decrease with increasing temperature?

5.7 Suppose that you have a hydrogenic donor impurity in GaAs ($m_{ii}^* = 0.07m_0$ and $\epsilon = 15$).

(a) Give an example of a substitutional impurity that will produce such donor states. At which site?

(b) At what donor concentration will the donor electron orbitals start to overlap and an impurity band will be formed?

(c) If the material is compensated, does this modify your answer to (b)? Explain!

Suggested Readings

A.C. Smith, J.F. Janak, R.B. Adler, *Electronic Conduction in Solids* (McGraw-Hill, New York, 1967)

C. Kittel, *Introduction to Solid State Physics*, 7th edn. (Wiley, New York, 1996)

N.W. Ashcroft, N.D. Mermin, Solid State Phys. (1976) (Saunders)

Chapter 6
Lattice Vibrations

6.1 Introduction

Condensed matter systems are not only probed by electrons for their electronic excitational structures, but also their structures can experience perturbations in response to photons which are carriers of electromagnetic radiation and which can also provide thermal energy in the form of waves, which are called phonons. At low frequencies these lattice vibrational waves are described by quantum oscillators called acoustic waves with the motion of harmonic oscillators. For crystal lattices with more than one atom per unit cell, optical branches with higher energy excitations are also created. For infrared excitation radiation in the terahertz (THz) range, the lattice energies are linearly dependent on wave vector. Most laboratory experiments start with photon energies above a few meV to probe the energies of weakly bonded impurity levels, out to the infrared range (0.1–1.5 eV) and to the visible range (1.5–2.8 eV). The vibrations of the atoms are, however, described by the motion of the carriers of thermal energy called phonons. The harmonic oscillators describing these systems are discussed in Sect. 6.2 from a quantum mechanical standpoint. These harmonic oscillators are introduced into a solid with periodic boundary conditions in Sect. 6.3, where photons are introduced to create and annihilate the phonons. Examples of phonon dispersion relations for some specific materials systems in three dimensions are given in Sect. 6.4. The probing of phonon lattice vibrations by electrons is discussed in Sect. 6.5 where the electron-phonon interaction is also discussed briefly.

6.2 Quantum Harmonic Oscillators

In this section we briefly review the solution of the harmonic oscillator problem in quantum mechanics using raising and lowering operators. We can think of the phonon as a vibration of a crystal lattice caused by thermal excitation. The Hamiltonian for this problem of the one dimensional harmonic oscillator is written as:

M. Dresselhaus et al., *Solid State Properties*, Graduate Texts in Physics,
https://doi.org/10.1007/978-3-662-55922-2_6

$$\mathscr{H} = \frac{p^2}{2m} + \frac{1}{2}\kappa x^2. \tag{6.1}$$

Classically, we know that the frequency of oscillation is given by $\omega = \sqrt{\kappa/m}$ so that it is natural to think of the quantum mechanical description to be a wave with the same frequency as described by the Hamiltonian

$$\mathscr{H} = \frac{p^2}{2m} + \frac{1}{2}m\omega^2 x^2. \tag{6.2}$$

We define the lowering and raising operators $\hat{a}$ and $\hat{a}^\dagger$ for the harmonic oscillator, respectively, by

$$\hat{a} = \frac{\hat{p} - i\omega m\hat{x}}{\sqrt{2\hbar\omega m}} \tag{6.3}$$

and

$$\hat{a}^\dagger = \frac{\hat{p} + i\omega m\hat{x}}{\sqrt{2\hbar\omega m}}. \tag{6.4}$$

where ω is the frequency of the wave. Since the displacement of the harmonic oscillator $\hat{x}$ and its momentum do not commute, the Heisenberg uncertainty principle gives $[\hat{p}, \hat{x}] = \hbar/i$, then it follows that

$$[\hat{a}, \hat{a}^\dagger] = 1 \tag{6.5}$$

so that

$$\hat{\mathscr{H}} = \frac{1}{2m}\left[(\hat{p} + i\omega m\hat{x})(\hat{p} - i\omega m\hat{x}) + m\hbar\omega\right] \tag{6.6}$$

$$= \hbar\omega[\hat{a}^\dagger\hat{a} + 1/2]. \tag{6.7}$$

Let

$$\hat{N} = \hat{a}^\dagger\hat{a} \tag{6.8}$$

denote the number operator and we denote the eigenstates of this operator $\hat{N}$ by $|n\rangle$, so that

$$\hat{N}|n\rangle = n|n\rangle \tag{6.9}$$

where n is any real integer. However

$$\langle n|\hat{N}|n\rangle = \langle n|\hat{a}^\dagger\hat{a}|n\rangle = \langle y|y\rangle = n \geq 0 \tag{6.10}$$

where $|y\rangle = a|n\rangle$ implies that n is a non-negative integer. We note with regard to (6.10) that the absolute value square of any wavefunction cannot be negative, because quantum mechanically, this quantity signifies a probability. Hence n is a positive number or zero.

The action of the lowering operator is found from consideration of

$$\hat{N}\hat{a}|n\rangle = \hat{a}^\dagger\hat{a}\hat{a}|n\rangle = (\hat{a}\hat{a}^\dagger - 1)\hat{a}|n\rangle = (n-1)\hat{a}|n\rangle. \tag{6.11}$$

Hence we find that

$$\hat{a}|n\rangle = c|n-1\rangle. \tag{6.12}$$

However from (6.10), we have

$$\langle n|\hat{a}^\dagger\hat{a}|n\rangle = |c|^2, \tag{6.13}$$

and also from (6.10) we have

$$\langle n|\hat{a}^\dagger\hat{a}|n\rangle = n, \tag{6.14}$$

so that

$$c = \sqrt{n} \tag{6.15}$$

and

$$\hat{a}|n\rangle = \sqrt{n}|n-1\rangle. \tag{6.16}$$

Since the operator $\hat{a}$ lowers the quantum number of the state, $\hat{a}$ is called the annihilation or lowering operator, which physically corresponds to the annihilation of one quantum of phonon energy. From this argument you can also see that n has to be an integer. The null state is obtained for $n = 0$ at which point the phonon has the energy of the vacuum.

To obtain the corresponding raising operator, which corresponds to increasing the phonon energy by one energy quantum, we write

$$\hat{N}\hat{a}^\dagger|n\rangle = \hat{a}^\dagger\hat{a}\hat{a}^\dagger|n\rangle = \hat{a}^\dagger(1 + \hat{a}^\dagger\hat{a})|n\rangle = (n+1)\hat{a}^\dagger|n\rangle. \tag{6.17}$$

Hence we obtain

$$\hat{a}^\dagger|n\rangle = \sqrt{n+1}|n+1\rangle \tag{6.18}$$

so that $\hat{a}^\dagger$ is called a creation operator or a raising operator. Finally, for the Hamiltonian in (6.7) we write

$$\hat{\mathscr{H}}|n\rangle = \hbar\omega[\hat{N} + 1/2]|n\rangle = \hbar\omega(n + 1/2)|n\rangle \tag{6.19}$$

so that the eigenvalues for the harmonic oscillator are written as:

$$E = \hbar\omega(n + 1/2) \qquad n = 0, 1, 2, \ldots. \tag{6.20}$$

where the ground state energy of the phonon is $\hbar\omega/2$.

6.3 Phonons in 1D Solids

6.3.1 A Monoatomic Chain

In this section we relate the lattice vibrations of the crystal to harmonic oscillators and formally identify the quanta of the lattice vibrations with phonons, as suggested in Sect. 6.2. Consider the 1-D crystalline solid model which is formed by a harmonic oscillator as vibrating atoms connected to one another with springs as shown in Fig. 6.1.

The Hamiltonian for this case is written as

$$\hat{\mathscr{H}} = \sum_{s=1}^{N} \left(\frac{p_s^2}{2m_s} + \frac{1}{2}\kappa[(q_{s+1})^2 - (q_s)^2] \right). \tag{6.21}$$

This equation does not look like a set of independent harmonic oscillators since q_s and q_{s+1} are coupled. Physically the atoms in the crystal are coupled by long range forces that allow the atoms to vibrate around their equilibrium positions in the solid. In the (6.21), the subscript s is used to include as many atoms as we consider relevant for describing a given phenomenon. In most of the cases, the interactions with the first neighbors are enough to get the main physical insights, but including up to three neighbor interactions accounts for many phenomena observed in real crystals. To obtain normal mode solutions we write

$$\begin{aligned} q_s &= \left(1/\sqrt{N}\right) \textstyle\sum_k Q_k e^{iksa} \\ p_s &= \left(1/\sqrt{N}\right) \textstyle\sum_k P_k e^{iksa}. \end{aligned} \tag{6.22}$$

These Q_k's and P_k's are called the phonon coordinates. Since the atoms obey the uncertainty principle, it is straightforward that their operators obey the relation

$$[\hat{p}_s, \hat{q}_{s'}] = (\hbar/i)\delta_{ss'}. \tag{6.23}$$

The operators related to phonon coordinates corresponding to the atoms will also follow the uncertainty principle, so that

$$[\hat{P}_k, \hat{Q}_{k'}] = (\hbar/i)\delta_{kk'}. \tag{6.24}$$

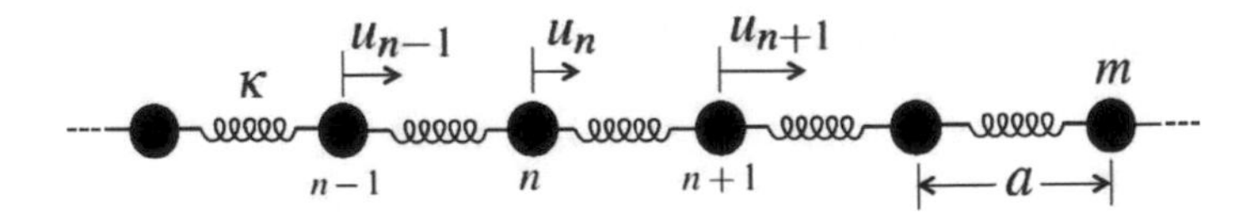

Fig. 6.1 Spring model for a 1D crystalline solid

The Hamiltonian operator for 1D lattice vibrations in phonon coordinates is then written as

$$\hat{\mathcal{H}} = \sum_k \left(\frac{1}{2} \hat{P}_k^\dagger \hat{P}_k + \frac{1}{2} \omega_k^2 \hat{Q}_k^\dagger \hat{Q}_k \right). \tag{6.25}$$

Similar to the harmonic oscillator, the Hamiltonian given by (6.21) can be written in terms of $\hat{a}$ and $\hat{a}^\dagger$ operators as

$$\hat{a}_k = \frac{i \hat{P}_k^\dagger + \omega_k \hat{Q}_k}{\sqrt{2\hbar\omega_k}}, \tag{6.26}$$

$$\hat{a}_k^\dagger = \frac{-i \hat{P}_k + \omega_k \hat{Q}_k^\dagger}{\sqrt{2\hbar\omega_k}}. \tag{6.27}$$

The Hamiltonian thus becomes using annihilation and creation operators:

$$\hat{\mathscr{H}} = \sum_k \hbar\omega_k (\hat{a}_k^\dagger \hat{a}_k + 1/2) \tag{6.28}$$

these yielding energy eigenvalues

$$E = \sum_k (n_k + 1/2)\hbar\omega_k. \tag{6.29}$$

The quantum excitation for the lattice vibration of the linear chain can be identified as a phonon, and the state vector of a system of phonons is written as $|n_1, n_2, \ldots, n_k, \ldots\rangle$. To annihilate or create a phonon in mode k with energy $(n_k + 1/2)\hbar\omega$, we then write

$$\hat{a}_k |n_1, n_2, \ldots, n_k, \ldots\rangle = \sqrt{n_k}\, |n_1, n_2, \ldots, n_k - 1, \ldots\rangle \tag{6.30}$$

$$\hat{a}_k^\dagger |n_1, n_2, \ldots, n_k, \ldots\rangle = \sqrt{n_k + 1}\, |n_1, n_2, \ldots, n_k + 1, \ldots\rangle \tag{6.31}$$

from which the probabilities n_k and $(n_k + 1)$ are obtained for the annihilation and creation processes, as described above.

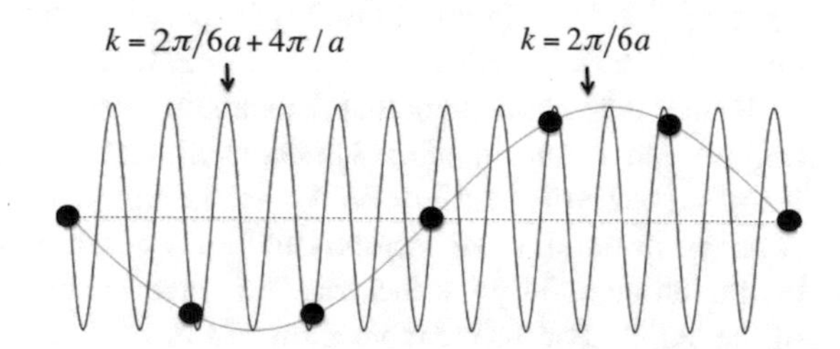

Fig. 6.2 Actual site lattices defining the waves in a 1D solid for different wave lengths corresponding to the solutions at the k points where the standing waves are the solutions of the dynamical motion

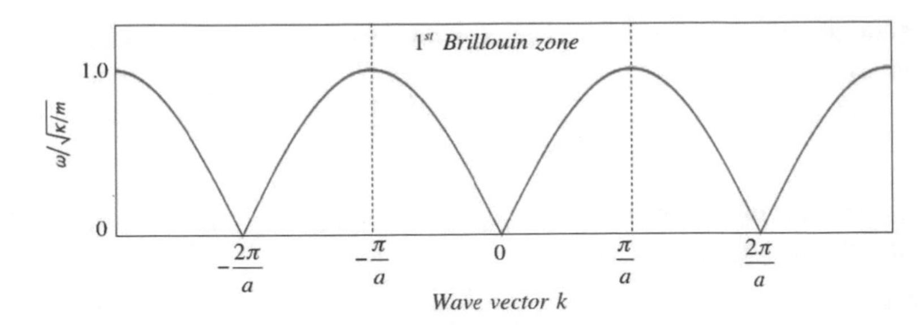

Fig. 6.3 Phonon dispersion relation for a 1D solid formed by a linear monoatomic chain

The solution of the Hamiltonian for a linear chain considering nearest neighbor interactions only gives rise to the 1-D phonon dispersion relation as shown in Fig. 6.3. A detailed derivation of this relation can be find in standard text books and here is left as an exercise in the problem set for this chapter.

$$\omega_k \equiv \sqrt{2\kappa(1 - \cos ka)} = \sqrt{4\kappa/m}|\sin(ka/2)|. \tag{6.32}$$

Equation (6.32) describes the waves that are allowed to propagate along the chain. Any wave whose frequency does not depend on phonon wave vector k as dictated by the dispersion relation in (6.32), does not propagate and is an evanescent wave which dies out in intensity after a while. By considering the small wave vector region ($k \to 0$) or equivalently by considering large wave length phonons, (6.32) can be written as

$$\omega_k = \sqrt{\kappa/m}\, ak = v_s k, \tag{6.33}$$

where we used the approximation $\sin(ka/2) \simeq ka/2$ and $v_s = \sqrt{\kappa/m}$. In (6.33), both phase velocity (ω/k) and group velocity ($\partial\omega/\partial k$) of the wave are equal to v_s, which is identified with the sound velocity in the solid. The physical meaning is that for very long wavelength phonons, the wave propagation does not depend on the details of the atomic structure and it behaves just like a propagating wave in a continuous elastic medium.

We should also discuss these other limit, the short wavelength limit, whose minimum value is just two lattice spacings ($\lambda_{min} = 2a$) which implies $k = \pi/a$. In this limit, where k is on the Brillouin zone boundary, the group velocity is zero, thus meaning that the solutions of (6.32) for periodic structures are standing waves at the points $k = \pi/(na)$, where $n = \pm1, \pm2, \pm3, \ldots$.

It should be pointed out that the standing wave has a special feature. It is defined only for the actual site of the lattice, as illustrated in Fig. 6.2 for two different wave lengths, i.e. $\lambda = 6a$ and $\lambda = 3a$. We emphasize that the positions of the atoms are the same for different wave vectors, and this is possible because of the $2\pi/a$ periodicity of the lattice and by the fact that the wave is defined for the actual physical sites of the lattice. For the example shown in Fig. 6.2, one wave is obtained by summing over a distance $4\pi/a$ in reciprocal space (2 reciprocal lattice vectors). The physical interpretation for this observation is that any waves with a shorter wavelength can be described by a wave with a longer wavelength, and we just need to consider the first Brillouin zone for describing the system in both cases.

6.3.2 Diatomic Linear Chain

We now discuss another model for a 1D solid for which the unit cell has two atoms with mass M and m as shown in Fig. 6.4. Since we have two atoms in the lattice, we can label the atom with a mass m by even ($2n$) and the atom with mass M by odd ($2n+1$) numbers. Similar to what was done with the monoatomic chain, the forces on each atom, can be written by considering just the nearest neighbor interactions to obtain

$$m\frac{d^2u_{2n}}{dt^2} = -\kappa(2u_{2n} - u_{2n+1} - u_{2n-1}) \tag{6.34}$$

$$M\frac{d^2u_{2n+1}}{dt^2} = -\kappa(2u_{2n+1} - u_{2n} - u_{2n+2}). \tag{6.35}$$

These two equations are coupled to each other and they can be solved by means of a wave-like solution, so that

$$u_{2n} = A_{2n}e^{i(kan-\omega t)} \tag{6.36}$$

$$u_{2n+1} = A_{2n+1}e^{i(ka(2n+1)-\omega t)} \tag{6.37}$$

where A_{2n} and A_{2n+1} denote the amplitudes of the vibrations of the two atoms. By inserting (6.36) and (6.37) into the equations of motion, we obtain

$$(m\omega^2 - 2\kappa)A_{2n} + \kappa(e^{ikan} + e^{-ikan})A_{2n+1} = 0 \tag{6.38}$$

$$\kappa(e^{ikan} + e^{-ikan})A_{2n} + (M\omega^2 - 2\kappa)A_{2n+1} = 0 \tag{6.39}$$

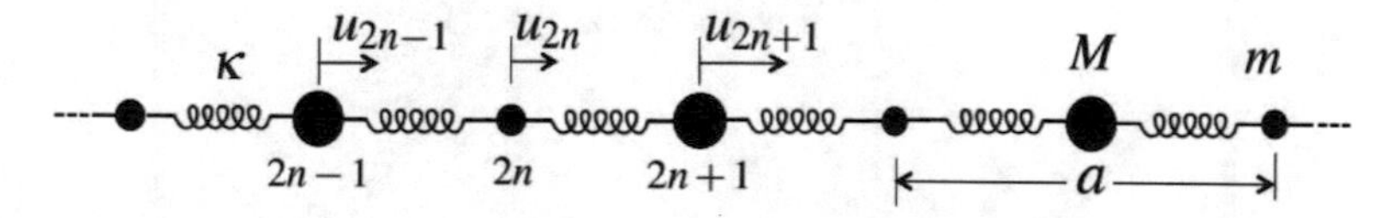

Fig. 6.4 1D model for solid based on a diatomic linear chain. The plots were constructed using arbitrary constants and $M = 1.6\,\mathrm{m}$

The above equations can be written in matrix form as

$$\begin{pmatrix} m\omega^2 - 2\kappa & \kappa(e^{ikan} + e^{-ikan}) \\ \kappa(e^{ikan} + e^{-ikan}) & M\omega^2 - 2\kappa \end{pmatrix} \begin{pmatrix} A_{2n} \\ A_{2n+1} \end{pmatrix} = \begin{pmatrix} 0 \\ 0 \end{pmatrix}. \tag{6.40}$$

The solutions for this eigenvalue equation are obtained when the determinant is zero, that is

$$\begin{vmatrix} m\omega^2 - 2\kappa & \kappa(e^{ikan} + e^{-ikan}) \\ \kappa(e^{ikan} + e^{-ikan}) & M\omega^2 - 2\kappa \end{vmatrix} = 0. \tag{6.41}$$

After a little algebra, we obtain

$$\omega^2 = \kappa\left(\frac{1}{m} + \frac{1}{M}\right) \pm \kappa\left[\left(\frac{1}{m} + \frac{1}{M}\right)^2 - \frac{4}{mM} sin^2\left(\frac{ka}{2}\right)\right]^{1/2}, \tag{6.42}$$

which represents the phonon dispersion relation for a linear chain with two atoms per unit cell. The plot of the (6.42) is shown in Fig. 6.5, where we can notice, as compared with the monoatomic linear chain with one atom per unit cell, the presence of a higher energy branch, which we call the optical phonon branch. Therefore, we now have two branches to the frequency versus wavevector plot in Fig. 6.5 because there are two atoms per unit cell.

Let us analyse the function and the Brillouin zone center ($k = 0$) and the Brillouin zone boundary ($k = \pi/a$). For $k = 0$, it is clear that one solution leads to $\omega(0) = 0$ and the other solution leads to $\omega(0) = \left[\kappa\left(\frac{1}{m} + \frac{1}{M}\right)\right]^{1/2}$. On the other hand, at the Brillouin zone boundary ($k = \pi/a$) we can note a gap of in the phonon dispersion and this energy gap is due to the different mass of the atoms. This gap closes if we make the atoms in the unit cell have the same mass.

We can now analyse the amplitudes A_{2n} and A_{2n+1} for the Brillouin zone center modes $k = 0$. By working with the (6.38), we find for the acoustic mode ($\omega(0) = 0$)

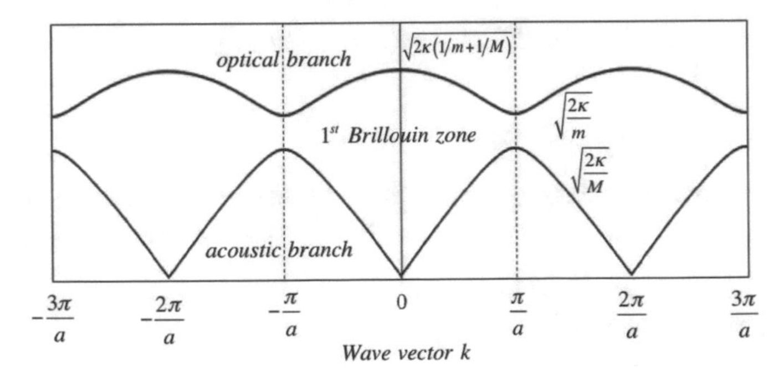

Fig. 6.5 Phonon dispersion for a 1D solid formed by a diatomic linear chain

for which we have that $A_{2n} = A_{2n+1}$ implies that the atoms are moving in phase. However, for the optical branch ($\omega(0) = \sqrt{2\kappa(\frac{1}{m} + \frac{1}{M})}$), we find that $-m/MA_{2n} = A_{2n+1}$, thus meaning that these atoms are moving out of phase. The discussion above is appropriate for the longitudinal modes but we can also include the transverse mode and calculate the phonon dispersion branches for the resulting linear chain in this case. This interesting case is left as an exercise in the Problem set listed at the end of this chapter.

We now use the concepts of Sect. 6.3 for phonons in real crystals in the next section, Sect. 6.4.

6.4 Phonons in 3D Crystals

In this section, we give some examples of phonons is 3D crystals. The first example is the zone center atomic displacements in graphite shown in Fig. 6.6. Graphite has 4 carbon atoms per unit cell, so that there are 12 zone center lattice modes. Twelve modes of these normal modes, 3 are acoustic modes and 9 are optic modes as shown.

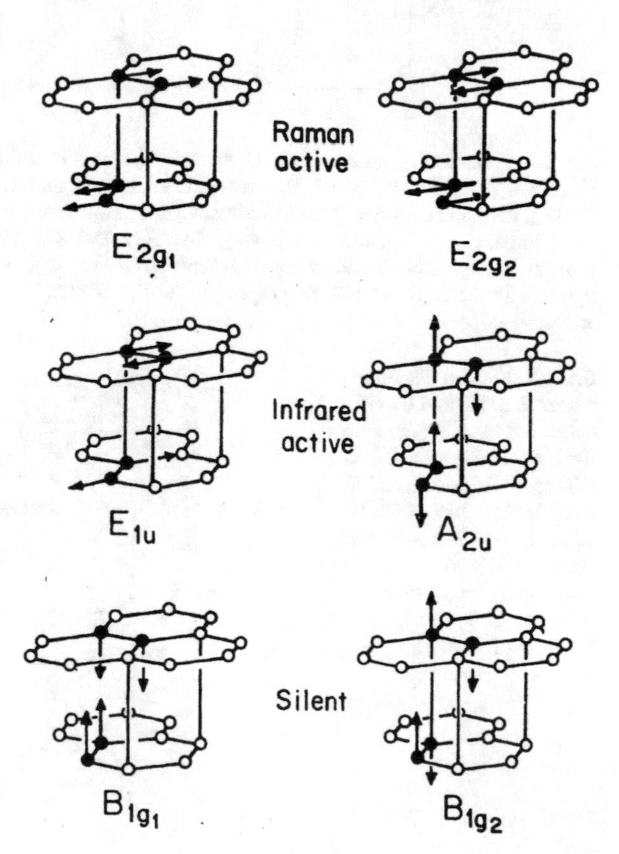

Fig. 6.6 The 9 zone center optical phonon modes in graphite. Here the A_{2u}, B_{1g1}, and B_{2g2} modes are non-degenerate while the E_{1u}, E_{2g} and E_{2g2} modes are two-fold degenerate

The next example is the phonon dispersion curves for diamond. The 3D structure for the diamond lattice is shown in Fig. 1.7a and the phonon dispersion relations are shown in Fig. 6.7 (curves are calculations and points are experimental). Diamond has 2 carbon atoms per fcc unit cell, thus 6 phonon branches. There are 3 acoustic and 3 are optic modes. The zone center optic modes are Raman active.

The next example is the phonon dispersion curves for Silicon and Germanium shown in Fig. 6.8a, b, respectively. Solid lines denote the phonon dispersion calculations by using ab initio models and points are experimental data from neutron scattering. It can be noticed that Si and Ge have very similar phonon dispersion as expected because they have the same basis structure but only different lattice parameters and atomic masses. The main difference relies on the highest frequency mode and this is in part due to the large mass of Ge as compared to Si. Silicon

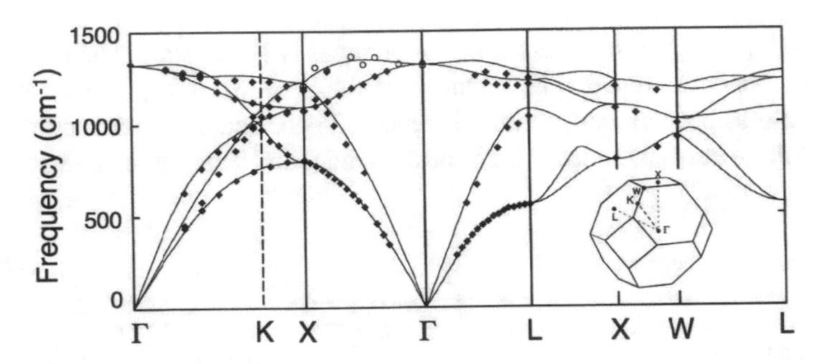

Fig. 6.7 Phonon dispersion curves in diamond calculated by using ab initio methods (P. Pavone, K. Karch, O. Schiitt, W. Windl, D. Strauch, P. Giannozzi, and S. Baroni, Phys. Rev. B 48, 3156 (1993)). The solid and open symbols are, respectively, experimental neutron scattering (J.L. Warren, J.L. Yarnell, G. Dolling, and R.A. Cowley, Phys. Rev. 158, 805 (1967)) and synchrotron scattering data. (E. Burkel, Inelastic Scattering of X Rays tooth Very High Energy Resolution, Vol. 125.) of Springer Tracts in Modern Physics (Springer, Berlin, 1991)). The first Brillouin zone is shown as an inset

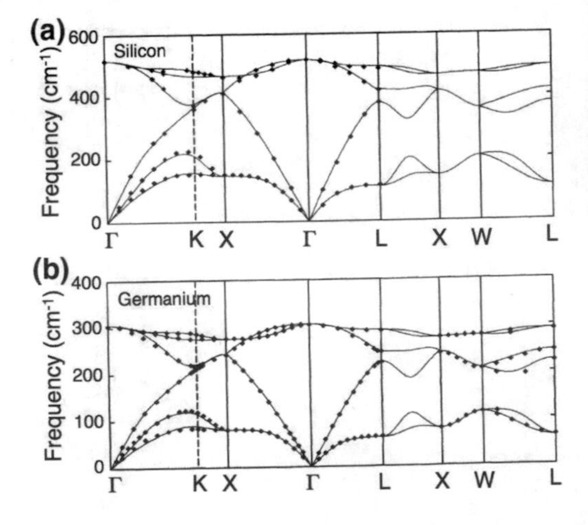

Fig. 6.8 Phonon dispersion curves in Si (**a**) and Ge (**b**) calculated by using ab initio methods (P. Giannozzi, S. De Gironcoli, P. Pavone, and S. Baroni, Phys. Rev. B 43(9), 7231 (1991)). The solid and open symbols are, respectively, experimental neutron scattering and synchrotron scattering data

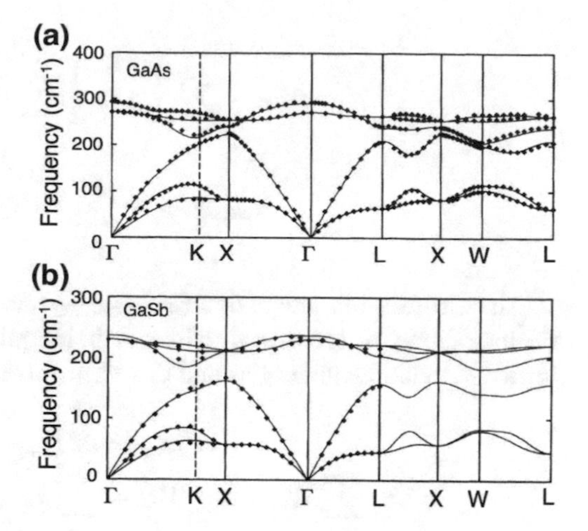

Fig. 6.9 Phonon dispersion curves in GaAs (**a**) and GaSb (**b**) calculated by using ab initio methods (P. Giannozzi, S. De Gironcoli, P. Pavone, and S. Baroni, Phys. Rev. B 43(9), 7231 (1991)). The solid and open symbols are, respectively, experimental neutron scattering and synchrotron scattering data

and Ge, like diamond, has 2 atoms per fcc unit cell (see Fig. 1.7a), and therefore has 6 phonon branches. The zone center optic modes are Raman active. There are 3 acoustic branches and 3 are optically active modes.

The next example is the phonon dispersion curves for GaAs and GaSb shown in Fig. 6.9a, b, respectively. These curves were calculated by using ab initio methods. GaAs and GaSb, like diamond, has 2 atoms per fcc unit cell, thus 6 branches. However, the two atoms are different and GaAs and GaSb, therefore, lacks inversion symmetry. The zone center optic modes are as a consequence both are infrared active and also Raman active. There are 3 acoustic branches and 3 are optic modes, all of which can be probed by optical spectroscopy techniques, as discussed further in Sect. 6.5. Phonon dispersion in 2D materials are left to the problems at the end of this chapter.

6.5 Electron-Phonon Interaction

The basic Hamiltonian for the electron-lattice system is

$$\mathscr{H} = \sum_k \frac{p_k^2}{2m} + \frac{1}{2}\sum_{kk'}{}' \frac{e^2}{|\mathbf{r}_k - \mathbf{r}_{k'}|} + \sum_i \frac{P_i^2}{2M} + \frac{1}{2}\sum_{ii'}{}' V_{\text{ion}}(\mathbf{R}_i - \mathbf{R}_{i'}) + \sum_{k,i} V_{\text{el-ion}}(\mathbf{r}_k - \mathbf{R}_i) \tag{6.43}$$

where

$$\mathscr{H} = \mathscr{H}_{\text{electron}} + \mathscr{H}_{\text{ion}} + \mathscr{H}_{\text{electron-ion}} \tag{6.44}$$

consisting of the

$$\mathscr{H}_{\text{electron}} = \sum_k \frac{p_k^2}{2m} + \frac{1}{2}\sum_{kk'}{}' \frac{e^2}{|\mathbf{r}_k - \mathbf{r}_{k'}|} \tag{6.45}$$

$$\mathscr{H}_{\text{ion}} = \sum_i \frac{P_i^2}{2M} + \frac{1}{2}\sum_{ii'}{}' V_{\text{ion}}(\mathbf{R}_i - \mathbf{R}_{i'}) \tag{6.46}$$

$$\mathscr{H}_{\text{electron-ion}} = \sum_{k,i} V_{\text{el-ion}}(\mathbf{r}_k - \mathbf{R}_i) \tag{6.47}$$

The electron-ion interaction ($\mathscr{H}_{\text{electron-ion}}$) term can be separated into two parts: the interaction of electrons with ions in their equilibrium positions, and an additional term due to lattice vibrations and this term is treated in perturbation theory:

$$\mathscr{H}_{\text{el-ion}} = \mathscr{H}^0_{\text{el-ion}} + \mathscr{H}_{\text{el-phonon}} \tag{6.48}$$

$$\sum_{k,i} V_{\text{el-ion}}(\mathbf{r}_k - \mathbf{R}_i) = \sum_{k,i} V_{\text{el-ion}}(\mathbf{r}_k - (\mathbf{R}_i^0 + \mathbf{s}_i)) \tag{6.49}$$

$$= \sum_{k,i} V_{\text{el-ion}}(\mathbf{r}_k - \mathbf{R}_i^0) - \sum_{k,i} \mathbf{s}_i \cdot \nabla V_{\text{el-ion}}(\mathbf{r}_k - \mathbf{R}_i^0) \tag{6.50}$$

$$= \mathscr{H}^0_{\text{el-ion}} + \mathscr{H}_{\text{el-phonon}}. \tag{6.51}$$

In solving the Hamiltonian $\mathscr{H}$ of (6.43), we seek a solution of the total problem in the form

$$\Psi = \psi(\mathbf{r}_1, \mathbf{r}_2, \ldots, \mathbf{R}_1, \mathbf{R}_2, \ldots)\varphi(\mathbf{R}_1, \mathbf{R}_2, \ldots) \tag{6.52}$$

such that

$$\mathscr{H}\Psi = E\Psi. \tag{6.53}$$

We then use an adiabatic approximation, which solves the electron part of the Hamiltonian by

$$(\mathscr{H}_{\text{electron}} + \mathscr{H}^0_{\text{el-ion}})\psi = E_{\text{el}}\psi. \tag{6.54}$$

Neglecting the $\mathscr{H}_{\text{el-phonon}}$ term at first, which we consider as a perturbation, we write:

$$\mathscr{H}_{\text{ion}}\varphi = (E - E_{\text{el}})\varphi = E_{\text{ion}}\varphi \tag{6.55}$$

and we have thus decoupled the electron-lattice system. This is valid for small perturbations by electromagnetic fields which are used as a probe.

Equation (6.55) gives us the phonon spectra and harmonic oscillator-like wave functions, as discussed in the previous section (Sect. 6.3). The term that was left out in the above discussion is the electron-phonon interaction

$$\mathscr{H}_{\text{el-phonon}} = -\sum_{k,i} \mathbf{s}_i \cdot \nabla V_{\text{el-ion}}(\mathbf{r}_k - \mathbf{R}_i^0) \tag{6.56}$$

which we now treat as a perturbation. We rewrite (6.55) by introducing normal coordinates, as before;

$$\mathbf{s}_i = \frac{1}{\sqrt{NM}} \sum_{\mathbf{q},j} Q_{\mathbf{q},j} e^{i\mathbf{q}\cdot\mathbf{R}_i^0} \hat{e}_j \tag{6.57}$$

where j is the polarization index and $\hat{e}_j$ is a unit displacement vector for mode j. Hence we obtain

$$\mathscr{H}_{\text{el-phonon}} = -\sum_{k,i} \frac{1}{\sqrt{NM}} \sum_{\mathbf{q},j} Q_{\mathbf{q},j} e^{i\mathbf{q}\cdot\mathbf{R}_i^0}\, \hat{e}_j \cdot \nabla V_{\text{el-ion}}(\mathbf{r}_k - \mathbf{R}_i^0) \tag{6.58}$$

where now this term in operator notation, reads

$$\hat{Q}_{\mathbf{q},j} = \left(\frac{\hbar}{2\omega_{\mathbf{q},j}}\right)^{\frac{1}{2}} (\hat{a}_{\mathbf{q},j} + \hat{a}^{\dagger}_{-\mathbf{q},j}). \tag{6.59}$$

Writing below the time dependence explicitly for the raising and lowering operators

$$\hat{a}_{\mathbf{q},j}(t) = \hat{a}_{\mathbf{q},j} e^{-i\omega_{\mathbf{q},j}t} \tag{6.60}$$

$$\hat{a}^{\dagger}_{\mathbf{q},j}(t) = \hat{a}^{\dagger}_{\mathbf{q},j} e^{i\omega_{\mathbf{q},j}t} \tag{6.61}$$

we obtain

$$\begin{aligned}\hat{\mathscr{H}}_{\text{el-phonon}} = &-\sum_{\mathbf{q},j} \left(\frac{\hbar}{2MN\omega_{\mathbf{q},j}}\right)^{\frac{1}{2}} (\hat{a}_{\mathbf{q},j} e^{-i\omega_{\mathbf{q},j}t} + \hat{a}^{\dagger}_{\mathbf{q},j} e^{i\omega_{\mathbf{q},j}t}) \\ &\times \sum_{k,i} (e^{i\mathbf{q}\mathbf{R}_i^0} + e^{i\mathbf{q}\mathbf{R}_i^0}) \hat{e}_j \cdot \nabla V_{\text{el-ion}}(\mathbf{r}_k - \mathbf{R}_i^0)\end{aligned} \tag{6.62}$$

which can be written as

$$\hat{\mathscr{H}}_{\text{el-phonon}} = -\sum_{\mathbf{q},j} \left(\frac{\hbar}{2NM\omega_{\mathbf{q},j}}\right)^{\frac{1}{2}} \left(\hat{a}_{\mathbf{q},j} \sum_{k,i} e^{i(\mathbf{q}\mathbf{R}_i^0 - \omega_{\mathbf{q},j}t)} \hat{e}_j \cdot \nabla(\mathbf{r}_k - \mathbf{R}_i^0) + c.c.\right)$$

If we are only interested in the interaction of one electron with a phonon on a particular phonon branch, say the longitudinal acoustic (LA) branch, then we drop the summation over j and k and write

$$\hat{\mathscr{H}}_{\text{el-phonon}} = -\sum_{\mathbf{q}} \left(\frac{\hbar}{2NM\omega_{\mathbf{q}}}\right)^{\frac{1}{2}} \left(\hat{a}_{\mathbf{q}} \sum_{i} e^{i(\mathbf{q}\cdot\mathbf{R}_i^0 - \omega_{\mathbf{q}}t)} \hat{e} \cdot \nabla V_{\text{el-ion}}(\mathbf{r} - \mathbf{R}_i^0) + c.c.\right) \tag{6.63}$$

where the 1st term in the brackets corresponds to phonon absorption and the c.c. term corresponds to phonon emission.

With $\mathscr{H}_{\text{el-phonon}}$ in hand, we can directly solve transport problems (e.g., τ due to phonon scattering), and optical problems (e.g., indirect transitions), since all these types of problems involve matrix elements $\langle f|\mathscr{H}_{\text{el-phonon}}|i\rangle$ that couple initial states i and final states f.

Problems

6.1 By considering (6.21) which defines the interaction between the atoms in a 1D solid crystalline model and the uncertainty relations for phonon coordinates, derive (6.32). In this derivation consider nearest neighbor interactions only.

6.2 We discuss in the chapter the diatomic linear chain model by considering only the nearest neighbor interactions. However, the linear chain model has been successfully used for describing the propagation of longitudinal optical modes in semiconductors, such as GaAs by considering the interactions with next nearest neighbors.

(a) Calculate the phonon dispersion for a diatomic linear chain by considering the interactions with the next nearest neighbors and compare with the results obtained for also considering the nearest neighbor interaction.
(b) By using this model show that the spring constant associated with nearest (κ) and next nearest (q_1 and q_2) neighbors can be written as

$$k = \frac{mM}{2(m+M)}\omega^2_{LO}(\Gamma)$$

$$q_1 = \frac{\omega^2_{LA}(X)m - 2\kappa}{4}$$

$$q_2 = \frac{\omega^2_{LO}(X)M - 2\kappa}{4}$$

6.3 Consider the Si and SiC bulk materials. In the "back-scattering"geometry the longitudinal optical modes are allowed, and along the c-axis direction and these crystals can be described by the linear chain model. The experimental frequencies for Si are 517 cm^{-1} for the LO (Γ), 409 cm^{-1} LO (X), and 409 cm^{-1} LA (X) modes, [LA means longitudinal acoustic and LO longitudinal optical]. For SiC the frequencies are : 972 cm^{-1} LO (Γ), 829 cm^{-1} LO (X), and 640 cm^{-1} LA (X). By using the results obtained in the previous problem, calculate k, q_1, and q_2.
The atomic masses of Si and C are respectively 28.09 and 12.01 a.m.u, and the data in Table 6.1 are multiplied by a factor of 1.672×10^{-18}.

6.4 Consider the diatomic linear chain model discussed in Sect. 6.3.2 regarding the transverse modes.

Table 6.1 Spring constants associated with nearest (k) and next nearest (q_1 and q_2) neighbors. Bezerra et al. 2000

	k	q_1	q_2
Si	112.41	13.95	13.95
SiC	236.18	52.85	4.53

(a) Show that for the transverse modes, the dynamical equations are given by

$$m\ddot{y}_{2n} = \frac{2\kappa_\alpha}{d}[6y_{2n} + y_{2n-2} + y_{2n+2} - 4y_{2n+1} - 4y_{2n-1}]$$

$$M\ddot{y}_{2n+1} = \frac{2\kappa_\alpha}{d}[6y_{2n+1} + y_{2n-1} + y_{2n+3} - 4y_{2n} - 4y_{2n+2}]$$

(b) By considering harmonic solutions for the transverse motion $y_{2n}=A_{2n}e^{-i[\omega t-nka]}$ and $y_{2n+1} = A_{2n+1}e^{-i[\omega t-(n+\frac{1}{2})ka]}$, show that the phonon dispersion for transverse optical and acoustic motions is given by

$$\omega^2 = \frac{2\kappa_\alpha}{mM}[(m+M)(3+coska) \pm [(m+M)^2(3+coska)^2 -$$

$$mM(cos^2ka - 2coska + 1)]^{1/2}]$$

where κ_α is the transverse effective spring constant.

6.5 Reflectivity measurements show that the LO and TO phonon (ω_{LO} and ω_{TO}) features for NaCl in the reflectivity spectra occur at 38 and 61 μm, respectively.

(a) From this information, estimate the force constant κ for the TO phonon mode assuming only nearest neighbor interactions.
(b) From the measured ω_{LO} and ω_{TO} splitting find the magnitude of the lattice polarization contribution to the dielectric constant for crystalline NaCl.
(c) Is the frequency difference ($\omega_{LO} - \omega_{TO}$) for NaCl expected to be temperature dependent? Why?
(d) Suppose that we have a material where there are 3 atoms per unit cell so that there are two different transverse optical frequencies ω_{TO1} and ω_{TO2}. Assume further that, for this material, the dielectric function has a frequency dependence given by

$$\varepsilon(\omega) = A + \frac{B_1}{\omega^2 - \omega_{TO1}^2} + \frac{B_2}{\omega^2 - \omega_{TO2}^2}$$

then generalize the Lyddane–Sachs–Teller relation when applying it to this material.

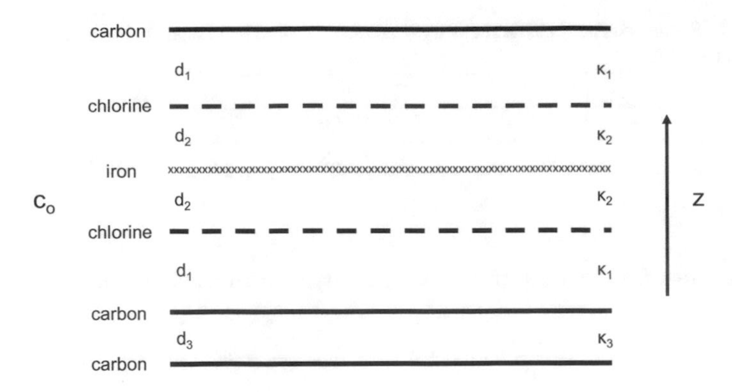

Fig. 6.10 The unit cell of $C_{24}FeCl_3$

6.6 Consider a stage 2 graphite–$FeCl_3$ layered compound (chemical formula $C_{24}FeCl_3$) with a unit cell consisting of 5 distinct layers, as indicated in the diagram, with the force constants κ_1, κ_2, and κ_3 coupling the carbon-chlorine layers, the chlorine-iron layers and the carbon-carbon layers, respectively. The masses M_C, M_{Cl}, and M_{Fe} denote the mass densities in the carbon, chlorine, and iron layers, respectively. Assume that d_1, d_2, and d_3 give the interlayer separation between carbon-chlorine layers, chlorine-iron layers and carbon-carbon layers, respectively. In solving this problem consider the number of layers, number of atoms per unit cell, and masses for each type of layer shown in Fig. 6.10.

(a) What are the mode frequencies for phonons propagating perpendicular to the layer planes (z-direction)?
(b) Which modes are excited by incident electromagnetic radiation at their respective mode frequencies?
(c) Which modes are Raman-active?
(d) With which experimental technique could you measure the entire phonon branch $\omega(q_z)$, for all q_z?

6.7 Starting from the raising and lowering operators for harmonic oscillator wave functions, find the following matrix elements taken between harmonic oscillator states.

(a) $\langle n|x|n'\rangle$
(b) $\langle n|x^2|n'\rangle$
(c) $\langle n|x^3|n'\rangle$
(d) $\langle n|p|n'\rangle$
(e) Show that the equipartition theorem applies to harmonic oscillator states: half the total energy goes into kinetic energy, and half into potential energy.

6.8 (a) Do you expect the LO phonon frequency for diamond to be higher or lower than that in GaAs? Why?

(b) Do you expect the LO–TO phonon splitting for diamond to be higher or lower than that for GaAs? Why?
(c) What experimental technique would you use to check these predictions? Why? Note: In a cubic material with optic modes, the TO modes are two-fold degenerate near the center of the Brillouin zone.

Reference

E.F. Bezerra et al., Appl. Phys. Letters **77**, 4316 (2000)

Part II
Transport Properties

Chapter 7
Basic Transport Phenomena

7.1 Introduction

In this section we study some of the transport properties for metals and semiconductors. An intrinsic semiconductor, meaning the absence of doping by foreign atoms, has no carriers at $T = 0\,\mathrm{K}$ and therefore there is no transport of carriers under the influence of external fields. However at finite temperatures there are thermally generated carriers. Impurities also can serve to generate carriers and interesting transport properties. For insulators, there is very little charge transport and in this case, the defects and the ions themselves can participate in charge transport under the influence of external applied fields. Metals make use of the Fermi–Dirac distribution function for describing the electron and hole carrier densities, but metals are otherwise similar to semiconductors, for which the Maxwell–Boltzmann distribution function is usually applicable because of their much lower carrier densities so that electron-electron (hole-hole) interactions are less important.

At finite fields, the electrical conductivity will depend on the product of the carrier density and the carrier mobility. For a one carrier type system, the Hall effect gives a measure of the carrier density and the magnetoresistance likewise gives the mobility, the key parameters governing the transport properties of a semiconductor. From the standpoint of device applications, the carrier density and the carrier mobility are the parameters of greatest importance.

To the extent that electrons can be considered as particles, the electrical conductivity, the electronic contribution to the thermal conductivity and the magnetoresistance are all found by solving the Boltzmann equation. For the case of nano-scale systems, where the wave aspects of the electron must be considered (called mesoscopic physics), more sophisticated approaches to the transport properties must be considered. To review the standard procedures for the transport properties of classical electrons (and holes), we briefly review the Boltzmann equation and its solution in the Sect. 7.2, and make use of the Boltzmann equation throughout the chapter.

M. Dresselhaus et al., *Solid State Properties*, Graduate Texts in Physics,
https://doi.org/10.1007/978-3-662-55922-2_7

7.2 The Boltzmann Equation

The Boltzmann transport equation is a statement that in the steady state, there is no net change in the distribution function $f(\mathbf{r}, \mathbf{k}, t)$ which determines the probability of finding an electron at position $\mathbf{r}$, with crystal momentum $\mathbf{k}$ and at time t. Therefore we obtain a zero sum for the changes in $f(\mathbf{r}, \mathbf{k}, t)$ due to the 3 processes of diffusion, the effect of forces and fields, and of collisions:

$$\left.\frac{\partial f(\mathbf{r}, \mathbf{k}, t)}{\partial t}\right|_{\text{diffusion}} + \left.\frac{\partial f(\mathbf{r}, \mathbf{k}, t)}{\partial t}\right|_{\text{fields}} + \left.\frac{\partial f(\mathbf{r}, \mathbf{k}, t)}{\partial t}\right|_{\text{collisions}} = 0. \tag{7.1}$$

It is customary to substitute the following differential form for the diffusion process

$$\left.\frac{\partial f(\mathbf{r}, \mathbf{k}, t)}{\partial t}\right|_{\text{diffusion}} = -\mathbf{v}(\mathbf{k}) \cdot \frac{\partial f(\mathbf{r}, \mathbf{k}, t)}{\partial \mathbf{r}} \tag{7.2}$$

which expresses the continuity equation in real space in the absence of forces, fields and collisions. For the forces and fields, we write correspondingly

$$\left.\frac{\partial f(\mathbf{r}, \mathbf{k}, t)}{\partial t}\right|_{\text{fields}} = -\frac{\partial \mathbf{k}}{\partial t} \cdot \frac{\partial f(\mathbf{r}, \mathbf{k}, t)}{\partial \mathbf{k}} \tag{7.3}$$

and by combining (7.1), (7.2), and (7.3), we obtain the Boltzmann equation:

$$\frac{\partial f(\mathbf{r}, \mathbf{k}, t)}{\partial t} + \mathbf{v}(\mathbf{k}) \cdot \frac{\partial f(\mathbf{r}, \mathbf{k}, t)}{\partial \mathbf{r}} + \frac{\partial \mathbf{k}}{\partial t} \cdot \frac{\partial f(\mathbf{r}, \mathbf{k}, t)}{\partial \mathbf{k}} = \left.\frac{\partial f(\mathbf{r}, \mathbf{k}, t)}{\partial t}\right|_{\text{collisions}} \tag{7.4}$$

for which the three derivatives for all the variables of the distribution function on the left hand side of the equation balance the collision terms appearing on the right hand side of (7.4). The first term in (7.4) gives the explicit time dependence of the distribution function $f(\mathbf{r}, \mathbf{k}, t)$ and is needed for the solution of the time dependent driving forces or for impulse perturbations.

Boltzmann's equation is usually solved using two approximations:

1. The perturbation due to external fields and forces is assumed to be small so that the distribution function can be linearized and written as:

$$f(\mathbf{r}, \mathbf{k}) = f_0(E) + f_1(\mathbf{r}, \mathbf{k}) \tag{7.5}$$

   where $f_0(E)$ is the equilibrium distribution function (the Fermi function), which depends only on the energy E, while $f_1(\mathbf{r}, \mathbf{k})$ is the perturbation term giving the departure from equilibrium.
2. The collision term in the Boltzmann equation is written in the **relaxation time approximation** so that the system returns to equilibrium uniformly:

$$\left.\frac{\partial f}{\partial t}\right|_{\text{collisions}} = -\frac{(f - f_0)}{\tau} = -\frac{f_1}{\tau} \tag{7.6}$$

where τ denotes the relaxation time and in general is a function of crystal momentum, i.e., $\tau = \tau(\mathbf{k})$ and the scattering process. The physical interpretation of the relaxation time is the time associated with the rate of return to the equilibrium distribution when the external fields or thermal gradients are switched off. Solution to (7.6) when the fields are switched off at $t = 0$ leads to

$$\frac{\partial f}{\partial t} = -\frac{(f - f_0)}{\tau} \tag{7.7}$$

which has solutions

$$f(t) = f_0 + \left[f(0) - f_0 \right] e^{-t/\tau} \tag{7.8}$$

where f_0 is the equilibrium distribution function and $f(0)$ is the distribution function at time $t = 0$. The relaxation process described by (7.8) follows a Poisson distribution, indicating that collisions relax the distribution function $f(t)$ exponentially to f_0 with a time constant τ.

With these approximations, the Boltzmann equation is solved to find the distribution function which in turn determines the number density $n(\mathbf{r}, t)$ and current density. The current density $\mathbf{j}(\mathbf{r}, t)$ is given by

$$\mathbf{j}(\mathbf{r}, t) = \frac{e}{4\pi^3} \int \mathbf{v}(\mathbf{k}) f(\mathbf{r}, \mathbf{k}, t) d^3k \tag{7.9}$$

in which the crystal momentum $\hbar\mathbf{k}$ plays the role of the momentum $\mathbf{p}$ in specifying a volume in phase space d^3k. Every element of size h (Planck's constant) in phase space can accommodate one spin $\uparrow$ and one spin $\downarrow$ electron. The carrier density $n(\mathbf{r}, t)$ is thus simply given by integration of the distribution function over k-space

$$n(\mathbf{r}, t) = \frac{1}{4\pi^3} \int f(\mathbf{r}, \mathbf{k}, t) d^3k \tag{7.10}$$

where d^3k is an element of 3D wavevector space. The velocity of a carrier with crystal momentum $\hbar\mathbf{k}$ is related to the $E(\mathbf{k})$ dispersion expression by

$$\mathbf{v}(\mathbf{k}) = \frac{1}{\hbar} \frac{\partial E(\mathbf{k})}{\partial \mathbf{k}} \tag{7.11}$$

and $f_0(E)$ is the Fermi distribution function

$$f_0(E) = \frac{1}{1 + e^{(E - E_F)/k_B T}} \tag{7.12}$$

which defines the equilibrium state in which E_F is the Fermi energy and k_B is the Boltzmann constant for a Fermi distribution.

7.3 Electrical Conductivity

To calculate the static electrical conductivity, we consider an applied electric field **E** which, for convenience, we will take to be along the x-direction. We will assume for the present that there is no magnetic field and that there are no thermal gradients present. The electrical conductivity is expressed in terms of the conductivity tensor $\overleftrightarrow{\sigma}$ which is evaluated explicitly from the relation

$$\mathbf{j} = \overleftrightarrow{\sigma} \cdot \mathbf{E}, \tag{7.13}$$

from solution of (7.9), using $\mathbf{v}(\mathbf{k})$ from (7.11) and the distribution function $f(\mathbf{r}, \mathbf{k}, t)$ from solution of the Boltzmann equation represented by (7.4). The first term in (7.4) vanishes since the dc applied field **E** has no time dependence.

For the second term in the Boltzmann equation (7.4), $\mathbf{v}(\mathbf{k}) \cdot \partial f(\mathbf{r}, \mathbf{k}, t)/\partial \mathbf{r}$, we note that

$$\frac{\partial f}{\partial \mathbf{r}} \simeq \frac{\partial f_0}{\partial \mathbf{r}} = \frac{\partial f_0}{\partial T}\frac{\partial T}{\partial \mathbf{r}}. \tag{7.14}$$

Since there are no thermal gradients present in the simplest calculation of the electrical conductivity given in this section, the term $\frac{\partial f}{\partial \mathbf{r}}$ does not contribute to (7.4). For the third term in (7.4), which we write as

$$\dot{\mathbf{k}} \cdot \frac{\partial f(\mathbf{r}, \mathbf{k}, t)}{\partial \mathbf{k}} = \sum_{\alpha} \dot{k}_{\alpha} \frac{\partial f(\mathbf{r}, \mathbf{k}, t)}{\partial k_{\alpha}} \tag{7.15}$$

where the right hand side shows the summation over the vector components. We do get a contribution to the sum over α in (7.15), since the equations of motion $(F = ma)$ give

$$\hbar \dot{\mathbf{k}} = e\mathbf{E} \tag{7.16}$$

and

$$\frac{\partial f(\mathbf{r}, \mathbf{k}, t)}{\partial \mathbf{k}} = \frac{\partial (f_0 + f_1)}{\partial \mathbf{k}} = \frac{\partial f_0}{\partial E}\frac{\partial E}{\partial \mathbf{k}} + \frac{\partial f_1}{\partial \mathbf{k}}. \tag{7.17}$$

In considering the linearized Boltzmann equation of (7.17), we retain only the leading terms in the perturbing electric field, so that $(\partial f_1/\partial \mathbf{k})$ can be neglected and only the term $(\partial f_0/\partial E)\hbar \mathbf{v}(\mathbf{k})$ need be retained. We thus obtain the linearized Boltzmann equation for the case of an applied static electric field and no thermal gradients:

$$\dot{\mathbf{k}} \cdot \frac{\partial f(\mathbf{r}, \mathbf{k}, t)}{\partial \mathbf{k}} = \frac{\phi}{\tau}\frac{\partial f_0}{\partial E} = -\frac{f_1}{\tau} \tag{7.18}$$

where it is convenient to write:

$$f_1 = -\phi\left(\frac{\partial f_0}{\partial E}\right) \tag{7.19}$$

in order to show the $(\partial f_0/\partial E)$ dependence explicitly. Substitution of (7.16) and (7.17) into (7.18) yields

$$\left[\frac{e\mathbf{E}}{\hbar}(\frac{\partial f_0}{\partial E})\right] \cdot [\hbar\mathbf{v}(\mathbf{k})] = \frac{\phi(\mathbf{k})}{\tau}(\frac{\partial f_0}{\partial E}) \tag{7.20}$$

so that

$$\phi(\mathbf{k}) = e\tau\mathbf{E} \cdot \mathbf{v}(\mathbf{k}). \tag{7.21}$$

Thus we can relate $\phi(\mathbf{k})$ to $f_1(\mathbf{k})$ by

$$f_1(\mathbf{k}) = -\phi(\mathbf{k})\frac{\partial f_0(E)}{\partial E} = -e\tau\mathbf{E} \cdot \mathbf{v}(\mathbf{k})\frac{\partial f_0(E)}{\partial E}. \tag{7.22}$$

The current density is then found from the distribution function $f(\mathbf{k})$ by calculation of the average value of $\langle ne\mathbf{v}\rangle$ over all k-space

$$\mathbf{j} = \frac{1}{4\pi^3}\int e\mathbf{v}(\mathbf{k})f(\mathbf{k})d^3k = \frac{1}{4\pi^3}\int e\mathbf{v}(\mathbf{k})f_1(\mathbf{k})d^3k \tag{7.23}$$

since

$$\int e\mathbf{v}(\mathbf{k})f_0(\mathbf{k})d^3k = 0. \tag{7.24}$$

Equation (7.24) states that no net current flows in the absence of an applied electric field, which is another statement of the equilibrium condition. Substitution for $f_1(\mathbf{k})$ given by (7.22) into (7.23) for $\mathbf{j}$ yields

$$\mathbf{j} = -\frac{e^2\mathbf{E}}{4\pi^3} \cdot \int \tau\mathbf{v}\mathbf{v}\frac{\partial f_0}{\partial E}d^3k \tag{7.25}$$

where in general $\tau = \tau(\mathbf{k})$ and reflects a variety of scattering mechanisms, each having a different k dependence and $\mathbf{v}$ is given by (7.11). A comparison of (7.25) and (7.13) thus yields the desired result for the conductivity tensor $\overleftrightarrow{\sigma}$

$$\overleftrightarrow{\sigma} = -\frac{e^2}{4\pi^3}\int \tau\mathbf{v}\mathbf{v}\ \frac{\partial f_0}{\partial E}d^3k \tag{7.26}$$

where $\overleftrightarrow{\sigma}$ is a symmetric second rank tensor ($\sigma_{ij} = \sigma_{ji}$). The evaluation of the integral in (7.26) over all k-space depends on the $E(\mathbf{k})$ relations through the $\mathbf{v}\mathbf{v}$ terms and the temperature dependence comes through the $\partial f_0/\partial E$ term. We will in Sect. 7.4

evaluate (7.26) for the simple example of a metal, and in Sect. 7.5 do the same for an intrinsic semiconductor.

7.4 Electrical Conductivity of Metals

To exploit the energy dependence of $(\partial f_0/\partial E)$ in applying (7.26) to metals, it is more convenient to evaluate $\overleftrightarrow{\sigma}$ if we replace $\int d^3k$ with an integral over the constant energy surfaces

$$\int d^3k = \int d^2S dk_\perp \equiv \int d^2S dE/|\partial E/\partial \mathbf{k}|. \tag{7.27}$$

Thus (7.26) is written as

$$\overleftrightarrow{\sigma} = -\frac{e^2}{4\pi^3}\int \frac{\tau \mathbf{v}\mathbf{v}}{|\partial E/\partial \mathbf{k}|}\frac{\partial f_0}{\partial E} d^2S\, dE. \tag{7.28}$$

From the Fermi–Dirac distribution function $f_0(E)$ shown in Fig. 7.1, we see that the derivative $(-\partial f_0/\partial E)$ can approximately be replaced by a δ-function for the case of a metal, so that (7.28) can be written as

$$\overleftrightarrow{\sigma} = \frac{e^2}{4\pi^3\hbar}\int_{\text{Fermi surface}} \tau \mathbf{v}\mathbf{v}\,\frac{d^2S}{v}. \tag{7.29}$$

For a cubic crystal, $[v_x v_x] = v^2/3$ and thus the conductivity tensor $\overleftrightarrow{\sigma}$ has only diagonal components σ that are all the diagonal components of $\overleftrightarrow{\sigma}$ are equal to each other:

$$\sigma = \frac{e^2}{4\pi^3\hbar}\int_{\text{Fermi surface}} \tau v\,\frac{d^2S}{3} = \frac{ne^2\tau}{m^*} \tag{7.30}$$

since

$$n = (1/4\pi^3)(4\pi/3)k_F^3 \tag{7.31}$$

and

$$v_F = \hbar k_F/m. \tag{7.32}$$

The result

$$\sigma = ne^2\tau/m^* \tag{7.33}$$

is called the Drude formula for the dc electrical conductivity. Generalization of this methodology to metals with anisotropic Fermi surfaces or with more than one type of carrier can be done directly and requires detailed numerical calculations in most cases for real metals.

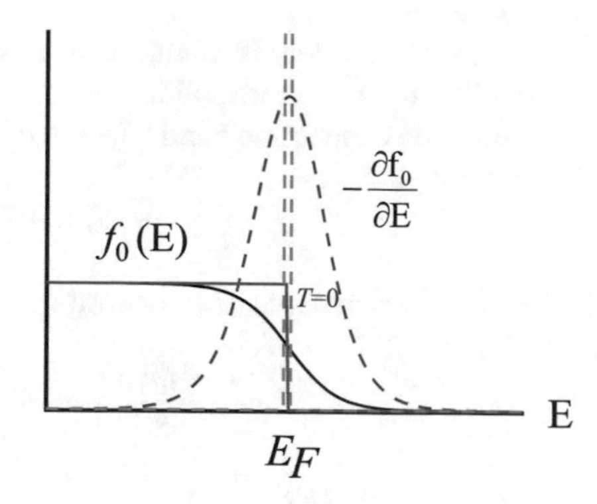

Fig. 7.1 Schematic plot of $f_0(E)$ and $-\partial f_0(E)/\partial E$ for a metal showing the δ-function like behavior near the Fermi level E_F for the derivative

7.5 Electrical Conductivity of Semiconductors

We show in this section that the simple Drude model $\sigma = ne^2\tau/m^*$ can also be recovered for a semiconductor from the general relation given by (7.26), using a simple parabolic band model and a constant relaxation time approximation. When a more complete theory is used or is needed to describe specific physical phenomena, departures from the simple Drude model will result.

In deriving the Drude model for a semiconductor we make three approximations:

- Approximation #1
  In the case of electron states in intrinsic semiconductors having no donor or acceptor impurities, we have the condition $(E - E_F) \gg k_BT$ since E_F is in the band gap and E is the energy of an electron in the conduction band, as shown in Fig. 7.2. Thus, the first approximation is equivalent to writing

$$f_0(E) = \frac{1}{1 + \exp[(E - E_F)/k_BT]} \simeq \exp[-(E - E_F)/k_BT] \qquad (7.34)$$

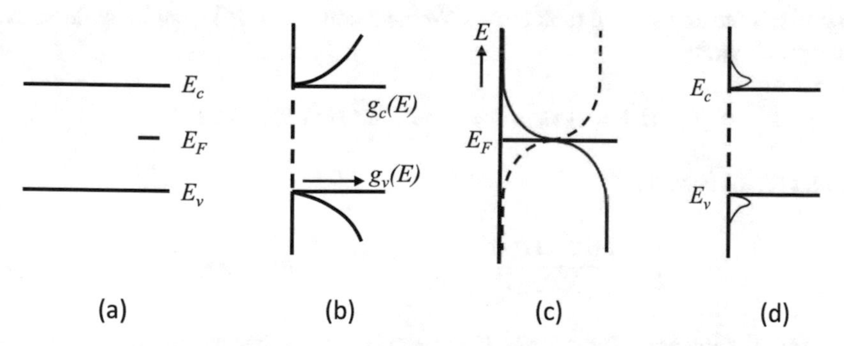

Fig. 7.2 Electron and hole states in the conduction and valence bands of an intrinsic semiconductor. **a** Location of E_F with approximately similar properties for electrons and holes. **b** The corresponding density of states for electrons and holes. **c** The Fermi functions for electrons (solid curve) and holes (dashed curve). **d** The occupation (or density) of electron and hole states in an intrinsic semiconductor at a finite temperature

which is equivalent to using the Maxwell–Boltzmann distribution in place of the full Fermi–Dirac distribution. Since E is usually measured with respect to the bottom of the conduction band, E_F is a negative energy and it is therefore convenient to write $f_0(E)$ as

$$f_0(E) \simeq e^{-|E_F|/k_B T} e^{-E/k_B T} \tag{7.35}$$

so that the derivative of the Fermi function becomes

$$\frac{\partial f_0(E)}{\partial E} = -\frac{e^{-|E_F|/k_B T}}{k_B T} e^{-E/k_B T}. \tag{7.36}$$

- Approximation #2
  For simplicity we assume a constant relaxation time τ that is independent of $\mathbf{k}$ and E. This approximation is made for simplicity and may not be valid for specific physical situations. Some common scattering mechanisms yield an energy-dependent relaxation time $\tau = \tau_0(E/k_B T)^r$, for which $r = -1/2$ and $r = +3/2$, respectively, for acoustic deformation potential scattering or ionized impurity scattering.
- Approximation #3
  To illustrate the explicit evaluation of the integral in (7.26), we consider the simplest case, assuming an isotropic, parabolic band $E = \hbar^2 k^2/2m^*$ for the evaluation of $\mathbf{v} = \partial E/\hbar \partial \mathbf{k}$ about the conduction band extremum.

Using this third approximation we can write

$$\begin{aligned}
\mathbf{vv} &= \tfrac{1}{3} v^2 \overleftrightarrow{1} \\
k^2 &= 2m^* E/\hbar^2 \\
2k dk &= 2m^* dE/\hbar^2 \\
v^2 &= 2E/m^* \\
v &= \hbar k/m^*
\end{aligned} \tag{7.37}$$

where $\overleftrightarrow{1}$ is the unit second rank tensor. We next convert (7.26) to an integration over energy and write

$$d^3k = 4\pi k^2 dk = 4\pi \sqrt{2} (m^*/\hbar^2)^{3/2} \sqrt{E} dE \tag{7.38}$$

so that (7.26) becomes

$$\sigma = \frac{e^2 \tau}{4\pi^3} \left(\frac{8\sqrt{2}\pi \sqrt{m^*}}{3\hbar^3 k_B T} \right) e^{-|E_F/k_B T} \int_0^\infty E^{3/2} dE e^{-E/k_B T} \tag{7.39}$$

in which the integral over energy E is extended to ∞ because there is negligible contribution for large E and because the definite integral

$$\int_0^\infty x^p dx e^{-x} = \Gamma(p+1) \tag{7.40}$$

can be evaluated exactly, $\Gamma(p)$ being the Γ function which has the property

$$\begin{aligned} \Gamma(p+1) &= p\Gamma(p) \\ \Gamma(1/2) &= \sqrt{\pi}. \end{aligned} \tag{7.41}$$

Substitution of (7.40) into (7.39) thus yields

$$\sigma = \frac{2e^2\tau}{m^*}\left(\frac{m^* k_B T}{2\pi\hbar^2}\right)^{3/2} e^{-|E_F|/k_B T} \tag{7.42}$$

which gives the temperature dependence of σ. Now the carrier density calculated using the same approximations becomes

$$\begin{aligned} n &= (4\pi^3)^{-1} \int f_0(E)\, d^3k \\ &= (4\pi^3)^{-1}\, e^{-|E_F|/k_B T} \int e^{-E/k_B T} 4\pi\, k^2 dk \\ &= (\sqrt{2}/\pi^2)\left(m^*/\hbar^2\right)^{3/2} e^{-|E_F|/k_B T} \int_0^\infty \sqrt{E} dE e^{-E/k_B T} \end{aligned} \tag{7.43}$$

where

$$\int_0^\infty \sqrt{E} dE e^{-E/k_B T} = \frac{\sqrt{\pi}}{2}(k_B T)^{3/2} \tag{7.44}$$

which gives the final result for the temperature dependence of the carrier density

$$n = 2\left(\frac{m^* k_B T}{2\pi\hbar^2}\right)^{3/2} e^{-|E_F|/k_B T} \tag{7.45}$$

so that by substitution into (7.42), the Drude formula is recovered

$$\sigma = \frac{ne^2\tau}{m^*} \tag{7.46}$$

for a semiconductor with constant τ and isotropic, parabolic dispersion relations.

To find σ for a semiconductor with more than one spherical carrier pocket, the conductivities per carrier pocket are added

$$\sigma = \sum_i \sigma_i \tag{7.47}$$

where i is the carrier pocket index. We use these simple formulae to make rough estimates for the carrier density and electrical conductivity of semiconductors. For more quantitative analysis, the details of the $E(\mathbf{k})$ relation must be considered, as well as an energy dependent τ and use of the complete Fermi function.

The electrical conductivity and carrier density of a semiconductor with one carrier type exhibits an exponential temperature dependence so that the slope of $\ln\sigma$ vs $1/T$ yields an activation energy (see Fig. 7.3). The plot of $\ln\sigma$ vs $1/T$ is called an "Arrhenius plot". If a plot of $\ln\sigma$ vs $1/T$ exhibits one temperature range with activation energy E_{A1} and a second temperature range with activation energy E_{A2}, then two carrier behavior is suggested. Also in such cases, the activation energies can be extracted from an Arrhenius plot as shown in the schematic of Fig. 7.3.

7.5.1 Ellipsoidal Carrier Pockets

The conductivity results given above for a spherical Fermi surface can easily be generalized to an ellipsoidal Fermi surface which is commonly found in degenerate semiconductors. Semiconductors are degenerate at $T = 0$ when the Fermi level is in the valence or conduction band rather than in the energy band gap.

For an ellipsoidal Fermi surface, we write the electronic dispersion relation as

$$E(\mathbf{k}) = \frac{\hbar^2 k_x^2}{2m_{xx}} + \frac{\hbar^2 k_y^2}{2m_{yy}} + \frac{\hbar^2 k_z^2}{2m_{zz}} \tag{7.48}$$

where the effective mass components m_{xx}, m_{yy} and m_{zz} are appropriate to the band curvatures in the x, y, z directions, respectively. Substitution of

$$k'_\alpha = k_\alpha \sqrt{m_0/m_\alpha} \tag{7.49}$$

for $\alpha = x, y, z$ brings (7.48) into spherical form

$$E(\mathbf{k}') = \frac{\hbar^2 k'^2}{2m_0} \tag{7.50}$$

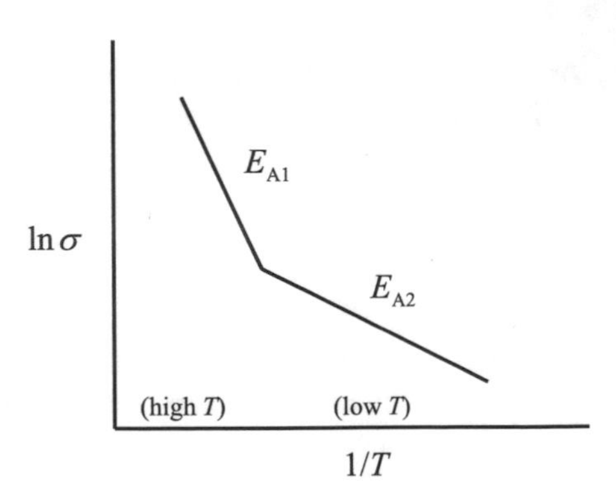

Fig. 7.3 Schematic diagram of an Arrhenius plot of $\ln\sigma$ vs $1/T$ showing two carrier types with different activation energies, indicating that one carrier type dominates transport at low temperature and another carrier type is dominant at high temperatures

where $k'^2 = k'^2_x + k'^2_y + k'^2_z$. For the volume element d^3k in (7.26) we have

$$d^3k = \sqrt{m_{xx}m_{yy}m_{zz}/m_0^3}\, d^3k' \tag{7.51}$$

and the carrier density associated with a single carrier pocket becomes

$$n_i = 2\sqrt{m_{xx}m_{yy}m_{zz}} \left(\frac{k_B T}{2\pi\hbar^2}\right)^{3/2} e^{-|E_F|/k_B T}. \tag{7.52}$$

For an ellipsoidal constant energy surface (see Fig. 7.4), the directions of the electric field, electron velocity and electron acceleration will in general be different. Let (x, y, z) be the coordinate system for the major axes of the constant energy ellipsoid and (X, Y, Z) be the laboratory coordinate system. Then in the laboratory system the current density $\mathbf{j}$ and electric field $\mathbf{E}$ are related by

$$\begin{pmatrix} j_X \\ j_Y \\ j_Z \end{pmatrix} = \begin{pmatrix} \sigma_{XX} & \sigma_{XY} & \sigma_{XZ} \\ \sigma_{YX} & \sigma_{YY} & \sigma_{YZ} \\ \sigma_{ZX} & \sigma_{ZY} & \sigma_{ZZ} \end{pmatrix} \begin{pmatrix} E_X \\ E_Y \\ E_Z \end{pmatrix} \tag{7.53}$$

As an example, suppose that the electric field is applied in the XY plane along the X axis at an angle θ with respect to the x axis of the constant energy ellipsoid (see Fig. 7.4). The conductivity tensor is easily written in the xyz crystal coordinate system where the xyz axes are along the principal axes of the ellipsoid:

$$\begin{pmatrix} j_x \\ j_y \\ j_z \end{pmatrix} = ne^2\tau \begin{pmatrix} 1/m_{xx} & 0 & 0 \\ 0 & 1/m_{yy} & 0 \\ 0 & 0 & 1/m_{zz} \end{pmatrix} \begin{pmatrix} E\cos\theta \\ E\sin\theta \\ 0 \end{pmatrix} \tag{7.54}$$

A coordinate transformation from the crystal axes to the laboratory frame allows us to relate $\overleftrightarrow{\sigma}_{\text{crystal}}$ which we have written easily by (7.54) to $\overleftrightarrow{\sigma}_{\text{Lab}}$ which we measure by (7.53). In general

$$\overleftrightarrow{\sigma}_{\text{Lab}} = R\, \overleftrightarrow{\sigma}_{\text{crystal}}\, R^{-1} \tag{7.55}$$

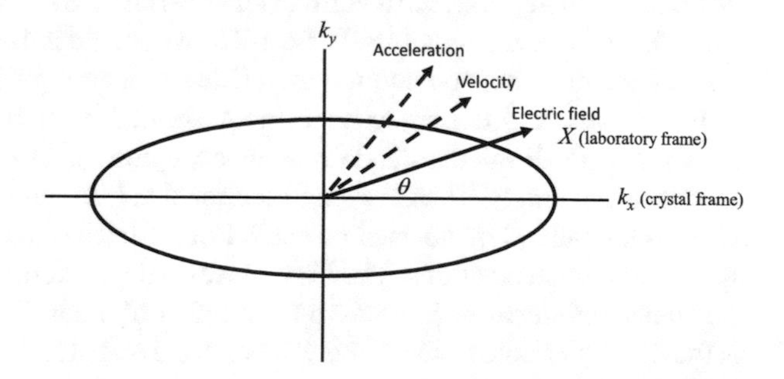

Fig. 7.4 Schematic diagram of an ellipsoidal constant energy surface

where R is the rotational transformation matrix

$$R = \begin{pmatrix} \cos\theta & \sin\theta & 0 \\ -\sin\theta & \cos\theta & 0 \\ 0 & 0 & 1 \end{pmatrix} \tag{7.56}$$

and R^{-1} is the inverse of the rotational transformation matrix

$$R^{-1} = \begin{pmatrix} \cos\theta & -\sin\theta & 0 \\ \sin\theta & \cos\theta & 0 \\ 0 & 0 & 1 \end{pmatrix} \tag{7.57}$$

so that the conductivity tensor $\overleftrightarrow{\sigma}_{\text{Lab}}$ in the lab frame becomes:

$$\overleftrightarrow{\sigma}_{\text{Lab}} = ne^2\tau \begin{pmatrix} \cos^2\theta/m_{xx} + \sin^2\theta/m_{yy} & \cos\theta\sin\theta(1/m_{yy} - 1/m_{xx}) & 0 \\ \cos\theta\sin\theta(1/m_{yy} - 1/m_{xx}) & \sin^2\theta/m_{xx} + \cos^2\theta/m_{yy} & 0 \\ 0 & 0 & 1/m_{zz} \end{pmatrix} \tag{7.58}$$

Semiconductors with ellipsoidal Fermi surfaces usually have several such surfaces located in crystallographically equivalent locations in the Brillouin zone. In the case of cubic symmetry, the sum of the conductivity components results in an isotropic conductivity even though the contribution from each ellipsoid is anisotropic. Thus measurement of the electrical conductivity provides no information on the anisotropy of the Fermi surfaces of cubic materials. However, measurement of the magnetoresistance does provide such information, since the application of a magnetic field gives special importance to the magnetic field direction, thereby lowering the effective crystal symmetry.

7.6 Electrons and Holes in Intrinsic Semiconductors

Intrinsic semiconductors refer to semiconductors with no environmental or intentional doping and no departures from perfect stoichiometry. In this section we consider the symmetry between electron and holes and we show how the Fermi energy is found for such semiconductors. In Sect. 7.7, we consider the corresponding issues for doped semiconductors containing impurities or departures from ideal stoichiometry.

In the absence of doping, carriers are generated by thermal or optical excitations. Thus at $T = 0$, all valence band states are occupied and all conduction band states are unoccupied or empty. Thus, for each electron that is excited into the conduction band, a hole is left behind in the valence band. For intrinsic semiconductors, conduction is by both holes and electrons. The Fermi level is thus determined by the condition that the number of electrons is equal to the number of holes. Writing $g_v(E_h)$ and $g_c(E_e)$ as the density of hole states in the valence band and electron states in the conduction band, respectively, we obtain

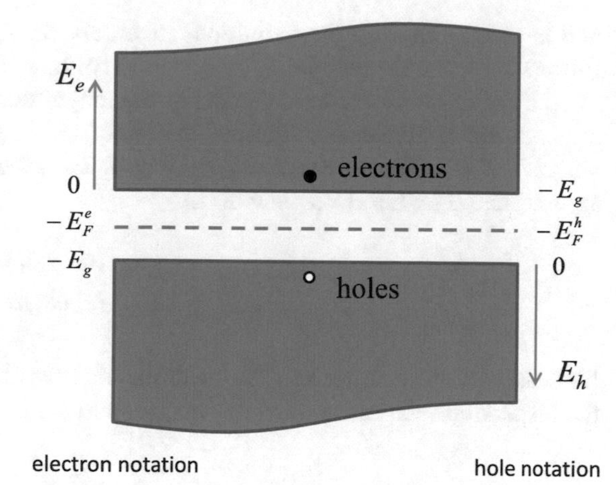

Fig. 7.5 Schematic diagram of the band gap in a semiconductor showing the symmetry of electrons and holes. The diagram is drawn for the case that the electron and hole dispersion relations are the same and the effective masses are the same

$$n_h = \int_0^\infty g_v(E_h)\hat{f}_0(E_h + E_F^h)dE_h = \int_0^\infty g_c(E_e)\hat{f}_0(E_e + E_F^e)dE_e = n_e \quad (7.59)$$

where the notation we have used is shown in Fig. 7.5. Here the energy gap E_g is written as the sum of the Fermi energies for electrons and holes, both taken as positive numbers

$$E_F^e + E_F^h = E_g, \quad (7.60)$$

and the Fermi functions $\hat{f}_0$ are written so as to include explicitly the Fermi energy. The condition $n_e = n_h$ for intrinsic semiconductors is used to determine the position of the Fermi levels for electrons and holes within the band gap. If the band curvatures of the valence and conduction bands are the same, then their effective masses have the same magnitude and E_F lies at midgap. We also derive in this section the general result for the placement of E_F when $m_e^* \neq m_h^*$.

On the basis of this interpretation, the holes obey Fermi statistics as do the electrons, only we must measure the hole energies downward, while electron energies are measured upwards, as indicated in Fig. 7.5. This approach clearly builds on the symmetry relation between electrons and holes. It is convenient to measure electron energies E_e with respect to the bottom of the conduction band E_c so that $E_e = E - E_c$ and to measure hole energies E_h with respect to the top of the valence band E_v so that $E_h = -(E - E_v)$. The Fermi level for the electrons is $-E_F^e$ (measured from the bottom of the conduction band which is taken as $E = 0$) and for holes it is $-E_F^h$ (measured from the top of the valence band which is taken as $E = 0$ for holes), so that E_F^e and E_F^h have positive values. Referring to (7.59), $\hat{f}_0(E_e + E_F^e)$ denotes the Fermi function for electrons where $E - E_F$ is written explicitly

$$\hat{f}_0(E_e + E_F^e) = \frac{1}{1 + \exp[(E_e + E_F^e)/k_B T]} \quad (7.61)$$

and is consistent with the definitions given above. A similar expression to (7.61) follows for the corresponding $\hat{f}_0(E_h + E_F^h)$ for holes.

In a typical intrinsic semiconductor, the magnitudes of the energies E_F^e and E_F^h are both much greater than thermal energies, i.e., $|E_F^e| \gg k_B T$ and $|E_F^h| \gg k_B T$, where $k_B T$ at room temperature is $\sim$25 meV. Thus the distribution functions can be approximated by the Boltzmann form

$$\begin{aligned} \hat{f}_0(E_e + E_F^e) &\simeq e^{-(E_e+E_F^e)/k_B T} \\ \hat{f}_0(E_h + E_F^h) &\simeq e^{-(E_h+E_F^h)/k_B T}. \end{aligned} \quad (7.62)$$

If m_e and m_h are, respectively, the electron and hole effective masses and if we write the dispersion relations around the valence and conduction band extrema as

$$\begin{aligned} E_e &= \hbar^2 k^2/(2m_e) \\ E_h &= \hbar^2 k^2/(2m_h) \end{aligned} \quad (7.63)$$

then the density of states for electrons at the bottom of the conduction band and for holes at the top of the valence band can be written in their respective nearly free electron forms (see (7.64))

$$\begin{aligned} g_c(E_e) &= \tfrac{1}{2\pi^2}\left(2m_e/\hbar^2\right)^{3/2} E_e^{1/2} \\ g_v(E_h) &= \tfrac{1}{2\pi^2}\left(2m_h/\hbar^2\right)^{3/2} E_h^{1/2}. \end{aligned} \quad (7.64)$$

These expressions follow from

$$n = \frac{1}{4\pi^3}\frac{4\pi}{3}k^3 \quad (7.65)$$

and substitution of k via the simple parabolic relation

$$E = \frac{\hbar^2 k^2}{2m^*} \quad (7.66)$$

so that

$$n = \frac{1}{3\pi^2}\left(\frac{2m^* E}{\hbar^2}\right)^{3/2} \quad (7.67)$$

and

$$g(E) = \frac{dn}{dE} = \frac{1}{2\pi^2}\left(\frac{2m^*}{\hbar^2}\right)^{3/2} E^{1/2}. \quad (7.68)$$

Substitution of this density of states expression into (7.45) results in a carrier density

$$n_e = 2\left(\frac{m_e k_B T}{2\pi\hbar^2}\right)^{3/2} e^{-E_F^e/k_B T}. \tag{7.69}$$

Likewise for holes we obtain

$$n_h = 2\left(\frac{m_h k_B T}{2\pi\hbar^2}\right)^{3/2} e^{-E_F^h/k_B T}. \tag{7.70}$$

Thus the famous product rule is obtained

$$n_e n_h = 4\left(\frac{k_B T}{2\pi\hbar^2}\right)^{3} (m_e m_h)^{3/2} e^{-E_g/k_B T} \tag{7.71}$$

where $E_g = E_F^e + E_F^h$. But for an intrinsic semiconductor $n_e = n_h$. Thus by taking the square root of the above expression, we obtain both n_e and n_h separately

$$n_e = n_h = 2\left(\frac{k_B T}{2\pi\hbar^2}\right)^{3/2} (m_e m_h)^{3/4} e^{-E_g/2k_B T}. \tag{7.72}$$

Comparison with the expressions given in (7.69) and (7.70) for n_e and n_h allows us to solve for the Fermi levels E_F^e and E_F^h

$$n_e = 2\left(\frac{m_e k_B T}{2\pi\hbar^2}\right)^{3/2} e^{-E_F^e/k_B T} = 2\left(\frac{k_B T}{2\pi\hbar^2}\right)^{3/2} (m_e m_h)^{3/4} e^{-E_g/2k_B T} \tag{7.73}$$

so that

$$\exp(-E_F^e/k_B T) = (m_h/m_e)^{3/4} \exp(-E_g/2k_B T) \tag{7.74}$$

and

$$E_F^e = \frac{E_g}{2} - \frac{3}{4} k_B T \ln(m_h/m_e). \tag{7.75}$$

If $m_e = m_h$, we obtain the simple result that $E_F^e = E_g/2$ which says that the Fermi level lies in the middle of the energy gap. However, if the masses are not equal, E_F will lie closer to the band edge with higher curvature, thereby enhancing the Boltzmann factor term in the thermal excitation process, to compensate for the lower density of states for the higher curvature band.

If however $m_e \ll m_h$, the Fermi level approaches the conduction band edge and the full Fermi functions have to be considered. In this case

$$n_e = \frac{1}{2\pi^2}\left(\frac{2m_e}{\hbar^2}\right)^{3/2} \int_{E_c}^{\infty} \frac{(E - E_c)^{1/2} dE}{\exp[(E - E_F^e)/k_B T] + 1} \equiv N_e F_{1/2}\left(\frac{E_F^e - E_c}{k_B T}\right) \tag{7.76}$$

where E_c denotes the bottom of the conduction band, E_F^e is the Fermi energy for electrons which here is allowed the possibility of moving up into the conduction band

(and therefore its sign cannot be predetermined), and N_e is the "effective electron density" which, in accordance with (7.69), is given by

$$N_e = 2\left(\frac{m_e k_B T}{2\pi\hbar^2}\right)^{3/2}. \tag{7.77}$$

The Fermi integral in (7.76) is written in standard form as

$$F_j(\eta) = \frac{1}{j!}\int_0^\infty \frac{x^j dx}{\exp(x-\eta)+1}. \tag{7.78}$$

We can take $F_j(\eta)$ from the tables in Blakemore, "Semiconductor Statistics" (Appendix B). For the semiconductor limit ($\eta < -4$), then $F_j(\eta) \to \exp(\eta)$. Clearly, when the full version of $F_j(\eta)$ is required to describe the carrier density, then $F_j(\eta)$ is also needed to describe the conductivity. These refinements are important for a detailed solution of the transport properties of semiconductors over the entire temperature range of interest, and eventually in the presence of doping.

7.7 Donor and Acceptor Doping of Semiconductors

In general a semiconductor has electron and hole carriers due to the presence of impurities, whether intentional or otherwise, as well as from thermal excitation processes. For many applications, impurities are intentionally introduced to generate carriers: donor impurities to generate electrons in n-type semiconductors, and acceptor impurities to generate holes in p-type semiconductors. Assuming for the moment that each donor contributes one electron to the conduction band, then the donors can contribute an excess carrier concentration up to N_d, where N_d is the donor impurity concentration. Similarly, if every acceptor contributes one hole to the valence band, then the excess hole concentration will be N_a, where N_a is the acceptor impurity concentration. In general, the semiconductor is **partly compensated**, which means that both donor and acceptor impurities are present, thereby giving a partial cancellation to the net carrier concentration. Furthermore, at finite temperatures, the donor and acceptor levels will be partially occupied, so that somewhat less than the maximum charge will be released as **mobile** charge into the conduction and valence bands. The density of electrons bound to a donor site n_d is found from the grand canonical ensemble in statistical mechanics as

$$\frac{n_d}{N_d} = \frac{\sum_j N_j e^{-(E_j-\mu N_j)/k_B T}}{\sum_j e^{-(E_j-\mu N_j)/k_B T}} \tag{7.79}$$

where E_j and N_j are, respectively, the energy and number of electrons that can be placed in state j, and μ is the chemical potential (Fermi energy). Referring to

Table 7.1 Occupation of impurity states in the grand canonical ensemble

States	N_j	Spin	E_j
1	0	–	0
2	1	↑	$-E_d$
3	1	↓	$-E_d$
4	2	↑↓	$-E_d + E_{\text{Coulomb}}$

Table 7.1, the system can be found in one of three states: one where no electrons are present in state j (hence no contribution is made to the energy), and two states where one electron is present (one with spin ↑, the other with spin ↓) corresponding to the donor energy E_d, where the sign is included directly so that E_d has a positive energy value. Placing two electrons in the same energy state would result in a very high energy because of the Coulomb repulsion between the two electrons; therefore this possibility is neglected in practical calculations. Writing either $N_j = 0, 1$ for the 3 states of importance, we obtain for the relative ion concentration of occupied donor sites

$$\frac{n_d}{N_d} = \frac{2e^{-(\varepsilon_d-\mu)/k_B T}}{1+2e^{-(\varepsilon_d-\mu)/k_B T}} = \frac{1}{1+\frac{1}{2}e^{(\varepsilon_d-\mu)/k_B T}} = \frac{1}{1+\frac{1}{2}e^{-(E_d-E_F^e)/k_B T}} \tag{7.80}$$

in which E_d and E_F^e are positive numbers, but lie below the zero of energy which is taken to be at the bottom of the conduction band. The energy ε_d denotes the energy for the donor level and is a negative number relative to the zero of energy.

Consequently, the concentration of electrons thermally ionized into the conduction band will be

$$N_d - n_d = \frac{N_d}{1+2e^{(E_d-E_F^e)/k_B T}} = n_e - n_h \tag{7.81}$$

where n_e and n_h are the mobile electron and hole concentrations. At low temperatures, where $E_d \sim k_B T$, almost all of the carriers in the conduction band will be generated by the ionized donors, so that $n_h \ll n_e$ and $(N_d - n_d) \simeq n_e$. The Fermi level will then adjust itself so that $N_d - n_d \simeq n_e$. From (7.45) and (7.81) the following equation determines E_F^e:

$$n_e = 2\left(\frac{m_e k_B T}{2\pi\hbar^2}\right)^{3/2} e^{-E_F^e/k_B T} \simeq \frac{N_d}{1+2e^{(E_d-E_F^e)/k_B T}}. \tag{7.82}$$

Solution of (7.82) shows that the presence of the ionized donor carriers moves the Fermi level up above the middle of the band gap and close to the bottom of the conduction band. For the donor impurity problem, the Fermi level will be close to the position of the donor level E_d, as shown in Fig. 7.6.

The position of the Fermi level also varies with temperature. Fig. 7.6 assumes that almost all the donor electrons (or acceptor holes) are ionized and are in the conduction

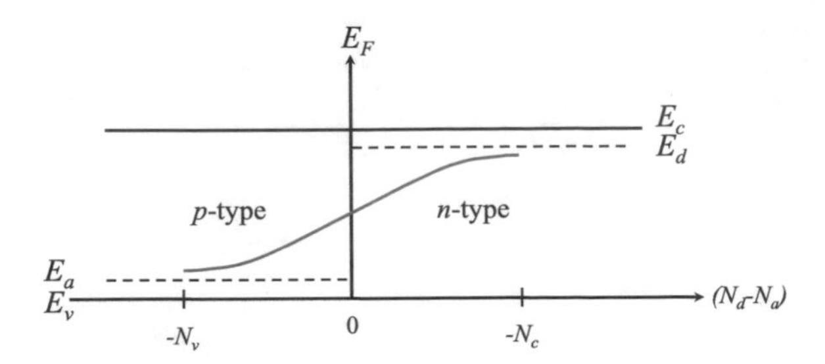

Fig. 7.6 Variation of the Fermi energy (E_F) with donor and acceptor concentrations, shown by the solid curve. For a heavily doped n-type semiconductor, E_F is close to the donor level E_d, while for a heavily doped p-type semiconductor, E_F is close to E_a. This plot is made assuming almost all the donor and acceptor states are ionized

band, which is typical of temperatures where the n-type (or p-type) semiconductor would be used for electron (or hole) transport.

Figure 7.7a shows the dependence of the Fermi level on temperature. Here T_1 denotes the temperature at which the thermal excitation of intrinsic electrons and holes become important, and T_1 is normally a high temperature. In contrast, T_2 is normally a very low temperature and denotes the temperature below which donor-generated electrons begin to freeze out in impurity level bound states and no longer contribute to conduction. This carrier freeze-out is illustrated in Fig. 7.7b. In the temperature range $T_2 < T < T_1$, the Fermi level in Fig. 7.7a falls as T increases according to

$$E_F = E_c - k_B T \ln(N_c/N_d) \tag{7.83}$$

where $N_c = 2m_e k_B T/(2\pi\hbar^2)$. In Fig. 7.7b we see the temperature dependence of the carrier concentration in the intrinsic range ($T > T_1$), the saturation range ($T_2 < T < T_1$), and finally the low temperature range ($T < T_2$) where carriers freeze out into bound states in the impurity band at E_d. The plot of the electron density n in Fig. 7.7b is presented as a function of (1000/T) and the corresponding temperature values are shown on the upper scale of the figure.

For the case of acceptor impurities, an ionized acceptor level releases a hole into the valence band, or alternatively, an electron from the valence band gets excited into an acceptor level, leaving a hole behind. At very low temperature, the acceptor levels are filled with holes under freeze-out conditions. Because of hole-hole Coulomb repulsion, we can place no more than one hole in each acceptor level. A singly occupied hole can have either spin up or spin down. Thus for the acceptor levels, a formula analogous to (7.80) for donors is obtained for the occupation of an acceptor level

$$\frac{n_a}{N_a} = \frac{1}{1 + \frac{1}{2}e^{-(E_a - E_F^h)/k_B T}} \tag{7.84}$$

so that the essential symmetry between holes and electrons is maintained. To obtain the hole concentration in the valence band, we use a formula analogous to (7.81).

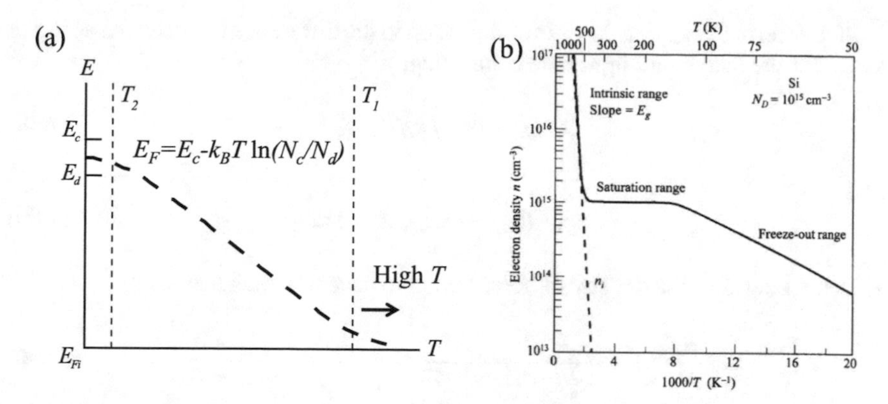

Fig. 7.7 **a** Temperature dependence of the Fermi energy for an n-type doped semiconductor. See the text for definitions of T_1 and T_2. Here E_{Fi} denotes the position of the Fermi level in the high temperature limit where the thermal excitation of carriers far exceeds the electron density contributed by the donor impurities. **b** Temperature dependence of the electron density for Si doped with 10^{15} cm^{-3} donors where the electron density is plotted as a function of 1000/T in units of K^{-1}. Reprinted with permission from John Wiley & Sons Inc, Sze and Ng, Physics of Semiconductor Devices

A situation which commonly arises for the acceptor levels relates to the degeneracy of the valence bands for group IV and III–V compound semiconductors. We will illustrate the degenerate valence band in the case where spin-orbit interaction is considered (which is usually the situation that is relevant for opto-electronic applications). Under strong spin-orbit interaction we have a degenerate heavy and light hole band and a lower lying split-off band. The two degenerate bands are only weakly coupled, so that we can approximate the impurity acceptor levels by hydrogenic acceptor levels for the heavy hole $\varepsilon_{a,h}$ and light hole $\varepsilon_{a,l}$ bands. In this case the split-off band does not contribute significantly because it lies much lower in energy. The density of holes bound to both types of acceptor sites is given by

$$\frac{n_a}{N_a} = \frac{\sum_j N_j e^{-(E_j - \mu N_j)/k_B T}}{\sum_j e^{-(E_j - \mu N_j)/k_B T}}, \tag{7.85}$$

following (7.79), where we note that the heavy hole and light hole bands can each accommodate one spin up and one spin down electron for each wavevector k. Using the same arguments as above, we obtain:

$$\frac{n_a}{N_a} = \frac{2e^{-(\varepsilon_{a,l} - \mu)/k_B T} + 2e^{-(\varepsilon_{a,h} - \mu)/k_B T}}{1 + 2e^{-(\varepsilon_{a,l} - \mu)/k_B T} + 2e^{-(\varepsilon_{a,h} - \mu)/k_B T}} \tag{7.86}$$

so that

$$\frac{n_a}{N_a} = \frac{1 + e^{-(\varepsilon_{a,h} - \varepsilon_{a,l})/k_B T}}{1 + \frac{1}{2}e^{(\varepsilon_{a,l} - \mu)/k_B T} + e^{-(\varepsilon_{a,h} - \varepsilon_{a,l})/k_B T}}. \tag{7.87}$$

If the thermal energy is large in comparison to the difference between the acceptor levels for the heavy and light hole bands, then

$$[(\varepsilon_{a,h} - \varepsilon_{a,l})/k_B T] \ll 1 \tag{7.88}$$

and

$$\exp[-(\varepsilon_{a,h} - \varepsilon_{a,l})/k_B T] \simeq 1 \tag{7.89}$$

so that the ratio of the density of holes bound to acceptor sites becomes

$$\frac{n_a}{N_a} \simeq \frac{1}{1 + \frac{1}{4} e^{-(\varepsilon_{a,l} - \mu)/k_B T}} = \frac{1}{1 + \frac{1}{4} e^{(E_a - E_F^h)/k_B T}} \tag{7.90}$$

where E_a and E_F^h are positive values corresponding to $\varepsilon_{a,l}$ and μ, respectively. From (7.81) and (7.90), the temperature dependence of E_F can be calculated for the case of doped semiconductors considering the doping by donor and acceptor impurities, either separately or at the same time. Figure 7.6 shows the doping dependence of E_F for p-doped semiconductors as well as for n-doped semiconductors.

7.8 Characterization of Semiconductors

In describing the electrical conductivity of semiconductors, it is customary to write the conductivity as

$$\sigma = n_e |e| \mu_e + n_h |e| \mu_h \tag{7.91}$$

in which n_e and n_h are the carrier densities for the carriers, and μ_e and μ_h are their **mobilities**. We have shown in (7.46) that for cubic materials the static conductivity can under certain approximations be written as

$$\sigma = \frac{ne^2 \tau}{m^*} \tag{7.92}$$

for each carrier type, so that the mobilities and effective masses are related by

$$\mu_e = \frac{|e| \langle \tau_e \rangle}{m_e} \tag{7.93}$$

and

$$\mu_h = \frac{|e| \langle \tau_h \rangle}{m_h} \tag{7.94}$$

which show that materials with small effective masses have high mobilities. By writing the electrical conductivity as a product of the carrier density with the mobility, it

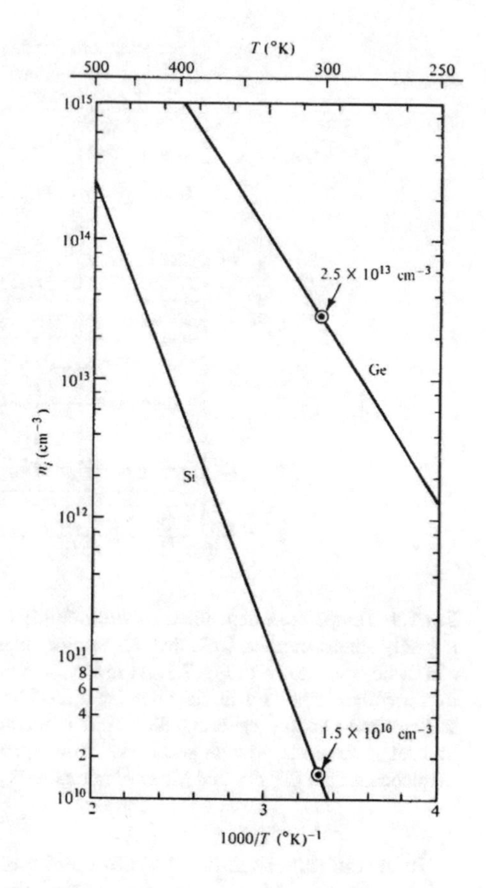

Fig. 7.8 Temperature dependence of the electron concentration for intrinsic Si and Ge in the range $250 < T < 500$ K. Circles indicate the doping levels that must be exceeded to have extrinsic carriers dominate over thermally excited carriers at 300 K. Reprinted with permission from John Wiley & Sons Inc, Sze and Ng, Physics of Semiconductor Devices

is easy to contrast the temperature dependence of σ for metals and semiconductors. For metals, the carrier density n is essentially independent of T, while μ is temperature dependent. In contrast, n for semiconductors is highly temperature dependent in the intrinsic regime [see Fig. 7.7b] and μ is relatively less temperature dependent. Fig. 7.8 shows the carrier concentration for intrinsic Si and Ge in the neighborhood of room temperature ($250 < T < 500$ K), demonstrating the rapid increase of the carrier concentration with increasing temperature. These values of n indicate the doping levels necessary to exceed the intrinsic carrier level to a desired level at a given temperature. Fig. 7.9 shows the mobility for n-type Si samples with various impurity levels. The observed temperature dependence can be explained by the different temperature dependences of the impurity scattering and phonon scattering mechanisms (see Fig. 7.10).

A table of typical mobilities for semiconductors is given in Table 7.2. By way of comparison, μ for copper at room temperature is 35 cm^2/V·s. When using conductivity formulae in esu units, remember that the mobility is expressed in cm^2/V·s and that all the numbers in Table 7.2 have to be multiplied by 300 to match the units given here in this book.

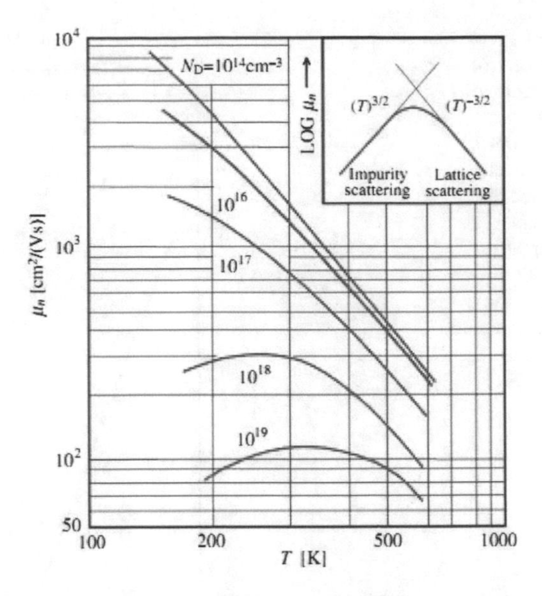

Fig. 7.9 Temperature dependence of the mobility for n-type Si for a series of samples with different impurity concentrations. Note that the mobility is not as strong a function of temperature as is the carrier density shown in Fig. 7.8. At low temperature, impurity scattering by the donor impurity ions becomes important, as shown in the inset. The different temperature dependences of impurity and electron-phonon scattering allows one to identify the important scattering mechanisms experimentally. Reproduced with permission from Springer-Verlag, Yu and Cardona, Fundamentals of Semiconductors: Physics and Materials Properties

In the characterization of a semiconductor for device applications, researchers are expected to provide information on the carrier density and mobility, preferably as a

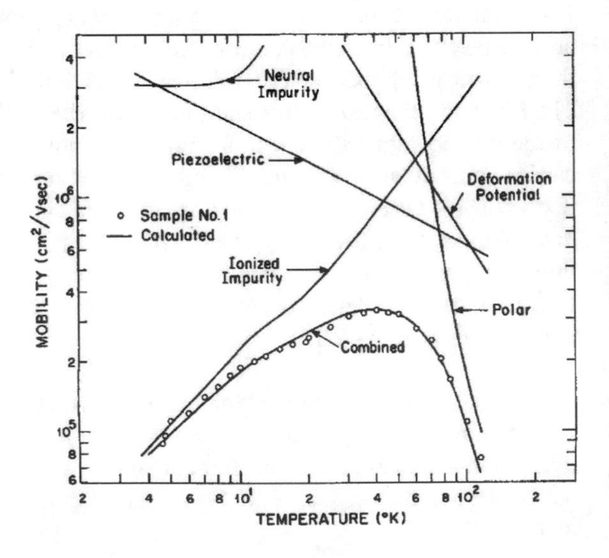

Fig. 7.10 Temperature dependence of the mobility for n-type GaAs, showing the separate and combined carrier scattering processes. Reprinted with permission from J. Appl. Phys. 41, 3088 (1970)

Table 7.2 Mobilities for some typical semiconductors at room temperature in units of $cm^2/V \cdot s$

Crystal	Electrons	Holes	Crystal	Electrons	Holes
Diamond	1800	1200	GaAs	8000	300
Si	1350	480	GaSb	5000	1000
Ge	3600	1800	PbS	550	600
InSb	77,000	750	PbSe	1020	930
InAs	30,000	460	PbTe	2500	1000
InP	4600	100	AgCl	50	–
AlAs	280	–	KBr (100K)	100	–
AlSb	900	400	SiC	100	10–20

Table 7.3 Semiconductor effective masses of electrons and holes in direct gap semiconductors

Crystal	Electron	Holes			Spin-orbit
	m_e/m_0	m_{hh}/m_0	m_{lh}/m_0	m_{soh}/m_0	Δ(eV)
InSb	0.015	0.39	0.021	(0.11)	0.82
InAs	0.026	0.41	0.025	0.08	0.43
InP	0.073	0.4	(0.078)	(0.15)	0.11
GaSb	0.047	0.3	0.06	(0.14)	0.80
GaAs	0.066	0.5	0.082	0.17	0.34

function of temperature. Such plots are shown in Figs. 7.8 and 7.9. When presenting characterization data in condensed form, the carrier density and mobility of semiconductors are traditionally given at 300 and 77 K. Other information on the values of the parameters commonly used in semiconductor physics are values of the effective masses (Table 7.3) and of the energy gaps (Table 7.4).

Table 7.4 Semiconductor energy gaps between the valence and conduction bands

Crystal	Gap[a]	E_g, eV		Crystal	Gap[a]	E_g, eV	
		0 K	300 K			0 K	300 K
Diamond	*i*	5.4	–	HgTe[b]	*d*	–	–
Si	*i*	1.17	1.11	PbS	*d*	0.286	0.34-0.37
Ge	*i*	0.744	0.66	PbSe	*i*	0.165	0.27
αSn	*d*	0.00	0.00	PbTe	*i*	0.190	0.29
InSb	*d*	0.23	0.17	CdS	*d*	2.582	2.42
InAs	*d*	0.43	0.36	CdSe	*d*	1.840	1.74
InP	*d*	1.42	1.27	CdTe	*d*	1.607	1.44
GaP	*i*	2.32	2.25	ZnO	–	3.436	3.2
GaAs	*d*	1.52	1.43	ZnS	–	3.91	3.6
GaSb	*d*	0.81	0.68	SnTe	*d*	0.3	0.18
AlSb	*i*	1.65	1.6	AgCl	–	–	3.2
SiC(hex)	*i*	3.0	–	AgI	–	–	2.8
Te	*d*	0.33	–	Cu_2O	*d*	2.172	–
ZnSb	–	0.56	0.56	TiO_2	–	3.03	–

[a]The indirect gap is labeled by i, and the direct gap is labeled by d

[b]HgTe is a zero gap semiconductor, and because of non-ideal stoichiometry, the Fermi level may be in the valence or conduction band

Problems

7.1 Suppose that you have a semimetal with three equivalent electron pockets along the $(1, \bar{1}, \bar{1})$, $(\bar{1}, 1, \bar{1})$ and $(\bar{1}, \bar{1}, 1)$ directions with

$$E(k) = \frac{\hbar^2 k_{el}^2}{2m_{el}} + \frac{\hbar^2 k_{et}^2}{2m_{et}}$$

(where $m_{el} = m_0$ and $m_{et} = 0.01m_0$), and a single hole pocket along the (1,1,1) direction

$$E(k) = \frac{\hbar^2 k_{hl}^2}{2m_{hl}} + \frac{\hbar^2 k_{ht}^2}{2m_{ht}}$$

(where $m_{hl} = m_0$ and $m_{ht} = 0.1m_0$). We note here that this semimetal (which is a simplification of the Bi electronic structure) does not strictly have cubic symmetry, although we use a cubic coordinate system for simplicity. Use these values for the longitudinal and transverse effective mass components for electrons and holes, and assume the band overlap for electrons and holes is 100 meV:

(a) Find the carrier density for electrons and holes for this semimetal at $T = 0$ and at a low but finite temperature T at $T = 10$ K and $T = 100$ K. State your definition of low T.
(b) Find the Fermi energy for electrons and holes.

(c) Find the electrical conductivity in the (111) direction at temperature T, using your results from (a).
(d) What are the qualitative differences between the electrical conductivity for a semimetal (band overlap of 100 meV) and a semiconductor (band gap of 100 meV) keeping all other band parameters the same?

7.2 (a) Suppose that we apply stress to a highly degenerate Si sample where the Fermi level lies at an energy E_F above the conduction band minimum. Assume that the stress Σ is applied along the (1,0,0) direction such that the conduction band along the stress direction is lowered, while maintaining the same indirect band gap E_g for the $(0, \Delta, 0)$ and $(0, 0, \Delta)$ conduction band minima as before the stress was applied. Assume that the change in band gap with stress for the $(\Delta, 0, 0)$ minima is linear with stress ($\partial E_g/\partial \Sigma = \alpha \Sigma$) and assume for simplicity that the total carrier density $n = 6 \times 10^{18}/\text{cm}^3$ is independent of stress. Write an expression for the stress level at which all the carriers will be in the $(\Delta, 0, 0)$ conduction band minima at $T = 0\,\text{K}$. Explain qualitatively the effect of increasing temperature on this system.
(b) What is the electrical conductivity for the Si sample in part (a) when all the electrons are in the (100) and $(\bar{1}00)$ carrier pockets and when the electric field is oriented along the $\mathbf{E} \parallel (011)$ direction? Assume that the longitudinal and transverse effective mass components are $m_l = 0.92m_0$ and $m_t = 0.19m_0$, respectively, and for simplicity assume that the effective mass components do not change with stress.

7.3 The valence band edge of lead telluride (PbTe) consists of 4 degenerate ellipsoidal hole pockets at the L-points (masses $m_t = 0.03m_0$ and $m_\ell = 0.24m_0$). For this simplified version of PbTe, assume that the conduction band edge is a mirror image of the valence band edge (i.e., the same masses and degeneracies and that the L-points in the Brillouin zone are all equivalent initially). The direct band gap E_g and the static dielectric constant ε_0 are 0.310 eV and 412, respectively at 300 K, and assume for simplicity that E_g and ε_0 are independent of temperature. Evaluate the effect of T and dopant concentration on E_F:

(a) First find E_d, the energy of the donor level. At which temperature range are most of the donor states ionized?
(b) Find $E_F(0\,\text{K})$, $E_F(300\,\text{K})$, and $n_i(300\,\text{K})$ for intrinsic PbTe.
(c) Find $E_F(0\,\text{K})$, $E_F(300\,\text{K})$, and $n_i(300\,\text{K})$ for lightly doped ($N_d = 10^{16}$ carriers/cm^3) PbTe. Justify your choice of the distribution function for this doping level.
(d) Repeat (c) for heavily doped ($N_d = 10^{19}$ carriers/cm^3) PbTe, assuming that the carrier masses do not change with doping.

7.4 Silicon crystallizes in the diamond structure. The lowest conduction band is at a Δ–point of the F.C.C. Brillouin zone 0.85 of the distance from Γ to X along a (100) direction, and its equivalent directions in the Brillouin zone. Six (6) ellipsoidal constant energy surfaces (ellipsoids of revolution) are formed, following the dispersion relation:

$$E(\mathbf{k}) = \frac{\hbar^2 k_\ell^2}{2m_\ell} + \frac{\hbar^2(k_{t1}^2 + k_{t2}^2)}{2m_t}$$

where m_ℓ is the longitudinal effective mass component along the ΓX direction and m_t is the transverse effective mass component perpendicular to this direction. Assume that each carrier pocket contains 10^{18} electrons/cm^3 and $\tau = 10^{-14}$ s is the relaxation time. For $(m_\ell/m_0) = 0.98$ and $(m_t/m_0) = 0.19$, find the contribution to the electrical conductivity from a single carrier pocket (labeled [1] in Fig. 7.11) for the following cases:

(a) The electric field $\mathbf{E} \parallel (001)$.
(b) The electric field $\mathbf{E} \parallel (111)$.
(c) Find the contribution to the electrical conductivity from all 6 electron pockets for $\mathbf{E} \parallel (001)$ and $\mathbf{E} \parallel (111)$ [see (a) and (b)]. It can generally be shown that, for cubic materials, σ is independent of the direction of the applied electric field.

7.5 Consider the direct band gap semiconductor GaAs (at room temperature $E_g =$ 1.43 eV) with 10^{16} hydrogenic donor impurities/cm^3.

(a) Write an expression for the temperature dependence of the Fermi level. Take $E_g = 1.42$ eV for the direct band gap, $m_e^* = 0.07m_0$ for the conduction band mass and $m_{\text{hh}} = 0.68m_0$, $m_{\text{lh}} = 0.12m_0$ for the heavy and light holes bands and ignore the effect of the split–off band $\Delta = 0.33$ eV below the top of the valence band. The static dielectric constant for GaAs is 15.
(b) Find the value of the Fermi energy at 300 K and at 30 K.
(c) What are the electron and hole carrier concentrations at room temperature (300 K)? at 30 K?
(d) Estimate the hole concentration in the split–off band ($m_{\text{soh}}^* = 0.20m_0$). This calculation should justify the neglect of the split–off band in the calculation in (a).
(e) Why must the spin orbit interaction be included when considering acceptor doping in GaAs?

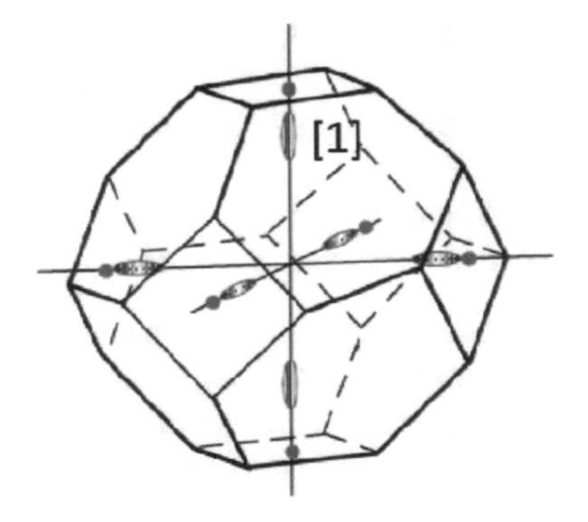

Fig. 7.11 Carrier pockets for silicon within the Brillouin zone

(f) At what doping concentration would you expect a substitutional impurity in GaAs to start forming an impurity band for the electron? How does an impurity band differ from an isolated impurity level?

(g) The bandgap for the indirect gap semiconductor AlAs (with the same crystal structure and nearly identical lattice constants as the direct gap semiconductor GaAs) is 2.13 eV. Why is it not correct to use Vegard's law to find the bandgap for $Ga_{0.9}Al_{0.1}As$? Hint: Vegard's law is an interpolation formula for property X given by

$$X(x) = X(0) + x[X(1) - X(0)]$$

where x is the concentration of Al in the alloy $Ga_{1-x}Al_xAs$.

7.6 The direct band gap semiconductor GaAs has a band gap of 1.43 eV at 300 K, and a second higher-lying conduction band extremum at the X-point in the Brillouin zone is 0.35 eV above the Γ point conduction band minimum. Assume effective masses $m_e(\Gamma) = 0.065m_0$ for the Γ-point electrons, $m_l(X) = 1.2m_0$ and $m_t(X) = 0.3m_0$ for the X-point electrons, and $m_{hh}(\Gamma) = 0.5m_0$ and $m_{lh}(\Gamma) = 0.12m_0$ for the Γ-point heavy and light holes. Assume all bands are parabolic.

(a) With intrinsic material and only thermal excitation, at what temperature and carrier concentration will the number of Γ point electrons equal the number of X-point electrons?

(b) For n-type material at $T = 0$ K, what is the carrier concentration just before the X-point electrons start to fill? Assume a step function Fermi distribution.

(c) Where does the Fermi level lie for cases (a) and (b)?

(d) Physically, what happens to the temperature dependence of the Fermi level as the X-point electrons start to fill? Indicate the conditions that must be satisfied when both Γ-point and X-point electrons are present?

(e) Compare the electrical conductivity of case (a) and case (b). Include contributions from all carriers in both cases.

7.7 Suppose that you have just discovered a new cubic compound semiconductor called novelite which crystallizes in the zincblende structure, a common crystal structure for III–V compound semiconductors such as GaAs. Suppose that the valence band maxima are at the Γ point ($k = 0$) are associated with constant energy surfaces that are approximately spherical and consist of a heavy hole band ($m^*_{hh} = m_0$), a degenerate light hole band ($m^*_{lh} = m_0/3$), and a split-off band ($m^*_{soh} = m_0/5$) which is 0.1 eV below the valence band maximum. Suppose that the conduction band minima are at all Σ points crystallographically equivalent to $(0.85,0.85,0)\pi/a$. These equivalent points are all located along all the equivalent $\{110\}$ directions (ΓK). Assume the following dispersion relations for the electrons

$$E_c(\mathbf{k}) = \frac{\hbar^2}{2}\left(\frac{k_1^2}{m_1^*} + \frac{k_2^2}{m_2^*} + \frac{k_3^2}{m_3^*}\right)$$

where the $\mathbf{k}_1$ and $\mathbf{k}_2$ vectors are, respectively, along the (110) and $(1\bar{1}0)$ directions, and $\mathbf{k}_3$ is along the (001) direction. Assume that along the longitudinal (110) direction $m_1^* = m_0$, and along the directions transverse to this direction take $m_2^* = m_0/3$ and $m_3^* = m_0/9$.

(a) If each electron carrier pocket has n electrons, find an expression for the minimum k_F for a single electron ellipsoid in terms of n and the effective mass components.
(b) What is the length of k_F along the (111) direction relative to the minimum k_F?
(c) How many equivalent electron ellipsoids are there?
(d) Find an expression for the density of states for the electrons.
(e) Find an expression for the density of states for the holes.
(f) Find an expression for the Fermi energy at temperature T if no dopants are added and all the carriers are generated thermally.
(g) Assuming equal relaxation times for all the holes and electrons, find the relative contributions of each electron and hole carrier type to the electrical conductivity (assume $\mathbf{E} \parallel (110)$).

7.8 Suppose that you have an fcc semimetal with 2 atoms per unit cell. Suppose that the electrons are at the L points, $\pi/a(1,1,1)$, ($m_l^* = 0.3m_0$ and $m_t^* = 0.1m_0$) in the Brillouin zone and the holes are in a single carrier pocket at the Γ point ($k = 0$) with $m_h^* = 0.3m_0$, and assume that the energy overlap for this semimetal is 10 meV.

(a) Find the position of the Fermi level for the 3 dimensional semimetal at $T = 0$.
(b) What is the smallest phonon wavevector required for inter-pocket scattering between the electron pockets for this 3D semimetal?
(c) What is the smallest vector for inter-pocket scattering if umklapp processes were to occur?
(d) Suppose that the semimetal is now prepared as a thin layer (quantum well) between alkali halide insulating barriers with the (001) crystalline direction normal to the thin layer of the semimetal (layer thickness = 50 Å). Find an expression for the energy of the lowest subband of the L point electrons as a function of the semimetal layer thickness.
(e) At what layer thickness does the thin film experience a semimetal-semiconductor transition?
(f) Suppose that the semimetal thin film is grown along a (111) direction. Is the layer thickness for the semimetal-semiconductor transition larger, smaller or the same as in (d)? Why?
(g) Design an experiment that will tell you that the semimetal-semiconductor transition has occurred.

7.9 Suppose that we apply a stress along the $(11\bar{1})$ direction of germanium so that all the electron carriers will be thermally excited into a single $(11\bar{1})$ carrier pocket. For germanium use $m_l = 1.58m_0$ and $m_t = 0.082m_0$ for the conduction band ellipsoids and $m_{hl} = 0.04m_0$ and $m_{hh} = 0.3m_0$ for the light and heavy holes.

(a) If an electric field is now applied along the (112) direction, find the magnitude of the conductivity for thermally excited carriers in intrinsic germanium. Consider the contribution of both electrons and holes (ignore the effect of stress on the hole bands).

(b) Along which direction does the current flow?

(c) What is the smallest wave vector for scattering an electron from one electron carrier pocket to another in germanium in the absence of stress?

Suggested Readings

J.M. Ziman, *Principles of the Theory of Solids* (Cambridge University Press, 1972). Chaps. 7 and 9

N.W. Ashcroft, N.D. Mermin, *Solid State Physics* (Holt, Rinehart and Winston, 1976). Chap. 13

A.C. Smith, J.F. Janak, R.B. Adler, *Electronic Conduction in Solids* (McGraw-Hill, 0, 1967). Chaps. 7, 8, and 9

Chapter 8
Thermal Transport

8.1 Introduction

The electrons in solids not only conduct electricity but also conduct heat, as they transfer energy from a hot junction to a cold junction. Just as the electrical conductivity characterizes the response of a material to an applied voltage, the thermal conductivity likewise characterizes the material with regard to heat flow in the presence of a temperature gradient. In fact, the electrical conductivity and thermal conductivity are coupled, since thermal conduction also transports charge and electrical conduction also transports energy. This coupling between electrical and thermal transport gives rise to thermo-electricity. In this chapter, we discuss first the thermal conductivity for metals, semiconductors and insulators and then consider the coupling between electrical and thermal transport which gives rise to thermoelectric phenomena. In Chap. 9, we discuss scattering mechanisms for electrons and phonons which relates closely to thermal transport.

8.2 Thermal Conductivity

8.2.1 General Considerations

Thermal transport in most materials systems, like electrical transport, follows from the Boltzmann equation. We will first derive a general expression for the electronic contribution to the thermal conductivity using Boltzmann's equation. We will then apply this general expression to find the thermal conductivity for metals and then for semiconductors. The total thermal conductivity $\overleftrightarrow{\kappa}$ of any material is, of course, the superposition of the electronic part $\overleftrightarrow{\kappa}_e$ with the lattice part $\overleftrightarrow{\kappa}_L$:

$$\overleftrightarrow{\kappa} = \overleftrightarrow{\kappa}_e + \overleftrightarrow{\kappa}_L \, . \tag{8.1}$$

M. Dresselhaus et al., *Solid State Properties*, Graduate Texts in Physics,
https://doi.org/10.1007/978-3-662-55922-2_8

We now consider the calculation of the electronic contribution to the thermal conductivity. The lattice contribution to the thermal conductivity is considered for the case of insulators in Sect. 8.2.4 where the contribution from the lattice vibration is discussed in terms of phonons which are the carriers of thermal energy. The application of a temperature gradient to a solid gives rise to a flow of heat. We define $\mathbf{U}$ as the thermal current that is driven by the thermal energy $E - E_F$, which in turn is defined as the excess energy of an electron above the equilibrium energy E_F, where is the Fermi energy. Neglecting time dependent effects, we define $\mathbf{U}$ as

$$\mathbf{U} = \frac{1}{4\pi^3} \int \mathbf{v}\,(E - E_F)\; f(\mathbf{r}, \mathbf{k}) d^3k \tag{8.2}$$

where the distribution function $f(\mathbf{r}, \mathbf{k})$ is related to the Fermi function f_0 by $f = f_0 + f_1$. Under equilibrium conditions, there is no thermal current density

$$\int \mathbf{v}(E - E_F) f_0 \, d^3k = 0 \tag{8.3}$$

so that the thermal current is driven by the thermal gradient which causes a departure from the equilibrium distribution occurring at the Fermi energy E_F:

$$\mathbf{U} = \frac{1}{4\pi^3} \int \mathbf{v}(E - E_F) f_1 \, d^3k. \tag{8.4}$$

Here the electronic contribution to the thermal conductivity tensor $\overleftrightarrow{\kappa}_e$ is defined by the relation

$$\mathbf{U} = -\overleftrightarrow{\kappa}_e \cdot \frac{\partial T}{\partial \mathbf{r}}. \tag{8.5}$$

Assuming no explicit time dependence for the distribution function, the function f_1 representing the departure of the distribution from equilibrium is found from solution of Boltzmann's equation

$$\mathbf{v} \cdot \frac{\partial f}{\partial \mathbf{r}} + \dot{\mathbf{k}} \cdot \frac{\partial f}{\partial \mathbf{k}} = -\frac{f_1}{\tau} \tag{8.6}$$

for a time independent temperature gradient. In the absence of an electric field, $\dot{\mathbf{k}} = 0$ and the drift velocity $\mathbf{v}$ is found from the equation

$$\mathbf{v} \cdot \frac{\partial f}{\partial \mathbf{r}} = -\frac{f_1}{\tau}. \tag{8.7}$$

Using the linear approximation for the term $\partial f / \partial \mathbf{r}$ in the Boltzmann equation, we obtain

$$\frac{\partial f}{\partial \mathbf{r}} \cong \frac{\partial f_0}{\partial \mathbf{r}} = \frac{\partial}{\partial \mathbf{r}}\left[\frac{1}{1+e^{(E-E_F)/k_BT}}\right]$$
$$= \left\{-\frac{e^{(E-E_F)/k_BT}}{[1+e^{(E-E_F)/k_BT}]^2}\right\}\left\{-\frac{1}{k_BT}\frac{\partial E_F}{\partial \mathbf{r}} - \frac{(E-E_F)}{k_BT^2}\frac{\partial T}{\partial \mathbf{r}}\right\}$$
$$= \left\{k_BT\frac{\partial f_0}{\partial E}\}\{-\frac{1}{k_BT}\frac{\partial T}{\partial \mathbf{r}}\right\}\left\{\frac{\partial E_F}{\partial T} + \frac{(E-E_F)}{T}\right\} \tag{8.8}$$
$$= -\frac{\partial f_0}{\partial E}\left\{\frac{\partial E_F}{\partial T} + \frac{(E-E_F)}{T}\right\}\frac{\partial T}{\partial \mathbf{r}}.$$

We will now give some typical values for these two terms for semiconductors and metals. For semiconductors, we evaluate the expression in (8.8) by referring to (7.75)

$$E_F = \frac{1}{2}E_g - \frac{3}{4}k_BT\ln(m_h/m_e) \tag{8.9}$$

from which

$$\frac{\partial E_F}{\partial T} \sim \frac{3}{4}k_B\ln(m_h/m_e) \tag{8.10}$$

showing that the temperature dependence of E_F arises from the inequality of the valence and conduction band effective masses. If $m_h = m_e$, which would be the case of strongly coupled "mirror" bands, then $\partial E_F/\partial T$ would vanish. However, for a significant mass difference such as $m_h/m_e = 2$, we obtain $\partial E_f/\partial T \sim 0.5k_B$ from (8.10). For a band gap of 0.5 eV and the Fermi level in the middle of the energy gap, we obtain for the other term in (8.8)

$$[(E - E_F)/T] \approx [0.5/(1/40)]k_B = 20k_B \tag{8.11}$$

where $k_BT \approx 1/40$ eV at room temperature. Thus for a semiconductor, the term $(E - E_F)/T$ is much larger than the term $(\partial E_F/\partial T)$.

For a metal with a spherical Fermi surface, the following relation

$$E_F = E_F^0 - \frac{\pi^2}{12}\frac{(k_BT)^2}{E_F^0} \tag{8.12}$$

is derived in standard textbooks on statistical mechanics, so that at room temperature and assuming that for a typical metal $E_F^0 = 5$ eV, we obtain from (8.12)

$$\left|\frac{\partial E_F}{\partial T}\right| = \frac{\pi^2}{6}\frac{(k_BT)}{E_F^0}k_B \approx \frac{10}{6}\frac{\left(\frac{1}{40}\right)}{5}k_B \approx 8\times 10^{-3}k_B. \tag{8.13}$$

Thus, for both semiconductors and metals, the term $(E - E_F)/T$ tends to dominate over $(\partial E_F/\partial T)$, though there can be situations where the term $(\partial E_F/\partial T)$ cannot be neglected. In this presentation, we will temporarily neglect the term ∇E_F in (8.8) when we calculate the electronic contribution to the thermal conductivity, but we will include this term formally in our derivation of thermoelectric effects in Sect. 8.3.

Typically, the electron energies of importance in any transport problem are those within $k_B T$ of the Fermi energy so that for many applications for metals, we can make the rough approximation,

$$\frac{E - E_F}{T} \approx k_B \tag{8.14}$$

though the results given in this section are derived without the above approximation for $(E - E_F)/T$. Rather, all integrations are carried out in terms of the variable $(E - E_F)/T$.

We return now to the solution of the Boltzmann's equation in the relaxation time approximation

$$\frac{\partial f}{\partial \mathbf{r}} = \left(-\frac{\partial f_0}{\partial E}\right)\left(\frac{E - E_F}{T}\right)\left(\frac{\partial T}{\partial \mathbf{r}}\right). \tag{8.15}$$

Solution of the Boltzmann's equation in the absence of an electric field yields

$$f_1 = -\tau \mathbf{v} \cdot \left(\frac{\partial f}{\partial \mathbf{r}}\right) = \tau \mathbf{v}\left(\frac{\partial f_0}{\partial E}\right)\left(\frac{E - E_F}{T}\right)\frac{\partial T}{\partial \mathbf{r}}. \tag{8.16}$$

Substitution of f_1 in (8.4) for the thermal current

$$\mathbf{U} = \frac{1}{4\pi^3}\int \mathbf{v}(E - F_F) f_1 d^3k \tag{8.17}$$

then results in

$$\mathbf{U} = \frac{1}{4\pi^3 T}\left(\frac{\partial T}{\partial \mathbf{r}}\right) \cdot \int \tau \mathbf{v}\mathbf{v}(E - E_F)^2 \left(\frac{\partial f_0}{\partial E}\right) d^3k. \tag{8.18}$$

Using the definition of the thermal conductivity tensor $\overleftrightarrow{\kappa}_e$ given by (8.5) we write the electronic contribution to the thermal conductivity $\overleftrightarrow{\kappa}_e$ as

$$\overleftrightarrow{\kappa}_e = \frac{-1}{4\pi^3 T}\int \tau \mathbf{v}\mathbf{v}(E - E_F)^2 \left(\frac{\partial f_0}{\partial E}\right) d^3k \tag{8.19}$$

where $d^3k = d^2S\ dk_\perp = d^2S\ dE/|\partial E/\partial \mathbf{k}| = d^2S dE/(\hbar v)$ is used to exploit our knowledge of the dependence of the distribution function on the energy, as discussed below.

8.2.2 *Thermal Conductivity for Metals*

In the case of a metal, the integral for $\overleftrightarrow{\kappa}_e$ given by (8.19) can be evaluated easily by converting the integral over phase space $\int d^3k$ to an integral over $\int dE\ d^2S_F$ in order to exploit the δ-function property of $-(\partial f_0/\partial E)$. We then make use of the following result that can be found in any standard statistical mechanics text (see for example, F. Reif, "Fundamentals of Statistical and Thermal Physics")

$$\int G(E)\left(-\frac{\partial f_0}{\partial E}\right)dE = G(E_F) + \frac{\pi^2}{6}(k_BT)^2\left[\frac{\partial^2 G}{\partial E^2}\right]_{E_F} + \dots \tag{8.20}$$

It is necessary to consider the expansion given in (8.20) for solving (8.19) since $G(E_F)$ vanishes at $E = E_F$ for the integral defined in (8.19) for $\overleftrightarrow{\kappa}_e$. To solve the integral equation of (8.19), we therefore make the identification of

$$G(E) = g(E)(E - E_F)^2 \tag{8.21}$$

where

$$g(E) = \frac{1}{4\pi^3}\int \tau \mathbf{v}\mathbf{v} d^2S/\hbar v \tag{8.22}$$

so that $G(E_F) = 0$ and $(\partial G/\partial E)|_{E_F} = 0$, while

$$\left[\frac{\partial^2 G}{\partial E^2}\right]_{E_F} = G''(E_F) = 2g(E_F). \tag{8.23}$$

These relations will be used again in connection with the calculation of the thermopower in Sect. 8.3.1. For the case of the thermal conductivity for a metal we then obtain

$$\overleftrightarrow{\kappa}_e = \frac{\pi^2}{3}(k_BT)^2 g(E_F) = \frac{(k_BT)^2}{12\pi\hbar}\int \tau\mathbf{v}\mathbf{v}\frac{d^2S_F}{v} \tag{8.24}$$

where the integration is over the Fermi surface. We immediately recognize that the integral appearing in (8.24) is the same as that for the electrical conductivity (see (7.26) and (7.29))

$$\overleftrightarrow{\sigma} = \frac{e^2}{4\pi^3\hbar}\int \tau\mathbf{v}\mathbf{v}\frac{d^2S_F}{v} \tag{8.25}$$

so that the electronic contribution to the thermal conductivity and the electrical conductivity tensors are proportional to each other

$$\overleftrightarrow{\kappa}_e = \overleftrightarrow{\sigma}\, T\left(\frac{\pi^2 k_B^2}{3e^2}\right) \tag{8.26}$$

and (8.26) is known as the Wiedemann–Franz Law. The physical basis for this relation is that in electrical conduction each electron carries a charge e and experiences an electrical force $e\mathbf{E}$ so that the electrical current per unit field is e^2. In thermal conduction, each electron carries a unit of thermal energy $k_B T$ and experiences a thermal force $k_B \partial T/\partial \mathbf{r}$ so that the heat current per unit thermal gradient is proportional to $k_B^2 T$. Therefore the ratio of $|\kappa_e|/|\sigma|$ must be on the order of $(k_B^2 T/e^2)$. The Wiedemann–Franz law suggests that the ratio $\kappa_e/(\sigma T)$ should be a constant (called the Lorenz number[1]), independent of materials properties

$$\left|\frac{\kappa_e}{\sigma T}\right| = \frac{\pi^2}{3}\left(\frac{k_B}{e}\right)^2 = 2.45 \times 10^{-8} \quad \text{watt ohm/deg}^2. \tag{8.27}$$

The ratio $(\kappa_e/\sigma T)$ is approximately constant for all metals at high temperatures $T > \Theta_D$ and at very low temperatures $T \ll \Theta_D$, where Θ_D is the Debye temperature. The derivation of the Wiedemann–Franz Law depends on the relaxation time approximation, which is valid at high temperatures $T > \Theta_D$ where the electron scattering is dominated by the quasi-elastic phonon scattering process. The Wiedemann–Franz Law is also valid at very low temperatures $T \ll \Theta_D$ where phonon scattering is unimportant and the dominant scattering mechanism is impurity and lattice defect scattering, both of which tend to be elastic scattering processes. These scattering processes are discussed in Chap. 9 where we discuss in more detail the temperature dependence for κ. When specific scattering processes are considered in detail, the value of the Lorenz number may then change. The Lorentz number has been reported as varying from $2.23 \times 10^{-8}\ \Omega\text{K}^{-2}$ for Cu at 0°C to $3.2 \times 10^{-8}\ \Omega\text{K}^{-2}$ for tungsten at 100°C. In fact, the Wiedemann–Franz Law works well for high and low (few Kelvins) temperatures but it is not accurate at intermediate temperatures. In the case the of graphene (a 2D material) the Lorenz number has been measured as being $3.23 \times 10^{-8}\ \Omega\text{K}^{-2}$.

The temperature dependence of the thermal conductivity of a metal is given in Fig. 8.1. From (8.27) we can write the following relation for the electronic contribution to the thermal conductivity κ_e when the Wiedemann–Franz law is satisfied

$$\kappa_e = \left(\frac{ne^2\tau}{m^*}\right) T \frac{\pi^2}{3}\left(\frac{k_B}{e}\right)^2. \tag{8.28}$$

At very low temperatures where scattering by impurities, defects, and crystal boundaries is dominant, σ is independent of T and therefore is independent from the Wiedemann–Franz law, where $\kappa_e \sim T$. At somewhat higher temperatures, but still in the regime $T \ll \Theta_D$, electron-phonon scattering starts to dominate and κ_e starts to decrease. In this regime, the electrical conductivity exhibits a T^{-5} dependence. However, only small q phonons participate in this regime. Thus it is only the phonon

[1] This number is $2.44 \times 10^{-8}\ W\Omega K^{-2}$ and was discovered by Ludvig Lorenz in 1872.

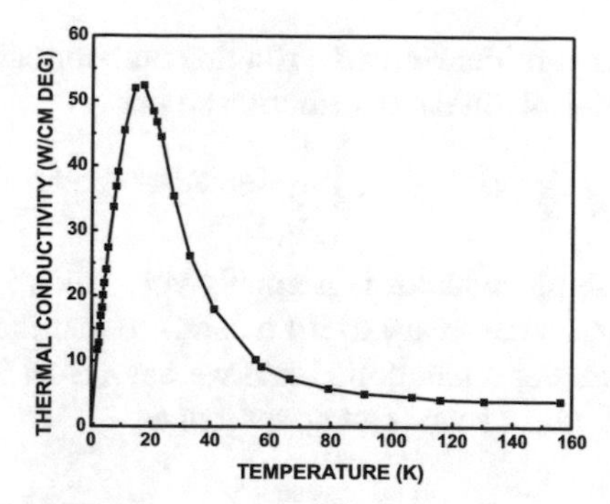

Fig. 8.1 The temperature dependence of the thermal conductivity of copper. Note that both κ and T are plotted on linear scales. At low temperatures where the phonon density is low, the thermal transport is by electrons predominantly, while at high temperatures, thermal transport by phonons becomes more important.

density which increases as T^3 that is relevant to the phonon-electron scattering, thereby yielding an electrical resistivity with a T^3 dependence and a conductivity with a T^{-3} dependence. Using (8.28), we thus find that in the low T range, where only low q phonons participate in thermal transport, κ_e should show a T^{-2} dependence, in agreement with Fig. 8.1. At high T where all the phonons contribute to thermal transport, we have $\sigma \sim 1/T$ so that κ_e becomes independent of T. Since $\Theta_D \sim 300$ K for Cu, the temperature range of Fig. 8.1 is well below the Debye temperature for copper.

The Wiedemann-Franz law does not hold for electrons confined in low dimensions, as theoretically predicted by Kane and Fisher 1996, and experimentally demonstrated by Wakeham et al. 2011 in the metallic phase of quasi-one-dimensional $Li_{0.9}Mo_6O_{17}$ where the ratio of thermal and electrical conductivity diverges with decreasing temperature.

8.2.3 Thermal Conductivity for Semiconductors

For the case of non-degenerate semiconductors, the integral for $\overleftrightarrow{\kappa}_e$ in (8.19) is evaluated by replacing $(E - E_F) \rightarrow E$, since in a semiconductor the electrons that can conduct heat must be in the conduction band, and the lowest energy an electron can have in the thermal conduction process is at the conduction band minimum which is taken as the zero of energy in these calculations. Then the thermal conductivity for a non-degenerate semiconductor can be written as

$$\overleftrightarrow{\kappa}_e = \frac{1}{4\pi^3 T} \int \tau \mathbf{v}\mathbf{v}\, E^2 \left(- \frac{\partial f_0}{\partial E} \right) d^3k. \tag{8.29}$$

For intrinsic semiconductors, the Fermi distribution function can normally be approximated by the Maxwell–Boltzmann distribution so that

$$(\partial f_0/\partial E) \rightarrow -(1/k_BT)e^{(-|E_F^e|/k_BT)}e^{(-E/k_BT)}. \tag{8.30}$$

The doping level in a semiconductor is normally very much lower than 2 electrons per atom (which would be necessary to fill a band) so that the Fermi level remains close to the band edges. For a parabolic band we have $E = \hbar^2k^2/2m^*$, so that the volume element in reciprocal space can be written as

$$\int d^3k = \int 4\pi k^2 dk = \int_0^\infty 2\pi(2m^*/\hbar^2)^{3/2}E^{1/2}dE, \tag{8.31}$$

and $\mathbf{v} = (1/\hbar)(\partial E/\partial \mathbf{k}) = \hbar\mathbf{k}/m$. Assuming a constant relaxation time, we then substitute all these terms into (8.29) for $\overleftrightarrow{\kappa}_e$ and integrate over the energy E to obtain for a diagonal component of the tensor

$$\begin{aligned}\kappa_{exx} &= (1/4\pi^3T)\ \int \tau v_x^2 E^2\left[(k_BT)^{-1}e^{-|E_F^e|(/k_BT)}\ e^{-E/(k_BT)}\right]2\pi(2m^*/\hbar^2)^{3/2}E^{1/2}dE \\ &= \left[k_B(k_BT)\tau/(3\pi^2m^*)\right](2m^*k_BT/\hbar^2)^{3/2}\ e^{-|E_F^e|/(k_BT)}\ \int_0^\infty x^{7/2}e^{-x}dx\end{aligned} \tag{8.32}$$

where $\int_0^\infty x^{7/2}e^{-x}dx = 105\sqrt{\pi}/8$, from which it follows that κ_{exx} has a temperature dependence of the form

$$T^{5/2}e^{-|E_F^e|/(k_BT)} \tag{8.33}$$

in which the exponential term is dominant for temperatures of physical interest, where $k_BT \ll |E_F^e|$. We note from (7.42) that for a semiconductor, the temperature dependence of the electrical conductivity is given by

$$\sigma_{xx} = \frac{2e^2\tau}{m^*}\left(\frac{m^*k_BT}{2\pi\hbar^2}\right)^{3/2}e^{-|E_F^e|/(k_BT)}. \tag{8.34}$$

Assuming cubic symmetry for simplicity, we can write the conductivity tensor as

$$\overleftrightarrow{\sigma} = \begin{pmatrix}\sigma_{xx} & 0 & 0\\ 0 & \sigma_{xx} & 0\\ 0 & 0 & \sigma_{xx}\end{pmatrix} \tag{8.35}$$

so that the electronic contribution to the thermal conductivity of a semiconductor can be written as

$$\kappa_{exx} = \left(\frac{35}{2}\right)\left(\frac{k_B^2}{e^2}\right)\sigma_{xx}T \tag{8.36}$$

where

$$\sigma_{xx} = ne^2\tau/m_{xx} = ne\mu_{xx} \tag{8.37}$$

and we note that the coefficient (35/2) for this calculation for semiconductors is different from the corresponding coefficient ($\pi^2/3$) for metals (see (8.27)). Except for numerical constants, the formal results relating the electronic contribution to the thermal conductivity κ_{exx} and σ_{xx} are similar for metals and semiconductors, with the electronic thermal conductivity and electronic electrical conductivity being proportional to one another. For low symmetry materials, the electrical and thermal conductivity tensors have off-diagonal elements.

A major difference between semiconductors and metals is the magnitude of the electrical conductivity, and hence of the electronic contribution to the thermal conductivity. Since σ_{xx} is much smaller for semiconductors than for metals, κ_e for semiconductors is relatively unimportant, and the thermal conductivity tends to be dominated by the lattice contribution κ_L.

8.2.4 Thermal Conductivity for Insulators

In the case of insulators, heat is only carried by phonons (lattice vibrations). The thermal conductivity in insulators therefore depends on the phonon scattering mechanisms (see Chap. 9). The lattice thermal conductivity is calculated from kinetic theory and is given by

$$\kappa_L = \frac{C_p \overline{v}_q \Lambda_{\mathrm{ph}}}{3} \tag{8.38}$$

where C_p is the heat capacity, $\overline{v}_q$ is the average phonon velocity and Λ_{ph} is the phonon mean free path. As discussed above, the total thermal conductivity of a solid is given as the sum of the lattice contribution κ_L and the electronic contribution κ_e. For metals the electronic contribution dominates, while for insulators and semiconductors the phonon contribution dominates. Let us now consider the temperature dependence of κ_{exx} (see Fig. 8.2), for heat conduction by phonons. At very low T in the defect scattering range, the heat capacity has a dependence $C_p \propto T^3$ while $\overline{v}_q$ and Λ_{ph} are almost independent of T. As T increases and we enter the phonon-phonon scattering regime due to normal scattering processes and involving only low q phonons, C_p is still increasing with T but the increase is slower than T^3, while $\overline{v}_q$ remains independent of T and Λ_{ph}. As T increases further, the thermal conductivity increases more and more gradually and eventually starts to decrease because of phonon-phonon scattering events, for which the density of phonons available for scattering depends on the Bose–Einstein factor $[\exp(\hbar\omega/k_B T) - 1]$. This causes a peak in $\kappa_L(T)$. The decrease in $\kappa_L(T)$ becomes more pronounced as C_p becomes independent of T and Λ_{ph} continues to be proportional to $[\exp(\hbar\bar{\omega}/k_B T) - 1]$ where $\bar{\omega}$ is a typical phonon

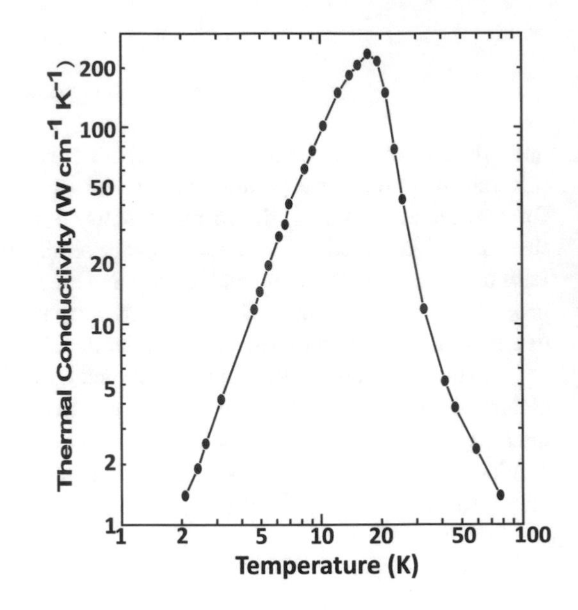

Fig. 8.2 Temperature dependence of the thermal conductivity of a highly purified insulating crystal of NaF. Note that both κ and T are plotted on a log scale, and that the peak in κ occurs at quite a low temperature (~17 K).

frequency (see Sect. 9.4.1). As T increases further, we eventually enter the $T \gg \Theta_D$ regime, where Θ_D is the Debye temperature. In this regime, the temperature dependence of Λ_{ph} simply becomes $\Lambda_{\text{ph}} \sim (1/T)$. Referring to Fig. 8.2 for $\kappa(T)$ for NaF we see that the peak in κ occurs at about 18 K where the complete Bose–Einstein factor $[\exp(\hbar\bar{\omega}/k_B T) - 1]$ must be used to describe the T dependence of κ_L. For much of the temperature range in Fig. 8.2, only low q phonons participate in the thermal conduction process. At higher temperatures where larger q phonons contribute to thermal conduction, umklapp processes become important in the phonon scattering process, as discussed in Sect. 9.3.1. The discussion in this section regarding phonons also applies to the lattice contribution to the thermal conductivity for metals, semimetals and semiconductors.

8.3 Thermoelectric Phenomena

In many metals and semiconductors there exists a significant coupling between the electrical current and the thermal current. This coupling can be appreciated by observing that when electrons carry thermal current, they are also transporting charge and therefore generating electric fields. This coupling between the charge transport and heat transport gives rise to thermoelectric phenomena. In our discussion of thermoelectric phenomena we start with a general derivation of the coupled equations for the electrical current density $\mathbf{j}$ and the thermal current density $\mathbf{U}$ defined in (8.4):

$$\mathbf{j} = \frac{e}{4\pi^3} \int \mathbf{v} f_1 d^3k \tag{8.39}$$

$$\mathbf{U} = \frac{1}{4\pi^3} \int \mathbf{v}(E - E_F) f_1 d^3k \tag{8.40}$$

and the perturbation to the distribution function f_1 is found from solution of Boltzmann's equation in the relaxation time approximation:

$$\mathbf{v} \cdot \frac{\partial f}{\partial \mathbf{r}} + \dot{\mathbf{k}} \cdot \frac{\partial f}{\partial \mathbf{k}} = -\frac{(f - f_0)}{\tau}, \tag{8.41}$$

which is written here for the case of time independent forces and fields. Substituting for $(\partial f/\partial \mathbf{r})$ from (8.8) and (8.15), for $\partial f/\partial \mathbf{k}$ from (7.17), and using the fact that $\partial f/\partial \mathbf{k} = (\partial f_0/\partial E)(\partial E/\partial \mathbf{k})$ yields

$$\mathbf{v} \cdot (\partial f_0/\partial E)\left[(e\mathbf{E} - \nabla E_F) - \frac{(E - E_F)}{T}(\nabla T)\right] = -\frac{f_1}{\tau}, \tag{8.42}$$

so that the solution to the Boltzmann equation in the presence of an electric field and a temperature gradient is

$$f_1 = \mathbf{v}\tau \cdot (\partial f_0/\partial E)\left\{[(E - E_F)/T]\nabla T - e\mathbf{E} + \nabla E_F\right\}, \tag{8.43}$$

in which e is negative for electrons and positive for holes. The electrical and thermal currents in the presence of both an applied electric field and a temperature gradient can thus be obtained by substituting f_1 into $\mathbf{j}$ and $\mathbf{U}$ in (8.39) and (8.40) to yield expressions of the form:

$$\mathbf{j} = e^2 \overleftrightarrow{\kappa}_0 \cdot \left(\mathbf{E} - \frac{1}{e}\nabla E_F\right) - (e/T) \overleftrightarrow{\kappa}_1 \cdot \nabla T \tag{8.44}$$

and

$$\mathbf{U} = e \overleftrightarrow{\kappa}_1 \cdot \left(\mathbf{E} - \frac{1}{e}\nabla E_F\right) - (1/T) \overleftrightarrow{\kappa}_2 \cdot \nabla T \tag{8.45}$$

where the thermal conductivity tensor $\overleftrightarrow{\kappa}_0$ is related to the electrical conductivity tensor $\overleftrightarrow{\sigma}$ by

$$\overleftrightarrow{\kappa}_0 = \frac{1}{4\pi^3} \int \tau \mathbf{v}\mathbf{v}(-\partial f_0/\partial E) d^3k = \frac{\overleftrightarrow{\sigma}}{e^2}, \tag{8.46}$$

and

$$\overleftrightarrow{\kappa}_1 = \frac{1}{4\pi^3} \int \tau \mathbf{v}\mathbf{v}(E - E_F)\left(-\frac{\partial f_0}{\partial E}\right) d^3k, \tag{8.47}$$

and $\overleftrightarrow{\kappa}_2$ is related to the thermal conductivity tensor $\overleftrightarrow{\kappa}_e$ by

$$\overleftrightarrow{\kappa}_2 = \frac{1}{4\pi^3} \int \tau \mathbf{v}\mathbf{v}(E - E_F)^2(-\partial f_0/\partial E) d^3k = T \overleftrightarrow{\kappa}_e . \tag{8.48}$$

Note that the integrands for $\overleftrightarrow{\kappa}_1$ and $\overleftrightarrow{\kappa}_2$ are both related to that for $\overleftrightarrow{\kappa}_0$ by introducing factors of $(E - E_F)$ and $(E - E_F)^2$, respectively. Note also that the same integral $\overleftrightarrow{\kappa}_1$ occurs in the expression for the electric current $\mathbf{j}$ induced by a thermal gradient ∇T and in the expression for the thermal current $\mathbf{U}$ induced by an electric field $\mathbf{E}$. The motion of charged carriers across a temperature gradient results in a flow of electric current expressed by the term $-(e/T)(\overleftrightarrow{\kappa}_1) \cdot \nabla T$. This term is the origin of thermoelectric effects.

The discussion up to this point has been general. If specific boundary conditions are imposed, we obtain a variety of thermoelectric effects such as the Seebeck effect, the Peltier effect and the Thomson effect. We now define the conditions under which each of these thermoelectric effects occur.

We define the thermopower $\overleftrightarrow{\mathscr{S}}$ (Seebeck coefficient) and the Thomson coefficient $\mathscr{T}_b$ under conditions of zero current flow. Then referring to (8.44), we obtain under open circuit conditions

$$\mathbf{j} = 0 = e^2 \overleftrightarrow{\kappa}_0 \cdot \left(\mathbf{E} - \frac{1}{e}\nabla E_F\right) - (e/T) \overleftrightarrow{\kappa}_1 \cdot \nabla T \tag{8.49}$$

so that the Seebeck coefficient $\overleftrightarrow{\mathscr{S}}$ is defined by

$$\mathbf{E} - \frac{1}{e}\nabla E_F = (1/eT)\overleftrightarrow{\kappa}_0^{-1} \cdot \overleftrightarrow{\kappa}_1 \cdot \nabla T \equiv \overleftrightarrow{\mathscr{S}} \cdot \nabla T, \tag{8.50}$$

and $\mathscr{S}$ is sometimes also called the thermopower. Using the relation $\nabla E_F = \frac{\partial E_F}{\partial T} \nabla T$ we obtain the definition for the Thomson coefficient $\mathscr{T}_b$

$$\mathbf{E} = \left(\frac{1}{e}\frac{\partial E_F}{\partial T} + \overleftrightarrow{\mathscr{S}}\right)\nabla T \equiv \overleftrightarrow{\mathscr{T}}_b \cdot \nabla T \tag{8.51}$$

where

$$\overleftrightarrow{\mathscr{T}}_b = T \frac{\partial}{\partial T} \overleftrightarrow{\mathscr{S}} . \tag{8.52}$$

For many thermoelectric systems of interest, $\overleftrightarrow{\mathscr{S}}$ has a linear temperature dependence, and in this case it follows from (8.52) that $\overleftrightarrow{\mathscr{T}}_b$ and the Seebeck coefficient $\overleftrightarrow{\mathscr{S}}$ for such systems are almost equivalent for practical purposes. Therefore the Seebeck and Thomson coefficients are used almost interchangeably in the literature, but one should always check that the conditions required for the use of (8.52) are satisfied before using this relation.

From (8.50) and neglecting the term in ∇E_F, as is usually done, we have

$$\overleftrightarrow{\mathscr{S}} = (1/eT)\, \overleftrightarrow{\kappa}_0^{-1} \cdot \overleftrightarrow{\kappa}_1 \tag{8.53}$$

which is simplified by assuming an isotropic medium, yielding the scalar quantities

$$\mathscr{S} = (1/eT)(\kappa_1/\kappa_0). \tag{8.54}$$

However in an anisotropic medium, the tensor components of $\overleftrightarrow{\mathscr{S}}$ are found from

$$\mathscr{S}_{ij} = (1/eT)(\kappa_0^{-1})_{i\alpha}(\kappa_1)_{\alpha j}, \tag{8.55}$$

where the Einstein summation convention is assumed in summing over α. Fig. 8.3 shows a schematic diagram for measuring the thermopower or Seebeck effect in an n-type semiconductor. At the hot junction the Fermi level is higher than at the cold junction. Electrons will move from the hot junction to the cold junction in an attempt to equalize the Fermi level, thereby creating an electric field which can be measured in terms of the open circuit voltage V shown in Fig. 8.3.

Another important thermoelectric coefficient is the Peltier coefficient $\overleftrightarrow{\Pi}$ which is a second rank tensor defined as the proportionality between the two vectors $\mathbf{U}$ and $\mathbf{j}$

$$\mathbf{U} \equiv \overleftrightarrow{\Pi} \cdot \mathbf{j} \tag{8.56}$$

in the absence of a thermal gradient.

For $\nabla T = 0$, (8.44) and (8.45) become

$$\mathbf{j} = e^2\, \overleftrightarrow{\kappa}_0 \cdot \left(\mathbf{E} - \frac{1}{e}\nabla E_F\right) \tag{8.57}$$

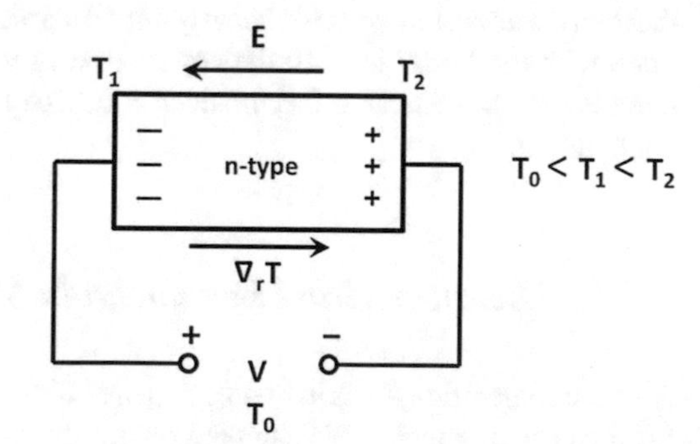

Fig. 8.3 Determination of the Seebeck effect for an n-type semiconductor. In the presence of a temperature gradient, electrons will move from the hot junction to the cold junction, thereby creating an electric field and a voltage V across the semiconductor.

$$\mathbf{U} = e \overset{\leftrightarrow}{\kappa}_1 \cdot \left(\mathbf{E} - \frac{1}{e} \nabla E_F \right) \tag{8.58}$$

so that

$$\mathbf{U} = (1/e) \overset{\leftrightarrow}{\kappa}_1 \cdot (\overset{\leftrightarrow}{\kappa}_0)^{-1} \cdot \mathbf{j} = \overset{\leftrightarrow}{\Pi} \cdot \mathbf{j} \tag{8.59}$$

where

$$\overset{\leftrightarrow}{\Pi} = (1/e) \overset{\leftrightarrow}{\kappa}_1 \cdot (\overset{\leftrightarrow}{\kappa}_0)^{-1}. \tag{8.60}$$

Comparing (8.53) and (8.60) we see that $\overset{\leftrightarrow}{\Pi}$ and $\overset{\leftrightarrow}{\mathcal{S}}$ are related by

$$\overset{\leftrightarrow}{\Pi} = T \overset{\leftrightarrow}{\mathcal{S}}, \tag{8.61}$$

where T is the temperature. For isotropic materials, the Peltier coefficient thus becomes a scalar, and is proportional to the thermopower $\mathcal{S}$:

$$\Pi = \frac{1}{e} \left(\kappa_1 / \kappa_0 \right) = T \mathcal{S}, \tag{8.62}$$

while for anisotropic materials the tensor components of $\overset{\leftrightarrow}{\Pi}$ can be found in analogy with (8.55). We note that both $\overset{\leftrightarrow}{\mathcal{S}}$ and $\overset{\leftrightarrow}{\Pi}$ exhibit a linear dependence on e and therefore depend explicitly on the sign of the carrier, and measurements of $\overset{\leftrightarrow}{\mathcal{S}}$ or $\overset{\leftrightarrow}{\Pi}$ can be used to determine whether transport is dominated by electrons or holes.

We have already considered the evaluation of $\overset{\leftrightarrow}{\kappa}_0$ in treating the electrical conductivity and $\overset{\leftrightarrow}{\kappa}_2$ in treating the thermal conductivity. To treat thermoelectric phenomena, we need now to evaluate $\overset{\leftrightarrow}{\kappa}_1$

$$\overset{\leftrightarrow}{\kappa}_1 = \frac{1}{4\pi^3} \int \tau \mathbf{v}\mathbf{v} (E - E_F)(-\partial f_0 / \partial E) d^3k. \tag{8.63}$$

In Sect. 8.3.1 we evaluate $\overset{\leftrightarrow}{\kappa}_1$ for the case of a metal and in Sect. 8.3.2 we evaluate $\overset{\leftrightarrow}{\kappa}_1$ for the case of the electrons in an intrinsic semiconductor. In practice, the thermopower is of interest for heavily doped semiconductors, which are either degenerate with the Fermi level in the conduction or valence band or very close to these band edges. In general, a thermoelectric device has both n-type and p-type legs or constituents.

8.3.1 Thermoelectric Phenomena in Metals

All thermoelectric effects in metals depend on the tensor $\overset{\leftrightarrow}{\kappa}_1$ which we evaluate below for the case of a metal. We can then obtain the thermopower

$$\overleftrightarrow{\mathscr{S}} = \frac{1}{eT}\left(\overleftrightarrow{\kappa}_1 \cdot \overleftrightarrow{\kappa}_0^{-1}\right) \tag{8.64}$$

or the Peltier coefficient

$$\overleftrightarrow{\Pi} = \frac{1}{e}\left(\overleftrightarrow{\kappa}_1 \cdot \overleftrightarrow{\kappa}_0^{-1}\right) \tag{8.65}$$

or the Thomson coefficient

$$\overleftrightarrow{\mathscr{T}}_b = T\frac{\partial}{\partial T}\overleftrightarrow{\mathscr{S}}. \tag{8.66}$$

To evaluate $\overleftrightarrow{\kappa}_1$ for metals we wish to exploit the δ-function behavior of $(-\partial f_0/\partial E)$. This is accomplished by converting the integration over d^3k to an integration over dE and over a constant energy surface, $d^3k = d^2S\ dE/\hbar v$. From Fermi statistics we have the general relation (see (8.20))

$$\int G(E)\left(-\frac{\partial f_0}{\partial E}\right)dE = G(E_F) + \frac{\pi^2}{6}\left(k_B T\right)^2\left[\frac{\partial^2 G}{\partial E^2}\right]_{E_F} + \cdots. \tag{8.67}$$

For the integral in (8.63) which defines $\overleftrightarrow{\kappa}_1$, we can write

$$G(E) = g(E)(E - E_F) \tag{8.68}$$

where

$$g(E) = \frac{1}{4\pi^3}\int \tau \mathbf{v}\mathbf{v}\ d^2S/v \tag{8.69}$$

and the integration in (8.69) is carried out over a constant energy surface at energy E. The differentiation of $G(E)$ then yields

$$\begin{aligned} G'(E) &= g'(E)(E - E_F) + g(E) \\ G''(E) &= g''(E)(E - E_F) + 2g'(E). \end{aligned} \tag{8.70}$$

Evaluation of (8.70) at $E = E_F$ yields

$$\begin{aligned} G(E_F) &= 0 \\ G''(E_F) &= 2g'(E_F). \end{aligned} \tag{8.71}$$

We therefore obtain

$$\overleftrightarrow{\kappa}_1 = \frac{\pi^2}{3}\left(k_B T\right)^2 g'(E_F). \tag{8.72}$$

We interpret $g'(E_F)$ in (8.72) to mean that the same integral $\overleftrightarrow{\kappa}_0$ which determines the conductivity tensor is evaluated on a constant energy surface E, and $g'(E_F)$ is the energy derivative of that integral evaluated at the Fermi energy E_F. The temperature dependence of $g'(E_F)$ is related to the temperature dependence of τ, since v is essentially temperature independent. For example, we will see in Chap. 9 that acoustic phonon scattering in the high temperature limit $T \gg \Theta_D$ yields a temperature dependence $\tau \sim T^{-1}$ so that $\overleftrightarrow{\kappa}_1$ in this important case for metals will be proportional to T.

For a spherical constant energy surface $E = \hbar^2 k^2/2m^*$ and assuming a relaxation time τ that is independent of energy, we can readily evaluate (8.72) to obtain

$$g(E) = \frac{\tau}{3\pi^2 m^*}\left(\frac{2m^*}{\hbar^2}\right)^{3/2} E^{3/2} \tag{8.73}$$

$$g'(E_F) = \frac{\tau}{2\pi^2 m^*}\left(\frac{2m^*}{\hbar^2}\right)^{3/2} E_F^{1/2} \tag{8.74}$$

and

$$\kappa_1 = \frac{\tau}{6m^*}\left(\frac{2m^*}{\hbar^2}\right)^{3/2} E_F^{1/2}(k_B T)^2. \tag{8.75}$$

Using the same approximations, we can write for κ_0:

$$\kappa_0 = \frac{\tau}{3\pi^2 m^*}\left(\frac{2m^*}{\hbar^2}\right)^{3/2} E_F^{3/2} \tag{8.76}$$

so that from (8.64) we have for the Seebeck coefficient

$$\mathscr{S} = \frac{\kappa_1}{\kappa_0 e T} = \frac{\pi^2 k_B}{2e}\frac{k_B T}{E_F} \tag{8.77}$$

for a simple metal. From (8.77) we see that $\mathscr{S}$ exhibits a linear dependence on T and a sensitivity to the sign of the carriers. We note from (8.64) that a low carrier density implies a large $\mathscr{S}$ value. Thus degenerate semiconductors (heavily doped with $n \sim 10^{18}$–$10^{19}/\mathrm{cm}^3$) tend to have higher thermopowers than metals. The derivation given here works as a good approximation for very heavily doped semiconductors with simple band structures. In general, practical thermoelectric material do not have simple band structures, so more sophisticated calculations need to be made. In this case, density functional theory is commonly used.

8.3.2 *Thermopower for Intrinsic Semiconductors*

In this section we evaluate $\overleftrightarrow{\kappa}_1$ for electrons in an intrinsic or lightly doped semiconductor for illustrative purposes. Intrinsic semiconductors are not important for

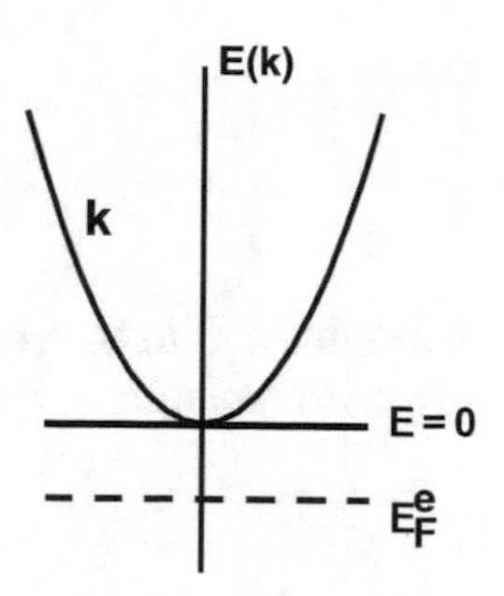

Fig. 8.4 Schematic E versus k diagram, showing that $E = 0$ is the lowest electronic energy for heat conduction.

practical thermoelectric devices since the contributions of electrons and holes to the matrix $\overleftrightarrow{\kappa}_1$ are of opposite signs and tend to cancel. Thus it is only heavily doped semiconductors with a single carrier type that are important for thermoelectric applications.

The evaluation of the general expression for the integral $\overleftrightarrow{\kappa}_1$

$$\overleftrightarrow{\kappa}_1 = \frac{1}{4\pi^3}\int \tau\mathbf{v}\mathbf{v}(E - E_F)\left(-\frac{\partial f_0}{\partial E}\right)d^3k \tag{8.78}$$

is different for semiconductors and metals. Referring to Fig. 8.4 for an intrinsic semiconductor we need to make the substitution $(E - E_F) \rightarrow E$ in (8.78), since only conduction electrons can carry heat. Of course phonons also carry heat and this is discussed in Sect. 8.5.

The equilibrium distribution function for an intrinsic semiconductor can be written as

$$f_0 = e^{-E/(k_BT)}\; e^{-|E_F^e|/(k_BT)} \tag{8.79}$$

so that

$$\frac{\partial f_0}{\partial E} = -\frac{1}{k_BT}\; e^{-E/(k_BT)}\; e^{-|E_F^e|/(k_BT)}. \tag{8.80}$$

To evaluate d^3k we need to assume a model for $E(\mathbf{k})$. For simplicity and for illustration, assume a simple parabolic band

$$E = \hbar^2k^2/2m^* \tag{8.81}$$

$$d^3k = 4\pi k^2 dk \tag{8.82}$$

so that

$$d^3k = 2\pi\left(\frac{2m^*}{\hbar^2}\right)^{3/2} E^{1/2}dE \tag{8.83}$$

and also

$$\mathbf{v} = \frac{1}{\hbar}(\partial E/\partial \mathbf{k}) = \hbar\mathbf{k}/m^*. \tag{8.84}$$

Substitution into the equation for $\overleftrightarrow{\kappa}_1$ for a semiconductor with the simple energy dispersion relation $E = \hbar^2 k^2/2m^*$ then yields upon integration from (8.78)

$$\kappa_{1xx} = \frac{5\tau k_B T}{m^*}\left(m^* k_B T/2\pi\hbar^2\right)^{3/2} e^{-|E_F^e|/(k_B T)}. \tag{8.85}$$

This expression is valid for a semiconductor with a simple parabolic band for which the Fermi level is far from the band edge $(E - E_F) \gg k_B T$. The thermopower is then found by substitution

$$\mathscr{S} = \frac{1}{eT}\left(\kappa_{1xx}/\kappa_{0xx}\right) \tag{8.86}$$

where the expression for $\kappa_{0xx} = \sigma_{xx}/e^2$ is given by (8.46). We thus obtain the result

$$\mathscr{S} = \frac{5}{2}\frac{k_B}{e} \tag{8.87}$$

which is a constant independent of temperature, independent of the band structure, but sensitive to the sign of the carriers. The calculation in this section is for the contribution of electrons. In an actual intrinsic semiconductor, the contribution of both electrons and holes to κ_1 must be found. Likewise the calculation for κ_{0xx} would also include contributions from both electrons and holes. Since the contribution to $(1/e)\kappa_{1xx}$ for holes and electrons are of opposite sign, we can from (8.87) expect that $\mathscr{S}$ for holes will cancel $\mathscr{S}$ for electrons for an intrinsic semiconductor, while the κ_0 for holes and electrons will add.

Materials with a high thermopower or Seebeck coefficient are heavily doped degenerate semiconductors for which the Fermi level is close to the band edge or even within the conduction band for electrons or the valence band for holes. In either of the two cases, the complete Fermi function must be used. Since $\mathscr{S}$ depends on the sign of the charge carriers, thermoelectric materials are doped either heavily doped n-type or heavily doped p-type semiconductors to prevent cancellation of the contribution from electrons and holes. For intrinsic semiconductors, carriers are created by thermal excitations so that these materials have approximately equal concentrations of electrons and holes and cannot be used for most practical thermoelectric applications.

8.3.3 *Effect of Thermoelectricity on the Thermal Conductivity*

From the coupled equations given by (8.44) and (8.45) it is seen that the proportionality between the thermal current **U** and the temperature gradient ∇T in the absence of electrical current ($\mathbf{j} = 0$) contains terms related to $\overset{\leftrightarrow}{\kappa}_1$. We now solve (8.44) and (8.45) to find the contribution of the thermoelectric terms to the electronic thermal conductivity. When $\mathbf{j} = 0$, (8.44) becomes

$$\left(\mathbf{E} - \frac{1}{e}\nabla E_F\right) = \frac{1}{eT}\,\overset{\leftrightarrow}{\kappa}^{-1}{}_0 \cdot \overset{\leftrightarrow}{\kappa}_1 \cdot \nabla T \tag{8.88}$$

so that

$$\mathbf{U} = -(1/T)\left[\overset{\leftrightarrow}{\kappa}_2 - \overset{\leftrightarrow}{\kappa}_1 \cdot \overset{\leftrightarrow}{\kappa}^{-1}{}_0 \cdot \overset{\leftrightarrow}{\kappa}_1\right] \cdot \nabla T \tag{8.89}$$

where $\overset{\leftrightarrow}{\kappa}_0$, $\overset{\leftrightarrow}{\kappa}_1$, and $\overset{\leftrightarrow}{\kappa}_2$ are given by (8.46), (8.47), and (8.48), respectively, or

$$\overset{\leftrightarrow}{\kappa}_0 = \frac{1}{4\pi^3\hbar}\int \tau\mathbf{v}\mathbf{v}\frac{d^2S_F}{v}, \tag{8.90}$$

$$\overset{\leftrightarrow}{\kappa}_1 = \frac{\pi^2}{3}\left(k_BT\right)^2\left(\frac{\partial\,\overset{\leftrightarrow}{\kappa}_0}{\partial E}\right)_{E_F}, \tag{8.91}$$

and

$$\overset{\leftrightarrow}{\kappa}_2 = \frac{(k_BT)^2}{12\pi\hbar}\int \tau\mathbf{v}\mathbf{v}\frac{d^2S_F}{v}. \tag{8.92}$$

We now evaluate the contribution to the thermal conductivity from the thermoelectric coupling effects for the case of a metal having a simple dispersion relation

$$E = \hbar^2k^2/2m^*. \tag{8.93}$$

In the case where τ is considered to be independent of E, (8.91) and (8.90), respectively, provide expressions for κ_1 and κ_0 from which

$$\frac{1}{T}\kappa_1\kappa_0^{-1}\kappa_1 = \frac{\pi^4 n\tau}{4m^*}k_B^2T\left(\frac{k_BT}{E_F}\right)^2 \tag{8.94}$$

so that from (8.28) and (8.94) the total electronic thermal conductivity for the metal becomes

$$\kappa_e = \frac{\pi^2 n\tau}{3m^*}k_B^2T\left[1 - \frac{3\pi^2}{4}\left(\frac{k_BT}{E_F}\right)^2\right]. \tag{8.95}$$

For typical metals $(T/T_F) \sim (1/30)$ at room temperature so that the thermoelectric correction term is less than 1%. For highly degenerate semiconductors as are of interest for thermoelectric applications, the complete Fermi function must be considered. Nevertheless, for most thermoelectric applications, the contribution of phonons strongly dominates over that of electrons, and the thermal transport of thermoelectric materials requires detailed consideration of phonons and electron-phonon interaction.

8.4 Thermoelectric Measurements

8.4.1 Seebeck Effect (Thermopower)

The thermopower $\mathcal{S}$ as defined in (8.50) and is the characteristic coefficient in the Seebeck effect, where a metal subjected to a thermal gradient ∇T exhibits an electric field $\mathbf{E} = \mathcal{S}\nabla T$. The measurements are made under an open-circuit voltage V and under conditions of no current flow.

In the application of the Seebeck effect to thermocouple operation, we usually measure the difference in thermopower $\mathcal{S}_A - \mathcal{S}_B$ between two different metals A and B by measuring the open circuit voltage V_{AB} as shown in Fig. 8.5. This voltage can be calculated from

$$\begin{aligned} V_{AB} &= -\oint \mathbf{E} \cdot d\mathbf{r} = -\oint \mathcal{S} \tfrac{\partial T}{\partial \mathbf{r}}\, d\mathbf{r} \\ &= \int_{T_0}^{T_1} \mathcal{S}_B dT + \int_{T_1}^{T_2} \mathcal{S}_A dT + \int_{T_2}^{T_0} \mathcal{S}_B dT \\ &= \int_{T_1}^{T_2} (\mathcal{S}_A - \mathcal{S}_B) dT. \end{aligned} \tag{8.96}$$

With $T_1 \neq T_2$, an open-circuit potential difference V_{AB} can be measured and (8.96) shows that V_{AB} is independent of the temperature T_0 for the simple case considered. Thus if T_1 is known and V_{AB} is measured, then the temperature T_2 can be found from

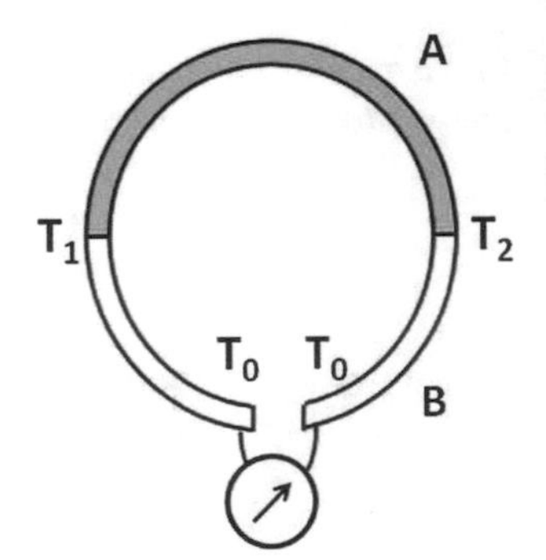

Fig. 8.5 Thermopower between two different metals showing the principle of operation of a thermocouple material or device under open circuit conditions (i.e., $j = 0$).

the calibration table of the thermocouple. From the simple expression of (8.77)

$$\mathscr{S} = \frac{\pi^2 k_B}{2e}\left(\frac{k_B T}{E_F}\right) \tag{8.97}$$

a linear dependence of $\mathscr{S}$ on T is predicted for simple metals. For actual thermocouples used for temperature measurements, the $\mathscr{S}(T)$ dependence is approximately linear, but is given by an accurate calibration table to account for small deviations from this linear relation that occur because of the simplifying approximations made in the actual devices where the electrons actually interact with phonons and with other electrons and such effects need to be considered for real materials and real devices. Thermocouples are calibrated at several fixed temperatures and the calibration table actually used data comes from a fit made for these thermal data to a polynomial function that is approximately linear in T.

8.4.2 Peltier Effect

The Peltier effect is the observation of a thermal current $\mathbf{U} = \overleftrightarrow{\Pi} \cdot \mathbf{j}$ in the presence of an electric current $\mathbf{j}$ with no thermal gradient ($\nabla T = 0$) so that

$$\overleftrightarrow{\Pi} = T \overleftrightarrow{\mathscr{S}}. \tag{8.98}$$

The Peltier effect measures the heat generated (or absorbed) at the junction of two dissimilar metals held at constant temperature, when an electric current passes through the junction. Sending electric current around a circuit of two dissimilar metals cools one junction and heats another and is the basis for the operation of thermoelectric coolers. This thermoelectric effect is represented schematically in Fig. 8.6. Because of the similarities between the Peltier coefficient and the Seebeck coefficient, materials exhibiting a large Seebeck effect also show a large Peltier effect. Since both $\overleftrightarrow{\mathscr{S}}$ and $\overleftrightarrow{\Pi}$ are proportional to $(1/e)$, the sign of $\overleftrightarrow{\mathscr{S}}$ and $\overleftrightarrow{\Pi}$ is negative for electrons and positive for holes in the case of degenerate semiconductors. Reversing the direction of $\mathbf{j}$, will interchange the junctions where heat is generated (absorbed).

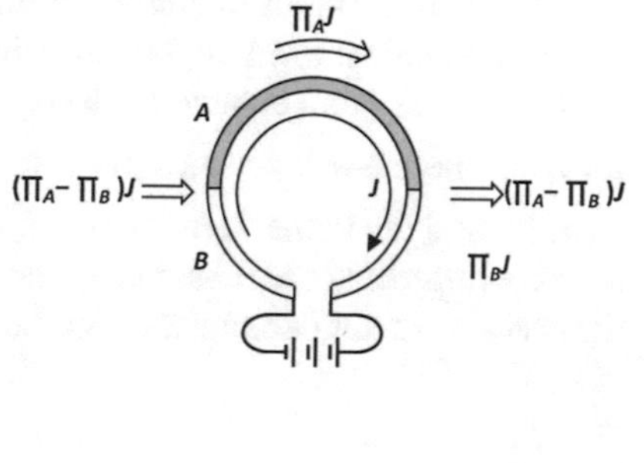

Fig. 8.6 A heat engine based on the Peltier Effect with heat $(\Pi_A - \Pi_B)j$ introduced at one junction and extracted at another under the conditions of no temperature gradient ($\nabla T = 0$).

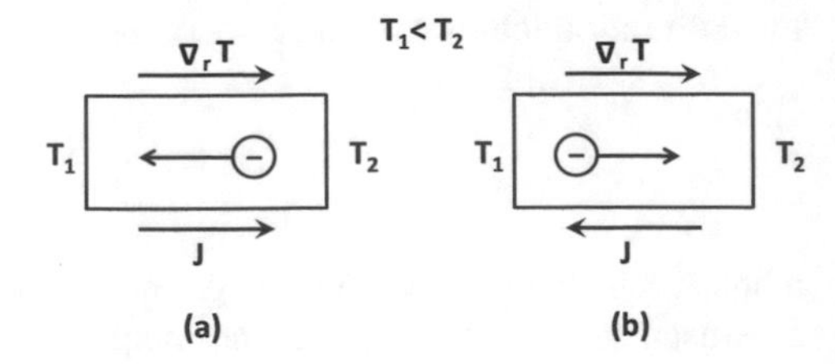

Fig. 8.7 The Thomson term in an n-type semiconductor produces **(a)** heating when **j** and ∇T are in the same direction and **(b)** cooling when **j** and ∇T are in opposite directions.

8.4.3 Thomson Effect

Assume that we have an electric circuit consisting of a single metal conductor. The power, generated in a sample, such as an n-type semiconductor, as shown in Fig. 8.7, is

$$P = \mathbf{j} \cdot \mathbf{E} \tag{8.99}$$

where the electric field can be obtained from (8.49) and (8.51) as

$$\mathbf{E} = (\overset{\leftrightarrow}{\sigma^{-1}}) \cdot \mathbf{j} - \overset{\leftrightarrow}{\mathcal{T}_b} \cdot \nabla T \tag{8.100}$$

where $\overset{\leftrightarrow}{\mathcal{T}_b}$ is the Thomson coefficient defined in (8.51) and is related to the Seebeck coefficient $\overset{\leftrightarrow}{\mathcal{S}}$ as discussed in Sects. 8.3 and 8.3.1. Substitution of (8.100) into (8.99) yields the total power dissipation

$$P = \mathbf{j} \cdot (\overset{\leftrightarrow}{\sigma^{-1}}) \cdot \mathbf{j} - \mathbf{j} \cdot \overset{\leftrightarrow}{\mathcal{T}_b} \cdot \nabla T. \tag{8.101}$$

The first term in (8.101) is the conventional joule heating term while the second term is the contribution from the Thomson effect. For an n-type semiconductor $\overset{\leftrightarrow}{\mathcal{T}_b}$ is negative. Thus when **j** and ∇T are parallel, heating will result, as in Fig. 8.7a. However if **j** and ∇T are antiparallel, as in Fig. 8.7b, cooling will occur. Thus reversal of the direction of **j** without changing the direction of ∇T will reverse the sign of the Thomson contribution. Likewise, a reversal in the direction of ∇T keeping the direction of **j** unchanged will also reverse the sign of the Thomson contribution.

Thus, if either (but not both) the directions of the electric current or the direction of the thermal gradient is reversed, an absorption of heat from the surroundings will take place. The Thomson effect is utilized in thermoelectric refrigerators which are useful as practical low temperature laboratory coolers. Referring to Fig. 8.8, we see a schematic diagram explaining the operation of a thermoelectric cooler. We see that for a degenerate n-type semiconductor where the Thomson term $\overset{\leftrightarrow}{\mathcal{T}_b}$, the Seebeck term $\overset{\leftrightarrow}{\mathcal{S}}$, and the Peltier term $\overset{\leftrightarrow}{\Pi}$ are all negative, and when **j** and ∇T are antiparallel, then cooling occurs and heat is extracted from the cold junction and transferred to the heat sink at temperature T_H. For the p-type leg, all the thermoelectric coefficients

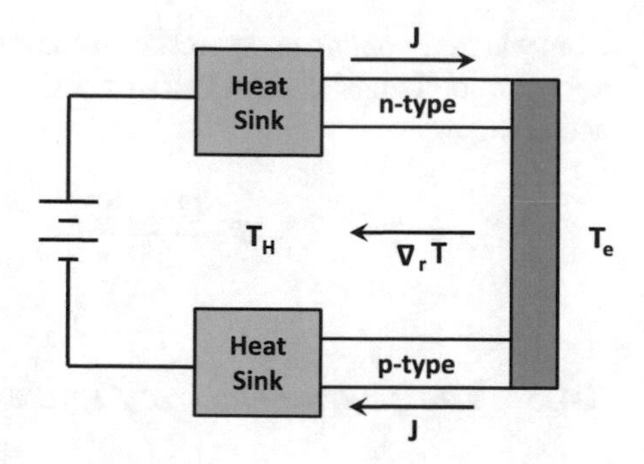

Fig. 8.8 Schematic diagram of a thermoelectric cooler. The heat sinks and cold junctions are metals that form ohmic contacts to the active thermoelectric n-type and p-type semiconductors.

are positive, so (8.101) shows that cooling occurs when $\mathbf{j}$ and ∇T are parallel. Thus both the n-type and p-type legs in a thermoelectric element contribute to cooling in a thermoelectric cooler.

8.4.4 The Kelvin Relations

The three thermoelectric effects are related and the relations between these coefficients were first derived by Lord Kelvin after he became a Lord and changed his family name from Thomson to Kelvin. The Kelvin relations are based on arguments of irreversible thermodynamics and relate Π, $\mathcal{S}$, and $\mathcal{T}_b$.

If we define the thermopower $\mathcal{S}_{AB} = \mathcal{S}_B - \mathcal{S}_A$ and the Peltier coefficient similarly $\Pi_{AB} = \Pi_A - \Pi_B$ for material A joined to material B, then we obtain the first Kelvin relation:

$$\mathcal{S}_{AB} = \frac{\Pi_{AB}}{T}. \tag{8.102}$$

The Thomson coefficient $\mathcal{T}_b$ is defined by

$$\mathcal{T}_b = T\frac{\partial \mathcal{S}}{\partial T} \tag{8.103}$$

which allows determination of the Seebeck coefficient at temperature T_0 by integration of (8.103)

$$\mathcal{S}(T_0) = \int_0^{T_0}\left[\frac{\mathcal{T}_b(T)}{T}\right]dT. \tag{8.104}$$

Furthermore, from the above definitions, we deduce the second Kelvin relation

$$\mathcal{T}_{b,A} - \mathcal{T}_{b,B} = T\frac{\partial \mathcal{S}_{AB}}{\partial T} = T\frac{\partial \mathcal{S}_A}{\partial T} - T\frac{\partial \mathcal{S}_B}{\partial T} \tag{8.105}$$

from which we obtain an expression relating all three thermoelectric coefficients, the Seebeck coefficient $\mathscr{S}$, the Peltier coefficient Π and the Thomson coefficient $\mathscr{T}_{b,A}$ according to

$$\mathscr{T}_{b,A} = T\frac{\partial \mathscr{S}_A}{\partial T} = T\frac{\partial(\Pi_A/T)}{\partial T} = \frac{\partial \Pi_A}{\partial T} - \frac{\Pi_A}{T} = \frac{\partial \Pi_A}{\partial T} - \mathscr{S}_A. \tag{8.106}$$

8.4.5 The Thermoelectric Figure of Merit

A good thermoelectric material for cooling applications must have a high thermoelectric figure of merit, ZT, which is a dimensionless parameter defined by

$$ZT = \frac{\mathscr{S}^2\sigma}{\kappa}T \tag{8.107}$$

where $\mathscr{S}$ is the thermoelectric power (Seebeck coefficient), σ is the electrical conductivity, and κ is the thermal conductivity. In order to achieve a high ZT, one requires a high thermoelectric power $\mathscr{S}$ to couple electrons transport and temperature differences, a high electrical conductivity σ to maintain high carrier mobility, and a low thermal conductivity κ to retain the applied thermal gradient. In general, it is difficult in practical systems to increase ZT for the following reasons: increasing $\mathscr{S}$ for simple materials also leads to a simultaneous decrease in σ, and an increase in σ leads to a comparable increase in the electronic contribution to κ because of the Wiedemann–Franz law. So with known conventional solids, a limit is rapidly obtained where a modification to any one of the three parameters $\mathscr{S}$, σ, or κ adversely affects the other transport coefficients, so that the resulting ZT does not vary significantly. Currently, the commercially available materials with the highest ZT values are Bi_2Te_3 alloys such as $Bi_{0.5}Sb_{1.5}Te_3$ with $ZT \sim 1$ at 300 K.

Only small increases in ZT were achieved in the 1960–1990 period. Since 1994, new interest has been revived in thermoelectricity with the discovery of (1) new materials: skutterudites – $CeFe_{4-x}Co_xSb_{12}$ or $LaFe_{4-x}Co_xSb_{12}$ for $0 < x < 4$, which offer promise for higher ZT values in bulk materials, and (2) low dimensional systems (quantum wells, quantum wires) which offer promise for enhanced ZT relative to their higher ZT bulk counterparts Z values in the same material. Thus thermoelectricity has again become an active research field with significant progress made since 1994 and quite a lot of this progress since 1994 involves the use of nanomaterials.

8.5 The Phonon Drag Effect

For a simple metal such as an alkali metal one would expect the thermopower $\mathcal{S}$ to be given by the simple expression in (8.77), and to be negative since the carriers are electrons. This is true at room temperature for all of the alkali metals except Li. Furthermore, $\mathcal{S}$ is positive for the noble metals Au, Ag and Cu. The anomalous sign of $\mathcal{S}$ in these metals can be understood by recalling the complex Fermi surfaces for these metals (see Fig. 4.6), where we note that copper in fact exhibits hole orbits in the extended zone. In general, with multiple carrier types as occur in semiconductors, the interpretation of thermopower data can become complicated.

Another complication which must also be considered, especially at low temperatures, is the *phonon drag effect*. In the presence of a thermal gradient, the phonons will diffuse and "drag" the electrons along with them because of the electron-phonon interaction (discussed in Chap. 6). For a simple explanation of phonon drag, consider a gas of phonons with an average energy density E_{ph}/V where V is the volume. Using kinetic theory, we find that the phonon gas exerts a pressure

$$P = \frac{1}{3}\left(\frac{E_{\mathrm{ph}}}{V}\right) \tag{8.108}$$

on the electron gas. In the presence of a thermal gradient, the electrons are subject to a force density

$$F_x/V = -dP/dx = -\frac{1}{3V}\left(\frac{dE_{\mathrm{ph}}}{dT}\right)\frac{dT}{dx}. \tag{8.109}$$

To prevent the flow of current, this force must be balanced by the electric force. Thus, for an electron density n, we obtain

$$-neE_x + F_x/V = 0 \tag{8.110}$$

giving a phonon-drag contribution to the thermopower. Using the definition of the Seebeck coefficient for an open circuit system, we can write

$$\mathcal{S}_{\mathrm{ph}} = \frac{E_x}{(dT/dx)} \approx -\left(\frac{1}{3enV}\right)\frac{dE_{\mathrm{ph}}}{dT} = \frac{C_{\mathrm{ph}}}{3en} \tag{8.111}$$

where C_{ph} is the phonon heat capacity per unit volume. Although this is only a rough approximate derivation, it predicts the correct temperature dependence, in that the phonon-drag contribution is important at temperatures where the phonon specific heat is large. The total thermopower is a sum of the diffusion contribution (considered in Sect. 8.4.1) and the phonon drag term $\mathcal{S}_{\mathrm{ph}}$ in (8.111).

The phonon drag effect depends on the electron-phonon coupling; at higher temperatures where the phonon-phonon coupling (Umklapp processes) becomes more important than the electron-phonon coupling, phonon drag effects become less important as is discussed in the next chapter (see Sect. 9.4.4).

Problems

For problems 8.1–8.7 below, assume the following properties for Bi_2Te_3

- Effective masses: $m_x = 0.02$, $m_y = 0.08$, $m_z = 0.32$
- Mobility: $\mu_x = 1200\,cm^2V^{-1}$
- Lattice thermal conductivity: $\kappa_L = 1.5\ Wm^{-1}K^{-1}$

Perform all calculations at room temperature, T = 300 K.

8.1 Write a function in Matlab or Mathematica (or other software of your choice) that calculates the dimensionless Fermi integral:

$$F_i = F_i(\zeta^*) = \int_0^{\infty} \frac{x^i dx}{e^{(x-\zeta^*)} + 1} \tag{8.112}$$

where

$$\zeta^* = \frac{E_F}{k_B T} \tag{8.113}$$

8.2 The transport coefficients of a bulk (3D) material in the constant relaxation time approximation are given by the following Boltzmann transport equations:

$$\sigma = \frac{e}{3\pi^2}\left(\frac{2k_BT}{\hbar^2}\right)^{\frac{3}{2}} (m_x m_y m_z)^{\frac{1}{2}} \mu_x \left(\frac{3}{2}F_{1/2}\right) \tag{8.114}$$

$$\mathscr{S} = -\frac{k_B}{e}\left(\frac{5F_{3/2}}{3F_{1/2}} - \zeta^*\right) \tag{8.115}$$

$$\kappa_e = \frac{k_B^2 T}{3\pi^2 e}\left(\frac{2k_BT}{\hbar^2}\right)^{\frac{3}{2}} (m_x m_y m_z)^{\frac{1}{2}} \mu_x \left(\frac{7}{2}F_{5/2} - \frac{25F_{3/2}^2}{6F_{1/2}}\right) \tag{8.116}$$

Using these equations, write functions in MatLab for the following quantities:

(a) Electrical conductivity (σ)
(b) Seebeck coefficient ($\mathscr{S}$) in units of μV/K
(c) Electrical component of the thermal conductivity (κ_e).

Plot these coefficients as a function of Fermi level E_F (or chemical potential).

8.3 The transport coefficients of a 2D material in the constant relaxation time approximation are given by the following Boltzmann transport equations:

$$\sigma = \frac{e}{2\pi a}\left(\frac{2k_B T}{\hbar^2}\right)(m_x m_y)^{\frac{1}{2}}\mu_x\,(F_0) \tag{8.117}$$

$$\mathcal{S} = -\frac{k_B}{e}\left(\frac{2F_1}{F_0} - \zeta^*\right) \tag{8.118}$$

$$\kappa_e = \frac{k_B^2 T}{2\pi a e}\left(\frac{2k_B T}{\hbar^2}\right)(m_x m_y)^{\frac{1}{2}}\mu_x\left(3F_2 - \frac{4F_1^2}{F_0}\right) \tag{8.119}$$

For a 5 nm quantum well (a = 5 nm), write functions in MatLab for the following quantities:

(a) Electrical conductivity (σ)
(b) Seebeck coefficient ($\mathcal{S}$) in units of μV/K
(c) Electrical component of the thermal conductivity (κ_e).

Plot these coefficients as a function of Fermi level E_F (or chemical potential).

8.4 In 1D, the transport coefficients are given by

$$\sigma = \frac{2e}{\pi a^2}\left(\frac{2k_B T}{\hbar^2}\right)^{\frac{1}{2}} m_x^{\frac{1}{2}}\mu_x\left(\frac{1}{2}F_{-\frac{1}{2}}\right) \tag{8.120}$$

$$\mathcal{S} = -\frac{k_B}{e}\left(\frac{3F_{1/2}}{3F_{-1/2}} - \zeta^*\right) \tag{8.121}$$

$$\kappa_e = \frac{2k_B^2 T}{\pi^2 a e}\left(\frac{k_B T}{\hbar^2}\right)^{\frac{1}{2}} m_x^{\frac{1}{2}}\mu_x\left(\frac{5}{2}F_{3/2} - \frac{9F_{1/2}^2}{2F_{-1/2}}\right) \tag{8.122}$$

For a nanowire with a 5 nm × 5 nm square cross section, write functions in Matlab for the following quantities

(a) Electrical conductivity (σ)
(b) Seebeck coefficient ($\mathcal{S}$) in units of μV/K
(c) Electrical component of the thermal conductivity (κ_e).

Plot these coefficients as a function of Fermi level E_F (or chemical potential).

8.5 The thermoelectric figure of merit is given by

$$Z = \frac{\mathcal{S}^2\sigma}{\kappa_e + \kappa_L} \tag{8.123}$$

Using the functions derived in the problems above, plot the dimensionless figure of merit ZT (at room temperature) as a function of the Fermi energy for the three cases

(a) 3D bulk
(b) 2D (5 nm width)
(c) 1D (5 nm × 5 nm cross section)

8.6 Find the optimum value for ZT (with respect to the Fermi energy) for each of the 3 cases above. Plot the optimum ZT value as a function of quantum well width and nanowire width for 1, 2, 3, 4 and 5 nm.

8.7 The Wiedemann-Franz Law states that

$$\mathbf{K} = \sigma T \frac{\pi^2 k_B^2}{3e^2} \tag{8.124}$$

On this basis, explain how it is possible for sapphire to be both an excellent thermal conductor and an excellent electrical insulator.

8.8 Suppose that you measure the thermal conductivity of a sample at 100°C. How would you estimate the fraction of the heat that is carried by electron (or hole) carriers?

8.9 Suppose that Si is doped with an isoelectronic impurity in column IV of the periodic table (such as Ge or Sn), will the doping effect be greater on the electrical conductivity or on the thermal conductivity, and why?

8.10 Could this thermoelectric cooler (see figure below) be modified to become a thermoelectric heater? If so, explain how it can be done.

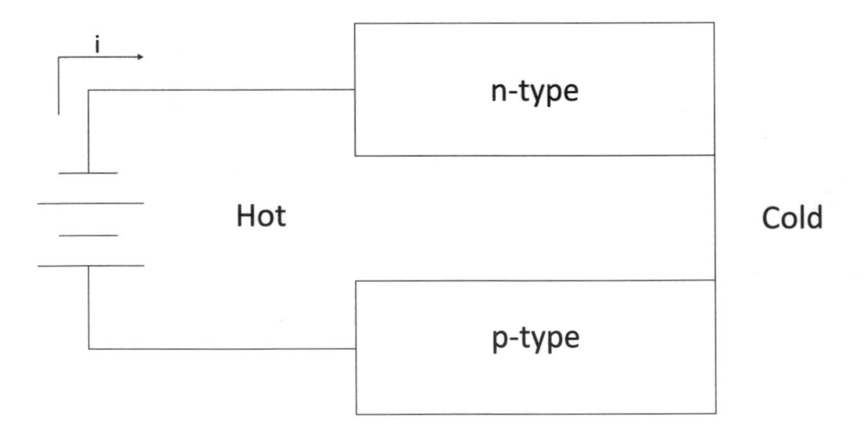

Suggested Readings

Ziman, *Principles of the Theory of Solids* (Cambridge University Press, Cambridge, 1972). Chapters 7

F. Reif, *Fundamentals of Statistical and Thermal Physics* (McGraw-Hill, New York, 2008)

C. Kittel, *Introduction to Solid State Physics* (Wiley, New York, 2005)

H. Rosenberg, *The Solid State* (Oxford University Press, New York, 2004)

Wolfe, Holonyak, Stillman, *Physical Properties of Semiconductors* (Prentice Hall, Englewood Cliffs, 1989). Chapter 5

Fong et al., Measurement of the electronic thermal conductance channels and heat capacity of graphene at low temperature. Phys. Rev. X **3**, 041008 (2013)

References

C.L. Kane, M.P.A. Fisher, Thermal transport in a Luttinger liquid. Phys. Rev. Lett. **76**, 3192 (1996)

Nicholas Wakeham, Alimamy F. Bangura, Xu Xiaofeng, Jean-Francois Mercure, Martha Greenblatt, Nigel E. Hussey, Gross violation of the Wiedemann Franz law in a quasi-one-dimensional conductor. Nat. Commun. **2**, 396 (2011)

Chapter 9
Electron and Phonon Scattering

9.1 Electron Scattering

The thermal properties of solid materials depend on the availability of carriers and on their scattering rates. In the previous chapters, we focused on the carriers and their generation. In this Chapter we focus on the relevant electron and phonon scattering mechanisms.

Electron scattering brings an electronic system which has been subjected to external perturbations back to equilibrium. Collisions also alter the momentum of all the carriers, as the electrons are brought back into equilibrium. Electron collisions can occur through a variety of mechanisms, such as electron-phonon, electron-impurity, electron-defect, electron-boundary and electron-electron scattering processes. Electron scattering is handled here by considering the collision term in the Boltzmann equation.

In principle, the collision rates can be calculated from using scattering theory in single form. To do this, we introduce a transition probability $S(\mathbf{k}, \mathbf{k}')$ for scattering the electron from a state $\mathbf{k}$ to a state $\mathbf{k}'$. Since electrons obey the Pauli principle, scattering will occur from an occupied to an unoccupied state. The process of scattering from $\mathbf{k}$ to $\mathbf{k}'$ decreases the distribution function $f(\mathbf{r}, \mathbf{k}, t)$ depending on the probability that $\mathbf{k}$ is occupied and that $\mathbf{k}'$ is unoccupied. The process of scattering an electron from $\mathbf{k}'$ to $\mathbf{k}$ increases the distribution function $f(\mathbf{r}, \mathbf{k}, t)$ and depends on the probability that state $\mathbf{k}'$ is occupied and state $\mathbf{k}$ is unoccupied. We will use the following notation for describing a general scattering process:

- f_k is the probability that an electron occupied with initial state $\mathbf{k}$
- $[1 - f_k]$ is the probability that state $\mathbf{k}$ is unoccupied
- $S(\mathbf{k}, \mathbf{k}')$ is the probability per unit time that an electron in state $\mathbf{k}$ will be scattered to state $\mathbf{k}'$
- $S(\mathbf{k}', \mathbf{k})$ is the probability per unit time that an electron in state $\mathbf{k}'$ will be scattered back into state $\mathbf{k}$.

M. Dresselhaus et al., *Solid State Properties*, Graduate Texts in Physics,
https://doi.org/10.1007/978-3-662-55922-2_9

Using these definitions, the rate of change of the distribution function in the Boltzmann equation (see (7.4)) due to collisions can be written as:

$$\left.\frac{\partial f(\mathbf{r}, \mathbf{k}, t)}{\partial t}\right|_{\text{collisions}} = \int d^3k' [f_{k'}(1 - f_k)S(\mathbf{k}', \mathbf{k}) - f_k(1 - f_{k'})S(\mathbf{k}, \mathbf{k}')] \quad (9.1)$$

where d^3k' is a volume element in $\mathbf{k}'$ space. The integration in (9.1) is over k space and the spherical coordinate system is shown in Fig. 9.1, together with the arbitrary force $\mathbf{F}$ responsible for the scattering event that introduces a perturbation described by

$$f_{\mathbf{k}} = f_{0\mathbf{k}} + \frac{\partial f_{0\mathbf{k}}}{\partial E}\frac{\hbar}{m^*}\mathbf{k} \cdot \mathbf{F} + \ldots \quad (9.2)$$

where $f_{0\mathbf{k}}$ denotes the equilibrium distribution. Using Fermi's Golden Rule for the transition probability per unit time between states $\mathbf{k}$ and $\mathbf{k}'$ we can write

$$S(\mathbf{k}, \mathbf{k}') \simeq \frac{2\pi}{\hbar}|\mathscr{H}_{\mathbf{kk}'}|^2\{\delta[E(\mathbf{k})] - \delta[E(\mathbf{k}')]\} \quad (9.3)$$

where the matrix element of the Hamiltonian coupling states $\mathbf{k}$ and $\mathbf{k}'$ is

$$\mathscr{H}_{\mathbf{kk}'} = \frac{1}{N}\int_V \psi^*_{\mathbf{k}}(\mathbf{r})\nabla V \psi_{\mathbf{k}'}(\mathbf{r})d^3r, \quad (9.4)$$

in which N is the number of unit cells in the sample and ∇V is the perturbation Hamiltonian term responsible for the scattering event associated with the force $\mathbf{F}$.

At equilibrium $f_k = f_0(E)$ and the principle of detailed balance applies

$$S(\mathbf{k}', \mathbf{k})f_0(E')[1 - f_0(E)] = S(\mathbf{k}, \mathbf{k}')f_0(E)[1 - f_0(E')] \quad (9.5)$$

so that the distribution function does not experience a net change via collisions when in the equilibrium state:

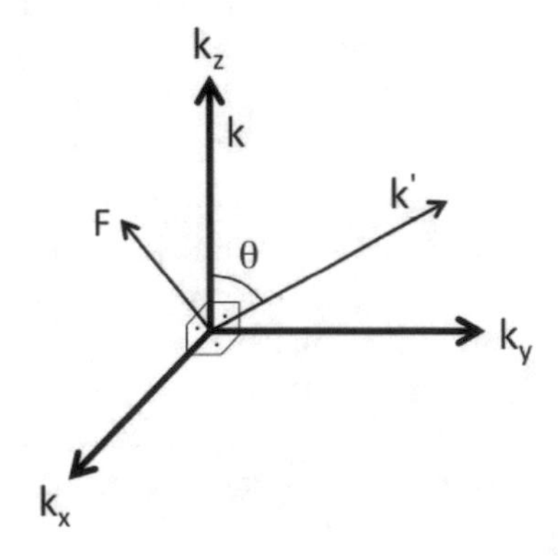

Fig. 9.1 Coordinate system in reciprocal space for an electron with wave vector $\mathbf{k}$ (along the k_z axis) scattering into a state with wavevector $\mathbf{k}'$ in an arbitrary force field $\mathbf{F}$. The scattering center is at the origin of the coordinate system. For simplicity the event is rotated so that $\mathbf{F}$ has no k_y component

$$(\partial f(\mathbf{r}, \mathbf{k}, t)/\partial t)|_{\text{collisions}} = 0. \tag{9.6}$$

We define collisions as *elastic collisions* when $E(\mathbf{k}') = E(\mathbf{k})$ and in this case $f_0(E') = f_0(E)$ so that $S(\mathbf{k}', \mathbf{k}) = S(\mathbf{k}, \mathbf{k}')$. Collisions for which $E(\mathbf{k}') \neq E(\mathbf{k})$ are termed *inelastic* collisions. The term *quasi-elastic* is used to characterize collisions where the percentage change in energy is small. For our purposes here, we shall consider $S(\mathbf{k}, \mathbf{k}')$ as a known function which can be calculated quantum mechanically by a detailed consideration of the scattering mechanisms which are important for a given practical case; this statement is true in principle, but in practice $S(\mathbf{k}, \mathbf{k}')$ is usually specified in an approximate way.

The return to equilibrium depends on the frequency of collisions and the effectiveness of a scattering event in randomizing the motion of the electrons. Thus, small angle scattering is not as effective in restoring a system to equilibrium as for the case of large angle scattering. For this reason we distinguish between τ_D, the time for the system to be restored to equilibrium, and τ_c, the time between collisions. These times are related by

$$\tau_D = \frac{\tau_c}{1 - \cos\theta} \tag{9.7}$$

where θ is the mean change of angle of the electron velocity on collision (see Fig. 9.1). The time τ_D is the quantity which enters into Boltzmann's equation as the relaxation time, while $1/\tau_c$ determines the actual scattering rate.

The mean free time between collisions, τ_c, is related to several other quantities of interest: the mean free path ℓ_f, the scattering cross section σ_d, and the concentration of scattering centers N_c by

$$\tau_c = \frac{1}{N_c \sigma_d v} \tag{9.8}$$

where v is the drift velocity given by

$$v = \frac{\ell_f}{\tau_c} = \frac{1}{N_c \sigma_d \tau_c} \tag{9.9}$$

and v is in the direction of the electron transport. From (9.9), we see that $\ell_f = 1/N_c\sigma_d$. The drift velocity is of course very much smaller in magnitude than the instantaneous velocity of the electron at the Fermi level, which is typically of magnitude $v_F \sim 10^8$ cm/s. Electron scattering centers include phonons, impurities, dislocations, vacancies, the crystal surface, etc.

The most important electron scattering mechanism for both metals and semiconductors is electron-phonon scattering (scattering of electrons by the thermal motion of the lattice), though the scattering processes for metals differs in detail from those in semiconductors. In the case of metals, much of the Brillouin zone is occupied by electrons, while in the case of semiconductors, most of the Brillouin zone is unoccupied, and represents states into which electrons can be scattered. In the case of metals, electrons are scattered from one point on the Fermi surface to another point on the Fermi surface, and a large change in momentum occurs, corresponding to a large change in **k**. In the case of semiconductors, changes in wave vector from **k** to $-\mathbf{k}$ normally correspond to a very small change in wave vector, and thus changes from **k** to $-\mathbf{k}$ can be accomplished much more easily in the case of semiconductors. By the same token, small angle scattering (which is not so efficient for returning the system to equilibrium) is especially important for semiconductors where the change in wavevector is small. Since the scattering processes in semiconductors and metals are quite different, they will be discussed separately in the next sections.

Scattering probabilities for more than one scattering process are taken to be additive and therefore so are the reciprocal scattering times and scattering rates. For the total reciprocal scattering time $(\tau^{-1})_{\text{total}}$ we write:

$$(\tau^{-1})_{\text{total}} = \sum_i \tau_i^{-1} \tag{9.10}$$

since $1/\tau$ is proportional to the scattering probability. Equation (9.10) is commonly referred to as "Matthiessen's rule" Metals have large Fermi wavevectors k_F, and therefore large momentum transfers Δk can occur as a result of electronic collisions. In contrast, for semiconductors, k_F is small and so also is Δk on collision.

9.2 Scattering Processes in Semiconductors

9.2.1 Electron-Phonon Scattering in Semiconductors

Electron-phonon scattering is the dominant scattering mechanism in crystalline semiconductors except at very low temperatures where the phonon density is low. Conservation of energy in the scattering process, which creates or absorbs a phonon of energy $\hbar\omega(\mathbf{q})$, is written as:

$$E_i - E_f = \pm\hbar\omega(\mathbf{q}) = \frac{\hbar^2}{2m^*}(k_i^2 - k_f^2), \tag{9.11}$$

where E_i is the initial energy, E_f is the final energy, k_i the initial wavevector, and k_f the final wavevector. Here, the "+" sign corresponds to the creation of phonons (the phonon emission process), while the "−" sign corresponds to the annihilation of phonons (the phonon absorption process). Conservation of momentum in the scattering of an electron by a phonon of wavevector **q** yields

$$\mathbf{k_i} - \mathbf{k_f} = \pm\mathbf{q}. \tag{9.12}$$

For semiconductors, the electrons involved in the scattering event generally remain in the vicinity of a single band extremum and involve only a small change in **k** and hence only low phonon **q** vectors participate. The probability that an electron makes a transition from an initial state i to a final state f is proportional to:

(a) the availability of final states for electrons,
(b) the probability of absorbing or emitting a phonon,
(c) the strength of the electron-phonon coupling or electron-phonon interaction.

The first factor, the availability of final states, is proportional to the density of final electron states $\rho(E_f)$ times the probability that the final state is unoccupied. This occupation probability for a semiconductor is assumed to be unity since the conduction band is essentially empty. For a simple parabolic band, $\rho(E_f)$ is (from (7.64)):

$$\rho(E_f) = \frac{(2m^*)^{3/2} E_f^{1/2}}{2\pi^2\hbar^3} = (2m^*)^{3/2} \frac{[E_i \pm \hbar\omega(\mathbf{q})]^{1/2}}{2\pi^2\hbar^3}, \tag{9.13}$$

where (9.11) has been employed and the "+" sign corresponds to absorption of a phonon and the "−" sign corresponds to phonon emission.

The probability of absorbing or emitting a phonon is proportional to the electron-phonon coupling $G(\mathbf{q})$ and to the phonon density $n(\mathbf{q})$ for absorption, and the phonon density $[1 + n(\mathbf{q})]$ for emission, where $n(\mathbf{q})$ is given by the Bose-Einstein factor

$$n(\mathbf{q}) = \frac{1}{e^{\hbar\omega(\mathbf{q})/k_B T} - 1}. \tag{9.14}$$

Combining the terms in (9.13) and (9.14) gives a scattering probability (or $1/\tau_c$) proportional to a sum over final states

$$\frac{1}{\tau_c} \sim \frac{(2m^*)^{3/2}}{2\pi^2\hbar^3} \sum_{\mathbf{q}} G(\mathbf{q}) \left[\frac{[E_i + \hbar\omega(\mathbf{q})]^{1/2}}{e^{\hbar\omega(\mathbf{q})/k_B T} - 1} + \frac{[E_i - \hbar\omega(\mathbf{q})]^{1/2}}{1 - e^{-\hbar\omega(\mathbf{q})/k_B T}} \right] \tag{9.15}$$

where the first term in the big bracket of (9.15) corresponds to phonon absorption and the second term to phonon emission. If $E_i < \hbar\omega(\mathbf{q})$, only the *phonon absorption* process is energetically allowed.

The electron-phonon coupling coefficient $G(\mathbf{q})$ in (9.15) depends on the electron-phonon coupling mechanism. There are three important coupling mechanisms in semiconductors which we briefly describe below: electromagnetic coupling, piezoelectric coupling, and deformation-potential coupling.

Electromagnetic Coupling

This coupling is important only for semiconductors where the charge distribution has different signs on neighboring ion sites when two species of atoms are involved. In this case, the oscillatory electric field can give rise to oscillating dipole moments associated with the motion of neighboring ion sites in the optical modes (see Fig. 9.2). The electromagnetic coupling mechanism is important in coupling electrons to optical phonon modes in III-V and II-VI compound semiconductors, but does not contribute in the case of silicon. To describe the optical modes we can use the Einstein approxi-

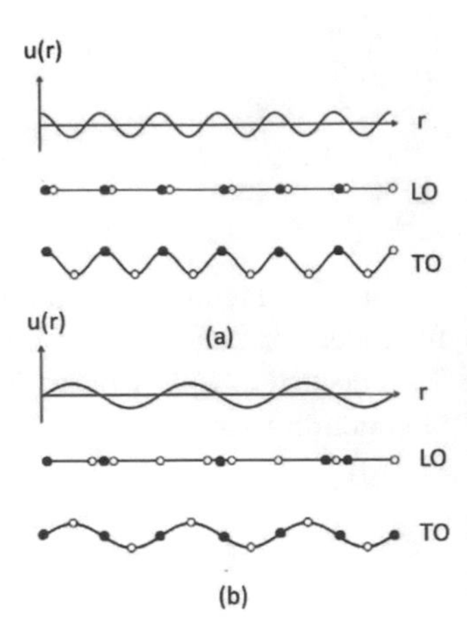

Fig. 9.2 Displacements $\mathbf{u}(\mathbf{r})$ of atoms in a diatomic chain for longitudinal optical (LO) and transverse optical (TO) phonons at **a** the center and **b** the edge of the Brillouin zone. The lighter mass atoms are indicated by open circles. For zone edge optical phonons, only the lighter atoms are displaced

mation, since $\omega(\mathbf{q})$ is only weakly dependent on $\mathbf{q}$ for the optical modes of frequency ω_0. In this case $\hbar\omega_0 \gg k_B T$ and $\hbar\omega_0 \gg E$ where E is the electron energy, so that from (9.15) the collision rate is proportional to

$$\frac{1}{\tau_c} \sim \frac{m^{*3/2}(\hbar\omega_0)^{1/2}}{e^{\hbar\omega_0/k_B T} - 1}. \tag{9.16}$$

Thus, the collision rate depends on the temperature T, the optical phonon frequency ω_0 and the electron effective mass m^*. The corresponding mobility for *optical phonon scattering* is

$$\mu = \frac{e\langle\tau\rangle}{m^*} \sim \frac{e(e^{\hbar\omega_0/k_B T} - 1)}{m^{*5/2}(\hbar\omega_0)^{1/2}} \tag{9.17}$$

Thus for optical phonon scattering, the mobility μ is independent of the electron energy E and decreases with increasing temperature.

Piezoelectric Coupling

As in the case of electromagnetic coupling, piezoelectric coupling is important in semiconductors which are ionic or partly ionic. If these crystals lack inversion symmetry, then *acoustic* mode vibrations generate regions of compression and rarefaction in a crystal which in here lead to the generation of electric fields (see Fig. 9.3). The piezoelectric scattering mechanism is thus associated with the coupling between electrons and phonons arising from these electromagnetic fields. The zincblende structure of the III–V compounds (e.g., GaAs) lacks inversion symmetry. In this case the perturbation potential is given by

$$\Delta V(\mathbf{r}, t) = \frac{-ie\varepsilon_{pz}}{\varepsilon_0 q} \nabla \cdot \mathbf{u}(\mathbf{r}, t) \tag{9.18}$$

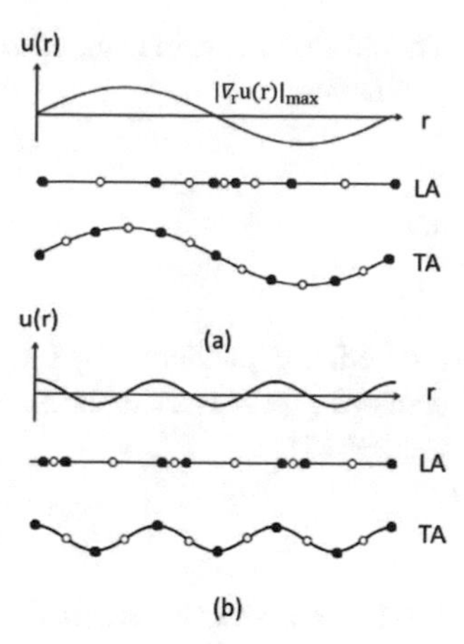

Fig. 9.3 Displacements $\mathbf{u}(\mathbf{r})$ of atoms on a diatomic chain for longitudinal acoustic (LA) and transverse acoustic (TA) phonons at **a** the center and **b** the edge of the Brillouin zone. The lighter mass atoms are indicated by open circles. For zone edge acoustic phonons, only the heavier atoms are displaced

where ε_{pz} is the piezoelectric coefficient and $\mathbf{u}(\mathbf{r}, t) = u \exp(i\mathbf{q} \cdot \mathbf{r} - \omega t)$ is the displacement during a normal mode oscillation. Note that the phase of $\Delta V(\mathbf{r}, t)$ in piezoelectric coupling is shifted by $\pi/2$ relative to the case of electromagnetic coupling.

Deformation-Potential Coupling

The deformation-potential coupling mechanism is associated with energy shifts of the energy band extrema caused by the compression and rarefaction of crystals during acoustic mode vibrations. The deformation potential scattering mechanism is important in crystals like silicon which have inversion symmetry (and hence no piezoelectric scattering coupling) and have the same species on each site (and hence no electromagnetic coupling). The longitudinal acoustic modes are important for phonon coupling in n-type Si and Ge where the conduction band minima occur away from $\mathbf{k} = 0$.

For deformation potential coupling, it is the LA acoustical phonons that are most important, though contributions by LO optical phonons still make some contribution. For the acoustic phonons, we have the condition $\hbar\omega \ll k_B T$ and $\hbar\omega \ll E$, while for the optical phonons it is usually the case that $\hbar\omega \gg k_B T$ at room temperature. For the range of acoustic phonon modes of interest, $G(\mathbf{q}) \sim q$, where q is the phonon wave vector and $\omega \sim q$ for acoustic phonons. Furthermore for the LA phonon branch, the phonon absorption process will depend on $n(q)$ in accordance with the Bose factor

$$\frac{1}{e^{\hbar\omega/k_B T} - 1} \simeq \frac{1}{\left[1 + \frac{\hbar\omega}{k_B T} + \cdots\right] - 1} \sim \frac{k_B T}{\hbar\omega} \sim \frac{k_B T}{q}, \tag{9.19}$$

while for phonon emission

$$\frac{1}{1 - e^{-\hbar\omega/k_B T}} \simeq \frac{1}{1 - \left[1 - \frac{\hbar\omega}{k_B T} + \cdots\right]} \sim \frac{k_B T}{\hbar\omega} \sim \frac{k_B T}{q}. \tag{9.20}$$

Therefore, in considering both phonon absorption and phonon emission, the respective factors

$$G(\mathbf{q})[e^{\hbar\omega/k_BT} - 1]^{-1}$$

and

$$G(\mathbf{q})[1 - e^{-\hbar\omega/k_BT}]^{-1}$$

are both independent of q for the LA branch. Consequently for the *acoustic phonon scattering* process, the carrier mobility μ decreases with increasing T according to (see (9.15))

$$\mu = \frac{e\langle\tau\rangle}{m^*} \sim m^{*-5/2} E^{-1/2} (k_B T)^{-1}. \tag{9.21}$$

For the optical LO contribution, we have a $G(\mathbf{q})$ independent of $\mathbf{q}$ but an $E^{1/2}$ factor is introduced by (9.15) for both phonon absorption and emission, leading to the same basic dependence as given by (9.21). Thus, we find that the temperature and energy dependence of the mobility μ is different for the various electron-phonon coupling mechanisms. These differences in the E and T dependences can thus be used to identify which scattering mechanism is dominant in specific semiconducting samples. Furthermore, when explicit account is taken of the energy dependence of τ, then departures from the strict Drude model $\sigma = ne^2\tau/m^*$ can be expected.

9.2.2 *Ionized Impurity Scattering*

As the temperature is reduced, phonon scattering becomes less important so that in this regime, ionized impurity scattering and other defect scattering mechanisms can become dominant. Ionized impurity scattering can also be important in heavily doped semiconductors over a wider temperature range because of the larger defect density. This scattering mechanism involves the deflection of an electron with velocity v by the Coulomb field of an ion with charge Ze, as modified by the dielectric constant ε of the medium and by the screening of the impurity ion by free electrons (see Fig. 9.4). Most electrons are scattered through small angles as they are scattered by ionized impurities. The perturbation potential is given by

$$\Delta V(\mathbf{r}) = \frac{\pm Ze^2}{4\pi\varepsilon_0 r} \tag{9.22}$$

and the $\pm$ signs denote the different scattering trajectories for electrons and holes (see Fig. 9.4). In (9.22) the screening of the electron by the semiconductor environment is handled by the static dielectric constant of the semiconductor ε_0. Because of the long-range nature of the Coulomb interaction, screening by other free carriers and by other ionized impurities could be important. Such screening effects are further discussed in Sect. 9.2.4.

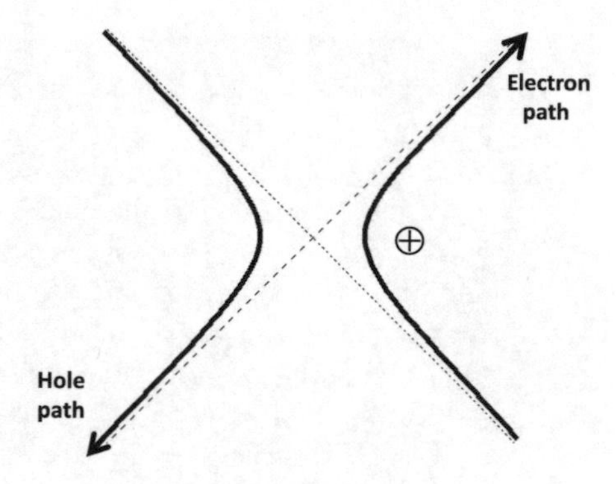

Fig. 9.4 Trajectories of electrons and holes in ionized impurity scattering. The scattering center is at the origin

The scattering rate $1/\tau_I$ due to ionized impurity scattering is given to a good approximation by the Conwell–Weisskopf formula

$$\frac{1}{\tau_I} \sim \frac{Z^2 N_I}{m^{*1/2} E^{3/2}} \, \ell n \left\{ 1 + \left[\frac{4\pi \varepsilon E}{Z e^2 N_I^{1/3}} \right]^2 \right\} \tag{9.23}$$

in which N_I is the ionized charged impurity density. The Conwell–Weisskopf formula works quite well for heavily doped semiconductors. We note here that $\tau_I \sim E^{3/2}$, so that it is the low energy electrons that are most affected by ionized impurity scattering (see Fig. 7.10).

Neutral impurities also introduce a scattering potential, but it is much weaker than that for the ionized impurity. Free carriers can polarize a neutral impurity and interact with the resulting dipole moment, or can undergo an exchange interaction. In the case of neutral impurity scattering, the perturbation potential is given by

$$\Delta V(\mathbf{r}) \simeq \frac{\hbar^2}{m^*} \left(\frac{r_B}{r^5} \right)^{1/2} \tag{9.24}$$

where r_B is the ground state Bohr radius of the electron in a doped semiconductor and r is the distance of the electron to the neutral impurity scattering center.

9.2.3 *Other Scattering Mechanisms*

Other scattering mechanisms in semiconductors include:

(a) neutral impurity centers — these make contributions at very low temperatures, and are mentioned in Sect. 9.2.2. Neutral impurity centers can also cause local strain effects which scatter carriers.

(b) dislocations — these defects give rise to anisotropic scattering at low temperatures.
(c) boundary scattering by crystal surfaces — this scattering becomes increasingly important, the smaller the crystal size. Boundary scattering can become a dominant scattering mechanism in nanostructures (e.g., quantum wells, quantum wires and quantum dots), when the sample size in the confinement direction is smaller that the bulk mean free path. These electrons do not reach equilibrium and are therefore called ballistic electrons (or holes).
(d) intervalley scattering from one equivalent conduction band minimum to another. This scattering process requires a phonon with large q and consequently results in a relatively large energy transfer.
(e) electron-electron scattering – similar to charged impurity scattering in being dominated by a Coulomb scattering mechanism, except that spin effects become important for spin–spin scattering. This mechanism can be important in distributing energy and momentum among the electrons in the solid and thus can act in conjunction with other scattering mechanisms in establishing equilibrium.
(f) electron-hole scattering — depends on having both electrons and holes present. Because the electron and hole motions induced by an applied electric field are in opposite directions, electron-hole scattering tends to reverse the direction of the incident electrons and holes. Radiative recombination, i.e., electron-hole recombination with the emission of a photon, must also be considered.
(g) ballistic carriers — charge carriers passing through sample without scattering and not coming to equilibrium with the lattice.

9.2.4 Screening Effects in Semiconductors

In the vicinity of a charged impurity or an acoustic phonon, charge carriers are accumulated or depleted by the scattering potential, giving rise to a charge density

$$\rho(\mathbf{r}) = e[n(\mathbf{r}) - p(\mathbf{r}) + N_a^-(\mathbf{r}) - N_d^+(\mathbf{r})] = en^*(\mathbf{r}) \tag{9.25}$$

where $n(\mathbf{r})$, $p(\mathbf{r})$, $N_a^-(\mathbf{r})$, $N_d^+(\mathbf{r})$, and $n^*(\mathbf{r})$ are, respectively, the electron, hole, ionized acceptor, ionized donor, and effective total carrier concentrations as a function of distance r to the scatterer. We can then write expressions for these quantities in terms of their excess charge above the uniform potential in the absence of the charge perturbation

$$\begin{aligned} n(\mathbf{r}) &= n + \delta n(\mathbf{r}) \\ N_d^+(\mathbf{r}) &= N_d^+ + \delta N_d^+(\mathbf{r}), \end{aligned} \tag{9.26}$$

and similarly for the holes and acceptors. The space charge $\rho(\mathbf{r})$ is related to the perturbing potential by Poisson's equation

$$\nabla^2\phi(\mathbf{r}) = -\frac{\rho(\mathbf{r})}{\varepsilon_0}. \tag{9.27}$$

Approximate relations for the excess concentrations are

$$\begin{aligned} \delta n(\mathbf{r})/n &\simeq -e\phi(\mathbf{r})/(k_B T) \\ \delta N_d^+(\mathbf{r})/N_d^+ &\simeq e\phi(\mathbf{r})/(k_B T) \end{aligned} \tag{9.28}$$

and similar relations for the holes. Substitution of (9.25) into (9.26) and (9.28) yield

$$\nabla^2\phi(\mathbf{r}) = -\frac{n^* e^2}{\varepsilon_0 k_B T}\phi(\mathbf{r}). \tag{9.29}$$

We define an effective Debye screening length λ such that

$$\lambda^2 = \frac{\varepsilon_0 k_B T}{n^* e^2}. \tag{9.30}$$

For a spherically symmetric potential (9.29) becomes

$$\frac{d^2}{dr^2}\Big(r\phi(r)\Big) = \frac{r\phi(r)}{\lambda^2} \tag{9.31}$$

which yields a solution

$$\phi(r) = \frac{Ze^2}{4\pi\varepsilon_0 r}e^{-r/\lambda}. \tag{9.32}$$

Thus, the screening effect produces an exponential decay of the scattering potential $\phi(r)$ with a characteristic length λ that depends through (9.30) on the effective electron concentration. When the concentration gets large, λ decreases and screening becomes more effective.

When applying screening effects to the ionized impurity scattering problem, we Fourier expand the scattering potential to take advantage of the overall periodicity of the lattice

$$\Delta V(\mathbf{r}) = \sum_G A_G \exp(i\mathbf{G}\cdot\mathbf{r}) \tag{9.33}$$

where the Fourier coefficients are given by

$$A_G = \frac{1}{V}\int_V \nabla V(\mathbf{r})\exp(-i\mathbf{G}\cdot\mathbf{r})d^3r \tag{9.34}$$

and the matrix element of the perturbation Hamiltonian in (9.4) becomes

$$\mathscr{H}_{\mathbf{k},\mathbf{k}'} = \frac{1}{N} \sum_G \int_V e^{-i\mathbf{k}\cdot\mathbf{r}} u_k^*(r) A_G e^{-i\mathbf{G}\cdot\mathbf{r}} e^{i\mathbf{k}'\cdot\mathbf{r}} u_{k'}(r) d^3r. \tag{9.35}$$

We note that the integral in (9.35) vanishes unless $\mathbf{k} - \mathbf{k}' = \mathbf{G}$ so that

$$\mathscr{H}_{\mathbf{k},\mathbf{k}'} = \frac{A_G}{N} \int_V u_k^*(r) u_{k'}(r) d^3r \tag{9.36}$$

within the first Brillouin zone so that for parabolic bands $u_k(\mathbf{r}) = u_{k'}(\mathbf{r})$ and

$$\mathscr{H}_{\mathbf{k},\mathbf{k}'} = A_{\mathbf{k}-\mathbf{k}'}. \tag{9.37}$$

Now substituting for the scattering potential in (9.34) we obtain

$$A_G = \frac{Ze^2}{4\pi\varepsilon_0 V} \int_V \exp(-i\mathbf{G}\cdot\mathbf{r}) d^3r \tag{9.38}$$

where $d^3r = r^2 \sin\theta d\theta d\phi dr$ so that, for $\phi(r)$ depending only on r, the angular integration gives 4π and the spatial integration gives

$$A_G = \frac{Ze^2}{\varepsilon_0 V |\mathbf{G}|^2} \tag{9.39}$$

and

$$\mathscr{H}_{\mathbf{k},\mathbf{k}'} = \frac{Ze^2}{\varepsilon_0 V |\mathbf{k} - \mathbf{k}'|^2}. \tag{9.40}$$

Equations (9.39) and (9.40) are valid for the scattering potential without screening. When screening is included in considering the ionized impurity scattering mechanism, the integration becomes

$$A_G = \frac{Ze^2}{4\pi\varepsilon_0 V} \int_V e^{-r/\lambda} e^{-i\mathbf{G}\cdot\mathbf{r}} d^3r = \frac{Ze^2}{\varepsilon_0 V [|\mathbf{G}|^2 + |1/\lambda|^2]} \tag{9.41}$$

and

$$\mathscr{H}_{\mathbf{k},\mathbf{k}'} = \frac{Ze^2}{\varepsilon_0 V [|\mathbf{k} - \mathbf{k}'|^2 + |1/\lambda|^2]} \tag{9.42}$$

so that screening clearly reduces the scattering due to ionized impurity scattering. The discussion given here also extends to the case of scattering in metals, which is treated below.

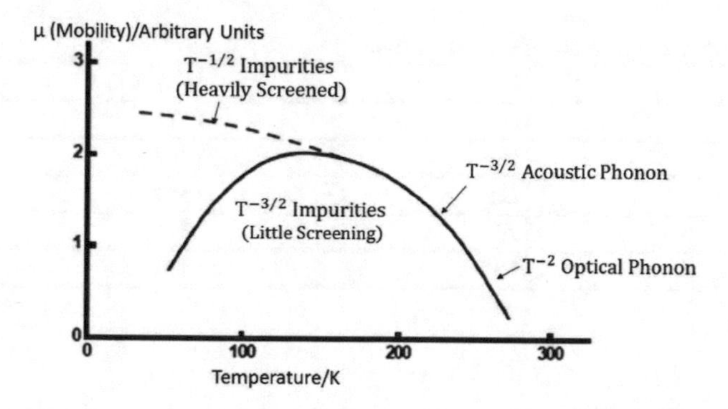

Fig. 9.5 Typical temperature dependence of the carrier mobility in semiconductors, showing the effect of the dominant scattering mechanisms and the temperature dependence of each

Combining the various scattering mechanisms discussed above for semiconductors, the picture given by Fig. 9.5 emerges. Here we see the temperature dependence of each of the important scattering mechanisms and the effect of each of these processes on the carrier mobility. Here it is seen that screening effects are important for carrier mobilities at low temperature.

9.3 Electron Scattering in Metals

Basically the same scattering mechanisms are present in metals as in semiconductors, but because of the large number of occupied states in the conduction bands of metals, the temperature dependences of the various scattering mechanisms are quite different.

9.3.1 Electron-Phonon Scattering in Metals

In metals as in semiconductors, the dominant scattering mechanism is usually electron-phonon scattering. In the case of metals, electron scattering is mainly associated with an electromagnetic interaction of ions with *nearby* electrons, the longer range interactions being **screened** by the numerous mobile electrons. For metals, we must therefore consider explicitly the probability that a state $\mathbf{k}$ is occupied $f_0(\mathbf{k})$ or unoccupied $[1 - f_0(\mathbf{k})]$. The scattering rate is found by explicit consideration of the scattering rate into a state $\mathbf{k}$ and the scattering out of that state. Using the same arguments as in Sect. 9.2.1, the collision term in Boltzmann's equation is given by

Table 9.1 Debye temperature of several metals

Symbol	Metal	Θ_D(K)
⊕	Au	175
∘	Na	202
△	Cu	333
⊓	Al	395
•	Ni	472

$$\left.\frac{\partial f}{\partial t}\right|_{\text{collisions}} \sim \frac{1}{\tau} \simeq$$
$$\sum_q G(\mathbf{q}) \left\{ \overbrace{[1-f_0(\mathbf{k})]\Big[\underbrace{f_0(\mathbf{k}-\mathbf{q})n(\mathbf{q})}_{\text{phonon absorption}} + \underbrace{f_0(\mathbf{k}+\mathbf{q})[1+n(\mathbf{q})]}_{\text{phonon emission}}\Big]}^{\text{scattering into }\mathbf{k}} \right.$$
$$\left. - \overbrace{[f_0(\mathbf{k})]\Big[\underbrace{[1-f_0(\mathbf{k}+\mathbf{q})]n(\mathbf{q})}_{\text{phonon absorption}} + \underbrace{[1-f_0(\mathbf{k}-\mathbf{q})][1+n(\mathbf{q})]}_{\text{phonon emission}}\Big]}^{\text{scattering out of }\mathbf{k}} \right\} \tag{9.43}$$

Here the first term in (9.43) is associated with scattering electrons into an element of phase space at **k** with a probability given by $[1 - f_0(\mathbf{k})]$ that state **k** is unoccupied and has contributions from both phonon absorption processes and phonon emission processes. The second term arises from electrons scattered out of state **k** and here, too, there are contributions from both phonon absorption processes and phonon emission processes. The equilibrium distribution function $f_0(\mathbf{k})$ for the electron is the Fermi distribution function while the function $n(\mathbf{q})$ for the phonons is the Bose distribution function (15.10). Phonon absorption depends on the phonon density $n(\mathbf{q})$, while phonon emission depends on the factor $\{1+n(\mathbf{q})\}$. These factors arise from the properties of the creation and annihilation operators for phonons (to be further discussed in the Problem set). The density of final states for metals is the density of states at the Fermi level which is consequently approximately independent of energy and temperature. In metals, the condition that electron scattering takes place to states near the Fermi level implies that the largest phonon wave vector in an electron collision is $2k_F$ where k_F is the electron wave vector at the Fermi surface.

Of particular interest is the temperature dependence of the phonon scattering mechanism in the limit of low and high temperatures. Experimentally, the temperature dependence of the resistivity of metals can be plotted on a universal curve (see Fig. 9.6) in terms of ρ_T/ρ_{Θ_D} vs. T/Θ_D where Θ_D is the Debye temperature. This plot includes data for several metals, and values for the Debye temperature of these metals are given with the figure (Table 9.1).

In accordance with the plot in Fig. 9.6, $T \ll \Theta_D$ defines the low temperature limit and $T \gg \Theta_D$ the high temperature limit. Except for the very low temperature

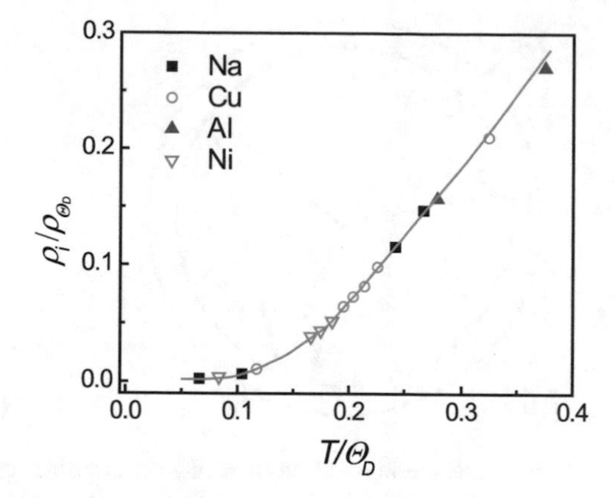

Fig. 9.6 Universal curve of the temperature dependence of the ideal resistivity of various metals normalized to the value at the Debye temperature as a function of the dimensionless temperature T/Θ_D

defect scattering limit, the electron-phonon scattering mechanism dominates, and the temperature dependence of the scattering rate depends on the product of the density of phonon states and the phonon occupation, since the electron-phonon coupling coefficient is essentially independent of T. The phonon concentration in the high temperature limit becomes

$$n(\mathbf{q}) = \frac{1}{\exp(\hbar\omega/k_B T) - 1} \approx \frac{k_B T}{\hbar\omega} \tag{9.44}$$

since $(\hbar\omega/k_B T) \ll 1$, so that from (9.44) we have $1/\tau \sim T$ and $\sigma = ne\mu \sim T^{-1}$. In this high temperature limit, the scattering is quasi-elastic and involves large-angle scattering, since phonon wave vectors up to the Debye wave vector q_D are involved in the electron scattering, where q_D is related to the Debye frequency ω_D and to the Debye temperature Θ_D according to

$$\hbar\omega_D = k_B \Theta_D = \hbar q_D v_q \tag{9.45}$$

where v_q is the velocity of sound.

We can interpret q_D as the radius of a Debye sphere in **k**-space which defines the range of accessible **q** vectors for scattering, i.e., $0 < q < q_D$. The magnitude of wave vector q_D is comparable to the Brillouin zone dimensions but the energy change of an electron (ΔE) on scattering by a phonon will be less than $k_B\Theta_D \simeq 1/40eV$ so that the restriction of $(\Delta E)_{max} \simeq k_B\Theta_D$ implies that the maximum electronic energy change on scattering will be small compared with the Fermi energy E_F. We thus obtain that for $T > \Theta_D$ (the high temperature regime), $\Delta E < k_B T$ and the scattering will be quasi-elastic as illustrated in Fig. 9.7a.

In the opposite limit, $T \ll \Theta_D$, we have $\hbar\omega_q \simeq k_B T$ (because only low frequency acoustic phonons are available for scattering) and in the low temperature limit there is the possibility that $\Delta E > k_B T$, which implies inelastic scattering. In the low temperature limit, $T \ll \Theta_D$, the scattering is also small-angle scattering, since only

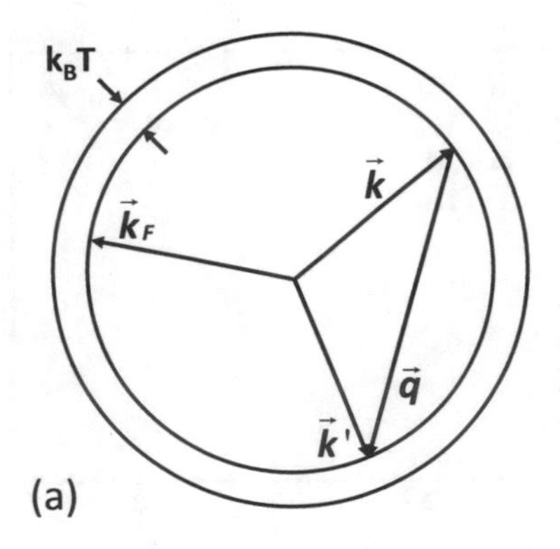

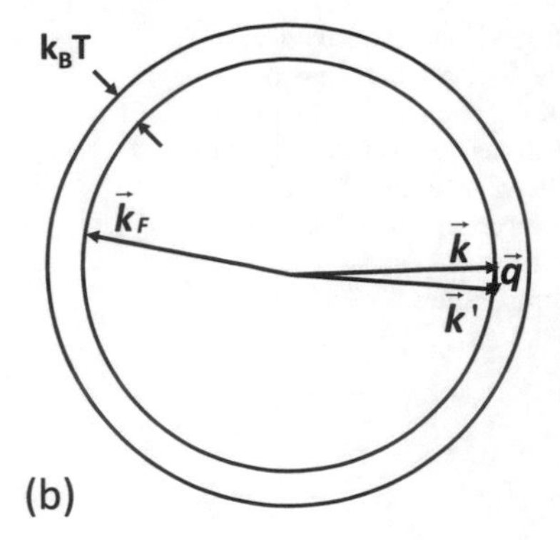

Fig. 9.7 **a** Scattering of electrons on the Fermi surface of a metal. Large angle scattering dominates at high temperature ($T > \Theta_D$) and this regime is called the "quasi-elastic" limit. **b** Small angle scattering is important at low temperature ($T < \Theta_D$) and is in general an inelastic scattering process

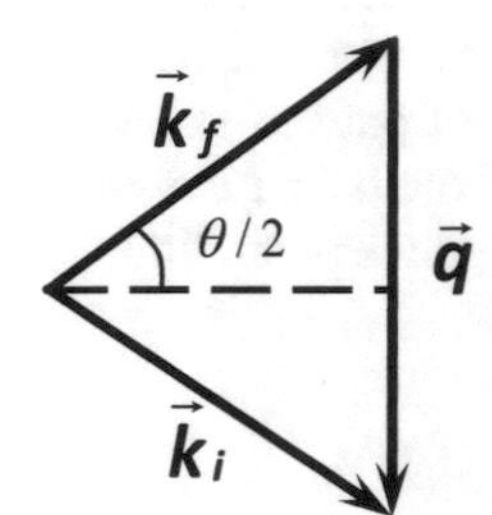

Fig. 9.8 Geometry of the scattering process, where θ is the scattering angle between the incident and scattered electron wave vectors $\mathbf{k}_i$ and $\mathbf{k}_f$, respectively, and q is the phonon wave vector

low energy (low q wave vector) phonons are available for scattering (as illustrated in Fig. 9.7b). At low temperature, the phonon density contributes a factor of T^3 to the scattering rate (9.43) when the sum over phonon states is converted to an integral and q^2dq is written in terms of the dimensionless variable $\hbar\omega_q/k_BT$ with $\omega = v_q q$. Since small momentum transfer gives rise to small angle scattering, the diagram in Fig. 9.8 involves Fig. 9.7. Because of the small energy transfer we can write,

$$|\mathbf{k_i} - \mathbf{k_f}| \sim k_f(1 - \cos\theta) \approx \frac{1}{2}k_f\theta^2 \approx \frac{1}{2}k_f(q/k_f)^2 \tag{9.46}$$

so that another factor of q^2 appears in the integration over $\mathbf{q}$ when calculating $(1/\tau_D)$. Thus, the electron scattering rate at low temperature is predicted to be proportional to T^5 so that $\sigma \sim T^{-5}$ (Bloch–Grüneisen formula). Thus, when phonon scattering is the dominant scattering mechanism in metals, the following results are obtained:

$$\sigma \sim \Theta_D/T \qquad T \gg \Theta_D \tag{9.47}$$

$$\sigma \sim (\Theta_D/T)^5 \qquad T \ll \Theta_D \tag{9.48}$$

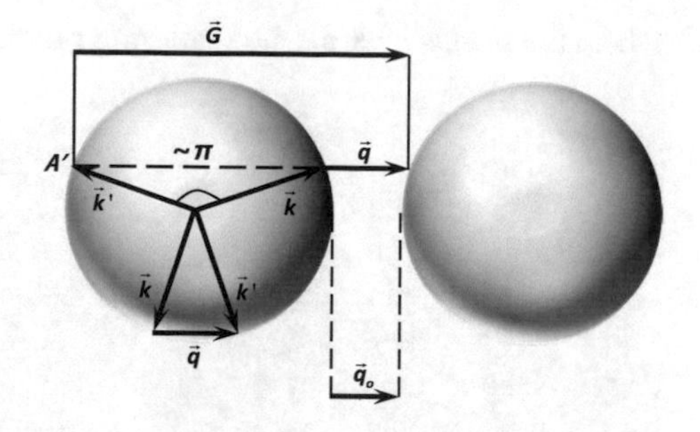

Fig. 9.9 Schematic diagram showing the relation between the phonon wave vector **q** and the electron wave vectors **k** and **k**′ in two Brillouin zones separated by the reciprocal lattice vector **G** (umklapp process)

In practice, the resistivity of metals at very low temperatures is dominated by other scattering mechanisms, such as impurities, boundary scattering, etc., and at very low T, electron-phonon scattering (see (9.48)) is relatively unimportant.

The possibility of *umklapp* processes further increases the range of phonon modes that can contribute to electron scattering in electron-phonon scattering processes. In an umklapp process, a non-vanishing reciprocal lattice vector can be involved in the momentum conservation relation, as shown in the schematic diagram of Fig. 9.9.

In this diagram, the relation between the wave vectors for the phonon and for the incident and scattered electrons $\mathbf{G} = \mathbf{k} + \mathbf{q} + \mathbf{k}'$ is shown when crystal momentum is conserved for a non-vanishing reciprocal lattice vector **G**. Thus, phonons involved in an umklapp process have large wave vectors with magnitudes of about 1/3 of the Brillouin zone dimensions. Therefore, substantial energies can be transferred on collision through an umklapp process. At low temperatures, normal scattering processes (i.e., normal as distinguished from umklapp processes) play an important part in completing the return to equilibrium of an excited electron in a metal, while at high temperatures, umklapp processes become more important.

The discussion presented up to this point is applicable to the creation or absorption of a single phonon in a particular scattering event. Since the restoring forces for lattice vibrations in solids are not strictly harmonic, *anharmonic* corrections to the restoring forces give rise to *multiphonon* processes where more than one phonon can be created or annihilated in a single scattering event. Experimental evidence for multiphonon processes is provided in both optical and transport studies. In some cases, more than one phonon at the *same* frequency can be created (harmonics), while in other cases, multiple phonons at *different* frequencies (overtones and combination modes comprising phonons with two different frequencies) are involved.

9.3.2 Other Scattering Mechanisms in Metals

At very low temperatures where phonon scattering is of less importance, other scattering mechanisms become important, and we can write

$$\frac{1}{\tau} = \sum_i \frac{1}{\tau_i} \tag{9.49}$$

where the sum is over all the scattering processes, according to Matthiessen's rule.

(a) Charged impurity scattering — The effect of charged impurity scattering (Z being the difference in the charge on the impurity site as compared with the charge on a regular lattice site) is of less importance in metals than in semiconductors, because of the strong screening effects by the free electrons in metals.
(b) Neutral impurities — This process pertains to scattering centers having the same charge as the host. Such scattering has less effect on the transport properties than scattering by charged impurity sites, because of the much weaker scattering potential.
(c) Vacancies, interstitials, dislocations, size-dependent effects — the effects for these defects on the transport properties are similar to those for semiconductors. Boundary scattering can become very important in metal nanostructures when the sample length in some direction becomes less than the mean free path in the corresponding bulk crystal. In this case ballistic transport can occur by electrons that remain out of equilibrium until reaching the boundary of the sample.

For most metals, phonon scattering is relatively unimportant at liquid helium temperatures, so that resistivity measurements at 4 K provide one sensitive method for the detection of impurities and crystal defects. In fact, in characterizing the quality of a high purity metal sample, it is customary to specify the resistivity ratio $\rho(300\,\text{K})/\rho(4\,\text{K})$. This quantity is usually called the *residual resistivity ratio* (RRR), or the residual *resistance* ratio. In contrast, a typical semiconductor is characterized by its conductivity and Hall coefficient at room temperature and at 77 K.

9.4 Phonon Scattering

Whereas electron scattering is important in electronic transport properties, phonon scattering is important in thermal transport, particularly for the case of insulators where heat is carried mainly by phonons. The major scattering mechanisms for phonons are phonon-phonon scattering, phonon-boundary scattering, defect-phonon scattering, and phonon-electron scattering which are briefly discussed in the following subsections.

9.4.1 Phonon-Phonon Scattering

The dominant phonon scattering process in crystalline materials is usually phonon-phonon scattering. Phonons are scattered by other phonons because of anharmonic terms in the restoring potential. This scattering process permits:

- two phonons to combine to form a third phonon or
- one phonon to break up into two phonons.

In these anharmonic processes, energy and wavevector conservation apply:

$$\mathbf{q}_1 + \mathbf{q}_2 = \mathbf{q}_3 \quad \text{normal processes} \tag{9.50}$$

or

$$\mathbf{q}_1 + \mathbf{q}_2 = \mathbf{q}_3 + \mathbf{Q} \quad \text{umklapp processes} \tag{9.51}$$

where $\mathbf{Q}$ corresponds to a phonon wave vector of magnitude equal to a non-zero reciprocal lattice vector. Umklapp processes are important when q_1 or q_2 are large, i.e., comparable to a reciprocal lattice vector (see Fig. 9.10). When umklapp processes (see Fig. 9.10) are present, the scattered phonon wavevector $\mathbf{q}_3$ can be in a direction opposite to the energy flow, thereby giving rise to thermal resistance. Because of the high momentum transfer and the large phonon energies that are involved, *umklapp* processes dominate the thermal conductivity at high T.

The phonon density is proportional to the Bose factor so that the scattering rate is proportional to

$$\frac{1}{\tau_{\text{ph}}} \sim \frac{1}{(e^{\hbar\omega/(k_B T)} - 1)}. \tag{9.52}$$

At high temperatures $T \gg \Theta_D$, the scattering time thus varies as T^{-1} since

$$\tau_{\text{ph}} \sim (e^{\hbar\omega/k_B T} - 1) \sim \hbar\omega/k_B T \tag{9.53}$$

while at low temperatures $T \sim \Theta_D$, an exponential temperature dependence for τ_{ph} is found

$$\tau_{\text{ph}} \sim e^{\hbar\omega/k_B T} - 1. \tag{9.54}$$

These temperature dependences are important in considering the lattice contribution to the thermal conductivity (see Sect. 2.4).

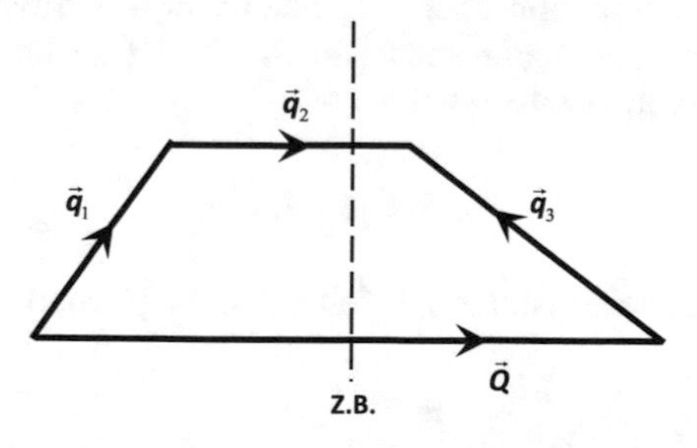

Fig. 9.10 Phonon-phonon umklapp processes. Here $\mathbf{Q}$ is a non-zero reciprocal lattice vector, and $\mathbf{q}_1$ and $\mathbf{q}_2$ are the incident phonon wavevectors involved in the scattering process, while $\mathbf{q}_3$ is the wavevector of the scattered phonon. The vertical dashed line denotes the Brillouin's zone boundary (Z.B.)

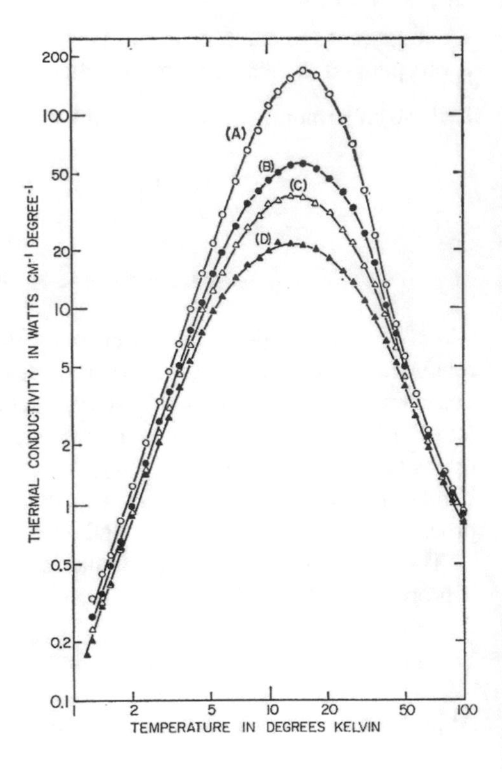

Fig. 9.11 For insulators, we often plot both the thermal conductivity κ and the temperature T on log scales. The various curves here are for LiF with different concentrations of Li isotopes ^{6}Li and ^{7}Li. For highly perfect crystals, it is possible to observe the scattering effects due to Li ions of different masses, which act as lattice defects but have little effect on the electronic properties. Reproduced with permission from Physical Review, vol. 156, pp. 975–988 Copyright (1967) American Physical Society

9.4.2 Phonon-Boundary Scattering

Phonon-boundary scattering is important at low temperatures where the phonon density is low. In this regime, the scattering time is independent of T. The thermal conductivity in this range is proportional to the phonon density which is in turn proportional to T^3. Phonon-boundary scattering is also very important for low dimensional systems where the sample size in some dimension is less than the corresponding phonon mean free path in the bulk 3D crystal. Phonon-boundary scattering combined with phonon-phonon scattering results in a thermal conductivity κ for insulators with the general shape shown in Fig. 9.11 (see Sect. 8.2.4). The lattice thermal conductivity follows the relation

$$\kappa_L = C_p v_q \Lambda_{\text{ph}}/3 \tag{9.55}$$

where the phonon mean free path Λ_{ph} is related to the phonon scattering probability $(1/\tau_{\text{ph}})$ by

$$\tau_{\text{ph}} = \Lambda_{\text{ph}}/v_q \tag{9.56}$$

in which v_q is the velocity of sound and C_p is the heat capacity at constant pressure. Phonon-boundary scattering becomes more important as the crystallite size decreases. The scattering conditions at the boundary can be specular (where after scattering only $q_\perp$ is reversed and $q_\parallel$ is unchanged) for a very smooth sample surface, or the scattering conditions can be diffuse (where after scattering the q is randomized) for a rough sample surface. Periodic corrugations on a surface can also give rise to interesting scattering effects for both electrons and phonons.

9.4.3 *Defect-Phonon Scattering*

Defect-phonon scattering includes a variety of crystal defects, charged and uncharged impurities and different isotopes of the host constituents. The thermal conductivity curves in Fig. 9.11 show the scattering effects due to different isotopes of Li. The low mass of Li makes it possible to see such effects clearly. Isotope effects are also important in graphite and diamond which have the highest thermal conductivity of any solid, and also have several isotopes with large fractional mass differences between one another.

9.4.4 *Electron-Phonon Scattering*

If electrons scatter from phonons, the reverse process also occurs. When phonons impart momentum to electrons, the electron distribution is affected. Thus, the electrons will also carry energy as they are dragged also by the stream of phonons. This phenomenon is called *phonon drag*. In the case of phonon drag, we must simultaneously solve the Boltzmann equations for the electron and phonon distributions which are coupled by the phonon drag interaction term.

9.5 Temperature Dependence of the Electrical and Thermal Conductivity

For the electrical conductivity, at very low temperatures, impurity, defect, and boundary scattering dominate. In this regime σ is independent of temperature. At somewhat higher temperatures but still far below Θ_D the electrical conductivity for metals exhibits a strong temperature dependence (see (9.48))

$$\sigma \propto (\Theta_D/T)^5 \qquad T \ll \Theta_D. \tag{9.57}$$

At higher temperatures where $T \gg \Theta_D$, scattering by phonons with any q vector is possible and the formula

$$\sigma \sim (\Theta_D/T) \qquad T \gg \Theta_D \tag{9.58}$$

applies. We now summarize the corresponding temperature ranges for the thermal conductivity.

Although the thermal conductivity was formally discussed in Chap. 8, a meaningful discussion of the temperature dependence of κ involves scattering processes because of the different temperature dependence of the various scattering processes. The total thermal conductivity κ in general depends on the lattice and electronic contributions, κ_L and κ_e, respectively. The temperature dependence of the lattice contribution is discussed in Sect. 8.2.4 with regard to the various phonon scattering processes and their temperature dependence. For the electronic contribution, we must consider the temperature dependence of the electron scattering processes discussed in Sect. 9.2.

At very low temperatures, in the impurity/defect/boundary scattering range, σ is independent of T, and the same scattering processes apply for both the electronic thermal conductivity and the electrical conductivity, thus $\kappa_e \propto T$ in the impurity scattering regime where $\sigma \sim$ constant and the Wiedemann–Franz law is applicable. From Fig. 8.1 we see that for copper, defect and boundary scattering are dominant below ~20 K, while phonon scattering becomes important at higher T.

At low temperatures $T \ll \Theta_D$, but with T in a regime where phonon scattering has already become the dominant scattering mechanism, the thermal transport depends on the electron-phonon collision rate which in turn is proportional to the phonon density. At low temperatures the phonon density is proportional to T^3. This follows from the proportionality of the phonon density of states arising from the integration of $\int q^2 dq$. From the dispersion relation for the acoustic phonons $\omega = qv_q$ we obtain

$$\omega/v_q = xkT/\hbar v_q \tag{9.59}$$

where $x = \hbar\omega/k_BT$. Thus in the low temperature range of phonon scattering where $T \ll \Theta_D$ and the Wiedemann–Franz law is no longer satisfied, the temperature dependence of τ is found from the product $T(T^{-3})$ so that now $\kappa_e \propto T^{-2}$. One reason why the Wiedemann–Franz law is not satisfied in this temperature regime is that κ_e depends on the collision rate τ_c, while σ depends on the time to reach thermal equilibrium, τ_D. At low temperatures where only low q phonons participate in scattering events, the times τ_c and τ_D are not the same, and τ_D can be very long.

At high T where $T \gg \Theta_D$ and the Wiedemann–Franz law applies, κ_e approaches a constant value corresponding to the regime where σ is proportional to $1/T$. This occurs at temperatures much higher than those shown in Fig. 8.1. The decrease in κ above the peak value at ~17 K follows a $1/T^2$ dependence quite well.

In addition to the electronic thermal conductivity, heat can be carried by the lattice vibrations or phonons. The phonon thermal conductivity mechanism is in fact the principal mechanism operative in semiconductors and insulators, since the electronic contribution in this case is negligibly small. Since κ_L contributes also to metals, the total measured thermal conductivity for metals should exceed the electronic contribution $(\pi^2 k_B^2 T\sigma)/(3e^2)$. In good metallic conductors of high purity, the electronic thermal conductivity dominates and the phonon contribution tends to be small. On the other hand, in conductors where the thermal conductivity due to phonons makes a significant contribution to the total thermal conductivity, it is necessary to separate the electronic and lattice contributions before applying the Wiedemann–Franz law to the total κ.

With regard to the lattice contribution, κ_L at very low temperatures is dominated by defect and boundary scattering processes. From the relation

$$\kappa_L = \frac{1}{3} C_p v_q \Lambda_{\text{ph}} \tag{9.60}$$

we can determine the temperature dependence of κ_L, since $C_p \sim T^3$ at low T, while the sound velocity v_q and phonon mean free path Λ_{ph} at very low T are independent of T. In this regime the number of scatterers is also independent of T.

In the regime where only low q phonons contribute to transport and to scattering, only normal scattering processes contribute. In this regime C_p is still increasing as T^3, v_q is independent of T, but $1/\Lambda_{\text{ph}}$ increases in proportion to the phonon density of states. With increasing T, the temperature dependence of C_p becomes less pronounced and that for Λ_{ph} becomes more pronounced as more scatters participate, leading eventually to a decrease in κ_L. We note that it is only the inelastic collisions that contribute to the decrease in Λ_{ph} and the inelastic collisions are of course due to anharmonic forces.

With increasing temperature, eventually phonons with wavevectors large enough to support umklapp processes are thermally activated. Umklapp processes give rise to thermal resistance and in this regime κ_L decreases as $\exp(-\Theta_{D/T})$. In the high temperature limit $T \gg \Theta_D$, the heat capacity and phonon velocity are both independent of T. Thus, the $\kappa_L \sim 1/T$ dependence arises from the $1/T$ dependence of the mean free path, since in this limit the scattering rate becomes proportional to $k_B T$.

Problems

9.1 By using simple physical arguments, demonstrate the relation given by (9.7).

$$\tau_D = \frac{\tau_c}{1 - \cos\theta} \tag{9.61}$$

9.2 The optical phonon energies of GaAs and AlAs are 36 and 50 meV, respectively, at the Brillouin zone center.

(a) What is the occupation probability of these optical phonons at 77 and 300 K?
(b) Estimate the relative importance of optical phonon absorption and phonon emission for scattering electrons in GaAs and AlAs at 100 K.
(c) Calculate the Debye temperature for GaAs where the sound velocity is 5.6×10^5 cm/s. Assume that the volume of the unit cell is 4.39×10^{-23} cm^{-3}.

9.3 In limiting the electrical transport in Si, the intervalley scattering is very important. In particular, two kinds of intervalley scatterings are important: in a g-scattering

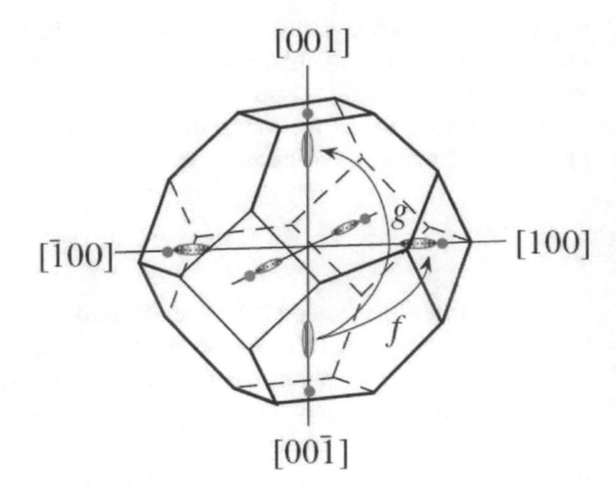

Fig. 9.12 Electron pockets in Si over the Brillouin zone. The schematic drawing highlights the g and f scattering processes listed in Sect. 9.2.3

event an electron goes from one valley (say a $(0,0,\Delta)$ valley) to an opposite valley $(0,0,-\Delta)$, while in an f-scattering event, the electron goes to a perpendicular valley ((00Δ) to $(0\Delta0)$, for example). The extra momentum for the transitions is provided by a phonon and may include a reciprocal lattice vector. Remember that Si valleys are not precisely at the X-point ($\Delta = 0.85$). The speed of sound in silicon is 8433 m/s.(See Fig. 9.12)

(a) Calculate the phonon wavevectors which allow these two scattering mechanisms to occur.
(b) Estimate the temperatures at which inter-valley scattering becomes important for electron conduction in Si, for both the g- and f-scattering mechanisms.

9.4 The phonon mean free path in bulk silicon is approximately $\lambda_{mfp} = 30\,\text{nm}$ at room temperature. This results in a lattice thermal conductivity of 1.38 W/cm·K, based on a heat capacity of C_v=1.66 J/K·cm^3 and a speed of sound v_g=8.3 × 10^5 cm/s. In silicon nanowires with diameters smaller than the bulk phonon mean free path, surface scattering can modify the phonon mean free path significantly. What is the lattice thermal conductivity of nanostructured silicon with a phonon mean free path of 10 nm?

9.5 Bi_2Te_3 has a lattice thermal conductivity of 1.5 W/m·K, heat capacity of 1.2×10^6 J/K·m^3, and speed of sound $v_g = 3 \times 10^3$ m/s. Based on these values, what is the approximate size of a Bi_2Te_3 nanostructure (i.e., nanowire) below which a substantial reduction in the thermal conductivity can be achieved through surface phonon scattering?

9.6 Isotopic doping can be used to increase the scattering of phonons in crystalline materials. Consider three graphite samples prepared with 1.1% ^{13}C (natural abundance), 50% ^{13}C, and 99% ^{13}C. Of these samples, which material has the lowest thermal conductivity? Explain why.

9.7 What is the minimum electron energy that is needed to create a 50 meV phonon of maximum wave vector (from the zone center to the zone boundary) in a semiconductor with the diamond structure where the nearest neighbor distance is 2 Å?

9.8 (a) Assuming a 1D carrier concentration of 10^6 electrons/cm, how many conduction electrons are contained per cm^3 in a quantum wire if $a = 25$ Å?

(b) Suppose that you have a Si sample and you would like to know the temperature below which defect scattering dominates over the intrinsic phonon scattering, how would you proceed?

(c) How would you distinguish between the relative importance between phonon absorption and phonon emission processes for scattering electrons in an electrical conductivity measurement in the 70–80 K temperature range in a high quality Si sample with few defects, other than the dopants used to generate the electron carrier concentration in making the sample n–type?

9.9 Why is optical phonon scattering not important for electron transport in copper?

9.10 Estimate the relative importance of phonon absorption to phonon emission for scattering electrons in an intrinsic GaAs sample at 100 K.

Suggested Readings

Ashcroft, Mermin, *Solid State Physics* (Holt, Rinehart and Winston, New York, 1976). Chapters 16 and 26

G. Chen, M.S. Dresselhaus, G. Dresselhaus, J.P. Fleurial, T. Caillat, Recent developments in thermoelectric materials. Int. Mater. Rev. **48**, 45–66 (2003)

Hang, *Theoretical Solid State Physics*, vol. 2 (Pergamon, New York, 1972). Chapter 4

C. Kittel, *Introduction to Solid State Physics*, 6th edn. (Wiley, New York, 1986). Appendix C

S. Link, M.A. El-Sayed, Spectral properties and relaxation dynamics of surface plasmon electronic oscillations in gold and silver nanodots and nanorods. Phys. Chem. B **103**, 8410–8426 (1999)

M. Lundstrom, *Fundamentals of Carrier Transport* (Cambridge University Press, Cambridge, 2000)

Chapter 10
Magneto-Transport Phenomena

10.1 Introduction

Since the electrical conductivity is sensitive to the product of the carrier density and the carrier mobility rather than to each of these quantities independently, as shown in (7.91), it is necessary to look for different transport techniques to provide information on the carrier density n, and the carrier mobility μ, separately. Magneto-transport provides us with such techniques, at least for simple cases, since the magnetoresistance is mostly sensitive to the carrier mobility, and the Hall effect is mostly sensitive to the carrier density. In this chapter, we consider magneto-transport in bulk solids. We return to the discussion of magneto-transport for lower dimensional systems later in this book, particularly with regard to the quantum Hall effect and giant magnetoresistance effects.

10.2 Magneto-Transport in the Classical Regime ($\omega_c \tau < 1$)

The magnetoresistance and Hall effect measurements, which are used to characterize semiconductors, are carried out in the weak magnetic field limit $\omega_c \tau \ll 1$ where the cyclotron frequency ω_c is given by

$$\omega_c = eB/m^*. \tag{10.1}$$

The cyclotron frequency ω_c is the angular frequency of rotation of a charged particle as it makes an orbit in a plane perpendicular to the magnetic field. In this chapter we explain the origin of magneto-transport effects and provide some insight into their measurement.

In the low field limit (defined by $\omega_c \tau \ll 1$) the carriers are scattered long before completing a single cyclotron orbit in real space, so that quantum effects are unimportant. In higher magnetic fields where $\omega_c \tau > 1$, quantum effects become important.

M. Dresselhaus et al., *Solid State Properties*, Graduate Texts in Physics,
https://doi.org/10.1007/978-3-662-55922-2_10

In this limit (discussed in Chap. 13), the electrons complete cyclotron orbits and the resonance achieved by tuning the microwave frequency of a resonant cavity to coincide with ω_c allows measurements of the effective mass of electrons in semiconductors.

A simplified version of the magnetoresistance phenomenon can be obtained in terms of the classical $\mathbf{F} = m\mathbf{a}$ approach and is presented in Sect. 10.2.1. The virtue of the simplified approach is to introduce the concept of the Hall magnetic field and the general form of the magneto-conductivity tensor. A more general version of these results will then be given using the Boltzmann equation formulation (Sect. 10.4). The advantage of the more general derivation is to put the derivation on a firmer quantitative foundation and to distinguish between the various effective masses which enter the transport equations: the cyclotron effective mass of (10.1), the longitudinal effective mass along the magnetic field direction, and the dynamical effective mass which describes transport in an electric field (see Sect. 10.6).

10.2.1 Classical Magneto-Transport Equations

For the simplified $\mathbf{F} = m\mathbf{a}$ treatment, let the magnetic field $\mathbf{B}$ be directed along the z direction. Then writing $\mathbf{F} = m\mathbf{a}$ for the electronic motion in the plane perpendicular to $\mathbf{B}$ we obtain

$$\mathbf{F} = e(\mathbf{E} + \mathbf{v} \times \mathbf{B}) = m^*\dot{\mathbf{v}} + m^*\mathbf{v}/\tau \tag{10.2}$$

where $m^*\mathbf{v}/\tau$ is introduced to account for damping or electron scattering. For static electric and magnetic fields, there is no time variation in the problem so that $\dot{\mathbf{v}} = 0$ and thus the equation of motion (10.2) is reduced to

$$\begin{aligned} m^*v_x/\tau &= e(E_x + v_y B) \\ m^*v_y/\tau &= e(E_y - v_x B) \end{aligned} \tag{10.3}$$

which can be written as

$$\frac{m^*}{\tau}(v_x + iv_y) = e(E_x + iE_y) - ieB(v_x + iv_y), \tag{10.4}$$

where i is the unit imaginary, so that $j_x + ij_y = ne(v_x + iv_y)$ becomes

$$(j_x + ij_y) = \left(\frac{ne^2\tau}{m^*}\right)\frac{(E_x + iE_y)}{1 + i\omega_c\tau} \tag{10.5}$$

where the cyclotron frequency ω_c is defined by (10.1). The unit imaginary i is introduced into (10.4) and (10.5) because of the circular motion of the electron orbit in a magnetic field, suggesting the use of circular polarization for fields and velocities.

Equating the real and imaginary parts of (10.5) yields

$$
\begin{aligned}
j_x &= \left(\frac{ne^2\tau}{m^*}\right)\left[\frac{E_x}{1+(\omega_c\tau)^2} + \frac{\omega_c\tau E_y}{1+(\omega_c\tau)^2}\right] \\
j_y &= \left(\frac{ne^2\tau}{m^*}\right)\left[\frac{E_y}{1+(\omega_c\tau)^2} - \frac{\omega_c\tau E_x}{1+(\omega_c\tau)^2}\right].
\end{aligned}
\tag{10.6}
$$

Since $\mathbf{v} = v_z\hat{z}$ is parallel to $\mathbf{B}$ or $(\mathbf{v} \times \mathbf{B}) = 0$, the motion of an electron along the magnetic field experiences no force due to the magnetic field, so that

$$
j_z = \frac{ne^2\tau}{m^*}E_z. \tag{10.7}
$$

Equations (10.6) and (10.7) yield the magnetoconductivity tensor defined by $\mathbf{j} = \overset{\leftrightarrow}{\sigma}_B \cdot \mathbf{E}$ in the presence of a magnetic field in the low field limit where $\omega_c\tau \ll 1$ and the classical approach given here is applicable. In this limit, an electron in a magnetic field is accelerated by an electric field and follows Ohm's law (as in the case of zero magnetic field):

$$
\mathbf{j} = \overset{\leftrightarrow}{\sigma}_B \cdot \mathbf{E} \tag{10.8}
$$

except that the magnetoconductivity tensor $\overset{\leftrightarrow}{\sigma}_B$ depends explicitly on magnetic field and in accordance with (10.6) and (10.7) assumes the form

$$
\overset{\leftrightarrow}{\sigma}_B = \frac{ne^2\tau/m^*}{1+(\omega_c\tau)^2}\begin{pmatrix} 1 & \omega_c\tau & 0 \\ -\omega_c\tau & 1 & 0 \\ 0 & 0 & 1+(\omega_c\tau)^2 \end{pmatrix}. \tag{10.9}
$$

The magnetoresistivity tensor (which is more closely related to laboratory measurements) is defined as the inverse of the magnetoconductivity tensor

$$
\overset{\leftrightarrow}{\rho}_B = [\overset{\leftrightarrow}{\sigma}_B]^{-1} = \frac{m^*}{ne^2\tau}\begin{pmatrix} 1 & -\omega_c\tau & 0 \\ \omega_c\tau & 1 & 0 \\ 0 & 0 & 1 \end{pmatrix}. \tag{10.10}
$$

10.2.2 Magnetoresistance

The magnetoresistance is defined in terms of the diagonal components of the magnetoresistivity tensor given by (10.10)

$$
\Delta\rho/\rho \equiv \Big(\rho(B) - \rho(0)\Big)/\rho(0) \tag{10.11}
$$

and, in general, depends on $(\omega_c\tau)^2$ or on B^2. Since $\omega_c\tau = (e\tau/m^*)B = \mu B$, the magnetoresistance provides information on the carrier mobility $\mu_s\tau$.

The *longitudinal magnetoresistivity* $\Delta\rho_{zz}/\rho_{zz}$ is measured with the electric field parallel to the magnetic field. On the basis of a spherical Fermi surface one carrier model, we have $E_z = j_z/\sigma_0$ from (10.10), so that there is no longitudinal magnetoresistivity in this case; that is, the resistivity is the same whether or not a magnetic field is present, since $\sigma_0 = ne^2\tau/m^*$. On the other hand, many semiconductors do exhibit longitudinal magnetoresistivity experimentally, and this effect arises from the non-spherical shape of their constant energy surfaces.

The *transverse magnetoresistivity* $\Delta\rho_{xx}/\rho_{xx}$ is measured with the current flowing in some direction (x) perpendicular to the magnetic field. With the direction of current flow along the x direction and $j_y = 0$, we can write from (10.8) and (10.9) that

$$E_y = (\omega_c\tau)E_x \tag{10.12}$$

and

$$j_x = \sigma_0\left[\frac{E_x}{1+(\omega_c\tau)^2} + \frac{(\omega_c\tau)^2 E_x}{1+(\omega_c\tau)^2}\right] = \sigma_0 E_x. \tag{10.13}$$

Again, there is no transverse magnetoresistance for a material with a single carrier type having a spherical Fermi surface. Introduction of either a more complicated Fermi surface or more than one type of carrier results in a transverse magnetoresistance. When the velocity distribution of carriers at a finite temperature is taken into account, a finite transverse magnetoresistance is also obtained. In a similar way, multi-valley semiconductors (having several electron or hole constant energy surfaces, some of which are equivalent by symmetry) can also display a transverse magnetoresistance. In all of these cases the magnetoresistance exhibits a B^2 dependence. The effect of two carrier types on the transverse magnetoresistance is discussed in Sect. 10.5 in some detail.

We note that the σ_{xy} and σ_{yx} terms arise from the presence of a magnetic field. The significance of these terms is further addressed in our discussion of the Hall effect (Sect. 10.3). We note that for non-spherical constant energy surfaces, (10.6) and (10.7) must be rewritten to reflect the fact that m^* is a tensor so that the vectors **v** and **E** need not be parallel, even in the absence of a magnetic field. This point is clarified to some degree in the derivation of the magneto-transport effects given in Sect. 10.4 using the Boltzmann Equation.

10.3 The Hall Effect

If an electric current is flowing in a semiconductor transverse to an applied magnetic field, an electric field is generated perpendicular to the plane containing **j** and **B**. This is known as the Hall effect. Because the magnetic field acts to deflect the charge carriers transverse to their current flow, the Hall field is required to ensure that the transverse current vanishes. Let x be the direction of current flow and z the direction of the magnetic field. Then the boundary condition for the Hall effect is $j_y = 0$. From the magnetoconductivity tensor (10.9), we have

$$j_y = \frac{ne^2\tau}{m^*}\left(\frac{1}{1+(\omega_c\tau)^2}\right)(E_y - \omega_c\tau E_x) \tag{10.14}$$

so that a non-vanishing Hall field

$$E_y = \omega_c\tau E_x \tag{10.15}$$

must be present to ensure the vanishing of j_y (see Fig. 10.1). It is convenient to define the Hall coefficient $R_{\rm Hall}$ as

$$R_{\rm Hall} \equiv \frac{E_y}{j_x B_z} = \frac{\tau E_x(eB_z/m^*)}{j_x B_z}. \tag{10.16}$$

Substitution of the Hall field into the expression for j_x in (10.17) then yields

$$j_x = \frac{ne^2\tau}{m^*[1+(\omega_c\tau)^2]}[E_x + \omega_c\tau E_y] = \frac{ne^2\tau[1+(\omega_c\tau)^2]E_x}{m^*[1+(\omega_c\tau)^2]} \tag{10.17}$$

or

$$j_x = \frac{ne^2\tau}{m^*}E_x = \sigma_{dc}E_x. \tag{10.18}$$

Substitution of this expression into the Hall coefficient in (10.16) yields

$$R_{\rm Hall} = \frac{e\tau}{m^*(ne^2\tau/m^*)} = \frac{1}{ne}. \tag{10.19}$$

The Hall coefficient is important because:

1. $R_{\rm Hall}$ depends only on the carrier density n and universal constants.
2. The sign of $R_{\rm Hall}$ determines whether conduction is by electrons ($R_{\rm Hall} < 0$) or by holes ($R_{\rm Hall} > 0$).

If the carriers are of one type, we can relate the Hall mobility $\mu_{\rm Hall}$ to $R_{\rm Hall}$:

$$\mu = \frac{e\tau}{m^*} = \left(\frac{ne^2\tau}{m^*}\right)\left(\frac{1}{ne}\right) = \sigma R_{\rm Hall}. \tag{10.20}$$

We define the Hall mobility as

$$\mu_{\rm Hall} \equiv \sigma R_{\rm Hall} \tag{10.21}$$

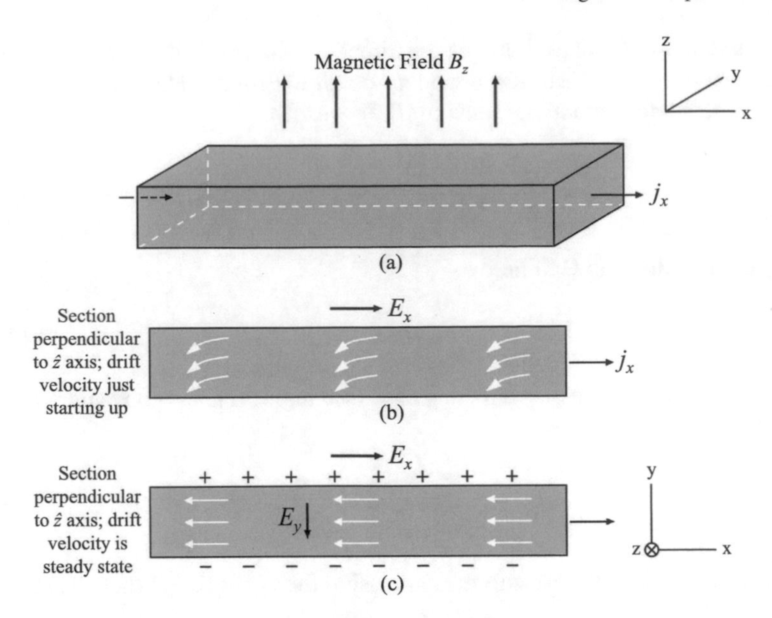

Fig. 10.1 The standard geometry for the Hall effect: a specimen of rectangular cross-section is placed in a magnetic field B_z as illustrated in **(a)**. An electric field E_x applied across the end electrodes causes an electric current density j_x to flow down the sample shown in **(a)**. The drift velocity of the electrons, immediately after the electric field is applied, is shown in **(b)**. The charge carrier deflection in the y direction is caused by the magnetic field. Electrons accumulate on one face of the sample and a positive ion excess is established on the opposite face until, as in **(c)**, the transverse electric field (Hall field) totally cancels the force due to the external magnetic field (Reprinted from C. Kittel)

and μ_{Hall} carries the same sign as R_{Hall}. The resistivity component $\rho_{xy} = ne\mu_{\mathrm{Hall}}$ is called the Hall resistivity. A variety of new effects can occur in R_{Hall} when there is more than one type of carrier, as is commonly the case in semiconductors, and this is discussed in Sect. 10.5.

10.4 Derivation of the Magneto-Transport Equations from the Boltzmann Equation

The corresponding results relating **j** and **E** will now be found using the Boltzmann equation (7.4) in the absence of temperature gradients. We first use the linearized Boltzmann equation given by (7.18) to obtain the distribution function f_1. Then we will use f_1 to obtain the current density **j** in the presence of an electric field **E** and a magnetic field $\mathbf{B} = B\hat{z}$ in the z-direction.

In the presence of a magnetic field, the equation of motion becomes

$$\hbar\dot{\mathbf{k}} = e\left(\mathbf{E} + \mathbf{v} \times \mathbf{B}\right). \tag{10.22}$$

We use, as in (7.17), $f = f_0 + f_1$ with

$$\frac{\partial f_0}{\partial \mathbf{k}} = \frac{\partial f_0}{\partial E}\frac{\partial E(\mathbf{k})}{\partial \mathbf{k}} = \hbar\mathbf{v}\frac{\partial f_0}{\partial E}. \tag{10.23}$$

Substituting into the linearized form of the Boltzmann equation (7.18) gives an equation for f_1:

$$\frac{e}{\hbar}(\mathbf{E} + \mathbf{v} \times \mathbf{B})\cdot\left(\hbar\mathbf{v}\frac{\partial f_0}{\partial E} + \frac{\partial f_1}{\partial \mathbf{k}}\right) = -\frac{f_1}{\tau}. \tag{10.24}$$

In analogy with the case of zero magnetic field, we assume a solution for f_1 of the form

$$f_1 = -e\tau\mathbf{v}\cdot\mathbf{V}\frac{\partial f_0}{\partial E} \tag{10.25}$$

(see (7.21)) where $\mathbf{V}$ is a vector to be determined in analogy with the solution for $B = 0$. The form of (10.25) is motivated by the form suggested by the magnetoconductivity tensor in (10.9).

For a simple parabolic band, $\mathbf{v} = \hbar\mathbf{k}/m^*$, and substitution of (7.21) into (10.24) gives

$$\frac{e}{\hbar}(\mathbf{v} \times \mathbf{B}) \cdot \frac{\partial f_1}{\partial \mathbf{k}} = -\frac{e^2\tau}{m^*}(\mathbf{v} \times \mathbf{B}) \cdot \mathbf{V}\frac{\partial f_0}{\partial E}. \tag{10.26}$$

The following equation for $\mathbf{V}$ is then obtained from (10.24)

$$\mathbf{v}\cdot\mathbf{E} - \frac{e\tau}{m^*}(\mathbf{v} \times \mathbf{B}) \cdot \mathbf{V} = \mathbf{v}\cdot\mathbf{V} \tag{10.27}$$

where we have neglected a term $\mathbf{E}\cdot\mathbf{V}$ which is small (of order $|\mathbf{E}|^2$ if $|\mathbf{E}|$ is small). Equation (10.27) is equivalent to

$$\begin{aligned} v_x E_x + \omega_c \tau v_x V_y &= v_x V_x \\ v_y E_y - \omega_c \tau v_y V_x &= v_y V_y \end{aligned} \tag{10.28}$$

which can be rewritten more compactly as:

$$\begin{aligned} \mathbf{V}_\perp &= \left(\mathbf{E}_\perp - (e\tau/m^*)[\mathbf{B} \times \mathbf{E}_\perp]\right)\left(1 + (e\tau B/m^*)^2\right)^{-1} \\ V_z &= E_z \end{aligned} \tag{10.29}$$

where the subscript notation "$\perp$" in (10.29) denotes the component in the $x - y$ plane, perpendicular to **B**. This solves the problem of finding f_1.

Now we can carry out the calculation of **j** in (7.23), using the new expression for f_1 given by (10.25), (7.13), and (10.29). With the more detailed calculation using the Boltzmann equation, it is clear that the cyclotron mass governs the cyclotron frequency, while the dynamic effective mass controls the coefficients $(ne^2\tau/m^*)$ in (10.6) and (10.7). In Sect. 10.6, we discuss how to calculate the cyclotron effective mass.

10.5 Two Carrier Model

In this section, we calculate both the Hall effect and the transverse magnetoresistance for a two-carrier model. Referring to Fig. 10.1, the geometry under which transport measurements are made ($\mathbf{j} \parallel \hat{x}$) imposes the condition $j_y = 0$. From the magneto-conductivity tensor of (10.9)

$$j_y = -\frac{\sigma_{01}\beta_1 E_x}{1+\beta_1^2} + \frac{\sigma_{01}E_y}{1+\beta_1^2} - \frac{\sigma_{02}\beta_2 E_x}{1+\beta_2^2} + \frac{\sigma_{02}E_y}{1+\beta_2^2} = 0 \tag{10.30}$$

where

$$\beta = \omega_c \tau \tag{10.31}$$

and the subscripts on σ_{0i} and β_i refer to the carrier index, $i = 1, 2$, so that $\sigma_{0i} = n_i e^2 \tau_i / m_i^*$ and $\beta_i = \omega_{0c}\tau_i$. Solving (10.30) yields a relation between E_y and E_x which defines the Hall field

$$E_y = E_x \left[\frac{\frac{\sigma_{01}\beta_1}{1+\beta_1^2} + \frac{\sigma_{02}\beta_2}{1+\beta_2^2}}{\frac{\sigma_{01}}{1+\beta_1^2} + \frac{\sigma_{02}}{1+\beta_2^2}} \right] \tag{10.32}$$

for a two carrier system. This basic equation for the Hall fields E_y is applicable to two kinds of electrons, two kinds of holes, or a combination of electrons and holes. The generalization of (10.30) to more than two types of carriers is immediate. The magnetoconductivity tensor is found by substitution of (10.32) into

$$j_x = \frac{\sigma_{01}E_x}{1+\beta_1^2} + \frac{\sigma_{01}\beta_1 E_y}{1+\beta_1^2} + \frac{\sigma_{02}E_x}{1+\beta_2^2} + \frac{\sigma_{02}\beta_2 E_y}{1+\beta_2^2}. \tag{10.33}$$

In general (10.33) is a complicated relation, but simplifications can be made in the low field limit $\beta \ll 1$, where we can neglect terms in β^2 relative to terms in β. Retaining the lowest power in terms in β then yields

$$E_y = E_x \left[\frac{\sigma_{01}\beta_1 + \sigma_{02}\beta_2}{\sigma_{01} + \sigma_{02}} \right] \tag{10.34}$$

and

$$j_x = (\sigma_{01} + \sigma_{02}) E_x. \tag{10.35}$$

We thus obtain the following important relation for the Hall coefficient which is independent of magnetic field in this low field limit

$$R_{\text{Hall}} \equiv \frac{E_y}{j_x B_z} = \frac{\beta_1\sigma_{01} + \beta_2\sigma_{02}}{(\sigma_{01} + \sigma_{02})^2 B} = \frac{\mu_1\sigma_{01} + \mu_2\sigma_{02}}{c(\sigma_{01} + \sigma_{02})^2} \tag{10.36}$$

where we have made use of the relation between β and the mobility μ

$$\beta = \mu B = e\tau B / m_c^* = \omega_c \tau. \tag{10.37}$$

This allows us to write R_{Hall} in terms of the Hall coefficients R_i for each of the two types of carriers

$$R_{\text{Hall}} = \frac{R_1\sigma_{01}^2 + R_2\sigma_{02}^2}{(\sigma_{01} + \sigma_{02})^2} \tag{10.38}$$

since

$$\frac{\beta}{B} = R_{\text{Hall}}\sigma, \tag{10.39}$$

where for each carrier type we have

$$\sigma_{0i} = \frac{n_i e^2 \tau_i}{m_i^*} \tag{10.40}$$

and

$$R_i = \frac{1}{n_i e_i} \tag{10.41}$$

where $i = 1, 2$. We note in (10.41) that $e_i = \pm|e|$ where $|e|$ is the magnitude of the charge on the electron. Thus electrons and holes contribute with opposite sign to R_{Hall} in (10.38). When more than one carrier type is present, it is not always the case that the sign of the Hall coefficient is the same as the sign of the majority carrier type. A minority carrier type may have a higher mobility, and the carriers with high mobility make a larger contribution per carrier to R_{Hall} than do the larger number of low mobility carriers.

The magnetoconductivity for two carrier types is obtained from (10.33) upon substitution of (10.34) into (10.33) and retaining terms in β^2. For the transverse magneto-conductance, we obtain

$$\frac{\sigma_B(B) - \sigma_B(0)}{\sigma_B(0)} = \frac{2\sigma_{01}^2\beta_1^2 + 2\sigma_{02}^2\beta_2^2 + \sigma_{01}\sigma_{02}(\beta_1 + \beta_2)^2}{(\sigma_{01} + \sigma_{02})^2} \tag{10.42}$$

which is an average of β_1 and β_2 appropriately weighted by conductivity components σ_{01} and σ_{02}. But since $\Delta\rho/\rho = -\Delta\sigma/\sigma$ we obtain the following result for the transverse magnetoresistance

$$\frac{\Delta\rho}{\rho} = -\frac{2\sigma_{01}^2\beta_1^2 + 2\sigma_{02}^2\beta_2^2 + \sigma_{01}\sigma_{02}(\beta_1 + \beta_2)^2}{(\sigma_{01} + \sigma_{02})^2}. \tag{10.43}$$

We note that the magnetoconductivity tensor (10.9) no longitudinal yields magnetoresistance for a spherical two-carrier model.

10.6 Cyclotron Effective Mass

To calculate the magnetoresistance and Hall effect explicitly for non-spherical Fermi surfaces, we need to derive a formula for the cyclotron frequency $\omega_c = eB/m_c^*$ which is generally applicable for non-spherical Fermi surfaces. The cyclotron effective mass can be determined in either of two ways. The first method is the *tube integral* method (see Fig. 10.2 for a schematic of the constant energy surfaces at energy E and $E+\Delta E$) which defines the cyclotron effective mass as

$$m_c^* = \frac{1}{2\pi}\oint \frac{\hbar d\kappa}{|v|} = \frac{\hbar^2}{2\pi}\oint \frac{d\kappa}{|\partial E/\partial k|} \tag{10.44}$$

where $d\kappa$ is an infinitesimal element of length along the contour and we can obtain m_c^* by direct integration. For the second method, we convert the line integral over an enclosed area, making use of $\Delta E = \Delta k(\partial E/\partial k)$ so that

$$m_c^* = \frac{\hbar^2}{2\pi}\frac{1}{\Delta E}\oint (\Delta k) d\kappa = \frac{\hbar^2}{2\pi}\frac{\Delta A}{\Delta E} \tag{10.45}$$

where ΔA is the area of the strip indicated in Fig. 10.2 by the separation ΔE. Therefore we obtain the relation

$$m_c^* = \frac{\hbar^2}{2\pi}\frac{\partial A}{\partial E} \tag{10.46}$$

which gives the second method for finding the cyclotron effective mass.

For a spherical constant energy surface, we have $A = \pi k^2$ and $E(\mathbf{k}) = (\hbar^2 k^2)/2m^*$ so that $m_c^* = m^*$. For an electron orbit described by an ellipse in reciprocal space (which is appropriate for the general orbit in the presence of a magnetic field on an ellipsoidal constant energy surface at wave vector k_B along the magnetic field) we write

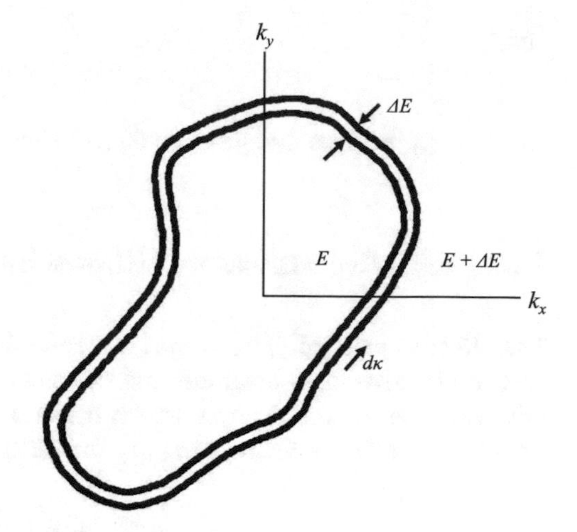

Fig. 10.2 Contour of the constant energy surface in k space used to calculate the path integral in the evaluation of the cyclotron effective mass

$$E(\mathbf{k}_\perp) = \frac{\hbar^2 k_1^2}{2m_1} + \frac{\hbar^2 k_2^2}{2m_2} = \frac{\hbar^2 k_0^2}{2m_0} \tag{10.47}$$

which defines the area A enclosed by the constant energy surface as

$$A = \pi k_1 k_2 = \pi k_0^2 \sqrt{m_1 m_2}/m_0 \tag{10.48}$$

where $(k_i^2/m_i) = (k_0^2/m_0)$. Then substitution in (10.46) gives

$$m_c^* = \sqrt{m_1 m_2}. \tag{10.49}$$

This expression for m_c^* gives a clear physical picture of the relation between m_c^* and the electron orbit on a constant ellipsoidal energy surface in the presence of a magnetic field. Since finding the electron orbit requires geometrical calculation for a general magnetic field orientation, it is more convenient to use the relation

$$m_c^* = \left(\frac{\det \overset{\leftrightarrow}{m^*}}{\hat{b} \cdot \overset{\leftrightarrow}{m^*} \cdot \hat{b}} \right)^{1/2} \tag{10.50}$$

for calculating m_c^* for ellipsoidal constant energy surfaces where $\hat{b} \cdot \overset{\leftrightarrow}{m^*} \cdot \hat{b}$ is the effective mass component along the magnetic field, while $\det \overset{\leftrightarrow}{m^*}$ denotes the determinant of the effective mass tensor $\overset{\leftrightarrow}{m^*}$, and where $\hat{b}$ is a unit vector along the magnetic field. One can then show that for this case the Hall mobility is given by

$$\mu_{\mathrm{Hall}} = \frac{e\tau}{m_c^*} \tag{10.51}$$

and

$$\mu_{\text{Hall}} B \equiv \omega_c \tau = \beta \tag{10.52}$$

so that μ_{Hall} involves the cyclotron effective mass.

10.7 Effective Masses for Ellipsoidal Fermi Surfaces

The effective mass of carriers in a magnetic field is complicated by the fact that several effective mass quantities are of importance. These include the cyclotron effective mass m_c^* for electron motion transverse to the magnetic field (Sect. 10.6) and the longitudinaleffective mass m_B^* for electron motion along the magnetic field

$$m_B^* = \hat{b} \cdot \overleftrightarrow{m^*} \cdot \hat{b} \tag{10.53}$$

obtained by projecting the effective mass tensor along the magnetic field. These motions are considered in finding f_1, the change in the electron distribution function due to forces and fields. And corresponding results are found for later.

Returning to (7.9) and (7.10) in the initial exposition for the current density calculated by the Boltzmann equation, we obtained the Drude formula

$$\overleftrightarrow{\sigma} = ne^2\tau \left(\overleftrightarrow{\frac{1}{m^*}}\right), \tag{10.54}$$

thereby defining the drift mass tensor in an electric field. Referring to the magnetoresistance and magnetoconductance tensors ((10.8) and (10.9)), we can see the drift term $(ne^2\tau/m^*)$ which utilizes the drift mass tensor and the terms in $(\omega_c\tau)$ which utilize the effective cyclotron mass m_c^* (see Sect. 10.6). Here we see that when the Fermi surface for a semiconductor consists of ellipsoidal carrier pockets, then the drift effective mass components are found in accordance with the procedure outlined in Sect. 7.5.1 for ellipsoidal carrier pockets. We can then conveniently use (10.50) to determine the cyclotron effective mass for ellipsoidal carrier pockets.

10.8 Dynamics of Electrons in a Magnetic Field

In this section we relate the electron motion on a constant energy surface in a magnetic field to real space orbits. Consider first the case of $B = 0$ shown in Fig. 10.3. At a given k value, $E(\mathbf{k})$ and $v(\mathbf{k})$ are specified and each of the quantities is a constant of the motion, where $\mathbf{v}(\mathbf{k}) = (1/\hbar)[\partial E(\mathbf{k})/\partial \mathbf{k}]$. If there are no forces acting on the system, then $E(\mathbf{k})$ and $v(\mathbf{k})$ are unchanged with time and are constants of the motion. Thus, at any instant of time there is an equal probability that an electron will

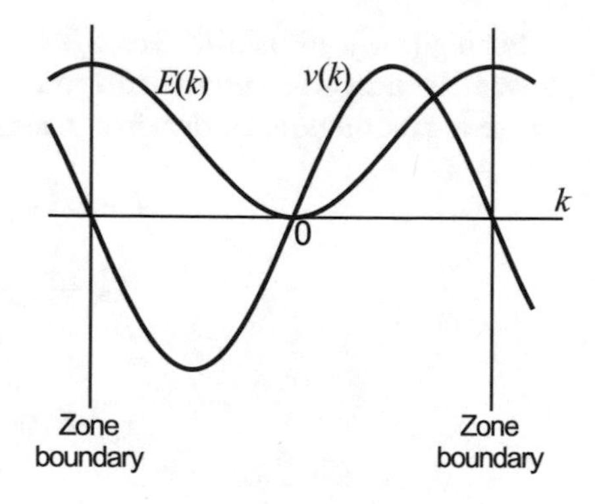

Fig. 10.3 Schematic diagram of $E(\mathbf{k})$ and of the velocity $v(\mathbf{k})$ which is proportional to the derivative $\partial E(\mathbf{k})/\partial \mathbf{k}$ for an electron in a nearly free electron model

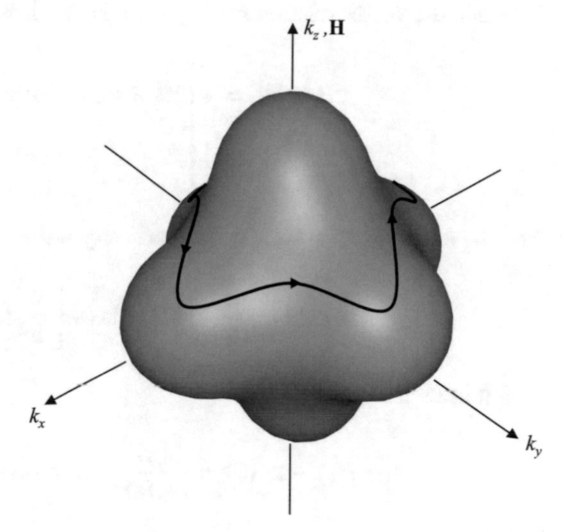

Fig. 10.4 Schematic diagram of the motion of an electron along a constant energy (and constant k_z) trajectory in the presence of a magnetic field **B** in the z-direction

be found anywhere on a constant energy surface. The role of an external electric field **E** is to change the **k** vector on this constant energy surface according to the equation of motion

$$\hbar \dot{\mathbf{k}} = e\mathbf{E} \tag{10.55}$$

so that under a force $e\mathbf{E}$, the energy of the system is changed.

In a constant magnetic field (but no electric field), the electron will move on a constant energy surface in **k** space in an orbit perpendicular to the magnetic field (see Fig. 10.4), and following an equation of motion

$$\hbar \dot{\mathbf{k}} = e(\mathbf{v} \times \mathbf{B}) \tag{10.56}$$

where we note that $|v_\perp|$ remains unchanged along the electron orbit. The electrons will execute the indicated orbit at a cyclotron frequency ω_c given by $\omega_c = eB/m_c^*$ where m_c^* is the cyclotron effective mass (see Sect. 10.6).

For high magnetic fields, when $\omega_c\tau \gg 1$, an electron circulates many times around its semiclassical orbit before undergoing a collision. In this limit, there is interest in describing the orbits of carriers in **r** space. The solution to the semiclassical equations

$$\dot{\mathbf{r}} = \mathbf{v} = (1/\hbar)\partial E/\partial \mathbf{k} \tag{10.57}$$

$$\hbar\dot{\mathbf{k}} = e[\mathbf{E} + \mathbf{v} \times \mathbf{B}] \tag{10.58}$$

is

$$\mathbf{r}_\perp = \mathbf{r} - \frac{\mathbf{B}}{B^2}(\mathbf{B}\cdot\mathbf{r}) \tag{10.59}$$

in which the vector relation $\mathbf{r} = \mathbf{r}_\parallel + \mathbf{r}_\perp$ is valid. We also note that

$$\begin{aligned} \mathbf{B} \times \hbar\dot{\mathbf{k}} &= e\Big[\{\mathbf{B} \times \mathbf{E}\} + \{\mathbf{B} \times (\mathbf{v} \times \mathbf{B})\}\Big] \\ &= e\Big[\mathbf{B} \times \mathbf{E} + (B^2\mathbf{v} - (\mathbf{B}\cdot\mathbf{v})\mathbf{B})\Big]. \end{aligned} \tag{10.60}$$

Making use of (10.59) and (10.60), we may write

$$\dot{\mathbf{r}}_\perp = \left[\frac{\hbar}{eB^2}\mathbf{B} \times \dot{\mathbf{k}}\right] + \left[\frac{1}{B^2}(\mathbf{E} \times \mathbf{B})\right] \tag{10.61}$$

which upon integration yields

$$\mathbf{r}_\perp(t) - \mathbf{r}_\perp(0) = \frac{\hbar}{\omega_c B m_c^*}\mathbf{B} \times [\mathbf{k}(t) - \mathbf{k}(0)] + \mathbf{w}\,t \tag{10.62}$$

where

$$\mathbf{w} = \frac{1}{B^2}(\mathbf{E} \times \mathbf{B}). \tag{10.63}$$

From (10.62) we see that the orbit in real space is $\pi/2$ out of phase with the orbit in reciprocal space. For the case of *closed orbits*, after a long time $t \approx \tau$. Then the second term $\mathbf{w}\,t$ of (10.62) will dominate, giving a transverse or Hall current

$$\mathbf{j}_\perp \rightarrow \frac{ne}{B^2}(\mathbf{E} \times \mathbf{B}) \tag{10.64}$$

where n is the electron density. Similarly the longitudinal current $\mathbf{j} \parallel \mathbf{E}$ will approach a constant value or the current will *saturate* since $\mathbf{E} \perp \mathbf{j}$ and $\mathbf{E} \perp \mathbf{B}$.

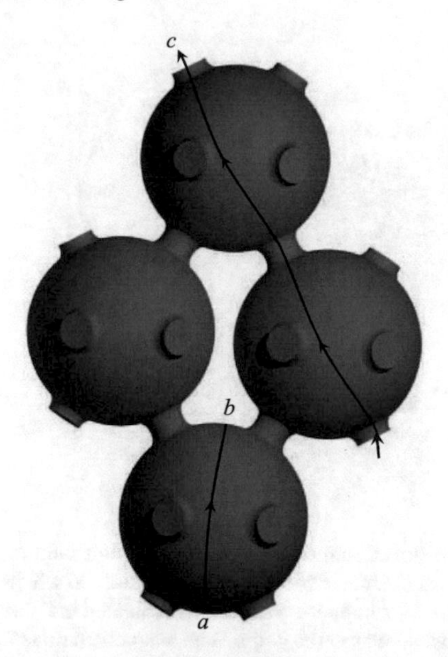

Fig. 10.5 This diagram for the electron orbits in metallic copper indicates only a few of the many types of orbits an electron can pursue in k-space when a uniform magnetic field is applied to a noble metal. (Recall that the orbits are given by slicing the Fermi surface with planes perpendicular to the magnetic field.) The figure here displays **a** a closed electron orbit; **b** a closed hole orbit; **c** an open hole orbit, which continues in the same general direction indefinitely in the repeated-zone scheme

The situation is very different for the magnetic and electric fields applied in special directions relative to the crystal axes. For these special directions called *open electron orbits* can occur, as illustrated in Fig. 10.5 for copper. In this case $\mathbf{k}(t) - \mathbf{k}(0)$ has a component proportional to $\mathbf{E}t$ which is *not* negligible. When the term $[\mathbf{k}(t) - \mathbf{k}(0)]$ must be considered, it can be shown that the magnetoresistance does not saturate but instead increases as B^2.

Figure 10.6 shows the angular dependence of the magnetoresistance in copper, which exhibits both closed and open orbits, depending on the direction of the magnetic field (see Fig. 10.6). The strong angular dependence is associated with the large difference in the magnitude of the magnetoresistance for closed and open orbits. (Compare orbits(a) and (b) in Fig. 10.5.) Large values of $\omega_c \tau$ are needed to distinguish clearly between the open and closed orbits, thereby requiring the use of samples of very high purity and measured at low temperature (e.g., 4.2 K) operation to observe open orbits.

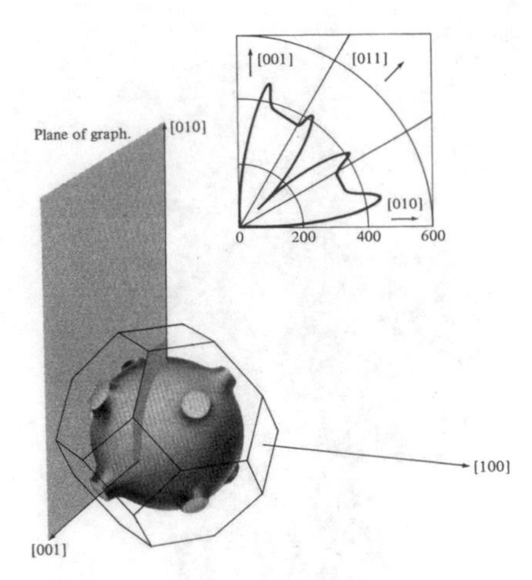

Fig. 10.6 The spectacular directional dependence of the high-field magnetoresistance in copper arises because of the special characteristics of the Fermi surface which supports open orbits. The [001] and [010] directions of the copper crystal are as indicated in the figure, and the current flows in the [100] direction perpendicular to the graph. The magnetic field is in the plane of the graph. Its magnitude is fixed at 18 kilogauss, and its direction is varied continuously from [001] to [010]. The graph is a polar plot of the transverse magnetoresistance $[\rho(H) - \rho(0)]/\rho(0)$ versus the orientation of the magnetic field

Problems

10.1 Consider n-type Si ($n = 10^{17}$ cm^{-3}) in a weak magnetic field, rotated between the (100) and (110) crystallographic directions.

(a) Calculate the cyclotron effective masses for all carrier pockets for the applied magnetic field along an arbitrary angle between (100) and (110). Use values for the effective mass components $m^*_{le} = 0.92m_0$ and $m^*_{te} = 0.19m_0$.
(b) Write an expression for the Hall Coefficient (R_H) for **B** along (100) and **j** along (001), including all the carrier pockets for Si.
(c) Find R_H for **B** along (110) and **j** along (001), including all carrier pockets.
(d) Should the results of (b) and (c) be the same?

10.2 (a) By using the result obtained in Problem 10.8(a), find an expression for the transverse magnetoresistance for $\mathbf{B} \parallel (100)$ and $\mathbf{j} \parallel (001)$ assuming a total carrier density n and a constant relaxation time τ.
(b) What is the longitudinal magnetoresistance for this n-type sample with $\mathbf{j} \parallel (001)$?

10.3 (a) Show that the equations of motion of a damped electron in an electric field $\mathbf{E} \perp$ magnetic field $\mathbf{B} = B\hat{z}$ are given by:

$$m(\frac{dv_x}{dt} + \frac{v_x}{\tau}) = -e(E_x + Bv_y)$$

$$m(\frac{dv_y}{dt} + \frac{v_y}{\tau}) = -e(E_y - Bv_x)$$

$$m(\frac{dv_z}{dt} + \frac{v_z}{\tau}) = -eE_z$$

(c) Consider a metal with a free electron concentration of n and with an electric charge e placed in a uniform magnetic field $\mathbf{B} = B\hat{z}$. Show that the electric current density in the xy plane is related to the electric field by the relations:

$$j_x = \sigma_{xx}E_x + \sigma_{xy}E_y$$

$$j_x = \sigma_{yx}E_x + \sigma_{yy}E_y$$

Consider that the electric field frequency $\omega \gg \omega_c$ and $\omega \gg 1/\tau$, where ω_c is the cyclotron resonance given by $\omega_c = eB/m$ and τ is the time between electron collisions.

(d) Explain physically what happens when $\omega\tau \gg 1$ and when $\omega\tau \ll 1$ in terms of sample preparation. Which regime is more strongly dependent on sample preparation?

10.4 By solving the equations developed in Problem 10.3, show that the components of the magnetoconductivity tensor are given by:

$$\sigma_{xx} = \sigma_{yy} = i\omega_p^2/4\pi\omega$$

$$\sigma_{yx} = -\sigma_{xy} = \omega_c\omega_p^2/4\pi\omega^2$$

where $\omega_p = \sqrt{4\pi ne^2/m}$ is the plasmon resonance. (The screened plasma frequency has been derived, which contains the core dielectric constant ε_{core}. In this problem, consider $\varepsilon_{core} = 1$).

10.5 (a) Show that the cyclotron mass m_c^* of carriers in graphene is given by $m_c^* = \sqrt{\pi n}/v_F$, where v_F is the Fermi velocity of graphene and n is the electronic density of carriers.

(b) Discuss how the cyclotron mass m_c^* in 3D graphite will deviate from the effective mass for a planar 2D system, as calculated above.

10.6 A sample of an extrinsic p-type semiconductor is placed in a uniform and steady magnetic field B_0. Assume that the effective mass tensor is isotropic, and that the mean free path is independent of velocity.

(a) First, show that at a finite temperature, the time-averaged drift velocity of electrons in the small-field limit ($\omega_r\tau \ll 1$) can be written as

$$\overline{v_x} = \frac{e}{m^\star}\left(\overline{\tau}E_x - \omega_c\overline{\tau^3}E_x - \omega_c\overline{\tau^2}E_y\right)$$

and

$$\overline{v_y} = \frac{e}{m^\star}\left(\overline{\tau} E_y - \omega_c \overline{\tau^2} E_x\right)$$

where $\bar{\tau^n}$ is averaged over a kinetic energy-weighted Maxwellian (Boltzmann) distribution.

(b) Show that the proper averaging over such a direction leads to an average of relaxation times over k space of the form

$$\overline{\tau^n} = \frac{\langle v^2 r^n \rangle}{\langle v^2 \rangle}$$

where $\langle x \rangle$ is given in terms of the Boltzmann distribution, f_0, as

$$\langle x \rangle = \int x f_0(v) d^3 v$$

(c) Evaluate the integral of part (b) using the fact that $\tau = \lambda/v$ where λ is the carrier mean free path and show that the magnetoconductivity obeys a quadratic dependence on the steady magnetic field B_0 (the so-called Lorentz term), despite the fact that the system under consideration is a one carrier, spherical Fermi surface system. Why should carriers be scattered by a magnetic field under steady-state conditions if the Hall field (set up by the accumulated charge of the deflected carriers) acts to 'cancel' the effect of the magnetic field?

(d) Show that the conductivity of the sample decreases with increasing magnetic field and in particular

$$\sigma(B_0) = \sigma_0 \left[1 - \frac{e^2 B_0^2}{m_e^\star} \frac{\lambda^2}{k_B T} \left(\frac{4-\pi}{8}\right)\right],$$

where σ_0 is the conductivity in the absence of a magnetic field and λ is the carrier mean free path. Assume $(\omega_r \tau \ll 1)$ where $\omega_c = \frac{eB_0}{m_e^\star}$.

10.7 Cyclotron resonance is observed when an electromagnetic wave is incident on a sample at $\omega = \omega_c$ where $\omega_c = eB/m$.

(a) What happens to an electromagnetic wave that is incident on a sample at $\omega = \omega_c$?
(b) The cyclotron resonance experiment is normally carried out by varying the magnitude of B. Why does this cyclotron resonance in 3D germanium occur at different values of B for electrons and holes ($m_e = 0.55 m_0$ and $m_h = 0.37 m_0$)?
(c) How can the differing resonance magnetic fields be utilized to distinguish electron doping from hole doping of germanium and for the amount of doping present in a given sample?

10.8 (a) Contrast the values for holes and electrons regarding the Hall coefficient and the transverse magnetoresistance for the holes in silicon for degenerate p-type material with 10^{17} hole carriers/cm^3 in the valence bands and for another material where the carrier in a degenerate n-type material with 10^{17} electron carriers/cm^3 in the conduction bands.

(b) Model the valence bands in (a) as a doubly degenerate heavy hole band and a non-degenerate light hole band with spherical Fermi surfaces and effective mass components $m_h = 0.50m_0$ and $m_\ell = 0.16m_0$.

(c) Model the electrons to occupy conduction band states associated with the 6 electron carrier pockets of Si using the mass components $m_t = 0.19m_0$ and $m_\ell = 0.98m_0$ and take $\mathbf{B} \parallel (001)$ and $\mathbf{j} \parallel (100)$.

10.9 Assume that you have an intrinsic direct gap III-V semiconductor sample heated to a temperature T, with spherical carrier pockets for the electrons and holes (an idealized version of GaAs). Suppose that the sample has an electron carrier concentration n in the conduction band. Denote the masses for the electrons, light holes and heavy holes by m_e, m_{lh} and m_{hh}, respectively.

(a) Find an expression for the Fermi level.
(b) Derive an expression for the electrical conductivity.
(c) Derive an expression for the transverse magnetoresistance. (Assume current flow along (100) and the magnetic field along (001).)

10.10 Both the Hall effect and thermoelectric power (Seebeck coefficient) are sensitive to the carrier density. Suppose that we have a simple semiconductor with spherical carrier pockets with a single parabolic non-degenerate band for the valence and conduction bands. Assume that the ratio of the effective masses for the electrons and holes is $m_e^*/m_h^* = 1/2$.

(a) For a total carrier density of n, find the ratio of the Hall coefficients for the case where the semiconductor is n-type by doping (no holes), and where the semiconductor is intrinsic ($n = n_e + n_h$).
(b) Is the Hall coefficient more sensitive or less sensitive than the Seebeck coefficient to the carrier scattering mechanism and why?
(c) Under what conditions would you expect an increase in the donor binding energy for a shallow donor level in a quantum well made from this material.

10.11 (a) Suppose that a sample with three electron carrier pockets in the k_x, k_y plane (and a three fold symmetry axis) is put in a weak magnetic field, find an expression for the cyclotron effective mass for each of the 3 electron carrier pockets as the field is rotated by an angle θ in the x – y plane from a (100) direction $\theta = 0°$ through an angle of 60° ($2\pi/6$).

(b) Using the result in (a), find the transverse magnetoresistance for current flow along an (001) direction as $\mathbf{H}$ is rotated in the x – y plane from a (100) direction through an angle of 60°. Assume a total carrier density n and a relaxation time τ. Also assume that the applied electric field is along the (001) direction.

(c) Repeat (b) for $\mathbf{j} \parallel (100)$ and the applied $\mathbf{E}$ field $\parallel$ (100). As in (a), the $\mathbf{H}$ field is allowed to rotate in the x – y plane from $\theta = 0$ to $\theta = 2\pi/6$.

Suggested Readings

Ashcroft, Mermin, *Solid State Physics* (Holt, Rinehart and Winston, 1976). Chapter 12
A.H. Castro Neto, F. Guinea, N.M.R. Peres, K.S. Novoselov, A.K. Geim, The electronic properties of graphene. Rev. Mod. Phys. **81**, 109 (2009)
M. Fox, *Optical Properties of Solids* (Oxford University Press, Oxford, 2009)
P. Hofmann, *Solid State Physics: An Introduction*, 1st edn. (Wiley, New York, 2008)
C. Kittel, *Introduction to Solid State Physics*, 7th edn. (Wiley, New York, 1996). Chapter 6
A.B. Pippard, *Magnetoresistance in Metals* (Cambridge University Press, Cambridge, 1989)

Chapter 11
Transport in Low Dimensional Systems

11.1 Introduction

Transport phenomena in low dimensional systems such as in quantum wells (2D), quantum wires (1D), and quantum dots (0D) are dominated by quantum effects not included in the classical treatments based on the Boltzmann equation and discussed in Chaps. 7–10. With the availability of experimental techniques to fabricate and synthesize materials with nanometer sizes and nanometer dimensions (0D, 1D, and 2D), transport studies in low dimensional systems have become an active research area. In this chapter, we consider some highlights on the subject of transport in low dimensional systems.

11.2 Observation of Quantum Effects in Reduced Dimensions

Quantum effects dominate the transport in quantum wells and other low dimensional systems, such as quantum wires and quantum dots, when the de Broglie wavelength of the electron

$$\lambda_{\mathrm{dB}} = \frac{\hbar}{(2m^{*}E)^{1/2}} \tag{11.1}$$

exceeds the size of a quantum structure of characteristic length L_z ($\lambda_{\mathrm{dB}} > L_z$). In this limit ($\lambda_{\mathrm{dB}} > L_z$), a new quantum effect becomes dominant, namely tunneling through a potential barrier of length L_z. To get some order of magnitude estimates of the electron kinetic energies E below which quantum effects become important, we show a log-log plot of λ_{dB} vs E for GaAs and InAs in Fig. 11.1. From this plot we see that an electron energy of $E \sim 0.1$ eV for GaAs corresponds to a de Broglie wavelength of λ_{dB}= 100 Å or 10 nm. Thus wave properties for electrons can be expected to become important for structures with L_z smaller than λ_{dB}. To observe

M. Dresselhaus et al., *Solid State Properties*, Graduate Texts in Physics,
https://doi.org/10.1007/978-3-662-55922-2_11

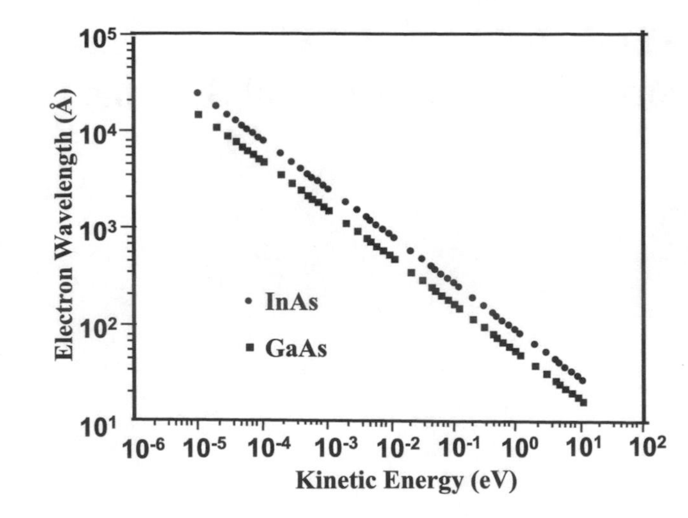

Fig. 11.1 Log-Log plot of the electron de Broglie wavelength λ_{dB} vs the electron kinetic energy E for GaAs (■) and InAs (•)

quantum tunneling effects, the thermal energy must also be less than the energy level separation ΔE between quantum levels within the quantum well, $k_B T < \Delta E$, where we note that room temperature (20 °C) corresponds to 25 meV. Since quantum effects depend on the phase coherence of electrons, scattering processes which destroy phase coherence can also destroy their quantum effects. The observation of quantum effects thus requires that the carrier mean free path be much larger than the dimensions of the quantum structures (quantum wells, wires, or dots).

The limit where quantum effects become important has been given the name of *mesoscopic physics*. Carrier transport in this limit simultaneously exhibits both particle and wave characteristics. In this ballistic transport limit, carriers can in some cases also transmit charge or energy without scattering. The small dimensions required for the observation of quantum effects can be achieved by the direct fabrication of semiconductor elements of small dimensions (quantum wells, quantum wires and quantum dots) and in this regime electrons can reach the boundaries of small objects before being scattered. Another approach is the use of gates on a field effect transistor to define an electron gas of reduced dimensionality. In this context, negatively charged metal gates can be used to control the source to drain current of a 2D electron gas formed near the GaAs/AlGaAs interface, as shown in Fig. 11.2. Between the dual gates shown on this figure, a thin conducting wire is formed out of the 2D electron gas. Controlling the gate voltage controls the amount of charge in the depletion region under the gates, as well as the charge flowing in the quantum wire. Thus, lower dimensional channels can be made in a 2D electron gas by using metallic gates. In the following sections of this chapter, a number of important applications are made that make use of this concept.

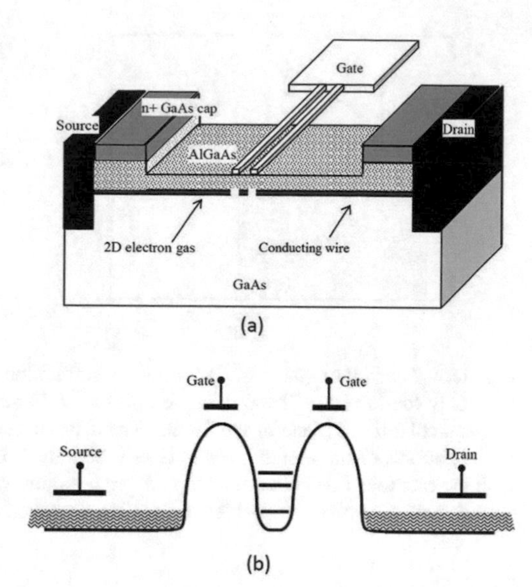

Fig. 11.2 **a** Schematic diagram of a lateral resonant tunneling field-effect transistor which has two closely spaced fine finger metal gates; **b** schematic of an energy band diagram for the device. A 1D quantum wire is formed in the 2D electron gas between the gates, which generally are metals

11.3 Density of States in Low Dimensional Systems

We showed in (8.40) that the density of states for a 2D electron gas that might form under the gate in Fig. 11.2a is a constant for each 2D subband

$$g_{\mathrm{2D}} = \frac{m^*}{\pi \hbar^2}, \tag{11.2}$$

and the 2D density of states g_{2D} for an electron gas is shown in Fig. 11.3a as a series of steps, where the inset to Fig. 11.3a shows the quantum well formed near a modulation doped GaAsAlGaAs interface. In the diagram of Fig. 11.3a, only the lowest bound state of the inset is partially occupied, with the upper levels being unoccupied for $(E_2 - E_1) \gg kT$. However, in Fig. 11.3b, four levels lie below the 4th peak in g(E).

Using the same arguments that are given above for g_{2D}, we now derive a simple formula for the density of states for a 1D electron gas. The total number of electronic states up to wavevector k in Fig. 11.3b is given by

$$N_{\mathrm{1D}} = \frac{2}{2\pi}(k) = \frac{1}{\pi}(k) \tag{11.3}$$

which for a parabolic band $E = E_n + \hbar^2 k^2/(2m^*)$ becomes

$$N_{\mathrm{1D}} = \frac{2}{2\pi}(k) = \frac{1}{\pi}\left(\frac{2m^*(E - E_n)}{\hbar^2}\right)^{1/2} \tag{11.4}$$

yielding an expression for the density of states $g_{\mathrm{1D}}(E) = \partial N_{\mathrm{1D}}/\partial E$

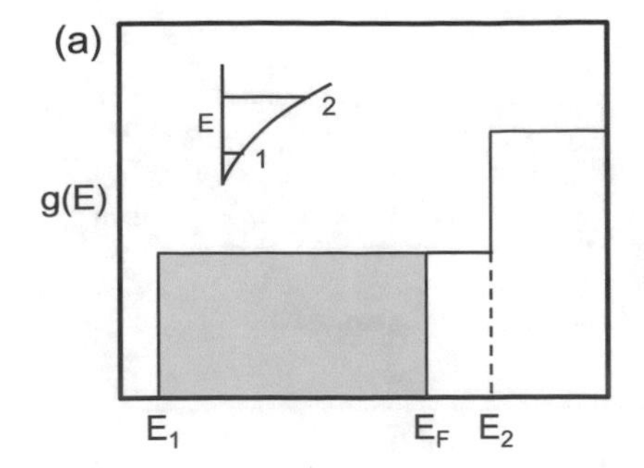

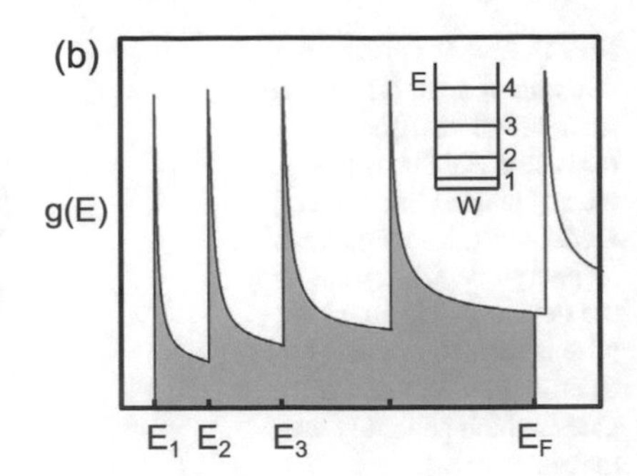

Fig. 11.3 Density of states $g(E)$ plotted as a function of energy. **a** Quasi-2D density of states, with only the lowest subband occupied up to the Fermi level E_F. Inset: Confinement potential perpendicular to the plane of the 2DEG. The discrete energy levels shown in the inset correspond to the bottoms of the first and second 2D subbands. **b** Quasi-1D density of states versus energy, with the first four 1D subbands occupied in a quantum well of width W. Inset: Square-well lateral confinement potential with discrete energy levels indicating the 1D subband extrema

$$g_{1D}(E) = \frac{1}{2\pi}\left(\frac{2m^*}{\hbar^2}\right)^{1/2}(E - E_n)^{-1/2}. \tag{11.5}$$

The interpretation of this expression is that, at each doubly confined bound state level E_n, there is a singularity in the density of states, as shown in Fig. 11.3b where the four lowest energy levels in the quantum well are occupied at a temperature $T = 0$ K, where no thermal excitation occurs.

11.3.1 Quantum Dots

A quantum dot consists of a small cluster of atoms such that the diameter of the quantum dot is small compared to the length scale relevant to the phenomena under investigation. In this limit, the cluster exhibits the properties of a zero dimensional system. Since the levels of a quantum dot are all discrete, any averaging would involve a sum over quantum levels and not an integral over energy. If, however, one chooses to think about quantum dots in terms of their density of states, then the DOS of a quantum dot would be a delta function singularity positioned at the energy of the localized state, with the integral over the quantum dot performed at $T = 0$ K, thereby giving the number of electrons in the quantum dot. At finite temperature, some tunneling of carriers becomes possible if the thermal energy is large enough to support the transport of electrical energy by a tunneling or collision process between quantum dots.

11.4 Ballistic Transport and the Landauer Formula

As the size of a conducting material is made small compared to the mean free path of the electrons, the Boltzmann transport model (and the Drude model in particular) breaks down and is not capable of describing the transport properties for this system accurately. The measured resistance to electron transport in such a low-dimensional system can, however, be described well by the Landauer transport model, which calculates the electron conduction by treating the possible scattering sources in the material as barriers to electron transport with a certain probability of electron transmission through the barriers. This approach allows for the possibility of ballistic transport, i.e., transport through a system without scattering.

As an illustration of a 1D system, the current I through a conductor with only one quantum channel is considered (i.e., an electronic band whose wavevector is single-valued in the transverse dimension but continuous in the longitudinal dimension). Such a system is described by

$$I = \left(\frac{2g_v e}{\pi}\right)\int_{-\infty}^{\infty}[f(E, eV_s) - f(E, eV_d)]\cdot g(E)\cdot \mid v \mid \cdot dE \tag{11.6}$$

In this generalized equation, g_v is the valley degeneracy which takes into account the possibility of multiple carrier pockets or valleys in the Brillouin zone and also take into account the spin degrees of freedom. Furthermore, $g(E)$ under the integral is the density of states for each carrier pocket, v is the carrier group velocity, and f denotes the Fermi function appropriate to that carrier pocket. For one-dimensional systems, $g(E) =\mid dE(k)/dk \mid^{-1}$, and since $v = (dE(k)/dk)/\hbar$, then the product of $g(E)\cdot v = 1/\hbar$. To calculate the electrical conductance G in (11.7), we then simply divide the current I in (11.6) by the bias voltage $V_b = V_s - V_d$ in (11.7) between the source and drain as shown in Fig. 11.2 to obtain as shown in (11.7)

$$G = \frac{I}{V_b} = G_0 \int \frac{[f(E, eV_s) - f(E, eV_d)]}{eV_b} dE. \tag{11.7}$$

The value in front of the integral in (11.7), $G_0 = 4e^2/h$, is the quantum conductance for a perfect resistance-piece 1D system, ($R_0 = 1/G_0 = 6.5\,\mathrm{k}\Omega$) for which $G = G_0 = 4e^2/h$ in (11.7). If there are two parallel degenerate quantum channels in the Brillouin zone, then $\nu = 2$. Some common examples of $\nu = 2$, would arise from 2 carrier pockets (e.g., at the K and K′ points in the Brillouin zone for carbon nanotubes) contributing equally to the measured current. Other examples of $\nu = 2$ could be due to spin degeneracy. For most materials, other than sp^2 carbon, with just one quantum conducting channel, $G_0 = 2e^2/h$. We can approximate the integrand of (11.7) as df/dE, which is valid for bias voltages not larger than the thermal energy, or for larger bias voltages under diffusive transport conditions (i.e., not ballistic conditions). Since the density of states is zero in the band gap, the integral in (11.7) is only taken over the bands containing carriers, and over these bands, we thus obtain

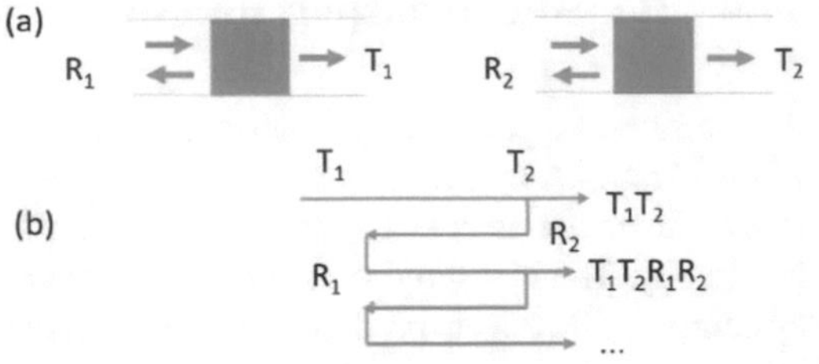

Fig. 11.4 Illustration of multiple scattering sources in a 1D system

$$G = G_0 \int_{carrier\ bands} \frac{df}{dE} dE. \tag{11.8}$$

Equation 11.8 assumes that the position of the Fermi level with respect to the bands is known. One method for calculating the Fermi level is discussed in Sect. 11.4.4. To incorporate the effects of scattering in the system, such as by a lattice defect or by atomic vibrations, we simply add the effects of the scattering sources together in (11.9) so that the transmission coefficient $\mathscr{T}$, simply determines the conductance G

$$G = G_0 \int_{carrier\ bands} \mathscr{T} \frac{df}{dE} dE. \tag{11.9}$$

Once scattering is taken into consideration, we can then describe transport equivalently in terms of a resistance R to carrier transmission as

$$R = R_0 \cdot \frac{1}{\int_{bands} \mathscr{T}(E)(df/dE)dE}. \tag{11.10}$$

To start our illustration, we will consider a metallic 1D system (i.e., without a band gap). In this case (assuming a constant, energy independent, value for $\mathscr{T}(E)$), the resistance of a system that allows scattering satisfies

$$R = R_0 \frac{1}{\mathscr{T}} = R_0 \left(1 + \frac{1 - \mathscr{T}}{\mathscr{T}} \right) \tag{11.11}$$

where we have separated the quantum resistance R_0 from the effects of the carrier scattering sources. To calculate the transmission through multiple barriers, we have to consider multiple reflections, as illustrated in Fig. 11.4. The total transmission coefficient in this case is given by

$$\mathscr{T}_{tot} = \mathscr{T}_1 \mathscr{T}_2 + \mathscr{T}_1 \mathscr{T}_2 R_1 R_2 + \mathscr{T}_1 \mathscr{T}_2 R_1^2 R_2^2 + ... = \frac{\mathscr{T}_1 \mathscr{T}_2}{1 - R_1 R_2}. \tag{11.12}$$

From (11.12), the resistances of an arbitrary number of scattering sources add according to

$$\frac{1-\mathcal{T}_{tot}}{\mathcal{T}_{tot}}=\frac{1-\mathcal{T}_1}{\mathcal{T}_1}+\frac{1-\mathcal{T}_2}{\mathcal{T}_2}+... \tag{11.13}$$

Using (11.11), we can calculate the resistance of a system with multiple scattering sources, such as: (1) a carbon nanotube with contact barriers, (2) diffusive scattering, and local defects.

11.4.1 *Relationship Between the Mean Free Path and the Transmission Coefficient*

In order to calculate the relationship between the mean free path ℓ_f and the transmission coefficient $\mathcal{T}$, we will use a weak scatterer approximation so that the scattering process acts like a weak perturbation to the physical system, where $\mathcal{T}_s \sim 1$. In this approximation,

$$\frac{L_s}{\ell_f}\approx 1-\mathcal{T}_s\approx\frac{1-\mathcal{T}_s}{\mathcal{T}_s} \tag{11.14}$$

where the average distance between scatterers L_s is much smaller than the mean free path of ℓ_f. This can be seen in Fig. 11.5, which shows in the transmission probability with no scattering (probability of ballistic transmission without a single scattering event) for a particle as a function of distance traveled (solid curve showing an exponential decay). The initial slope of the decay is $-(1-\mathcal{T}_s)/L_s$, which determines the decay constant or mean free path ℓ_f between scattering events.

The total scattering coefficient for all the scatterers (v_s) in the system (green dashed curve in Fig. 11.5), including multiple reflections, can then be calculated using the following equation:

$$\frac{1-\mathcal{T}_{total}}{\mathcal{T}_{total}}=N_s\frac{1-\mathcal{T}_s}{\mathcal{T}_s}=N_s\frac{L_s}{\ell_f}=\frac{L}{\ell_f}\Rightarrow\mathcal{T}_{total}=\frac{\ell_f}{\ell_f+L}. \tag{11.15}$$

Using the result of (11.15), the resistance with perfect contacts is

$$R=R_0\frac{\ell_f+L}{\ell_f}. \tag{11.16}$$

Now we consider a system like a semiconductor with a band gap. The integrals over energy in (11.9) and (11.10) both have a maximum value of unity, since $\mathcal{T}\leq 1$ and the integral over df is unity. So, in the presence of the band gap, the integral is less than unity, which physically means that charge carriers are being depleted. Therefore, it is useful to separate out the effects of carrier depletion due to the band gap from those of the reduced mobility due to carrier scattering sources. We can separate these two effects by modifying (11.10) as

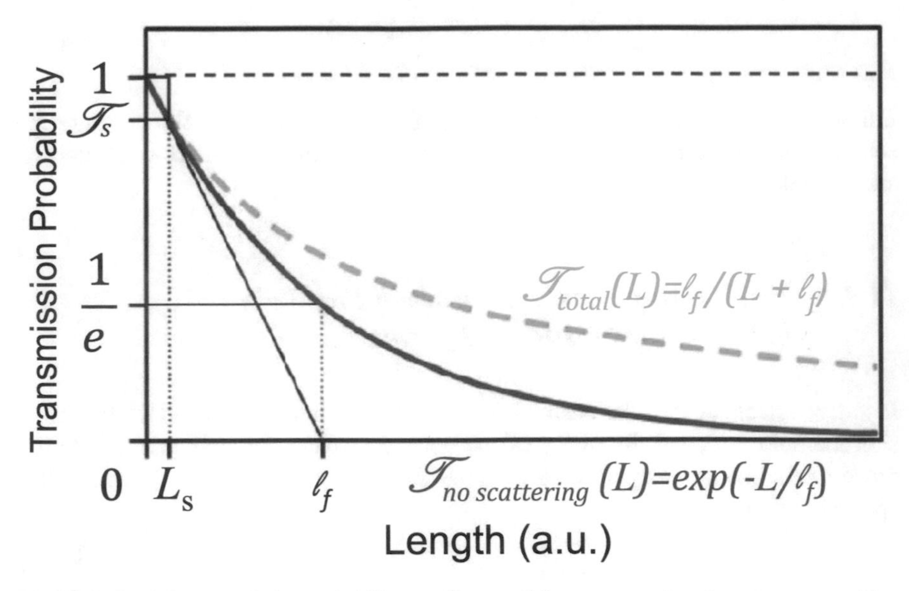

Fig. 11.5 Particle transmission probability to a distance L for a system of weak scatterers, considering the case of no scattering (ballistic) transmission (red solid curve) and total (multiple reflection) transmission (green dashed curve); $\mathcal{T}_s$ is the transmission probability to L_s after one scattering evernt, and ℓ_f is the mean free path

$$R = R_0 \cdot \frac{1}{\int_{bands} \mathcal{T}(E)\frac{df}{dE}dE} = \left(\frac{R_0}{\int_{bands} \frac{df}{dE}dE}\right)\cdot\left(\frac{\int_{bands} \frac{df}{dE}dE}{\int_{bands} \mathcal{T}(E)\frac{df}{dE}dE}\right) = R_0^* \frac{1}{\mathcal{T}^*}$$

$$= R_0^* \left(1 + \frac{1-\mathcal{T}_1^*}{\mathcal{T}_1^*} + \frac{1-\mathcal{T}_2^*}{\mathcal{T}_2^*} + ...\right) \tag{11.17}$$

Here, R_0^* denotes the *depleted quantum channel resistance*, or the resistance of the gapped system in the absence of other scattering sources, and $\mathcal{T}^*$ is the effective or weighted average transmission coefficient in the quantum channel. If $\mathcal{T}(E)$ is constant, then $\mathcal{T}^* = \mathcal{T}$.

We show here a simple example to demonstrate the importance of the separation of the calculations of the two terms in (11.17). Consider a resonant scattering source in the channel, which only scatters electrons within a narrow energy range. If that energy range happens to lie inside the band gap, then the scatterer should not affect the conductance of the system, because none of the conduction electrons are of the proper energy to scatter. This result is found with the separation above. However, if we neglect this carrier separation, then an additional resistance due to the carrier depletion needs to be added for each such scattering source that we add to the system, which is a non-physical result. The total resistance in a single walled nanotube, including contacts and diffusive electron-phonon scattering, can be then written as

$$R = R_0^* \left(1 + \frac{1 - \mathcal{T}_{ph}^*}{\mathcal{T}_{ph}^*} + 2\frac{1 - \mathcal{T}_c^*}{\mathcal{T}_c^*} \right) \tag{11.18}$$

where $\mathcal{T}_{ph}^*$ and $\mathcal{T}_c^*$ are the transmission coefficients for phonon scattering and for the two contacts, respectively.

11.4.2 Relationship to the Boltzmann Transport

As another instructive example, we can also calculate the electrical resistance of a one dimensional system using Boltzmann transport theory, which gives a current of j_{1D}, when

$$j_{1D} = \frac{2}{\pi} \int e \cdot v_g(k) \cdot f_1 dk = \frac{2}{\pi} \int e \cdot v_g \cdot f_1 \frac{dk}{dE} dE = \frac{2}{\pi} \int \frac{e \cdot v_g \cdot f_1}{\hbar(d\omega/dk)} dE \tag{11.19}$$

Since the carrier group velocity v_g is just $v_g = \frac{d\omega}{dk}$, we can substitute for v_g to get the conductance

$$\frac{j_{1D}}{E} = \sigma = \left(\frac{4e^2}{h}\right) \left(\int \tau v_g\right) \left(\frac{df_0}{dE} dE\right). \tag{11.20}$$

Considering the carrier mean free path $\ell_f = v_g \tau$ and the length of the conducting channel L, we can calculate the conductivity of the conducting channel as

$$G = \frac{\sigma}{L} = G_0 \int \frac{\ell_f}{L} \frac{df_0}{dE} dE \tag{11.21}$$

where G_0 is the quantum conductance. The only difference between this result and the result of the Landauer model above is that, in the Landauer model, $\frac{\ell_f}{L} \rightarrow \frac{\ell_f}{\ell_f + L}$, which explicitly accounts for ballistic conduction through the entire conducting channel in the case when $\ell_f \gg L$.

11.4.3 Relationship to Mobility Calculations

Traditionally, the electrical conductivity in semiconductors is calculated using the carrier mobility μ, by the equation

$$\sigma = en\mu \tag{11.22}$$

where

$$n = \int_{conduction-band} f(E) g(E) dE \tag{11.23}$$

is the n-type charge carrier density of the material, ignoring p-type conduction. This formula breaks down for degenerately doped semiconductors and metals, because not all charge carriers contribute to the conductivity equally, since carriers near the Fermi energy make the dominant contribution to conduction. Electrons with energies far below the Fermi energy do not contribute because of the Pauli exclusion principle. This issue does not arise for most semiconductor calculations; however, because only the tail of the Fermi distribution (light doping regime) extends into the conduction band.

The Landauer model can then be shown to be equivalent to the mobility equation $G = en\mu/L$ in the light doping approximation. First, using the Einstein mobility relation,

$$\mu = \frac{eD}{k_B T} \tag{11.24}$$

where $k_B T$ is the thermal energy of the system and D is the carrier diffusion coefficient, given by $D = \ell_f \cdot v/\pi$. So assuming two quantum channels for the system, we have

$$G = \frac{en\mu}{L} = \frac{2e^2}{\pi L} \int_{conduction-band} f(E)g(E)\frac{\ell_f \cdot v}{k_B T} dE \tag{11.25}$$

As in (11.6), the contribution from the density of states factor cancels the contribution from the group velocity in 1D, and we get

$$G = \frac{4e^2}{h} \int \frac{\ell_f}{L} \cdot \frac{f(E)}{k_B T} dE. \tag{11.26}$$

In the limit $(E - E_F)/k_B T = \Delta \gg 1$, then

$$\frac{df}{dE} = \frac{e^\Delta}{(1+e^\Delta)^2} \frac{1}{k_B T} \approx \frac{e^\Delta}{(1+e^\Delta)e^\Delta} \frac{1}{k_B T} = \frac{f(E)}{k_B T}. \tag{11.27}$$

Thus, we obtain

$$G = G_0 \int \frac{\ell_f}{L} \cdot \frac{df}{dE} dE \tag{11.28}$$

which is the same as (11.21), because the mobility equation is also based on a diffusive transport model. Under ballistic conditions ($\ell_f > L$), the mobility equation and Boltzmann transport model both break down because the electrical conductance must not exceed G_0.

11.4.4 Dependence of the Fermi Energy on Gate Voltage

The electron transport properties of nanostructures are often measured in a field effect transistor (FET) configuration, in which an electrostatic gate can be used to dope the material to vary the Fermi energy of the material without adding impurities. The application of a gate voltage changes the electrochemical potential (ζ) according to the relation

$$\zeta = eV_{gate} \tag{11.29}$$

where e is the elementary charge and V_{gate} is the applied gate voltage. The electrochemical potential is the sum of the electrostatic potential (ζ_{elect}) and of the chemical potential (ζ_{chem}). Thus, the electrochemical potential is given by

$$\zeta = \zeta_{elect} + \zeta_{chem} = e\phi + E_F \tag{11.30}$$

where $\phi = Q/C_{geom}$ is the electrostatic potential, which can be obtained by dividing the electric charge Q by the geometric capacitance C_{geom}. The E_F term in (11.30) can typically be neglected for bulk materials, which have a large number of electronic states. Replacing the electrochemical potential ζ in (11.29) and (11.30) by $e\phi = eQ/C_{geom}$ yields

$$eV_{gate} = eQ(E_F)/C_{geom} + E_F \tag{11.31}$$

Here, $Q(E_F)$ is the charge on the nanostructure, which can be found by integrating the density of states (remembering to include both p- and n-type charging terms)

$$Q(E_F) = \int g(E) f(E) dE. \tag{11.32}$$

11.4.5 Ballistic Phonon Transport

We can also use the Landauer transport model to calculate the thermal power conducted by all ballistic phonons. This thermal power flow is given by

$$\dot{Q} = \sum_m \int_0^\infty \frac{dk}{2\pi} \hbar\omega_m(k) v_m(k) \eta(\omega_m, T_{hot}) \mathscr{T}(\omega_m) \tag{11.33}$$

Here, $\hbar\omega_m(k)$ is the phonon energy, which is integrated over all wave vectors k and bands m, while η is the temperature-dependent Bose–Einstein distribution function, v_m is the group velocity of the mth phonon mode, T_{hot} is the temperature variable in the Bose–Einstein distribution, and $\mathscr{T}$ is the transmission coefficient, typically set to 1 in order to calculate the upper bound for heat transport. The thermal conductance is then calculated as $G_{th} = \dot{Q}_{ph}/(A\Delta T)$, where A is the cross-sectional area of the

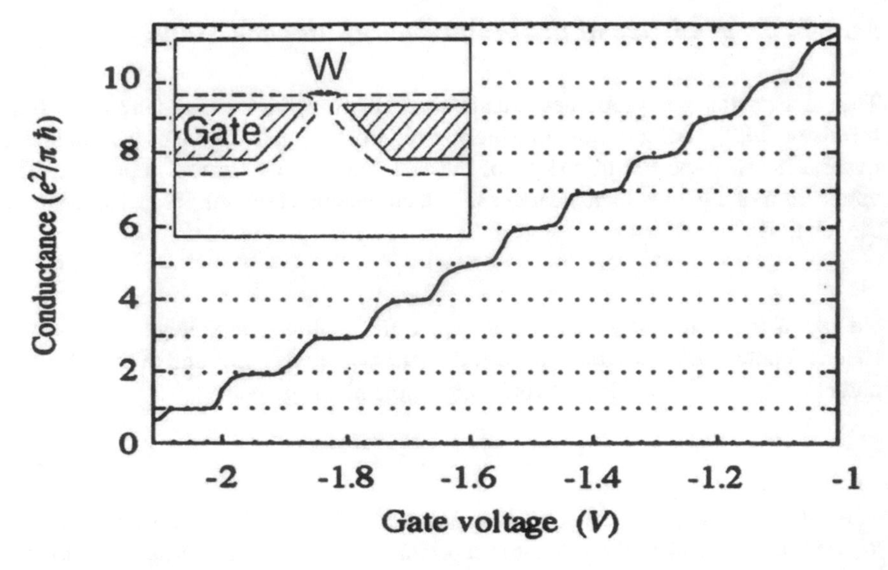

Fig. 11.6 Conductance-gate voltage data taken from a GaAs quantum point contact showing quantized behavior (Physical Review Letters, **60**, 848 (1988))

sample or device. Phonon scattering can also be included in this formalism by adding individual scattering events into the transmission coefficient $\mathscr{T}$, as described above for electrons.

11.5 Quantum Point Contacts (QPC) Effects

A quantum point contact serves as perhaps the simplest device that demonstrates ballistic conduction of electrons and their wave-like nature.The existense of QPCs were demonstrated in a top-gated GaAs/AlGaAs two-dimensional electron gas (2DEG) more than 25 years ago (see van Wees, et al., Physical Review Letters, **60**, 848 (1988)). The characteristic signature of a QPC is conductance quantized in units of $e^2/(\pi\hbar)$ or $(2e^2/h)$, as shown in Fig. 11.6. As the channel width W becomes wider with increasing gate voltage, the number of conducting channels (transverse modes in the channel) increases by integer values. The early point contact experiments were done using ballistic point contacts on a gate structure placed over a two-dimensional electron gas as shown schematically in the inset of Fig. 11.6. The width W of the gate (in this case 2500 Å) defines the effective width W' of the conducting electron channel, and the applied gate voltage is varied in order to control the effective width W'. Superimposed on the raw data for the resistance vs gate voltage is a collection of periodic steps as shown in Fig. 11.6, after subtracting off the background resistance of 400 Ω.

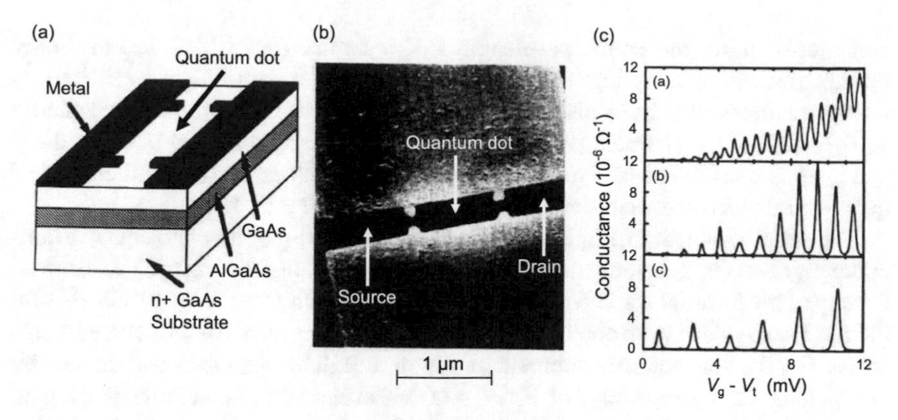

Fig. 11.7 Observation of the Coulomb Blockade: **a** Schematic diagram, **b** SEM image, and **c** conductance-gate voltage characteristics of a single electron transistor showing quantized behavior (Physical Review Letters, **65**, 771 (1990))

11.6 Coulomb Blockade and Single Electron Transistors (SETs)

A single electron transistor is another device that can be fabricated using this general fabrication scheme. Like QPCs, single electron transistors (SETs) were demonstrated in GaAs 2DEGs more than 25 years ago (see Meirav et al. 1990). In a SET, a quantum dot is formed by the top gated geometry shown in Fig. 11.7. Current tunneling between the source and drain through the quantum dot can be modulated by many orders of magnitude with the addition of each additional electron on the quantum dot. This produces periodic conductance resonances known as the Coulomb blockade (CB), as shown in Fig. 11.7c. Here, the peak width is determined by temperature ($k_B T$) rather than tunneling. In order to observe the CB effect, the tunneling resistance must be larger than h/e^2 in order to make the quantum dot an effectively isolated capacitor. Two experimental observations of these phenomena were simultaneously published (see Wharam et al. 1988 and van Wees et al. 1988).

There are several conditions necessary to observe perfect ($2e^2/h$) quantization of the 1D conductance. One requirement is that the electron mean free path ℓ_f be much greater than the length of the channel L. This limits the values of channel lengths to $L < 5{,}00\,\text{nm}$ even though the mean free path values in this work are much larger, $\ell_f = 8.5\,\mu\text{m}$. It is important to note, however, that $\ell_f = 8.5\,\mu\text{m}$ is the mean free path for the 2D electron gas. When the electron channel is formed, the screening effect of the 2D electron gas is no longer present and the effective mean free path becomes much shorter. A second condition is that there are adiabatic transitions at the inputs and outputs of the channel. These transitions minimize reflections at these two points, an important condition for the validity of the Landauer formula. A third condition requires the Fermi wavelength $\lambda_F = 2\pi/k_F$ (or $k_F L > 2\pi$) to satisfy the relation $\lambda_F < L$ by introducing a sufficient carrier density ($3.6 \times 10^{11}\,\text{cm}^{-2}$) into the channel. Finally, as discussed earlier, it is necessary that the thermal energy $k_B T \ll [E_j - E_{j-1}]$ where $[E_j - E_{j-1}]$ is the subband separation between the j and $[j-1]$ one dimensional energy levels. Therefore, these early measurements of quantum conductance were done at low temperatures ($T < 1$ K).

The point contacts in Fig. 11.6 were made on high-mobility molecular-beam-epitaxy-grown GaAs heterostructures using electron beam lithography. The electron density of the material was $3.6 \times 10^{11}/\text{cm}^2$ and the mobility was $8.5 \times 10^5\,\text{cm}^2/\text{V s}$ (at 0.6 K). These values were obtained directly from measurements of the devices themselves. For the transport measurements, a standard Hall bar geometry was defined by wet etching. At a gate voltage of $V_g = -0.6$ V the electron gas underneath the gate was depleted, so that conduction takes place through the point contact only. At this voltage, the point contacts have their maximum effective width W'_{max}, which is about equal to the opening W between the gates. By a further decrease (more negative) of the gate voltage, the width W of the point contacts in Fig. 11.6 can be reduced, until they are fully pinched off at $V_g = -2.2$ V.

Early measurements of QPC and SET behavior were performed at low temperatures (0.6 K) on relatively large structures ($W \sim 250$ nm). More recently, room temperature-operating SETs and single electron memories have been demonstrated in Si MOSFET structures fabricated using nanoimprint lithography (see Zhuang et al. 1998 and Wu et al. 2003). The single electron memory had a nanoimprint defined sub-10 nm quantum dot as the floating gate on top of a 20 nm wide channel. The information was then stored by charging/discharging each electron to/from the quantum dot. The very steep transconductance values in SET devices have attracted a lot of attention for logic and memory applications; however, their extreme sensitivity to any changes in the background charge has rendered large scale use of these new scientific advances unfeasible for the present. SETs, however, have found a niche for themselves in charge-sensing applications. For example, Chiu et al. have utilized this aspect of SETs to increase the sensitivity of micromechanical mass sensors to achieve a sensitivity of 0.066 zeptograms (i.e., 0.066×10^{-21} g)(see Chiu et al. 2008), and these SET devices have had a significant impact on nanotechnology.

Problems

11.1 Calculate the first five quantized energy levels and plot the density of states of an n-type silicon quantum well with the following parameters:

(a) (001) orientation, well width (d_w) = 10 Å
(b) (001) orientation, well width (d_w) = 50 Å
(c) (111) orientation, well width (d_w) = 10 Å

(d) (111) orientation, well width $(d_w) = 50\,\text{Å}$

Assume an infinite potential. How do these energies compare with the thermal energy at room temperature ($k_B T$=25 meV)? n-type silicon has $m_l = 0.98$ and $m_t = 0.19$. Hint: You first have to calculate the x, y, and z components of the effective mass for each carrier pocket. Because there are multiple carrier pockets, you will have to sort these energies in ascending order. You should give your answers in eV.

11.2 Calculate the first 10 quantized energy levels and plot the density of states of a p-type GaAs quantum wire with the following parameters:

(a) (001) orientation, wire diameter $(d_w) = 10\,\text{Å}$
(a) (001) orientation, wire diameter $(d_w) = 50\,\text{Å}$
(c) (111) orientation, wire diameter $(d_w) = 10\,\text{Å}$
(d) (111) orientation, wire diameter $(d_w) = 50\,\text{Å}$

Assume an infinite potential and a square wire cross-section. How do these energies compare with the thermal energy at room temperature ($k_B T$=25 meV)? Because there are two quantum numbers for each energy level, you will have to sort these in ascending order.

11.3 A metallic sphere with a radius of r is embedded in an infinitely large insulating medium with a relative permittivity of ϵ_r.

(a) What is the capacitance of the sphere?
(b) What is the difference of the energy needed to add the Nth electron and the $(N+1)^{th}$ electron into the sphere?
(c) If the insulating medium is SiO_2 and you want to observe the Coulomb blockade effect at room temperature with this system, which means the above energy difference needs to be larger than $3k_B T$, what then is the required size of the sphere?

11.4 Suppose you are building a single electron device based on the Coulomb blockade effect of a quantum dot. In order to make the device work practically, the dot cannot be an isolated dot, and it has to be located near some conducting electrodes. Will the capacitance of the dot be larger or smaller than the isolated dot mentioned in Problem 11.3? What then is the implication of the dot size needed to make the device work properly at room temperature?

11.5 The electrostatic capacitance of a carbon nanotube on a SiO_2/Si substrate is given by the relation:

$$C = \frac{2\pi\varepsilon L}{\ln(\frac{4h}{d})} \tag{11.34}$$

where d and L are the nanotube diameter and length, h is the oxide thickness respectively and the dielectric permittivity $\varepsilon = 3.9\varepsilon_0$ for SiO_2.

(a) Calculate the gate capacitance of a carbon nanotube 1 nm in diameter and 1 μm in length, and a SiO_2 thickness of 300 nm.

(b) Assuming a constant density of states of 0.15 states/graphene unit cell/eV, what gate voltage would be required to shift the Fermi energy by 10meV? Assume $T = 0K$ (i.e., $f(E) = 0$ or 1).
(c) Plot the relation between the Fermi energy and the gate voltage.
(d) Now assume that the nanotube has a small band gap of 50meV (quasi-metallic). Plot the relation between the Fermi energy and the gate voltage. How is this physically different from relation in (c)?
(e) Now plot the relation between the Fermi energy and the gate voltage for a semiconductor nanotube with a band gap of 1eV. In which range does the Fermi energy change most significantly with gate voltage?

11.6 (a) If the scattering length in a carbon nanotube is 1 μm, what is the effective mobility of a 100 μm long nanotube? (Assume a Fermi velocity of 10^8 cm/s.)
(b) What value of resistance does this nanotube have?
(c) As the length of the nanotube is made much shorter than 1 μm, what value of resistance do you expect the nanotube to have, assuming perfect contacts. Explain your answer.

11.7 For monolayer graphene on a SiO_2/Si substrate, the gate capacitance can be calculated using the parallel plate capacitor equation. Derive an equation relating the Fermi energy to the applied gate voltage for graphene on a SiO_2/Si substrate. The following relations for graphene many be helpful: DOS = $\frac{k^2}{2\pi}$ and $E = v_F \hbar k$.

References

Meirav et al., Phys. Rev. Lett. **65**, 771 (1990)
D.A. Wharam, T.J. Thornton, R. Newbury, M. Pepper, H. Ahmed, J.E.F. Frost, D.G. Hasko, D.C. Peacock, D.A. Ritchie, G.A.C. Jones, J. Phys. C: Solid State Phys. **21**, L209 (1988)
B.J. van Wees, H. van Houten, C.W.J. Beenakker, J.G. Williamson, L.P. Kouvenhoven, D. van der Marel, C.T. Foxon, Phys. Rev. Lett. **60**, 848 (1988)
Zhuang et al., Appl. Phys. Lett. **72**, 1205 (1998)
Wu et al., Appl. Phys. Lett. **83**, 2268 (2003)
Chiu et al., Nano Lett. **8**, 4342 (2008)

Chapter 12
Two Dimensional Electron Gas, Quantum Wells and Semiconductor Superlattices

12.1 Two-Dimensional Electronic Systems

One of the most important recent developments in semiconductors, both from the point of view of physics and for the purpose of device developments, has been the achievement of structures in which the electronic behavior is essentially two-dimensional (2D). This means that, at least for some phases of operation of the device, the carriers are confined in a potential well such that electron motion in one direction is restricted and thus is quantized, leaving plane wave motion for the electron with a two-dimensional wave vector k, which characterizes motion in a plane normal to the confining potential. The major systems where such 2D behavior has been studied are MOS (metal-oxide-semiconductor) structures, quantum wells and superlattices. More recently, quantization has been achieved in 2D layered materials, such as graphene and the transition metal dichalcogenide MoS_2.

12.2 MOSFETS

One of the most useful and versatile of these structures is the metal-insulator-semiconductor (MIS) layered structures, the most important of these being the metal-oxide-semiconductor (MOS) structures. As shown in Fig. 12.1, the MOS device is fabricated from a substrate of usually moderately-doped p-type or n-type silicon which together with its grounded electrode is called the base and labeled B in the figure. On the top of the base, an insulating layer of silicon dioxide is grown, followed by a metal layer; this structure is the gate (labeled G in the figure) and is used to apply an electric field through the oxide to the silicon. For the MOS device shown in the figure, the base region is p-type and the source (S) and drain (D) regions are n-type. Measurements of the changes in the properties of the carriers in the silicon layer immediately below the gate are made (by measuring the conductance in the source-drain channel), in response to changes in the applied electric field at the

M. Dresselhaus et al., *Solid State Properties*, Graduate Texts in Physics,
https://doi.org/10.1007/978-3-662-55922-2_12

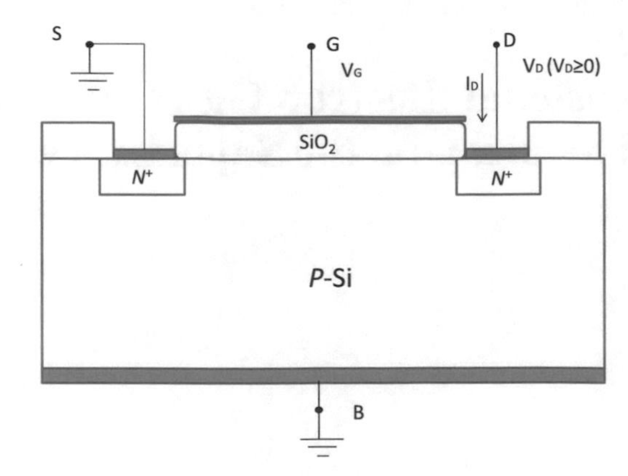

Fig. 12.1 Cross-sectional view of the basic MOSFET structure showing the terminal designations and standard biasing conditions

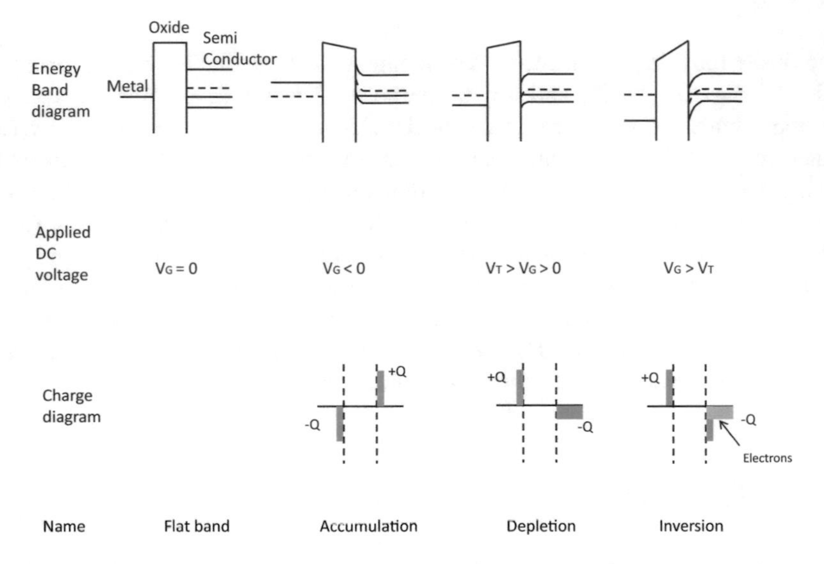

Fig. 12.2 Energy band and block charge diagrams for a *p*–type device operating under flat band, accumulation, depletion, and inversion conditions. V_G is the gate voltage, and V_T is the threshold voltage

gate electrode, and such measurements are called field-effect measurements. As we show below, the electric field dramatically changes the conducting properties of the carriers beneath the gate. Use is made of this effect in the so-called metal-oxide-semiconductor field-effect transistor (MOSFET).

To understand the operation of this device, we first consider the schematic energy band diagram of the MOS structure as shown in Fig. 12.2, for four different values of V_G, the gate potential relative to that of the substrate. For each V_G value, the diagram shows from left to right the metal (M) - oxide (O) - semiconductor (S) regions. In the semiconductor regions each of the diagrams show from top to bottom: the Si conduction band edge E_c, the "intrinsic" Fermi level for undoped Si as the dashed line, the Fermi level E_F in the p-type Si, and the valence band edge E_v. In each of the four diagrams, the central oxide region shows the valence band edge for the oxide. On the left hand side of each diagram, the Fermi level for the metal is shown as a solid horizontal line, and the dashed line gives the extension of the Si Fermi level. In the lower part of the figure, the applied DC voltage conditions are indicated, and the corresponding diagrams for the charge layers of the interfaces for each case are illustrated.

We now explain the diagrams in Fig. 12.2 as a function of the gate voltage V_G. For $V_G = 0$ (the flat-band case), there are (ideally) no charge layers, and the energy levels of the metal (M) and semiconducting (S) regions line up to yield the same Fermi level (chemical potential). The base region is doped p-type (p-Si). For a negative gate voltage ($V_G < 0$, the accumulation case), an electric field is set up in the oxide. The negative gate voltage causes the Si bands to bend up at the oxide interface (see Fig. 12.2) so that the Fermi level is closer to the valence-band edge. Thus extra holes accumulate at the semiconductor-oxide interface and electrons accumulate at the metal-oxide interface (see lower part of Fig. 12.2). In the third (depletion) case, the gate voltage is positive but less than some threshold value V_T. The voltage V_T is defined as the gate voltage where the intrinsic Fermi level and the actual Fermi level are coincident at the interface (see lower part of Fig. 12.2). For the "depletion" regime, the Si bands bend down at the interface resulting in a depletion of holes, and a negatively charged layer of localized states is formed at the semiconductor-oxide interface. The size of this "depletion region" (area of the rectangle labeled $-Q$) increases as V_G increases. The corresponding positively charged region at the metal-oxide interface is also shown. Finally, for $V_G > V_T$, the intrinsic Fermi level at the interface drops below the actual Fermi level, forming the "inversion layer", where mobile electrons (shown in orange color) reside. It is the electrons in this inversion layer which are of interest, both because they can be confined so as to exhibit two-dimensional transport behavior, and because they can be controlled by the gate voltage in the MOSFET (see Fig. 12.3)

The operation of a metal-oxide semiconductor field-effect transistor (MOSFET) is illustrated in Fig. 12.3, which shows the electron inversion layer under the gate for $V_G > V_T$ (for a p-type substrate), with the source region grounded, for various values of the drain voltage V_D. The inversion layer forms a conducting "channel" between the source and the drain (as long as the gate voltage is above threshold $V_G > V_T$). The dashed line in Fig. 12.3 shows the boundaries of the depletion region which forms in the p-type substrate adjoining the n^+ and p regions.

For $V_D = 0$ there is obviously no current between the source and the drain since both are at the same potential (Fig. 12.3a). For $V_D > 0$, the inversion layer or channel acts like a resistor, inducing the flow of electric current I_D between the source and drain. As shown in Fig. 12.3b, increasing V_D imposes a reverse bias on the n^+-p drain-substrate junction, thereby increasing the width of the depletion region and both decreasing the number of carriers and narrowing the channel in the inversion layer, as shown in Fig. 12.3. Finally as V_D increases further, the channel reaches

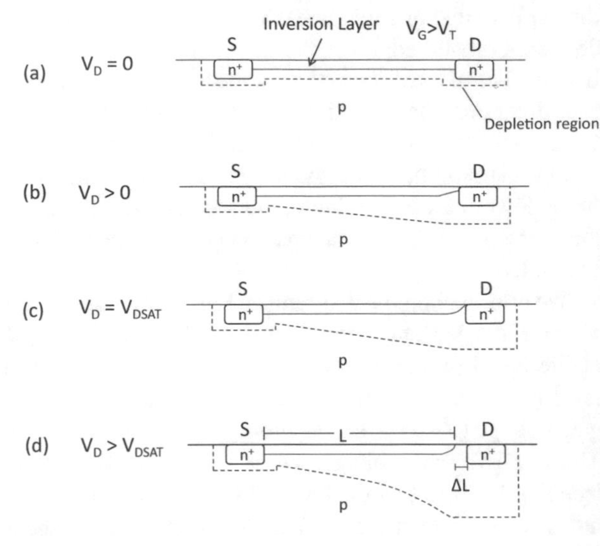

Fig. 12.3 Visualization of various phases of $V_G > V_T$ MOSFET operation. **a** Drain voltage $V_D = 0$, **b** channel (inversion layer) narrowing under moderate V_D positive voltage biasing, **c** pinch–off denoted by V_{Dsat}, and **d** post-pinch-off ($V_D > V_{Dsat}$) operation. (Note that the inversion layer widths, depletion widths, etc. are not drawn to scale.)

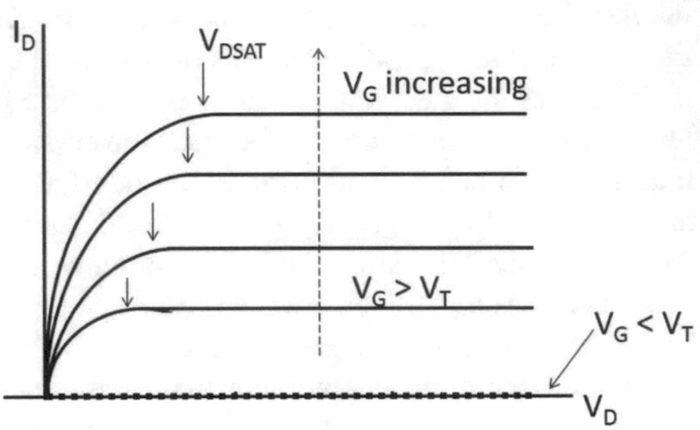

Fig. 12.4 General form of the $I_D - V_D$ characteristics expected from a long channel ($\Delta L \ll L$) MOSFET, where V_{Dsat} is the saturation voltage, L is the channel length, and ΔL is the change in the channel length due to the pinch off phenomenon

the "pinched-off" condition V_{Dsat} shown in Fig. 12.3c. Further increase in V_D does not increase I_D but rather causes "saturation" at V_{Dsat}. We note that at saturation, $V_{Dsat} = V_G - V_T$. Saturation is caused by a decrease in the carrier density in the channel due to the pinch-off phenomena. When the gate voltage V_G exceeds the threshold voltage V_T, a new regime is entered as is discussed next.

In Fig. 12.4, I_D versus V_D curves are plotted for fixed values of $V_G > V_T$. We note that V_{Dsat} increases with increasing V_G. These characteristic curves are qualitatively similar to the curves for the bipolar junction transistor. The advantage of MOSFET devices lie in the speed of their operation and in the ease with which they can be fabricated into ultra-small devices.

The MOSFET device, or an array of a large number of MOSFET devices, is fabricated starting with a large Si substrate or "wafer". At each stage of fabrication, areas of the wafer which are to be protected are masked off using a light-sensitive substance called photoresist, which is applied as a thin film, exposed to light (or an electron or x-ray beam) through a mask of the desired pattern, then chemically developed to remove the photoresist from only the exposed (or, sometimes only the un–exposed) area. On the formerly protected areas, first the source and drain regions are formed by either diffusing or implanting (bombarding) donor ions into the p-type substrate. Then a layer of SiO_2 (which is an excellent and stable insulator) is grown by exposing the desired areas to an atmosphere containing oxygen; usually only a thin oxide layer is grown over the gate regions and, in a separate step, thicker oxide layers are grown between neighboring devices to provide electrical isolation. Finally, the metal gate electrode, the source and drain contacts are formed by sputtering or evaporating a metal, such as aluminum, onto the desired regions of the substrate, where metallic contacts are made from the MOSFET device input and output connections to appropriate connections of the external system.

12.3 Two-Dimensional Behavior

Other systems where two-dimensional behavior has been observed include heterojunctions of III-V compounds such as GaAs/$Ga_{1-x}Al_xAs$, layer compounds such as GaSe, $GaSe_2$ and related III-VI compounds, graphite and intercalated graphite, and electrons on the surface of liquid helium. The GaAs/$Ga_{1-x}Al_xAs$ heterojunctions are important for device applications because the lattice constants and the coefficient of expansion of GaAs and $Ga_{1-x}Al_xAs$ are very similar. This lattice matching permits the growth of high mobility thin films of $Ga_{1-x}Al_xAs$ on a GaAs substrate (For further reading, see Weisbuch and Vinter 2014).

The interesting physical properties of the MOSFET device are connected to the two-dimensional behavior of the electrons in the channel inversion layer at low temperatures. Studies of these electrons have provided important tests of modern theories of localization, electron-electron interactions and many-body effects. In addition, the MOSFETs have exhibited a highly unexpected property that, in the presence of a magnetic field normal to the inversion layer, the transverse or Hall resistance ρ_{xy} is quantized in integer values of e^2/h. This quantization is accurate to parts in 10^7 or 10^8 and provides the best measure to date of the fine structure constant $\alpha = e^2/hc$, when combined with the precisely-known velocity of light c. We will further discuss the quantized Hall effect in Chap. 14.

We now discuss the two-dimensional behavior of MOSFET devices in the absence of a magnetic field. The two-dimensional behavior is associated with the nearly plane wave electron states in the inversion layer. The potential $V(z)$ is associated with the electric field $V(z) = eEz$ (where z is the direction perpendicular to the plane) and because of the negative charge on the electron, a potential well is formed containing bound states described by quantized levels. A similar situation occurs in the two–dimensional behavior for the case of electrons in quantum wells produced by molecular beam epitaxy. Explicit solutions for the bound states in quantum wells are given in Sect. 12.4. In the present section, we discuss the form of the differential equation and of the resulting eigenvalues and eigenfunctions associated with such quantum states.

A single electron in a one-dimensional potential well $V(z)$ will, from elementary quantum mechanics, have discrete allowed energy levels E_n corresponding to bound states and usually a continuum of levels at higher energies corresponding to states which are not bound. An electron in a bulk semiconductor is in a three-dimensional periodic potential. In addition, the potential causing the inversion layer of a MOSFET or a quantum well in GaAs/$Ga_{1-x}Al_xAs$ can both be described by a one-dimensional confining potential $V(z)$ and can be written as a first approximation using the effective-mass theorem

$$[E(-i\nabla) + \mathscr{H}']\Psi = i\hbar\left(\frac{\partial \Psi}{\partial t}\right) \tag{12.1}$$

where $\mathscr{H}' = V(z)$. The energy eigenvalues $E(\mathbf{k})$ near the band edge, denoted by the wavevector k_0, can be most simply approximated by

$$E(\mathbf{k}) = E(\mathbf{k}_0) + \frac{1}{2}\sum_{i,j}\left(\frac{\partial^2 E}{\partial k_i \partial k_j}\right)k_i k_j \tag{12.2}$$

so that the operator $E(-i\nabla)$ in (12.1) can be written as

$$E(-i\nabla) = \sum_{i,j}\frac{p_i p_j}{2m_{i,j}} \tag{12.3}$$

where the p_i and p_j are the operators

$$p_j = \frac{\hbar}{i}\frac{\partial}{\partial x_j} \tag{12.4}$$

which are substituted into Schrödinger's equation. The effect of the periodic potential is contained in the reciprocal of the effective mass tensor

$$\frac{1}{m_{ij}} = \frac{1}{\hbar^2}\frac{\partial^2 E(\mathbf{k})}{\partial k_i \partial k_j}\bigg|_{\mathbf{k}=\mathbf{k_0}} \tag{12.5}$$

where the components of $1/m_{ij}$ are evaluated at the band edge denoted by $\mathbf{k}_0$.

If $1/m_{ij}$ is a diagonal matrix, the effective-mass equation $\mathscr{H}\Psi = E\Psi$ is solved by a function of the form

$$\Psi_{n,k_x,k_y} = e^{ik_x x}e^{ik_y y}f_n(z) \tag{12.6}$$

where $f_n(z)$ is a solution of the equation

$$-\frac{\hbar^2}{2m_{zz}}\frac{d^2 f_n}{dz^2} + V(z)f_n = E_{n,z}f_n \tag{12.7}$$

and the total energy is

$$E_n(k_x, k_y) = E_{n,z} + \frac{\hbar^2}{2m_{xx}}k_x^2 + \frac{\hbar^2}{2m_{yy}}k_y^2. \tag{12.8}$$

Since the $E_{n,z}$ energies ($n = 0, 1, 2, \ldots$) are discrete, the energies states $E_n(k_x, k_y)$ for each n value form a "sub-band". We give below (in Sect. 12.3.1) a simple derivation for the discrete energy levels by considering a particle in various potential wells (i.e., quantum wells). The electrons in these "sub–bands" form a 2D electron gas.

12.3.1 *Quantum Wells and Superlattices*

Many of the quantum wells and superlattices that are commonly studied at present do not occur in nature, but rather are deliberately structured materials (see Fig. 12.5), and are helpful for fundamental understanding. In the case of superlattices formed by molecular beam epitaxy, the quantum wells result from the different bandgaps of the two constituent materials. The additional periodicity is in one–dimension (1–D) which we take along the z–direction, and the electronic behavior is usually localized on the basal planes (x–y planes) normal to the z–direction, giving rise to two–dimensional behavior.

A schematic representation of a semiconductor heterostructure superlattice is shown in Fig. 12.5a, where d is the superlattice periodicity composed of a distance d_1, of semiconductor S_1, and d_2 of semiconductor S_2, as shown in Fig. 12.5b. Because of the different band gaps in the two semiconductors, potential wells and barriers are formed. For example in Fig. 12.5b, the barrier heights in the conduction and valence bands are ΔE_c and ΔE_v, respectively. In Fig. 12.5 we see that the difference in bandgaps between the two semiconductors gives rise to band offsets ΔE_c and ΔE_v for the conduction and valence bands, as shown in Fig. 12.5b. In principle, these band

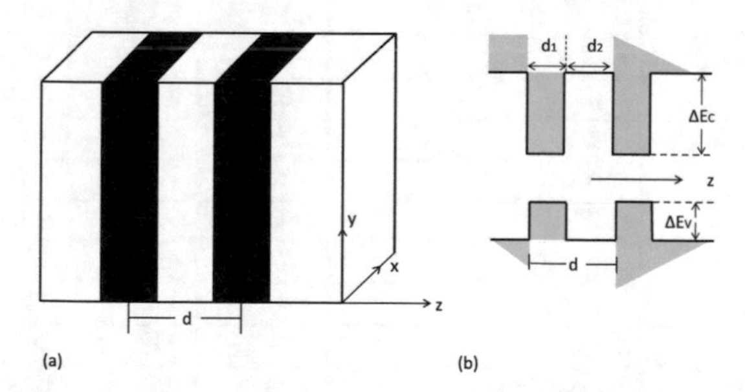

Fig. 12.5 **a** A heterojunction superlattice of periodicity d. **b** Each superlattice unit cell consists of a thickness d_1 of material #1 and d_2 of material #2. Because of the different band gaps, a periodic array of potential wells and potential barriers is formed. When the band offsets are both positive as shown in this figure, the structure is called a type I superlattice

Table 12.1 Material parameters of GaAs, GaP, InAs, and InP[a]

Property	Parameter (units)	GaAs	GaP	InAs	InP
Lattice constant	a(Å)	5.6533	5.4512	6.0584	5.8688
Density	g(g/cm^3)	5.307	4.130	5.667	4.787
Thermal expansion	$\alpha_{th}(\times 10^{-6}/°C)$	6.63	5.91	5.16	4.56
Γ point band gap plus spin orbit	E_0(eV)	1.42	2.74	0.36	1.35
	$E_0 + \Delta_0$(eV)	1.76	2.84	0.79	1.45
L point band gap plus spin orbit	E_1(eV)	2.925	3.75	2.50	3.155
	$E_1 + \Delta_1$(eV)	3.155	...	2.78	3.305
Γ point band gap	E_0'(eV)	4.44	4.78	4.44	4.72
Δ axis band gap plus spin orbit	E_2(eV)	4.99	5.27	4.70	5.04
	$E_2 + \delta$(eV)	5.33	5.74	5.18	5.60
Gap pressure coefficient	$\partial E_0/\partial P(\times 10^{-6}$eV/bar)	11.5	11.0	10.0	8.5
Gap temperature coefficient	$\partial E_0/\partial T(\times 10^{-4}$eV/°C)	−3.95	−4.6	−3.5	−2.9
Electron mass light hole heavy hole spin orbit hole	m^*/m_0	0.067	0.17	0.023	0.08
	$m_{\ell h}{}^*/m_0$	0.074	0.14	0.027	0.089
	$m_{hh}{}^*/m_0$	0.62	0.79	0.60	0.85
	$m_{so}{}^*/m_0$	0.15	0.24	0.089	0.17
Dielectric constant: static	ε_s	13.1	11.1	14.6	12.4
Dielectric constant: optic	ε_∞	11.1	8.46	12.25	9.55
Ionicity	f_1	0.310	0.327	0.357	0.421
Polaron coupling	α_F	0.07	0.20	0.05	0.08
Elastic constants	$c_{11}(\times 10^{11}$dyn/cm^2)	11.88	14.120	8.329	10.22
	$c_{12}(\times 10^{11}$dyn/cm^2)	5.38	6.253	4.526	5.76
	$c_{44}(\times 10^{11}$dyn/cm^2)	5.94	7.047	3.959	4.60
Young's modulus	$Y(\times 10^{11}$dyn/cm^2)	8.53	10.28	5.14	6.07
	P	0.312	0.307	0.352	0.360
Bulk modulus	$B(\times 10^{11}$ dyn/cm^2)	7.55	8.88	5.79	7.25
	A	0.547	0.558	0.480	0.485
Piezo–electric coupling	e_{14}(C/m^2)	−0.16	−0.10	−0.045	−0.035
	$K_{[110]}$	0.0617	0.0384	0.0201	0.0158
Deformation potential	a(eV)	2.7	3.0	2.5	2.9
	b(eV)	−1.7	−1.5	−1.8	−2.0
	d(eV)	−4.55	−4.6	−3.6	−5.0
Deformation potential	Ξ_{eff}(eV)	6.74	6.10	6.76	7.95
Donor binding	G(meV)	4.4	10.0	1.2	5.5
Donor radius	a_B(Å)	136	48	406	106
Thermal conductivity	κ(watt/deg-cm)	0.46	0.77	0.273	0.68
Electron mobility	μ_n(cm^2/V-sec)	8000	120	30000	4500
Hole mobility	μ_p(cm^2/V-sec)	300	–	450	100

[a] Table from Blakemore (1982) and Strauch et al. (2001)

offsets are determined by matching the Fermi levels for the two semiconductors. In actual materials, the Fermi levels are highly sensitive to impurities, defects and charge transfer at the heterojunction interface.

The two semiconductors of a heterojunction superlattice could be different semiconductors such as InAs with GaP (see Table 12.1 for parameters related to these compounds) or a binary semiconductor with a ternary alloy semiconductor, such as GaAs with $Al_xGa_{1-x}As$ (sometimes referred to by their slang names "Gaas" and "Algaas"). In the typical semiconductor superlattices, the periodicity $d = d_1 + d_2$ is repeated many times (e.g., 100 times). The period thicknesses typically vary between a few layers and many layers (10–500 Å).

The electronic states corresponding to the heterojunction superlattices are of two fundamental types–bound states in quantum wells and nearly free electron states in zone–folded energy bands. We here limit our discussion to the bound states in a single infinite quantum well. For generalizations to multiple quantum wells, see Davies (1997).

One important issue that comes up when two semiconductors are brought together to form an abrupt interface is how the conduction (and valence) band edges line up in the two materials. Several possible scenarios are illustrated in Fig. 12.7, where semi-

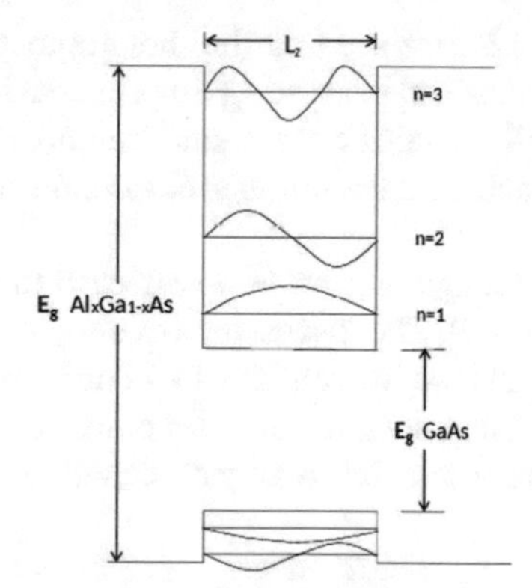

Fig. 12.6 The eigenfunctions and bound state energies of an infinitely deep potential well used here as an approximation to the states in two finite wells. The upper quantum well applies to electrons and the lower one to holes. This diagram is a schematic representation of a quantum well in the GaAs region formed by the adjacent wider gap semiconductor $Al_xGa_{1-x}As$

Fig. 12.7 Various band alignments of semiconductor heterostructures

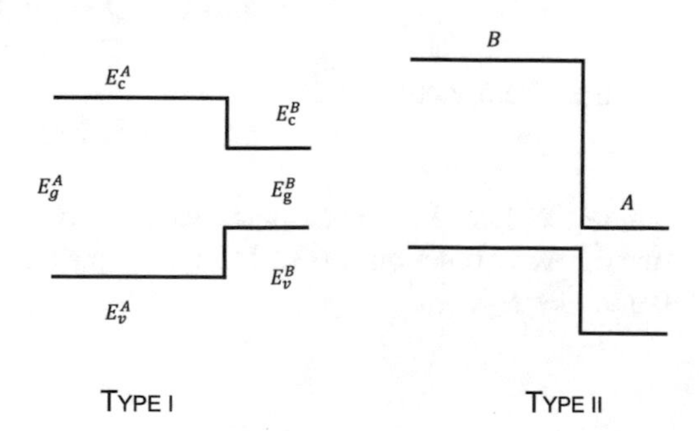

conductor A with band gap E_g^A and edges E_v^A and E_c^A is grown on semiconductor B with band gap E_g^B and edges E_v^B and E_c^B. In principle, it is possible to calculate these offsets from the electron affinities (or work functions) of the two materials. However, in practice, these band offsets must be determined empirically. In type I heterostructures, the conduction and valence bands of the smaller band gap material lie withing the band gap of the larger band gap material. In this case, the lowest conduction band and highest valence band exist in the same physical location (i.e., in the narrow gap material). These are the most widely used type of heterostructures, which include GaAs/AlGaAs, InGaAs/InP, and GaN/AlGaN. In type II heterostructures, the lowest conduction band and the highest valence band are in two different materials. In this case, the effective band gap of the heterostructure, that is the energy difference between the lowest conduction band and the highest valence band, can be rather small. Thus, type II heterostructures are useful for long wavelength optoelectronics. InAs/GaSb heterostructures show type II behavior.

12.4 Bound Electronic States

From the diagram in Fig. 12.5 we see that the heterojunction superlattice consists of an array of potential wells. The interesting limit to consider is the case where the width of the potential well contains only a small number of crystallographic unit cells ($L_z < 10\,\text{nm}$), in which case the number of bound states in the well is a small number.

From a mathematical standpoint, the simplest case to consider is an infinitely deep rectangular potential well. The discussions of simple cases and simple approximations are useful to understand the fundamental concepts involved. In the present case, a particle of mass m^* in a deep rectangular potential well of width L_z in the z direction satisfies the free particle Schrödinger's equation

$$-\frac{\hbar^2}{2m^*}\frac{d^2\psi}{dz^2} = E\psi \tag{12.9}$$

with eigenvalues

$$E_n = \frac{\hbar^2}{2m^*}\left(\frac{n\pi}{L_z}\right)^2 = \left(\frac{\hbar^2\pi^2}{2m^* L_z^2}\right)n^2 \tag{12.10}$$

and the eigenfunctions

$$\psi_n = A\sin(n\pi z/L_z) \tag{12.11}$$

where $n = 1, 2, 3 \ldots$ are the plane wave solutions that satisfy the boundary conditions that the wave functions in (12.11) must vanish at the walls of the quantum wells ($z = 0$ and $z = L_z$).

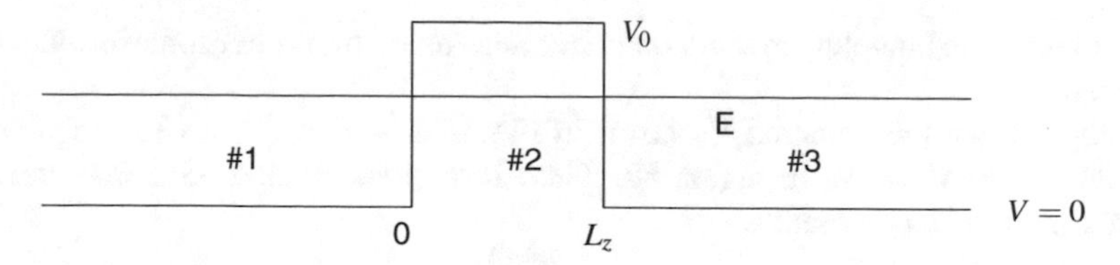

Fig. 12.8 Schematic of a potential barrier in region #2

We note that the energy levels are not equally spaced, but have energies $E_n \sim n^2$, though the spacings $E_{n+1} - E_n$ are proportional to n. We also note that $E_n \sim L_z{}^{-2}$, so that as L_z becomes large, the levels become very closely spaced as expected for a 3D semiconductor. However when L_z decreases, the number of states in the quantum well decreases, so that for a well depth E_d it would seem that there is a critical width $L_z{}^c$ below which there would be no bound states

$$L_z{}^c = \frac{\hbar\pi}{(2m^* E_d)^{\frac{1}{2}}}. \tag{12.12}$$

An estimate for $L_z{}^c$ is obtained by taking $m^* = 0.1m_0$ and $E_d = 0.1\,\text{eV}$ to yield $L_z{}^c = 61\,\text{Å}$. There is actually a theorem in quantum mechanics that says that there will be at least one bound state for an arbitrarily small potential well (Shankar 2011). More exact calculations considering quantum wells of finite thickness have been carried out, and show that the infinite well approximation gives qualitatively correct results for many cases of practical interest.

The closer level spacing of the valence band bound states in Fig. 12.6 reflects the heavier masses in the valence band of the GaAs system. Since the states in the potential well are quantized, the structures in Figs. 12.5 and 12.6 are called quantum well structures.

If the potential energy of the well V_0 is not infinite but finite, the wave functions are similar to those given in (12.11), but will have decaying exponentials on either side of the potential well walls. The effect of the finite size of the well on the energy levels and wave functions is most pronounced near the top of the well. When the particle has an energy greater than V_0, its eigenfunction corresponds to a continuum state approximated by $\exp(ik_z z)$.

In the case of MOSFETs, the quantum well is not of rectangular shape, as shown in Fig. 12.8, but rather is approximated as a triangular well. The solution for the bound states in a triangular well cannot be solved exactly, but can only be calculated approximately, as for example using the WKB approximation described in Sect. 12.6.

12.5 Review of Tunneling Through a Potential Barrier

When the potential well is finite, the wave functions do not completely vanish at the walls of the well, so that tunneling through the potential well becomes possible. We now briefly review the quantum mechanics of tunneling through a potential barrier.

We will return to tunneling in semiconductor heterostructures after some introductory material.

Suppose that the potential V shown in Fig. 12.8 is zero ($V = 0$) in regions #1 and #3, while $V = V_0$ in region #2. Then in regions #1 and #3 a free electron approximation can be used:

$$E = \frac{\hbar^2 k^2}{2m^*} \tag{12.13}$$

$$\psi = e^{ikz} \tag{12.14}$$

while in region #2 the wave function is exponentially decaying

$$\psi = \psi_0 e^{-\beta z} \tag{12.15}$$

so that substitution into Schrödinger's equation gives

$$\frac{-\hbar^2}{2m^*}\beta^2\psi + (V_0 - E)\psi = 0 \tag{12.16}$$

where

$$\beta^2 = \frac{2m^*}{\hbar^2}(V_0 - E). \tag{12.17}$$

The probability $\mathscr{P}$ that the electron tunnels through the rectangular potential barrier is then given by

$$\mathscr{P} = \exp\left\{-2\int_0^{L_z}\beta(z)dz\right\} = \exp\left\{-2\left(\frac{2m^*}{\hbar^2}\right)^{\frac{1}{2}}(V_0 - E)^{\frac{1}{2}}L_z\right\}. \tag{12.18}$$

As the width of the potential well L_z increases (see Fig. 12.8), the probability of tunneling decreases exponentially. Electron tunneling phenomena frequently occur in solid state physics (see Brown 1991; Lake 1997), and there is presently a large body of literature on this topic.

12.6 Quantum Wells of Different Shape and the WKB Approximation

With the sophisticated computer control available with state of the art molecular beam epitaxy and other computer-controlled deposition systems, it is now also possible to produce quantum wells with specified potential profiles $V(z)$ for semiconductor heterojunction superlattices. Potential wells with non–rectangular profiles also occur in the fabrication of other types of superlattices (e.g., by modulation doping). We therefore briefly discuss bound states in general potential wells, and then consider some examples.

In the general case where the potential well has an arbitrary shape, solution by the WKB (Wentzel–Kramers–Brillouin) approximation is very useful (see for example, Shankar 2011. According to this approximation, the energy levels satisfy the Bohr–Sommerfeld quantization condition

$$\int_{z_1}^{z_2} p_z dz = \hbar\pi(r + c_1 + c_2) \tag{12.19}$$

where $p_z = (2m^*[E - V])^{\frac{1}{2}}$ and the quantum number r is an integer $r = 0, 1, 2, \ldots$ while c_1 and c_2 are the phases which depend on the form of $V(z)$ at the turning points z_1 and z_2 where $V(z_i) = E$. If the potential has a sharp discontinuity at a turning point, then $c = 1/2$, but if V depends linearly on z at the turning point then $c = 1/4$.

For example, we consider the infinite rectangular well (see Fig. 12.9 for the finite potential well)

$$V(z) = 0 \quad \text{for } |z| < a \quad \text{(inside the well)} \tag{12.20}$$

$$V(z) = \infty \quad \text{for } |z| > a \quad \text{(outside the well)} \tag{12.21}$$

By the WKB rules, the turning points occur at the edges of the rectangular well and therefore $c_1 = c_2 = 1/2$. In this case p_z is a constant, independent of z so that $p_z = (2m^*E)^{\frac{1}{2}}$ and (12.19) yields

$$(2m^*E)^{\frac{1}{2}} L_z = \hbar\pi(r + 1) = \hbar\pi n \tag{12.22}$$

where $n = r + 1$ and

$$E_n = \frac{\hbar^2\pi^2}{2m^* {L_z}^2} n^2 \tag{12.23}$$

in agreement with the exact solution given by (12.10). The finite rectangular well shown in Fig. 12.9, when approximated as an infinite well, yields results consistent with solutions given by (12.10).

As a second example consider a harmonic oscillator potential well shown in Fig. 12.10, where $V(z) = m^*\omega^2 z^2/2$. The harmonic oscillator potential well is typical of quantum wells in periodically doped (i.e., nipi, which is short for n-type semiconductor; insulator; p-type semiconductor; insulator) superlattices. In this case

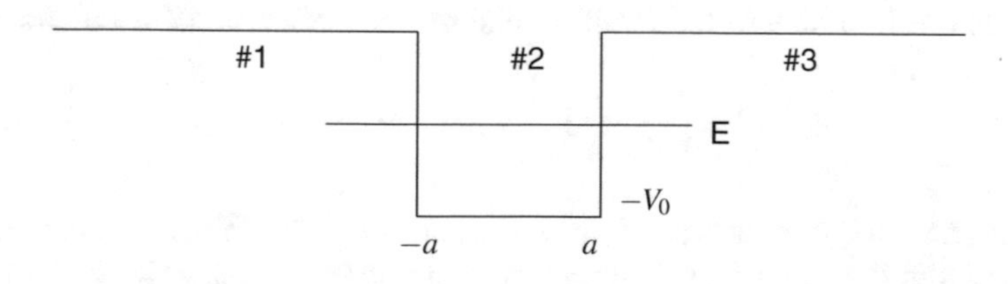

Fig. 12.9 Schematic of a rectangular quantum well plotted as $V(z)$

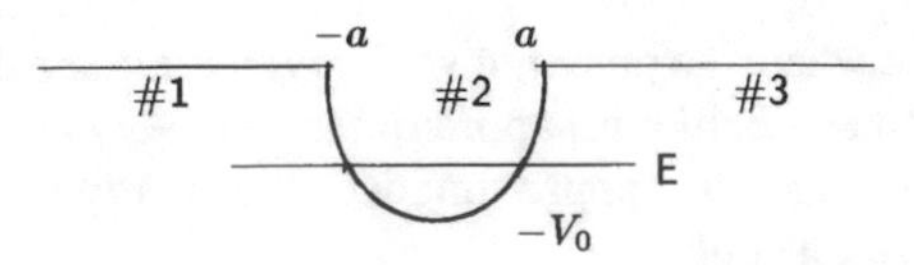

Fig. 12.10 Schematic plot of $V(z)$ of a harmonic oscillator potential well $V(z) = m^*\omega^2 z^2/2$

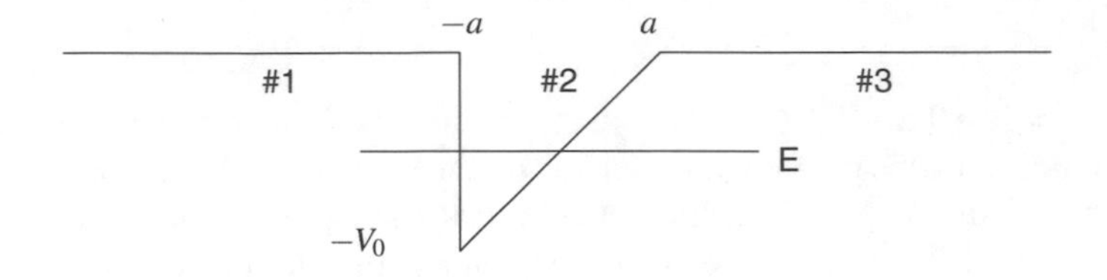

Fig. 12.11 Schematic plot of $V(z)$ of a triangular well

$$p_z = (2m^*)^{1/2}\left(E - \frac{m^*\omega^2}{2}z^2\right)^{\frac{1}{2}}. \tag{12.24}$$

The turning points for the harmonic oscillator quantum well occur when $V(z) = E$ so that the turning points are given by $z = \pm(2E/m^*\omega^2)^{\frac{1}{2}}$. Near the turning points $V(z)$ is approximately linear in z, so the phase factors become $c_1 = c_2 = \frac{1}{4}$. The Bohr–Sommerfeld quantization for this case thus yields

$$\int_{z_1}^{z_2} p_z dz = \int_{z_1}^{z_2} (2m^*)^{\frac{1}{2}}\left(E - \frac{m^*\omega^2}{2}z^2\right)^{\frac{1}{2}} dz = \hbar\pi\left(r + \frac{1}{2}\right). \tag{12.25}$$

Making use of the integral relation

$$\int \sqrt{a^2 - u^2}\, du = \frac{u}{2}\sqrt{a^2 - u^2} + \frac{a^2}{2}\sin^{-1}\frac{u}{a} \tag{12.26}$$

we obtain upon substitution of (12.26) into (12.25):

$$(2m^*)^{\frac{1}{2}}\left(\frac{m^*\omega^2}{2}\right)^{\frac{1}{2}}\left(\frac{E_r}{m^*\omega^2}\right)\pi = \frac{E_r\pi}{\omega} = \hbar\pi\left(r + \frac{1}{2}\right) \tag{12.27}$$

which can be simplified to the familiar relation for the harmonic oscillator energy levels:

$$E_r = \hbar\omega\left(r + \frac{1}{2}\right) \quad \text{where} \quad r = 0, 1, 2\ldots \tag{12.28}$$

thus yielding another example of an exact solution. The WKB method can also be used to find the energy levels for an asymmetric triangular well. Such quantum wells are typically used to model the semiconductor interface in metal–insulator–semiconductor (MOSFET) device structures (see Fig. 12.11).

12.7 The Kronig–Penney Model

We next review the Kronig–Penney model (see Kittel 1996), which gives an explicit solution for a one–dimensional array of finite potential wells as shown in Fig. 12.12. Starting with the one dimensional Hamiltonian with a periodic potential (see (3.40))

$$-\frac{\hbar^2}{2m^*}\frac{d^2\psi}{dz^2} + V(z)\psi = E\psi \tag{12.29}$$

we obtain solutions in the region $0 < z < a$ where $V(z) = 0$

$$\psi(z) = Ae^{iKz} + Be^{-iKz} \tag{12.30}$$

$$E = \frac{\hbar^2 K^2}{2m^*} \tag{12.31}$$

and in the region $-b < z < 0$ where $V(z) = V_0$ (the barrier region)

$$\psi(z) = Ce^{\beta z} + De^{-\beta z} \tag{12.32}$$

where

$$\beta^2 = \frac{2m^*}{\hbar^2}[V_0 - E]. \tag{12.33}$$

Continuity of $\psi(z)$ and $d\psi(z)/dz$ at $z = 0$ and $z = a$ determines the four coefficients A, B, C, D. At $z = 0$ we have:

$$\begin{aligned} A + B &= C + D \\ iK(A - B) &= \beta(C - D) \end{aligned} \tag{12.34}$$

At $z = a$ in Fig. 12.12, we apply Bloch's theorem by introducing a phase factor $\exp[ik(a + b)]$ to obtain $\psi(a) = \psi(-b)\exp[ik(a + b)]$

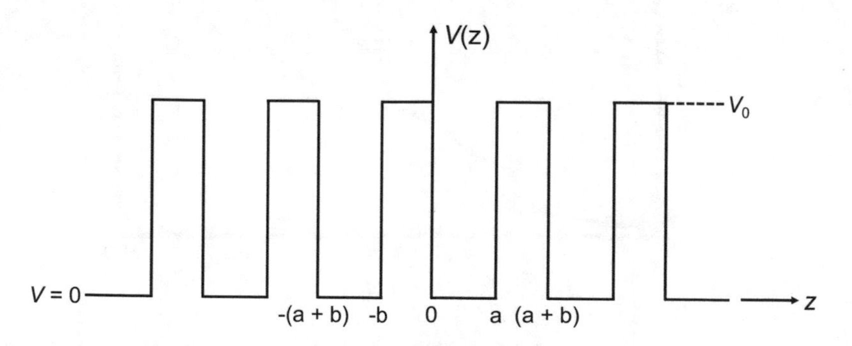

Fig. 12.12 Kronig–Penney square well periodic potential ($V(z)$)

$$\begin{aligned} Ae^{iKa} + Be^{-iKa} &= (Ce^{-\beta b} + De^{\beta b})e^{ik(a+b)} \\ iK(Ae^{iKa} - Be^{-iKa}) &= \beta(Ce^{-\beta b} - De^{\beta b})e^{ik(a+b)}. \end{aligned} \tag{12.35}$$

These 4 equations (12.34) and (12.35) all involve 4 unknowns, and determine the parameters A, B, C, D when solved simultaneously. The vanishing of the coefficient determinant restricts the conditions under which solutions to the Kronig–Penney model are possible, leading to the algebraic equation

$$\frac{\beta^2 - K^2}{2\beta K} \sinh \beta b \sin Ka + \cosh \beta b \cos Ka = \cos k(a + b) \tag{12.36}$$

which has solutions as given above for a limited range of β values, $E < V_0$. For the case $E > V_0$, continuous solutions exist.

Normally the Kronig–Penney model in textbooks (see Kittel 1996) is solved in the limit $b \to 0$ and $V_0 \to \infty$ in such a way that $[\beta^2 ba/2] = P$ remains finite. The restricted solutions in this limit lead to the energy band structure shown in Fig. 12.13.

For the superlattice problem, we are interested in solutions both within the quantum wells and in the continuum. This is one reason for discussing the Kronig–Penney model. Another reason for discussing this model is because it provides a review of

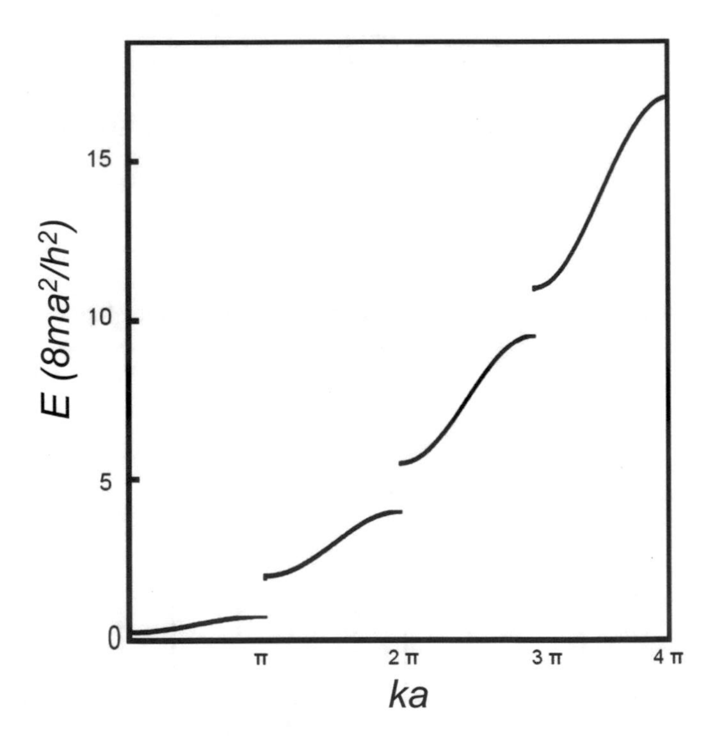

Fig. 12.13 Plot of energy E (in units of $8ma^2/h^2$) versus k for the Kronig–Penney model with $P = 3\pi/2$. (After Sommerfeld and Bethe 1933)

boundary conditions and the application of Bloch's theorem. In the quantum wells, the permitted solutions can give rise to narrow bands with large band gaps (corresponding to the wide gap semiconductors), while in the continuum regions, the solutions to the four linear equations in A, B, C, D correspond to wide bands and small band gaps.

12.8 3D Motion Within a 1-D Rectangular Well

The thin films used for the fabrication of quantum well structures (see Sect. 12.4) are very thin in the z–direction but have macroscopic size in the perpendicular x–y plane. An example of a quantum well structure would be a thin layer of GaAs sandwiched between two thicker $Al_xGa_{1-x}As$ layers, as shown in the Fig. 12.5. For the thin film, the motion in the x and y directions is similar to that of the corresponding bulk solid which can be treated qualitatively by the conventional 1–electron approximation and the Effective Mass Theorem. Thus the potential can be written as a sum of a periodic term $V(x, y)$ and the quantum well term $V(z)$. The electron energies thus are superimposed on the quantum well energies, and the periodic solutions obtained from solution of the 2–D periodic potential $V(x, y)$ are given by

$$E_n(k_x, k_y) = E_{n,z} + \frac{\hbar^2(k_x^2 + k_y^2)}{2m^*} = E_{n,z} + E_\perp \tag{12.37}$$

in which the quantized bound state energies $E_{n,z}$ are given by (12.10). A plot of the energy levels of the bound state subbands for a 2D electron gas is given in Fig. 12.14. The band of energies associated with each quantum state n is called a subband. At $(k_x, k_y) = (0,0)$ the energy is precisely the quantum well energy E_n for all n.

Of particular interest is the density of states for the quantum well structures. Associated with each two–dimensional subband is a constant density of states, as derived below. From elementary considerations, the number of electrons per unit area in a 2–dimensional circle is given by

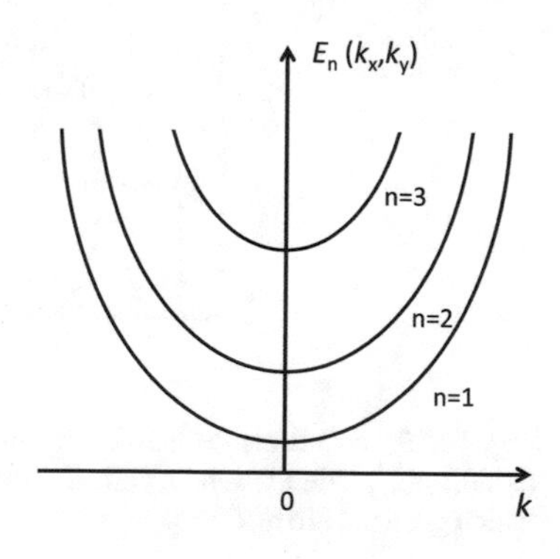

Fig. 12.14 Energy of the subbands labeled by integers n, associated with bound states for the 2D electron gas, are plotted versus wave vector k

$$N_{2D} = \frac{2}{(2\pi)^2}\pi k_\perp^2 \tag{12.38}$$

where $k_\perp^2 = k_x^2 + k_y^2$ and

$$E_\perp = \frac{\hbar^2 k_\perp^2}{2m^*} \tag{12.39}$$

so that for each subband the density of states $g_{2D}(E)$ contribution becomes

$$\frac{\partial N_{2D}}{\partial E} = g_{2D}(E) = \frac{m^*}{\pi\hbar^2}. \tag{12.40}$$

This 2D density of states $g_{2D}(E)$ concept discussed in this Chapter could be extended to few layer materials, which have been intensively studied since 2005. If we now plot the density of states corresponding to the 3D motion in a 1–D rectangular well, we have $g_{2D}(E) = 0$ until the lowest energy bound state energy E_1 is reached, when a step function contribution of $(m^*/\pi\hbar^2)$ is made. The density of states $g_{2D}(E)$ will then remain constant until the minimum of subband E_2 is reached when an additional step function contribution of $(m^*/\pi\hbar^2)$ is made, hence yielding the staircase density of states shown in Fig. 12.15. Two generalizations of (12.40) for the density of states for actual quantum wells are needed, as we discuss below. The first generalization takes into account the finite size L_z of the quantum well, so that the system is not completely two dimensional and some k_z dispersion must occur. Secondly, the valence bands of typical semiconductors are degenerate so that coupling between the valence band levels occurs, giving rise to departures from the simple parabolic bands as discussed below.

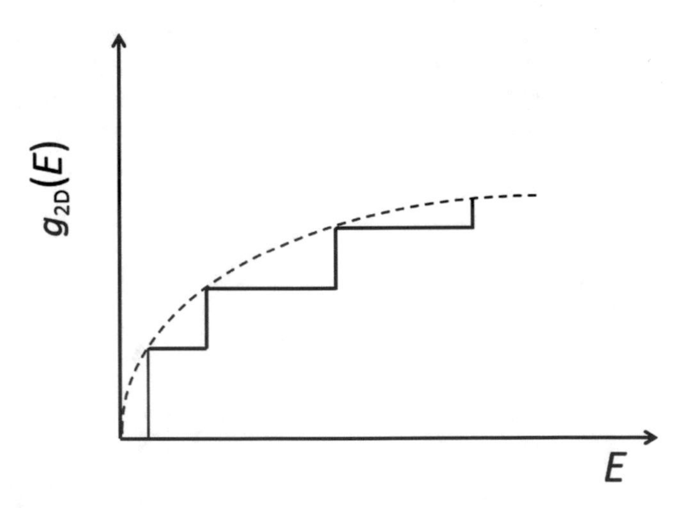

Fig. 12.15 Two dimensional density of states $g_{2D}(E)$ for a rectangular quantum well structure (solid line) plotted together with the three dimensional density of states $g_{3D}(E)$ (dashed line), which goes to zero at $E = 0$

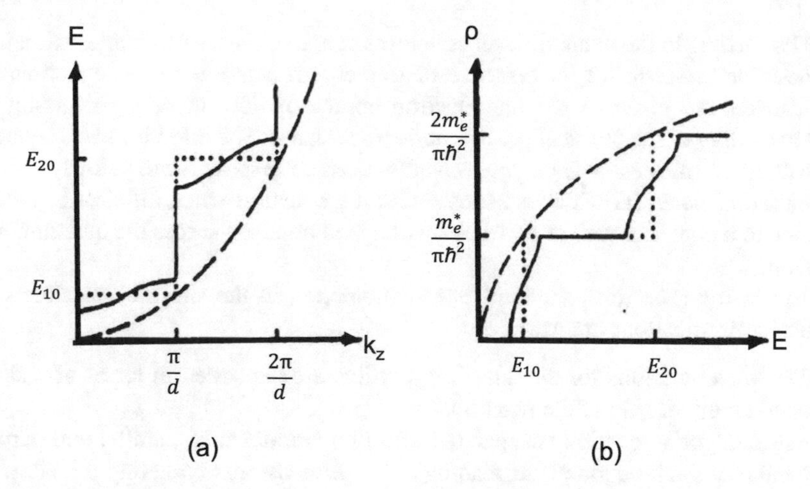

Fig. 12.16 Schematic diagrams of the **a** energy dispersion $E(k_z)$ and the **b** density of states $g_{2D}(E)$. Indicated are the two–dimensional (dotted), three–dimensional (dashed), and the intermediate (solid) cases

A comparison between the energy dispersion relation $E(\mathbf{k})$ and the density of states $g(E)$ in two dimensions and three dimensions is shown in Fig. 12.16 together with a quasi two–dimensional case, typical of actual quantum well samples. In the quasi–two dimensional case, the $E(\mathbf{k})$ relations exhibit a small degree of dispersion along k_z, leading to a corresponding width in the steps of the density of states function shown in Fig. 12.16b.

A generalization of the simple 2D density of states in Fig. 12.15 is also necessary to treat the complex valence bands of a typical III-V compound semiconductor. The $E(\mathbf{k})$ diagram (where $k_\perp$ is normal to k_z) for the heavy hole and light hole levels can be calculated to some level of approximation using $\mathbf{k} \cdot \mathbf{p}$ perturbation theory to be discussed later.

The most direct evidence for bound states in quantum wells comes from optical absorption measurements (to be discussed in Chap. 18) and resonant tunneling effects, which we discuss in Sect. 12.9.

12.9 Resonant Tunneling in Quantum Wells

Resonant tunneling (see Fig. 12.19) provides direct evidence for the existence of bound states in quantum wells. We review first the background material for tunneling across potential barriers in semiconductors and then apply these concepts to the resonant tunneling phenomenon. Such resonant phenomena are very important and of great practical interest in low dimensional systems. The large density of states of low dimensional resonant processes allows individual nanotubes and single layer graphene to be studied in great detail.

The carriers in the quantum well structures can, to a lowest order approximation, be described in terms of the effective mass theorem where the wave functions for the carriers are given by the one electron approximation, thereby promoting the understanding of the physical process under discussion. The effective mass equation is written in terms of slowly varying wavefunctions corresponding to a slowly varying perturbation potential which satisfies Poisson's equation when an electric field is applied to a quantum well (e.g., when a voltage is imposed across the quantum well structure).

To illustrate the resonant tunneling phenomenon in the simplest possible way, further simplifications are made:

1. The wavefunctions for the tunneling particle are expanded in terms of a single band on either side of the junction.
2. Schrödinger's equation is separated into two components, parallel and perpendicular to the junction plane, leading to a 1–dimensional tunneling problem.
3. The eigenstates of interest have energies sufficiently near those of critical points in the energy band structure on both sides of the interface so that the simplified form of the effective mass theorem can be used as a meaningful approximation.
4. The total energy, E, and the momentum parallel to the interface or perpendicular to the layering direction, $k_\perp$, are conserved in the tunneling process. Since the potential acts only in the z–direction, the 1–dimensional Schrödinger equation becomes a one dimensional problem:

$$\left[-\frac{\hbar^2}{2m}\frac{d^2}{dz^2} + V(z) - E\right]\psi_e = 0 \tag{12.41}$$

where $V(z)$ is the electrostatic potential, and ψ_e is an envelope function. The wave function ψ_e is subject, at an interface $z = z_1$ (see Fig. 12.17), to the following boundary conditions that guarantee current conservation. Continuity of the energy and momentum across the boundary leads to

$$\psi_e(z_1^-) = \psi_e(z_1^+) \tag{12.42}$$

$$\frac{1}{m_1}\frac{d}{dz}\psi_e\bigg|_{z_1^-} = \frac{1}{m_2}\frac{d}{dz}\psi_e\bigg|_{z_1^+}. \tag{12.43}$$

The current density for tunneling through a potential barrier becomes

$$J_z = \frac{e}{4\pi^3\hbar}\int dk_z d^2k_\perp f(E)T(E_z)\frac{dE}{dk_z} \tag{12.44}$$

where $f(E)$ is the Fermi–Dirac distribution for the current carriers, and $T(E_z)$ is the probability of tunneling through the potential barrier. Here $T(E_z)$ is expressed as the ratio between the transmitted and incident probability currents.

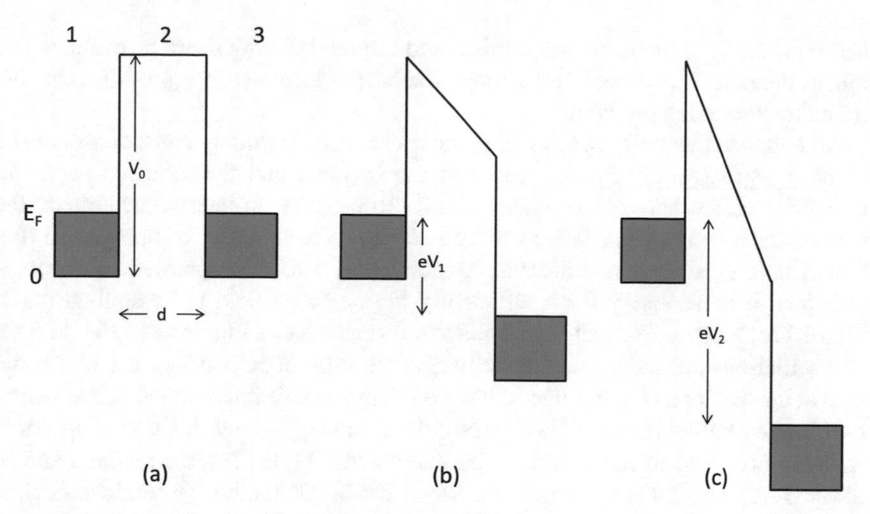

Fig. 12.17 Rectangular–potential model **a** used to describe the effect of an insulator, 2, between two metals, 1 and 3. When a negative bias is applied to 1, electrons, with energies up to the Fermi energy E_F, can tunnel through the barrier. For small voltages, **b**, the barrier becomes trapezoidal, but at high bias **c**, the barrier becomes effectively triangular

If an external bias V is applied to the barrier (see Fig. 12.17), the net current flowing through the barrier is the difference between the current from left to right and that from right to left. Thus, we obtain:

$$J_z = \frac{e}{4\pi^3\hbar}\int dE_z d^2k_\perp [f(E) - f(E+eV)]T(E_z) \tag{12.45}$$

where E_z represents the energy from the k_z component of crystal momentum, i.e., $E_z = \hbar^2 k_z^2/(2m)$. Since the integrand is not a function of $k_\perp$ in a plane normal to k_z, we can integrate over $d^2k_\perp$ by writing

$$dk_x dk_y = d^2k_\perp = \frac{2m}{\hbar^2} dE_\perp \tag{12.46}$$

where $E_\perp = \hbar^2 k_\perp^2/(2m)$ and after some algebra, the tunneling current can be written as,

$$J_z = \frac{em}{2\pi^2\hbar^3}\left[eV\int_0^{E_F-eV} dE_z T(E_z) + \int_{E_F-eV}^{E_F} dE_z (E_F - E_z)T(E_z)\right] \quad \text{if } eV \le E_F$$

$$J_z = \frac{em}{2\pi^2\hbar^3}\int_0^{E_F} dE_z (E_F - E_z)T(E_z) \quad \text{if } eV \ge E_F \tag{12.47}$$

(see Fig. 12.17 for the geometry pertinent to the model) which can be evaluated as long as the tunneling probability through the barrier is known. We now discuss how to find the tunneling probability.

An enhanced tunneling probability occurs for certain voltages as a consequence of the constructive interference between the incident and the reflected waves in the barrier region between regions 1 and 3. To produce an interference effect, the wavevector **k** in the plane wave solution e^{ikz} must have a real component so that an oscillating (rather than a decaying exponential) solution is possible. To accomplish this, it is necessary for a sufficiently high electric field to be applied (as in Fig. 12.17c) so that a virtual bound state is formed. As can be seen in Fig. 12.18a, the oscillations are most pronounced when the difference between the electronic mass at the barrier and at the electrodes is the largest. This interference phenomenon is frequently called resonant Fowler–Nordheim tunneling and this tunneling effect has been observed in metal–oxide–semiconductor (MOS) heterostructures and in GaAs/$Ga_{1-x}Al_xAs$/GaAs capacitors. Since the WKB method is semiclassical, it does not give rise to the resonant tunneling phenomenon, which is a quantum interference effect.

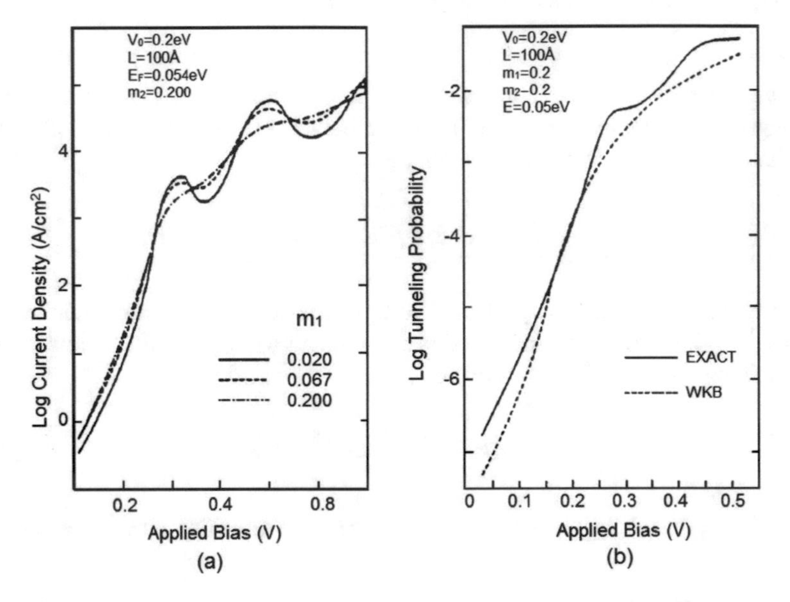

Fig. 12.18 **a** Tunneling current through a rectangular barrier (like the one of Fig. 12.17a) calculated as a function of bias for different values of m_1, in the quantum well. **b** Comparison of an "exact" calculation of the tunneling probability through a potential barrier under an external bias with an approximate result obtained using the WKB method. The barrier parameters are the same as in (**a**), and the energy of an incident electron, of mass $0.2m_0$, is 0.05 eV. (From the book of E.E. Mendez and K. von Klitzing, "Physics and Applications of Quantum Wells and Superlattices", NATO ASI Series, Vol. 170, p.159 (1987))

For the calculation of the resonant tunneling phenomenon, we must therefore use the quantum mechanical solution. In this case, it is convenient to use the transfer–matrix method to find the tunneling probability. In region (#1) of Fig. 12.17, the potential $V(z)$ is constant and solutions to (12.41) have the form

$$\psi_e(z) = A\exp(ikz) + B\exp(-ikz) \tag{12.48}$$

with

$$\frac{\hbar^2 k^2}{2m} = E - V. \tag{12.49}$$

When $E - V > 0$, then k is real and the wave functions are plane waves. When $E - V < 0$, then k is imaginary and the wave functions are growing or decaying waves. The boundary conditions (12.42) and (12.43) determine the coefficients A and B which can be described by a (2×2) matrix R such that

$$\begin{pmatrix} A_1 \\ B_1 \end{pmatrix} = R \begin{pmatrix} A_2 \\ B_2 \end{pmatrix} \tag{12.50}$$

where the subscripts on A and B refer to the region index and R can be written as

$$R = \frac{1}{2k_1 m_2} \begin{pmatrix} (k_1 m_2 + k_2 m_1)\exp[i(k_2 - k_1)z_1] & (k_1 m_2 - k_2 m_1)\exp[-i(k_2 + k_1)z_1] \\ (k_1 m_2 - k_2 m_1)\exp[i(k_2 + k_1)z_1] & (k_1 m_2 + k_2 m_1)\exp[-i(k_2 - k_1)z_1] \end{pmatrix} \tag{12.51}$$

and the terms in R of (12.51) are obtained by matching boundary conditions, as given in (12.42) and (12.43).

In general, if the potential profile consists of n regions, characterized by the potential values V_i and the masses m_i $(i = 1, 2, \ldots n)$, separated by $n - 1$ interfaces at positions z_i $(i = 1, 2, \ldots (n - 1))$, then

$$\begin{pmatrix} A_1 \\ B_1 \end{pmatrix} = (R_1 R_2 \ldots R_{n-1}) \begin{pmatrix} A_n \\ B_n \end{pmatrix}. \tag{12.52}$$

To illustrate the phenomena discussed above for more than one quantum well structure, let us consider tunneling through two potential barriers. The matrix elements of R_i are then written as

$$\begin{aligned}
(R_i)_{1,1} &= \left(\tfrac{1}{2} + \tfrac{k_{i+1} m_i}{2k_i m_{i+1}}\right) \exp[i(k_{i+1} - k_i)z_i] \\
(R_i)_{1,2} &= \left(\tfrac{1}{2} - \tfrac{k_{i+1} m_i}{2k_i m_{i+1}}\right) \exp[-i(k_{i+1} + k_i)z_i] \\
(R_i)_{2,1} &= \left(\tfrac{1}{2} - \tfrac{k_{i+1} m_i}{2k_i m_{i+1}}\right) \exp[i(k_{i+1} + k_i)z_i] \\
(R_i)_{2,2} &= \left(\tfrac{1}{2} + \tfrac{k_{i+1} m_i}{2k_i m_{i+1}}\right) \exp[-i(k_{i+1} - k_i)z_i]
\end{aligned} \tag{12.53}$$

where the k_i wave vectors are defined by (12.49). If an electron is incident from the left (region #1) only a transmitted wave will appear in the last region #n, and therefore $B_n = 0$. The transmission probability is then given by

$$T = \left(\frac{k_1 m_n}{k_n m_1}\right) \frac{|A_n|^2}{|A_1|^2}. \tag{12.54}$$

This is a general solution to the problem of transmission through multiple barriers. Under certain conditions, a particle incident on the left can appear on the right essentially without attenuation. This situation, called resonant tunneling, corresponds to a constructive interference between the two plane waves coexisting in the region between the barriers (quantum well).

The tunneling probability through a double rectangular barrier is illustrated in Fig. 12.19. In this figure, the mass of the particle is taken to be $0.067m_0$, the height of the barriers is 0.3 eV, their widths are 50 Åand their separations are 60 Å. As observed in the figures, for certain energies below the barrier height, the particle can tunnel without attenuation. These energies correspond precisely to the eigenvalues of the quantum well; this is understandable, since the solutions of Schrödinger's equation for an isolated well are standing waves. When the widths of the two barriers are different (see Fig. 12.19b), the tunneling probability does not reach unity, although the tunneling probability shows maxima for incident energies corresponding to the bound and virtual states.

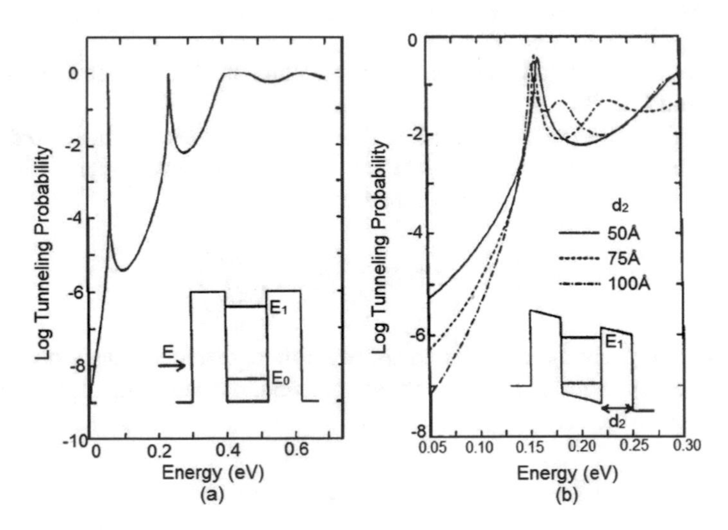

Fig. 12.19 **a** Probability of tunneling through a double rectangular barrier as a function of energy. The carrier mass is taken to be $0.1m_0$ in the barrier and $0.067m_0$ outside the barrier, and the width of the quantum well between the barriers is 60 Å. Note that the tunneling probability is plotted on a logarithmic scale. **b** Tunneling probability through a double–barrier structure, subject to an electric field of 1×10^5 V/cm. The width of the left barrier (1) is 50 Å, while that of the right barrier (2) is varied between 50 and 100 Å. The peak at ~0.16 eV corresponds to resonant tunneling through the first excited state (E_1) of the quantum well. The optimum transmission is obtained when the width of the right barrier (2) is ~75 Å

Problems

12.1 Calculate the bound state energies of a narrow width conductor, as illustrated in the Fig. 12.20 as a function of width (W) assuming an infinite square well potential for

(a) a material with an effective mass $m^* = m_o$.
(b) a material with an effective mass $m^* = 0.07m_o$.
(c) Discuss physically why the results are different depending on the value of m^*. Discuss special values of length L and width W, and discuss the characteristics that are different.

12.2 The confining potential in the z direction (perpendicular to the interfaces) for the 2D electron gas in a MOSFET (see Fig. 12.11) under inversion conditions) is modeled by a triangular well:

$$V(z) = \begin{cases} \infty & z \leq 0 \\ eEz & z > 0 \end{cases}$$

where the constant electric field is approximated by

$$E \simeq \frac{4\pi N_s e}{2\varepsilon}$$

in which $N_s = 10^{12}/\text{cm}^2$ is the 2D carrier concentration in the inversion layer and the dielectric constant has a value of $\varepsilon = 11.8$ for silicon. Use $m^*_{lh} = 0.16m_0$ and $m^*_{hh} = 0.50m_0$ for the light and heavy hole masses in Si, and $m^*_{le} = 0.92m_0$ and $m^*_{te} = 0.19m_0$ for the longitudinal and transverse effective mass components for the electron ellipsoids and an indirect band gap of $\sim$1 eV.

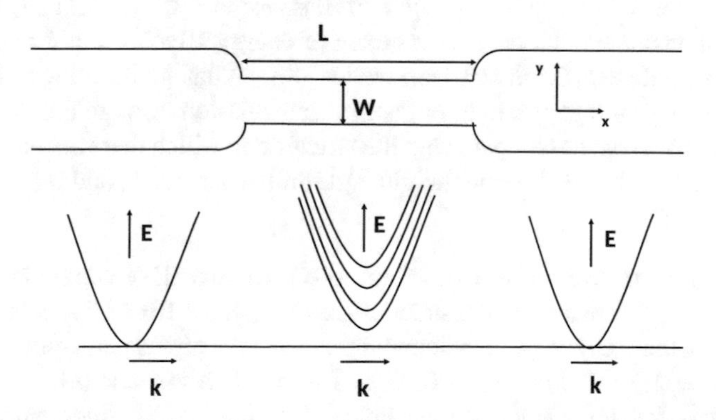

Fig. 12.20 Narrow conductor etched out of a wide conductor. In the wide regions, the transverse modes are essentially continuous. However, in the narrow region, the modes are well-separated in energy in the E(k) diagram

(a) Assume that a (001) crystal orientation is used in preparing the MOSFET (see Fig. 12.1). Using the WKB approximation, find the energies for the three lowest bound states for the 2D electrons in an infinite triangular quantum well for $V(z)$ as given above.
(b) What is the 2D density of electron states in (a) for energies up through the occupation of the third subband, and compare your results with that of a 3D electron gas for the same energy range.
(c) For a carrier concentration of $10^{12}/\text{cm}^2$ for the 2D electron gas, how many levels in (b) are occupied at $T \simeq 0\,\text{K}$?
(d) Suppose that a split gate electrode (see Fig. 12.2) is used to create a strip where a 1-dimensional electron gas is confined. Assume in this case an infinite two-dimensional square well potential $V(x, y) = 0$ for $-a \leq x, y < a$ and $V(x, y) = \infty$ elsewhere, and assume that periodicity is maintained along the z-direction. What are the energies for the lowest 3 bound states in this case? What is the 1D density of electron states?

12.3 Write a computer program in MatLab, Mathematica, or other similar software that uses the propagation matrix method to find the transmission resonance of a particle of mass $m^* = 0.07m_o$ in the following one-dimensional potentials:

(a) A double barrier potential with the barrier energies and widths indicated in Fig. 12.21.
(b) A parabolic energy barrier with $V(z) = (z^2/L^2)$ eV for $|\ z\ | \leq L = 5\,\text{nm}$ and $V(z) = 0\,\text{eV}$ for $|\ z\ | > L$. Your results should include a plot of the particle transmission probability as a function of incident energy for a particle incident from the left and list the energy level values and resonant line widths. You may find it useful to plot your results on a log-linear scale.
(c) What happens to the energy levels in problems (a) and (b) if $m^* = 0.14m_o$?

12.4 (a) Using your computer program from Problem 12.3, find the transmission as a function of energy for a particle of mass $m^* = 0.07m_o$ through 12 identical one-dimensional potential barriers each of energy 10 eV, width 0.1 nm, sequentially spaced every 0.5 nm (0.4 nm well width). What are the allowed (band) and disallowed (band gap) ranges of energy transmission through this structure?
(b) How do these bands compare with the situation in which there are only three barriers, each with 10 eV barrier height, 0.1 nm barrier width, and 0.4 nm quantum well width?

12.5 Suppose that we make a superlattice out of two III-V compound semiconductors (similar in concept to the superlattices fabricated from GaAs/$\text{Ga}_{1-x}\text{Al}_x\text{As}$). Assume that the narrow gap semiconductor (A) is a direct gap semiconductor at $k = 0$ ($E_g = 1.5\,\text{eV}$) with $m_e^* = 0.05m_0$ for the electrons and $m_{hh}^* = 1.0m_0$ and $m_{lh}^* = 0.3m_0$ for the heavy and light holes. Assume that the light and heavy hole masses for the valence band are the same for the wide band gap semiconductor (B), which is an indirect band gap semiconductor ($E_g = 2.4\,\text{eV}$) with the conduction

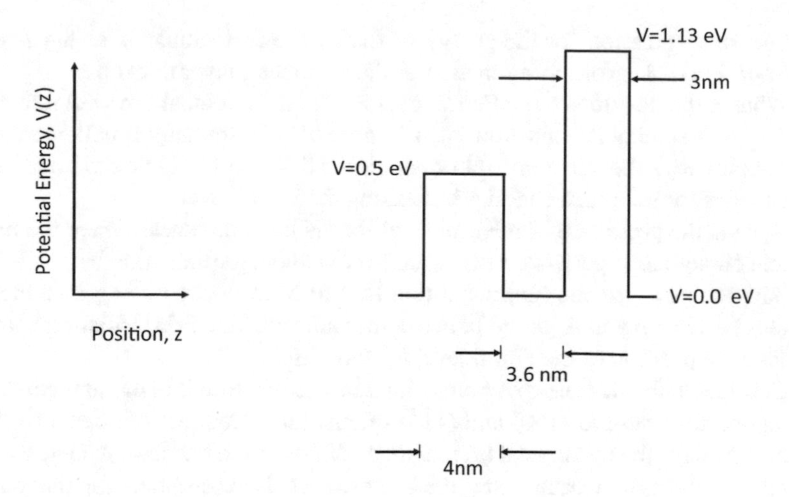

Fig. 12.21 Energy band diagram for a one-dimensional double barrier potential

band extrema at the X point, $\pi/a(100)$, with $m_t^* = 0.3m_0$ and $m_l^* = 1.0m_0$ for the transverse and longitudinal effective mass components. Assume that the width of the quantum wells and quantum barriers of the superlattice are each 50 Å, and that the parameters for the X-point are the same for semiconductors A and B.

(a) Assume an infinite barrier potential for calculation of the subband edge energies. At what photon energy (energies) would you expect the peak (peaks) for the optical absorption edge to occur? Are phonons necessary for these optical transitions? Why?
(b) Now assume that the X point conduction band energy in the bulk 3D narrow band gap and wide band gap semiconductors (A and B) is the same (2.5 eV above the valence band extremum), and that the band offset in the superlattice is twice as large in the conduction band as in the valence band. In which semiconductor (A or B) is the lowest X-point subband minimum located? Find the energies for the lowest X-point longitudinal and transverse subband minima relative to the band edge energy of the 3D conduction band.
(c) Find the energies of the optical transitions in this case. Are phonons necessary for these optical transitions?

12.6 Suppose that you have an fcc semimetal with 2 atoms per unit cell. Suppose that the electrons are at the L points, $\pi/a(1, 1, 1)$, ($m_l^* = 0.3m_0$ and $m_t^* = 0.1m_0$) in the Brillouin zone and the holes are in a single carrier pocket at the Γ point ($k = 0$) with $m_h^* = 0.3m_0$, and assume that the energy overlap for this semimetal is 10 meV. Suppose that the semimetal is now prepared as a thin layer (quantum well) between alkali halide insulating barriers with the (001) crystalline direction normal to the thin layer of the semimetal (layer thickness = d).

(a) Find an expression for the energy of the two lowest subbands of the L point $[\pi/a(111)]$ electrons as a function of the semimetal layer thickness.
(b) What is the position of the Fermi level E_F for a 5 nm quantum well relative to E_F for a semimetal thin film 1 μm in thickness (essentially a bulk semimetal sample) with the same crystal orientation. Hint: use the bulk semimetal edge energies for reference energies in locating the Fermi level.
(c) What is the position of the Fermi level for the layer thickness where the film is thin enough to experience a semimetal-semiconductor transition?
(d) Suppose now that the semimetal thin film of 5 nm thickness is grown along a (111) direction and is placed between the same alkali halide insulating barriers. Find the position of the Fermi level for this case.
(e) Consider a film thickness of 1 nm, which for the thin film is in the semiconductor regime for both the (100) and (111) orientations. Contrast the optical spectra observed for the two crystal orientations with regard to the photon energy where optical absorption occurs and the intensity of the absorption for the various optical transitions.

Suggested Readings

T. Ando, A. Fowler, F. Stern, Rev. Mod. Phys. **54**, 437 (1982)
R.F. Pierret, *Field Effect Devices*, vol. 4, Modular Series on Solid State Devices (Addison-Wesley, Reading, 1983)
B.G. Streetman, *Solid State Electronic Devices*, Series in Solid State Physical Electronics (Prentice-Hall, Englewood Cliffs, 1980)
C. Kittel, *Introduction to Solid State Physics*, 7th edn. (Wiley, New York, 1996)
E.R. Brown et al., Appl. Phys. Lett. **58**, 2291 (1991)
R. Lake et al., J. Appl. Phys. **81**, 7845 (1997)
A. Sommerfeld, H. Bethe, *Electronentheorie der Metalle*, 2nd edn., Handbuch Physik (Springer, Berlin, 1933)
S. Datta, *Electronic Transport in Mesoscopic Systems* (Cambridge University Press, Cambridge, 1995)
A.F.J. Levi, *Applied Quantum Mechanics*, 2nd edn. (Cambridge University Press, Cambridge, 2012)
J.H. Davies, *The Physics of Low-Dimensional Semiconductors* (Cambridge University Press, Cambridge, 1997)
R. Shankar, *Principles of Quantum Mechanics* (Plenum Press, New York, 2011). Chapter 6
J.S. Blakemore, J. Appl. Phys. **53**, 8777 (1982)
D. Strauch, H. Landolt, R. Börnstein, *Semiconductors: Group IV Elements, IV-IV and III-V Compounds. Lattice Properties. Subvol. A* (Springer, Berlin, 2001)
C. Weisbuch, B. Vinter, *Quantum Semiconductor Structures: Fundamentals and Applications* (Academic Press, San Diego, 2014)

Chapter 13
Magneto-Oscillatory and Other Effects Associated with Landau Levels

13.1 Introduction to Landau Levels

In Chap. 10, we discussed magneto-transport phenomena for nearly free electrons in a weak magnetic field (i.e., $\omega_c \tau \ll 1$). In this weak field limit, the carriers scatter many times before completing one cyclotron orbit, and the equations for motion can be derived from classical mechanics. In the high field limit (i.e., $\omega_c \tau \gg 1$), we are particularly interested in the case where the carriers complete their cyclotron orbit without scattering, and we have to treat this case quantum mechanically. Here, the orbital motion of electrons in strong magnetic fields gives rise to discrete magnetic energy levels called Landau levels, named after the famous Russian physicist Lev Davidovich Landau, who first studied this phenomenon theoretically back in the 1930s. In this chapter, we discuss the fundamental properties of Landau levels. Later in the chapter, we discuss magneto-oscillatory phenomena that are associated with Landau levels.

13.2 Quantized Magnetic Energy Levels in 3D

The Hamiltonian for a free electron in a magnetic field uses the basic Schrödinger's equation

$$\left[\frac{(\mathbf{p} - e\mathbf{A})^2}{2m}\right]\psi = E\psi \tag{13.1}$$

in which the square brackets on the first term denotes the scalar product of each of the factors $[\mathbf{p} - e\mathbf{A}]$. To represent a magnetic field along the z axis, we choose the asymmetric gauge (Landau gauge) for the vector potential $\mathbf{A}$:

M. Dresselhaus et al., *Solid State Properties*, Graduate Texts in Physics,
https://doi.org/10.1007/978-3-662-55922-2_13

$$
\begin{aligned}
A_x &= -By \\
A_y &= 0 \\
A_z &= 0.
\end{aligned} \tag{13.2}
$$

Since the only spatial direction in (13.1) and (13.2) is y, the form of the wave function $\psi(x, y, z)$ in (13.1) is chosen to make the differential equation separable into plane wave motion in the x and z directions. The wave function $\psi(x, y, z)$ is thus written as

$$\psi(x, y, z) = e^{ik_x x} e^{ik_z z} \phi(y). \tag{13.3}$$

Substitution of (13.3) in (13.1) results in the expression

$$\left[\frac{(\hbar k_x + eBy)^2}{2m} + \frac{p_y^2}{2m} + \frac{\hbar^2 k_z^2}{2m}\right]\phi(y) = E\phi(y) \tag{13.4}$$

We see immediately that the orbital portion of (13.4) is of the form of the harmonic oscillator (H.O.) equation

$$\left[\frac{p_x^2}{2m} + \frac{1}{2}mx^2\omega_c^2\right]\psi_{\text{H.O.}}^{(\ell)} = E_\ell \psi_{\text{H.O.}}^{(\ell)} \tag{13.5}$$

where $\psi_{\text{H.O.}}^{(\ell)}$ is a harmonic oscillator function and $E_\ell = \hbar\omega_c(\ell + 1/2)$ are the harmonic oscillator eigenvalues in which ℓ is an integer quantum number, $\ell = 0, 1, \ldots$. A comparison of (13.4) with the harmonic oscillator equation (13.5) shows that the characteristic frequency for the harmonic oscillator is the cyclotron frequency $\omega_c = eB/m$ and the harmonic oscillator is centered about the spatial coordinate y_0

$$y_0 = -\frac{\hbar k_x}{m\omega_c}. \tag{13.6}$$

These identifications yield the harmonic oscillator equation

$$\left[\frac{p_y^2}{2m} + \frac{\omega_c^2}{2m}\left(y - y_0\right)^2 + \frac{\hbar^2 k_z^2}{2m}\right]\phi(y) = E\phi(y). \tag{13.7}$$

Thus, the energy eigenvalues of (13.7) for a free electron moving in an orbit in the presence of a magnetic field in the z direction can be written down immediately as

$$E_\ell(k_z) = \frac{\hbar^2 k_z^2}{2m} + \hbar\omega_c\left(\ell + \frac{1}{2}\right) \tag{13.8}$$

recognizing that in the direction parallel to **B** we have plane wave motion, since there is no force acting along **B**, and in the plane perpendicular to **B** we have the harmonic oscillator motion described above.

For a band electron in a solid, the energy eigenvalues in a magnetic field are given in the "effective mass approximation" by an expression which is very similar to (13.8) except that the free electron mass is replaced by an effective mass tensor. Thus, Landau levels for carriers in a simple parabolic band in a semiconductor are given by

$$E_\ell(k_z) = \frac{\hbar^2 k_z^2}{2m_\parallel^*} + \hbar\omega_c^*\left(\ell + \frac{1}{2}\right) \tag{13.9}$$

where the various band parameters in (13.9) are defined as follows: $m_\parallel^*$ is the effective mass tensor component along the magnetic field, $\omega_c^* = eB/m_c^*$ is the cyclotron frequency, and m_c^* is the cyclotron effective mass for motion of the electron in the plane normal to the magnetic field.

The quantum numbers describing the energy eigenvalues $E_\ell(k_z)$ are as follows:

1. ℓ is the Landau level index (or the harmonic oscillator level index), $\ell = 0, 1, 2, 3, \ldots$.
2. k_z assumes values between $-\infty$ and $+\infty$ in free space and is a quasi-continuous variable in the first Brillouin zone of reciprocal space for a real solid.
3. k_x is the wave vector in the plane $\perp$ to **B** and does not enter into (13.9) for the energy levels.

Since the magnetic energy levels $E_\ell(k_z)$ are independent of k_x, the quantum number k_x contributes directly to the density of states in a magnetic field. This degeneracy factor is discussed in Sect. 13.2.1. The form of $E_\ell(k_z)$ is then discussed in Sect. 13.2.2 and finally the effective mass parameters $m_\parallel^*$ and m_c^* are discussed in Sect. 13.2.3.

13.2.1 Degeneracy of the Magnetic Energy Levels in k_x

The degeneracy of the magnetic energy levels $E_\ell(k_z)$ in k_x is found by considering the spatial center in coordinate space of the harmonic oscillator function, which from (13.6) is located at $y_0 = -\hbar k_x/(m\omega_c)$. Since y_0 lies in the interval

$$-L_y/2 < y_0 < L_y/2, \tag{13.10}$$

and since the center of the harmonic oscillator is located inside the sample, we have the requirement

$$-\frac{m\omega_c L_y}{2\hbar} < k_x < \frac{m\omega_c L_y}{2\hbar}. \tag{13.11}$$

Thus, the limits on the range of the quantum number k_x are between $k_x^{\min}$ and $k_x^{\max}$ which are given by

$$k_x^{\min}=-m\omega_c L_y/2\hbar$$
$$k_x^{\max}=m\omega_c L_y/2\hbar. \tag{13.12}$$

With the limits on k_z imposed by (13.12), the sum over states (using Fermi statistics) becomes

$$\mathscr{Z}=\sum_{\ell=0}^{\infty}\sum_{k_z=-\infty}^{\infty}\sum_{k_x=k_x^{\min}}^{k_x^{\max}}\ln\left(1+e^{[E_F-E_\ell(k_z)]/k_BT}\right). \tag{13.13}$$

Since the energy levels are independent of k_x, we can sum (13.13) over k_x to obtain a degeneracy factor which is important in all magnetic energy level phenomena

$$\sum_{k_x}\rightarrow\int_{k_x^{\min}}^{k_x^{\max}}dk_x\frac{L_x}{2\pi}=\frac{L_xL_ym\omega_c}{2\pi\hbar} \tag{13.14}$$

utilizing the uncertainty principle which requires that there is one k_x state per $2\pi/L_x$ since $N_xa=L_x$, in which a is the lattice constant. It is important to emphasize that the sum over k_x in (13.14) is proportional to the magnetic field since $\omega_c\propto B$.

Referring to Fig. 13.1d, we see how upon application of a magnetic field in the z direction the wave vector quantum numbers k_x and k_y in the plane normal to the magnetic field are transformed into the Landau level index ℓ and the quantum number k_x which has a high degeneracy factor per unit area of $(m\omega_c/h)$. It is convenient to introduce the characteristic magnetic length λ defined by

$$\lambda^2\equiv\frac{\hbar}{eB} \tag{13.15}$$

so that from (13.14) the degeneracy factor per unit area becomes $1/(2\pi\lambda^2)$.

From Fig. 13.1d we see a qualitative difference between the states in a magnetic field and the states in zero field. For fields too small to confine the carriers into a cyclotron orbit with a characteristic length less than λ, the electrons are best described in the zero field limit, or we can say that the Landau level description applies for magnetic fields large enough to define a cyclotron orbit within the sample dimensions and for electron relaxation times long enough for an electron not to be scattered before completing an electron orbit, $\omega_c\tau>1$.

13.2.2 Dispersion of the Magnetic Energy Levels Along the Magnetic Field

The dispersion of the magnetic energy levels is given by (13.9) and is displayed in Fig. 13.1a. In this figure it is seen that the dispersion relations $E_\ell(k_z)$ are parabolic

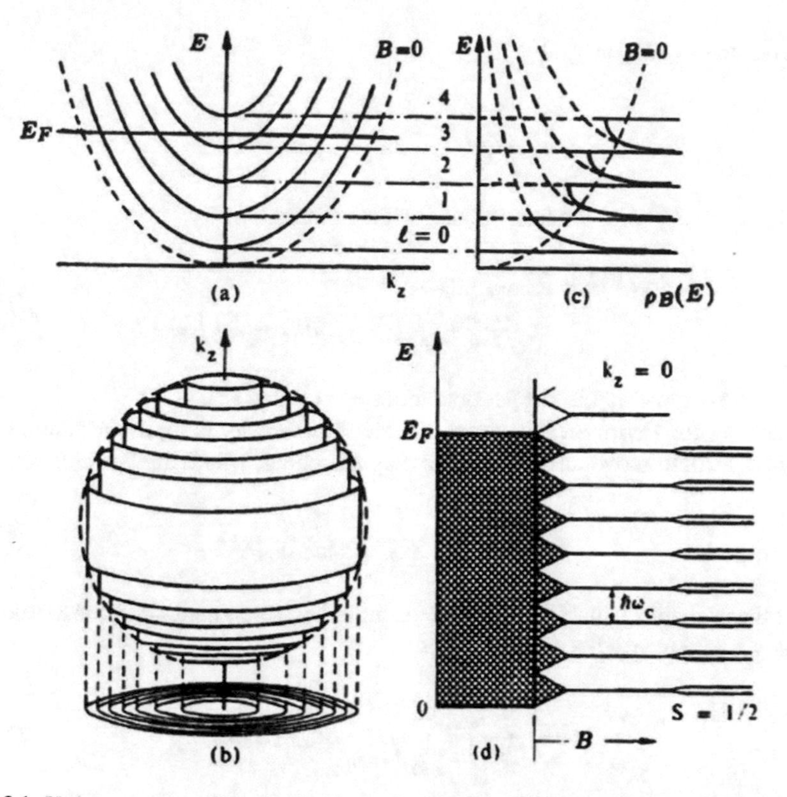

Fig. 13.1 Various aspects of Landau levels. **a** E vs k_z for the first few(lowest energy) Landau levels $\ell = 0, 1, \ldots, 4$. The $B = 0$ parabola (dashed curve) refers to the ordinary free electron case with zero magnetic field. **b** k–space showing Landau levels in 3D. The allowed k-values lie on the concentric cylinders, and the spherical Fermi surface cuts these cylinders. **c** The solid line is the density of states for all the Landau levels while the dashed-solid curves give the density of states in a magnetic field for each of the Landau levels. Singularities in the density of states occur whenever a Landau level pops through the Fermi level. The dashed curve labeled $B = 0$ refers to the density of states in zero field, and shows the expected $\sqrt{E}$ dependence. **d** A schematic diagram showing how the states in zero field go into Landau levels when the B field is applied. The diagram also shows the effect of electron-spin $s = 1/2$ splitting of the Landau levels

in k_z for each Landau level, each level ℓ being displaced from levels $\ell + 1$ and $\ell - 1$ by the Landau level separation $\hbar\omega_c$. The lowest Landau level ($\ell = 0$) is at an energy ($\hbar\omega_c/2$) above the energy of electrons in zero magnetic field. The occupation of each Landau level is found by integration up to the Fermi level E_F. Figure 13.1b shows special k_z values where either a Landau level crosses the Fermi level or a Landau level pops through the Fermi level E_F. As the magnetic field increases the Landau level separation increases until a Landau level pops through E_F, requiring a redistribution of electrons through the remaining Landau levels.

In this section we focus on the k_z dependence of the magnetic energy levels. First we obtain the sum of the number density over k_z which involves conversion of the sum on states to an integral

$$\sum_{k_z} \to 2\int_0^\infty dk_z \frac{L_z}{2\pi}. \tag{13.16}$$

Using Fermi statistics we then obtain for the number density:

$$\begin{aligned} n(L_x, L_y, L_z) &= \sum_{\text{states}} \frac{1}{1+e^{(E_\ell(k_z)-E_F)/k_BT}} \\ &= \frac{L_x L_y m\omega_c}{2\pi\hbar}\frac{2L_z}{2\pi}\sum_\ell \int_0^\infty dk_z \frac{1}{1+e^{(E_\ell(k_z)-E_F)/k_BT}}, \end{aligned} \tag{13.17}$$

so that the degeneracy factor per unit volume is $(1/2\pi^2\lambda^2)$.

Keeping the Fermi level constant, the electron density is found by summing the Fermi distribution over all states in the magnetic field, where the Fermi function

$$f(E_\ell(k_z)) = \frac{1}{1+e^{(E_\ell(k_z)-E_F)/k_BT}} \tag{13.18}$$

gives the probability that the state (ℓ) is occupied. In a magnetic field, the 3D electron density n of a nearly free electron solid is

$$n = \frac{2eB}{(2\pi)^2\hbar}\sum_{\ell=0}^{\ell_{\max}}\int_{-\pi/a}^{\pi/a} dk_z f(E_\ell(k_z)) \tag{13.19}$$

in which a factor of 2 for the electron spin degeneracy has been inserted.

For simplicity, we further consider the magnetic energy levels for a simple 3D parabolic band

$$E_\ell(k_z) = \frac{\hbar^2 k_z^2}{2m^*} + \hbar\omega_c^*(\ell+1/2) \tag{13.20}$$

so that

$$k_z = \left(\frac{2eB}{\hbar}\right)^{1/2}\left[\frac{E}{\hbar\omega_c^*} - (\ell+1/2)\right]^{1/2} \tag{13.21}$$

where we have written E to denote $E_\ell(k_z)$. Differentiating (13.21) gives

$$dk_z = \left(\frac{2eB}{\hbar}\right)^{1/2}\frac{dE}{2\hbar\omega_c^*}\left[\frac{E}{\hbar\omega_c^*} - (\ell+1/2)\right]^{-1/2}. \tag{13.22}$$

For the case of a 2D electron gas, the electrons are confined in the z direction and exhibit bound states. Thus, no integration over k_z (see (13.19)) is needed for a 2D electron gas. However, for the 3D electron gas, integration of (13.19) thus yields a carrier density at $T = 0$ of

$$n = \frac{1}{\pi^2\lambda^2}\sum_{\ell=0}^{\ell_F}\int_{E(k_z=0)}^{E_F}\left(\frac{2eB}{\hbar}\right)^{1/2}\frac{dE}{2\hbar\omega_c^*}\left[\frac{E}{\hbar\omega_c^*}-(\ell+1/2)\right]^{-1/2} \quad (13.23)$$

where the characteristic magnetic length λ is given by (13.15). Carrying out the integration in (13.23), we obtain the result

$$n = \frac{1}{\pi^2\lambda^2}\sum_{\ell=0}^{\ell_F}\left(\frac{2eB}{\hbar}\right)^{1/2}\left[\frac{E_F}{\hbar\omega_c^*}-(\ell+1/2)\right]^{1/2} \quad (13.24)$$

where ℓ_F is the highest occupied Landau level. The oscillatory effects associated with (13.24) are discussed in Sect. 13.3.

From differentiation of (13.24) with respect to energy, we obtain the density of states in a magnetic field $\rho_B(E) = (\partial n/\partial E)$

$$\rho_B(E) = \frac{\sqrt{2eB/\hbar}}{2\pi^2\hbar\omega_c^*(\hbar/eB)}\sum_{\ell}\left[\frac{E}{\hbar\omega_c^*}-(\ell+1/2)\right]^{-1/2} \quad (13.25)$$

which is plotted in Fig. 13.1c, showing singularities at each magnetic subband extrema. Because of the singular behavior of physical quantities associated with these extrema, the subband extrema contribute resonantly to the magneto-optical spectra, as discussed in Sect. 13.4.

To illustrate the oscillatory behavior of (13.24) in $1/B$, we write (13.24) as the 3D electron number density sum over Landau levels and can be written whree n is

$$n = \frac{\sqrt{2}}{\pi^2\lambda^3}\sum_{\ell=0}^{\ell_F}\left(\ell_F' - \ell\right)^{1/2} \quad (13.26)$$

and the resonance condition is

$$\ell_F' = \frac{E_F}{\hbar\omega_c^*} - \frac{1}{2} \quad (13.27)$$

which gives a measure of the occupation level as is illustrated in Fig. 13.1a. The oscillatory behavior of n and other related physical observables is the subject of Sects. 13.3 and 13.4.

13.2.3 Band Parameters Describing the Magnetic Energy Levels

The magnetic energy levels given by (13.9) depend on several band parameters $m_\parallel^*$ and m_c^*. In this section we summarize the properties of these band parameters. To observe the effects associated with the Landau levels we require that $\omega_c\tau \gg 1$, which implies that an electron can execute at least one cyclotron orbit before being scattered. Because of the small effective masses of carriers in many interesting semiconductors, the cyclotron frequency is high, and the spacing between magnetic energy levels also becomes large in comparison to the corresponding spacing for free electrons.

The effective mass parameters $m^*_\parallel$ and m^*_c which enter (13.9) can be simply written for semiconductors because of the simplicity of their Fermi surfaces. For arbitrary magnetic field directions, it is often convenient to use the formula

$$m^*_c = \left(\frac{\det[\overleftrightarrow{m}^*]}{\hat{b} \cdot \overleftrightarrow{m}^* \cdot \hat{b}} \right)^{1/2} \tag{13.28}$$

to find the cyclotron effective mass m^*_c for an ellipsoidal constant energy surface, where det $[\overleftrightarrow{m}^*]$ is the determinant of the effective mass tensor $\overleftrightarrow{m}^*$ and $\hat{b}$ is a unit vector in the direction of the magnetic field so that

$$m^*_\parallel = \hat{b} \cdot \overleftrightarrow{m}^* \cdot \hat{b}. \tag{13.29}$$

Equations (13.28) and (13.29) are particularly useful when the constant energy ellipsoidal surface does not have its major axes along the crystalline axes and the magnetic field is arbitrarily directed with respect to the major axes of the ellipsoidal constant energy surface. For ellipsoidal constant energy surfaces, neither $m^*_\parallel$ nor m^*_c depend on k_z. For more general Fermi surfaces, m^*_c is found by integration of $k/(\partial E/\partial k)$ around a constant energy surface normal to the magnetic field and m^*_c will depend on k_z in general. The effective mass component along the magnetic field $m^*_\parallel$ is unaffected by the applied field in the z direction.

13.3 Overview of Landau Level Effects

Studies of the density of states in a magnetic field and intraband (cyclotron resonance) and interband transitions between magnetic energy levels provide three of the most informative techniques listed in this subsection for study of the constant energy surfaces, Fermi surfaces, and effective mass parameters in solid state physics:

1. The de Haas–van Alphen effect and the other related magneto-oscillatory effects provide the main method for studying the shape of the constant energy surfaces of semiconductors and metals. This is the main focus of this chapter.
2. Cyclotron resonance (see Fig. 13.2) gives specific values for the effective mass tensor components by measurement of the transition between adjacent magnetic energy levels in a single energy band (intraband transitions)

$$\hbar\omega^*_c = E_\ell - E_{\ell-1} \tag{13.30}$$

where the cyclotron frequency for a carrier orbit normal to the magnetic field is given by $\omega_c^* = eB/m_c^*$ and E_ℓ denotes a Landau level with Landau level index ℓ. The magnetic energy level structure for two simple parabolic bands shown in Fig. 13.3 can be interpreted as the Landau level subbands for the valence and conduction bands for a model semiconductor. The dispersion of the energy levels along k_z that is discussed in this Chapter

$$E_\ell(k_z) = \frac{\hbar^2 k_z^2}{2m} + \hbar\omega_c\left(\ell + \frac{1}{2}\right) \quad (13.31)$$

is shown in Fig. 13.3 for each magnetic subband ℓ for a given spin state. As shown in Fig. 13.3, cyclotron resonance experiments can be carried out on both the electron and hole carrier pockets of semiconductors. Electrons and holes correspond to the two different circularly polarizations of the microwave excitation radiation. Because of the different band curvatures and effective masses associated with the various carrier types in metals and semiconductors, $\hbar\omega_c^*$ will be in resonance with $\hbar\omega$ of the resonant microwave cavity at different magnetic field values for electrons as compared to holes. As seen in Fig. 13.3, the optical selection rule for these intraband (cyclotron resonance) transitions is $\Delta\ell = \pm 1$. By varying the magnetic field direction relative to the crystal axes, the corresponding cyclotron effective masses can be determined, in this way the effective mass tensors for electrons and/or holes as a function of k_x, k_y, and k_z for electrons and holes are found. By varying the direction of the magnetic field, the shape of each carrier pocket for electrons and holes can, in principle, be determined.

3. Interband Landau level transitions (see Fig. 13.3) occur when the optical frequency is equal to the separation between the extrema ($k_z = 0$) of the Landau levels ℓ and ℓ'

$$\hbar\omega = E_{\ell,c} - E_{\ell',v} = E_g + \hbar\omega_{c,c}^*(\ell + 1/2) + \hbar\omega_{c,v}^*(\ell' + 1/2). \quad (13.32)$$

These interband transitions provide information on the effective masses for the valence and conduction bands and the bandgaps between them. The optical selection rule for these interband transitions is $\Delta\ell = 0$. In (13.32) the subscripts v and c refer to the valence and conduction bands, respectively. Table 13.1 gives values for the effective masses not only for the conduction band m_e, the valence bands are more complicated being derived from p-orbitals with angular momentum ℓ. This means that the state has $m_{\ell h}$ and m_{hh} components. The upper state is the heavy hole (hh) state with a relatively heavy components and a lower lying light hole state with a lighter mass component, as seen in Table 13.1. And light holes m_{lh} and for the split-off bands for several direct gap semiconductors, and data are included for the split-off band shown in Table 13.1.

In semiconductors it is relatively easy to observe quantum effects in a magnetic field because of the light mass, high mobility and long relaxation times of the carriers which make it easy to satisfy $\omega_c\tau \gg 1$, the requirement for observing quantum effects.

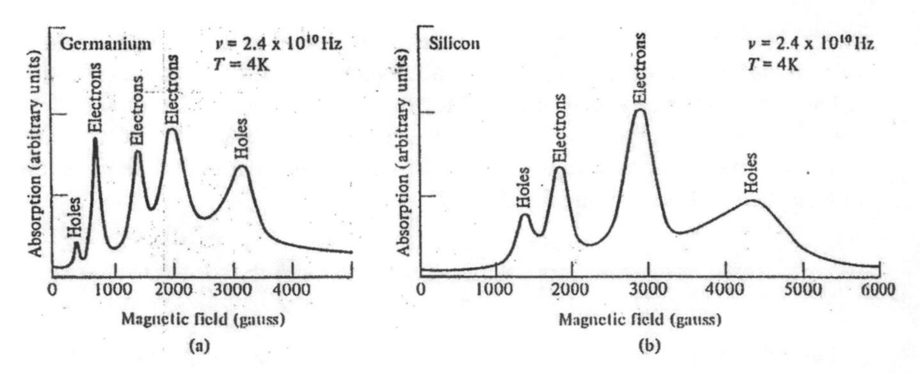

Fig. 13.2 Historic cyclotron resonance signals in the semiconductors **a** germanium and **b** silicon. The magnetic field lies in a (110) plane and makes an angle with the [001] axis of 60° for the spectrum shown for Ge, and 30° for the spectrum shown for Si. (From G. Dresselhaus, et al., Phys. Rev. **98**, 368 (1955).)

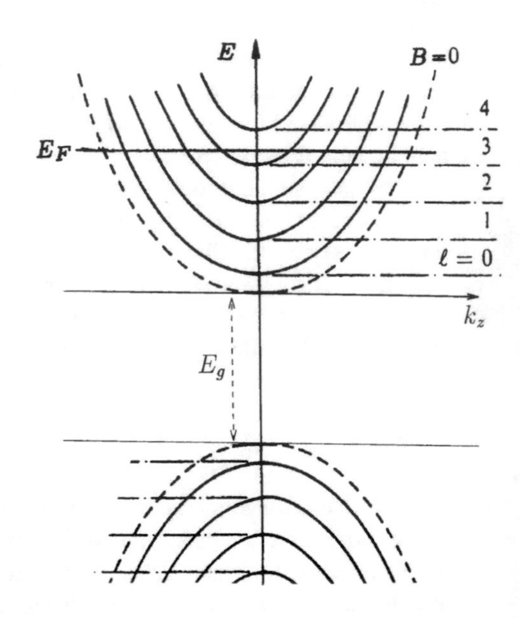

Fig. 13.3 Magnetic energy levels for simple parabolic valence and conduction bands. Intraband cyclotron resonance and interband Landau level transitions occur between these magnetic energy levels. The dashed curves represent the energy dispersion relation in zero magnetic field. In this diagram the spin on the electron is neglected

Table 13.1 Effective masses of electrons and holes in direct gap semiconductors

Crystal	Electron m_e/m_0	Heavy hole m_{hh}/m_0	Light hole m_{lh}/m_0	Split-off hole m_{soh}/m_0	Spin-orbit Δ (eV)	Gap E_g (eV)
InSb	0.015	0.39	0.021	(0.11)	0.82	0.23
InAs	0.026	0.41	0.025	(0.08)	0.43	0.43
InP	0.073	0.40	(0.078)	(0.15)	0.11	1.42
GaSb	0.047	0.30	0.06	(0.14)	0.80	0.81
GaAs	0.070	0.68	0.12	(0.20)	0.34	1.52

13.4 Quantum Oscillatory Magnetic Phenomena

Consider the magnetic energy levels such as those shown in Fig. 13.3 for a band electron in a solid. Assume, for example, that we have carriers in the conduction band and hence a Fermi level E_F as indicated in Fig. 13.4 where we plot the parabolic $E(\mathbf{k})$ relation in zero magnetic field and indicate the energy of each magnetic sub-band extremum by its Landau level index. For each magnetic subband, the density of states is singular at its subband extremum and the resonances in the magneto-oscillatory experiments occur when an energy extremum is at the Fermi energy. Now imagine that we increase the magnetic field. The Landau level spacing is $\hbar\omega_c^* = \hbar e B m_c^*$ and is proportional to B. Thus as we increase B, we eventually reach a value B_ℓ for which the highest occupied Landau level ℓ crosses the Fermi level and the electrons that formerly were in this level must redistribute themselves among the lower levels below the Fermi level.

Assume for the moment that the Fermi level is independent of magnetic field, which is a good approximation when many Landau levels are occupied. We will now show that the passage of Landau levels through the Fermi level produces an oscillatory dependence of the electron density upon the reciprocal of the magnetic field.

Since many physical quantities depend on the density of states, these physical quantities will also exhibit an oscillatory dependence on $1/B$. Thus, this oscillatory dependence on $(1/B)$ is observed in a large class of observables, such as the electrical resistivity (Shubnikov–de Haas effect), Hall effect, Seebeck coefficient (important for thermoelectric effects), ultrasonic attenuation, velocity of sound, optical dielectric constant, relaxation time, temperature dependence (magneto-thermal effect), magnetic susceptibility (the de Haas–van Alphen effect). We discuss below the oscillatory dependence of the carrier density on $1/B$ as representative of this whole class of magneto-oscillatory effects, some of which are explicitly discussed in this chapter.

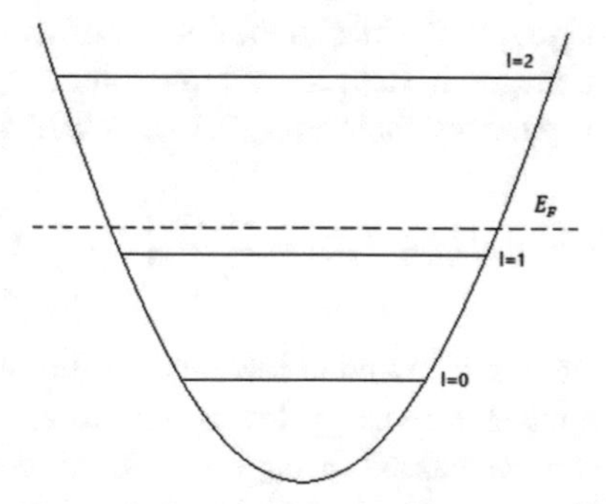

Fig. 13.4 Schematic diagram of the extrema of the energy of the Landau levels for the quantum limit $\ell = 0, 1, 2, \ldots$ showing occupation of the two lowest magnetic sub-bands for $k_z = 0$. The parabola indicates the parabolic energy(momentum) k_x (or k_y) dependence at $B = 0$

In order for the de Haas–van Alphen effect to be observable, we require that an electron complete an orbit before scattering. The time to complete an orbit is $2\pi/\omega_c^*$ and this time must be small compared with τ, the average time between electron scattering events (also called electron collisions). Thus the condition for observing the de Haas–van Alphen effect is usually written as $\omega_c^*\tau \gg 1$. For this reason the observation of magneto-oscillatory phenomena usually requires high magnetic fields (large ω_c) and low temperatures (long τ). Low temperatures are also necessary so that the Landau level separations can be large compared with thermal energies. Landau level separations generally are quite small in magnitude. For example, for $B = 10$ tesla or 100 kilogauss, and m equal to the free electron mass, the Landau level separation is $\sim 10^{-3}$ eV (or ~ 12 K), which is to be compared with k_BT at room temperature with a thermal energy of 0.025 eV or 25 meV (milli electron volts). Therefore it is desirable to carry out de Haas–van Alphen experiments in the vicinity of 1 K. For simplicity, we will take $T = 0$ K in our simple discussion of magneto-oscillatory effects so that the Fermi function is 1 for $E < E_F$ (occupied states) and is 0 for $E > E_F$ (unoccupied states).

The oscillatory behavior of the electron density in a magnetic field can there be simply understood physically from the following considerations. As we increase the magnetic field, two things happen:

1. The density of states degeneracy associated with k_x increases because this degeneracy is proportional to B (see (13.14)).
2. With increasing field B, the number of electrons in a magnetic energy level ℓ decreases as its magnetic energy level extremum approaches the Fermi level. This emptying of electrons from higher lying magnetic sub-bands is not a linear function of B. In particular, when a level crosses the Fermi level, the emptying of electron states is very rapid due to the high density of states at $k_z = 0$ (see Fig. 13.1c).

Consider, for example the emptying of the $\ell = 1$ Landau level as it passes through E_F, with increasing magnetic field. All electrons in this level must be emptied when the Landau level crosses E_F (see Fig. 13.4). We show below that the extrema in the Landau levels correspond to singularities in the density of states (see Fig. 13.1c). The 3D density of states in a magnetic field has a monotonic magnetic field-dependent background due to the degeneracy factor of (13.14), as well as a resonance at

$$E_F = \hbar\omega_c^*(\ell + 1/2) = \frac{\hbar e B_\ell}{m_c^*}\left(\ell + 1/2\right) \tag{13.33}$$

denoting the energy where the lth Landau level passes through the Fermi level. As B increases further, the magnetic energy levels tend to empty their states slowly just after the Landau level has passed through the Fermi level, and the monotonic linearly increasing degeneracy term (13.14) dominates. The interplay of these two factors leads to oscillations in the density of states and consequently in all physical observables depending on the density of states. The resonance condition in the density of states in a magnetic field is given by (13.33) which defines the resonant magnetic field B_ℓ as the field where the E_ℓ Landau level passes through E_F. Making use of (13.33), we see that the resonances in the density of states (13.25) are periodic in $1/B$ with a period defined by

$$\mathscr{P} \equiv \frac{1}{B_\ell} - \frac{1}{B_{\ell-1}} = \frac{e\hbar}{m_c^* E_F}\left[(\ell + 1/2) - (\ell - 1/2)\right] = \frac{e\hbar}{m_c^* E_F}. \tag{13.34}$$

Equation (13.34) shows that the period $\mathscr{P}$ is independent of the quantum number (Landau level index) ℓ, but depends on the product $m_c^* E_F$. It turns out that the temperature dependence of the amplitude of the de Haas–van Alphen resonances depends on the cyclotron effective mass m_c^* so that one can thus measure both m_c^* and the product $m_c^* E_F$ through study of these magneto-oscillatory phenomena, thereby yielding E_F and m_c^* independently, on the basis of this simple model.

It is often convenient to discuss the de Haas–van Alphen effect in terms of cross–sectional areas $\mathscr{A}$ of the Fermi surface. Since $E_F = \hbar^2 k_F^2/2m^*$ and $\mathscr{A} = \pi k_F^2$, we have $E_F = \hbar^2 \mathscr{A}/2\pi m^*$ and from (13.34) the de Haas–van Alphen period $\mathscr{P}$ becomes

$$\mathscr{P} = \frac{1}{B_\ell} - \frac{1}{B_{\ell-1}} = \frac{2\pi e}{\hbar \mathscr{A}}. \tag{13.35}$$

Equation (13.35) shows that the de Haas–van Alphen period $\mathscr{P}$ depends only on the Fermi surface cross sectional area $\mathscr{A}$ Fig. 13.5 except for universal constants. A more rigorous derivation of the de Haas–van Alphen period $\mathscr{P}$ shows that (13.35) is valid for an arbitrarily shaped Fermi surface and the area $\mathscr{A}$ that is associated with the resonance is the extremal cross-sectional area – either the maximum or the minimum, as illustrated in Fig. 13.5. A physical explanation for the dominance of the extremal cross section of the Fermi surface is that all cross sectional areas normal to the magnetic field contribute to the magneto-oscillatory effect, but upon integration over k_z, the cross-sections which do not vary as much with k_z (or varies very little with k_z) will contribute to the same de Haas–van Alphen period $\mathscr{P}$, while the non-extremal cross sections will each contribute to different values of $\mathscr{P}$ and therefore will not give a resonant oscillatory period. By varying the magnetic field orientation, different cross-sections will become extremal, and in this way the shape of the Fermi surface can be monitored. Ellipsoidal constant energy surfaces have only one extremal (maximum) cross-section and here the cyclotron effective mass m_c^* is independent of k_z.

As an example of the de Haas–van Alphen effect in a real material, we see in Fig. 13.6a oscillations observed in silver with $\mathbf{B} \parallel (111)$ direction. In this figure we see oscillations with a long period as well as fast or short period oscillations. From the Fermi surface diagram in the extended Brillouin zone shown for silver in Fig. 13.6b, we identify the fast periods with the large Fermi surface cross sections associated with the belly orbits and the slow oscillations with the small cross sectional necks. From Fig. 13.6b, it is clear that the necks can be clearly observed only for the $\mathbf{B} \parallel (111)$ directions. However, the anisotropy of the belly orbit can be monitored by varying the orientation of the applied magnetic field $\mathbf{B}$.

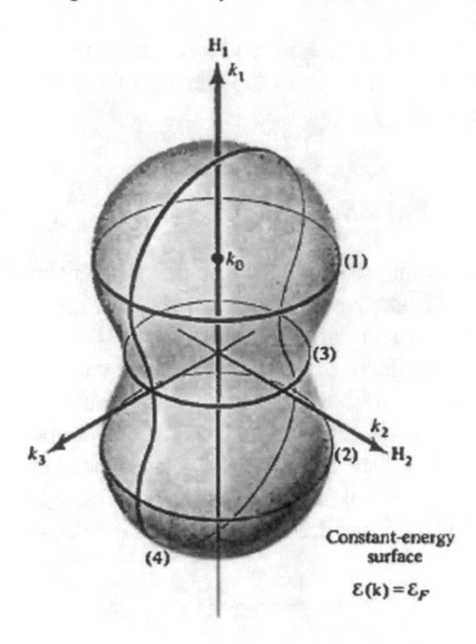

Fig. 13.5 Fermi surface showing extremal cross–sectional areas. The indicated maximum and minimum areas would each show distinct de Haas–van Alphen periods. The larger cross section would have a shorter period. Here k_0 denotes the center of the maximum cross sectional area $\mathscr{A}$ denoted by (1), and k_2 denotes the wave vector axis along the direction H_2, while k_3 denotes the wave vector on the same cross sectional area as k_2 but normal to k_2

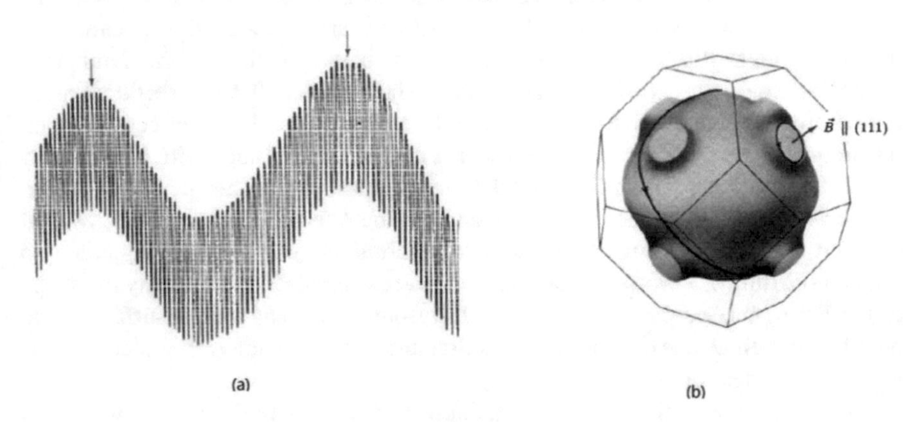

Fig. 13.6 **a** De Haas–van Alphen effect for silver with **B** $\|$ (111), allowing observation of the belly (fast oscillation) orbit and neck (slow oscillation) orbit shown in **b**. The Fermi surface for silver is inferred from measurement of the de Haas–van Alphen effect as a function of magnetic field orientation. The period for the neck orbits [see (b)] is given by the distance between the vertical arrows in **a**. The long period comes from the maximum belly orbits

Not only do electrons execute orbits in reciprocal space. They also can execute orbits in real space in the presence of a magnetic field. Because the length scales in real space and reciprocal space are inversely proportional to one another, large orbits in k-space correspond to small orbits in real space. Furthermore for ellipsoidal orbits (which commonly occur in semiconductor physics), a large k_y/k_x ratio in the k-space orbit would correspond to a small y/x ratio in the real space orbit but a large value for x/y, so that the semi-major axis in the real space orbit is rotated by 90°, relative to the semi-major axis of the reciprocal space orbit.

Although we have neglected the electron spin in the above discussion, it is nevertheless important. De Haas–van Alphen oscillations occur whenever a spin-up or a spin-down level crosses E_F. In fact, magneto-oscillatory observations provide an excellent tool for studying both the Landau level spacing as well as the effective g–factor g_{eff}, as can be seen from Fig. 13.1d. Values for m_c^* and g_{eff} can be obtained independently since the period between every second resonance yields the Landau level separation, while sequential resonances overall are separated by $g_{\text{eff}}\mu_B B$.

13.5 Selection Rules for Landau Level Transitions

Since the magnetic energy states are described in simple terms by harmonic oscillator wave functions, the matrix elements coupling different Landau levels are described by the selection rules for harmonic oscillators. Utilizing the matrix element of the coordinate taken between harmonic oscillator states, we write

$$\langle \ell | x | \ell' \rangle = \sqrt{\frac{\hbar}{2m_c^* \omega_c^*}} \left[\sqrt{\ell + 1}\, \delta_{\ell', \ell+1} + \sqrt{\ell}\, \delta_{\ell', \ell-1} \right]. \tag{13.36}$$

The corresponding matrix element for p_x is

$$\langle \ell | p_x | \ell' \rangle = \sqrt{\frac{\hbar m_c^* \omega_c^*}{2}} \left[\sqrt{\ell + 1}\, \delta_{\ell', \ell+1} + \sqrt{\ell}\, \delta_{\ell', \ell-1} \right]. \tag{13.37}$$

The matrix elements for x and p_x determine the matrix elements for intraband transitions, referred to in Sect. 13.3. It is also of interest to discuss the expectation value of $\langle \ell | x^2 | \ell' \rangle$ and $\langle \ell | p_x^2 | \ell' \rangle$ which are

$$\langle \ell | x^2 | \ell \rangle = \left(\frac{\hbar}{2m_c^* \omega_c^*} \right) \left(2\ell + 1 \right) = \frac{\hbar}{m_c^* \omega_c^*} \left(\ell + 1/2 \right) \tag{13.38}$$

$$\langle \ell | p_x^2 | \ell \rangle = \left(\frac{\hbar m_c^* \omega_c^*}{2} \right) \left(2\ell + 1 \right) = \hbar m_c^* \omega_c^* \left(\ell + 1/2 \right) \tag{13.39}$$

to yield the partition theorem that the kinetic and potential energies of the harmonic oscillator are each $(\hbar\omega_c^*/2)(\ell + 1/2)$. The "classical mean radius" for a harmonic oscillator state is defined by

$$\sqrt{\langle \ell | x^2 | \ell \rangle} = \lambda \sqrt{(\ell + 1/2)} \tag{13.40}$$

using (13.38), thus giving physical meaning to the characteristic length λ in a magnetic field which is $\lambda = (\hbar/m_c^*\omega_c^*)^{1/2} = (\hbar/eB)^{1/2}$ as given in (13.15). We see here that λ is independent of m_c^* and, except for universal constants, depends only on B. The classical mean radius thus has a value at 10 tesla (or 100 kG) of $\sim 10^{-6}$ cm which is about 30 lattice constants in extent. Thus to get a classical orbit within a unit cell we would require fields of $\sim$3,000 tesla or 30 megagauss. With present technology it is not yet possible to generate an external magnetic field with magnetic effects comparable in magnitude to crystal fields, though the highest available fields (300 tesla in the form of pulsed fields) permit entry into this important and interesting regime.

13.6 Landau Level Quantization for Large Quantum Numbers

The most general quantization condition for electrons in conduction bands was given by Onsager many years ago. Suppose that a magnetic field is applied parallel to the z–axis. Then the wave vector components k_x, k_y which are perpendicular to the magnetic field B should satisfy the commutation relation

$$[k_x, k_y] = \frac{s}{i}, \tag{13.41}$$

where $s = 1/\lambda^2$ is proportional to the magnetic field B and is defined as $s = eB/\hbar$ and where

$$k_x \rightarrow \frac{1}{i}\frac{\partial}{\partial x} - \frac{eB}{\hbar}y \tag{13.42}$$

and

$$k_y \rightarrow \frac{1}{i}\frac{\partial}{\partial y}. \tag{13.43}$$

The reason why k_x and k_y in a magnetic field do not commute, of course, relates to the fact that y and p_y do not commute for quantum mechanical systems (the uncertainty principle). We define the raising and lowering operators k_+ and k_- in terms of k_x and k_y

$$k_\pm = \frac{1}{\sqrt{2}}\left(k_x \pm ik_y\right), \tag{13.44}$$

and the operation of $k_\pm$ on the harmonic oscillator wavefunction ϕ_ℓ is given by

$$\begin{matrix} k_+\phi_\ell = [(\ell+1)s]^{\frac{1}{2}}\phi_{\ell+1} \\ k_-\phi_\ell = (\ell s)^{\frac{1}{2}}\phi_{\ell-1}. \end{matrix}, \quad \ell = 0, 1, 2, \ldots . \tag{13.45}$$

The general quantization condition gives $k^2 = k_+k_- + k_-k_+$ so that

$$(2\ell + 1)s \longrightarrow k^2 \tag{13.46}$$

and corresponds to the Bohr–Sommerfeld–Onsager relation:

$$\oint_{E(k)=\text{const}} |k|dk = 2\pi s\left(\ell + \frac{1}{2}\right), \quad (\ell \gg 1) \tag{13.47}$$

where the line integral is over an orbit on the constant energy surface. This semi-classical quantization gives the classical limit for large quantum numbers and can be applied to calculate orbits of carriers in a magnetic field on any constant energy surface.

Problems

13.1 Consider the six ellipsoidal conduction band carrier pockets in silicon with longitudinal and transverse effective masses of $m_l/m_0 = 1.0$ and $m_t/m_0 = 0.2$, respectively. Suppose that a magnetic field is applied in the (100) direction. The magnetic energy levels for an electron associated with one of these ellipsoidal constant energy surfaces is given by

$$E_n(k_B) = \hbar^2 k_B^2/2m_\parallel^* + \hbar\omega_c^*(n + 1/2)$$

where k_B and $m_\parallel^*$ are, respectively, the wavevector and the effective mass component along the magnetic field, and ω_c^* is the cyclotron frequency corresponding to motion in the plane perpendicular to the magnetic field.

(a) Suppose that there are a total of 10^{17} electrons/cm^3 in the conduction band of a silicon specimen. For a magnetic field along the (100) direction, find the magnetic field value B_0 for which all of the carriers have been emptied out of the light cyclotron mass carrier pockets.
(b) For the magnetic field B_0 in part (a), how many Landau levels are occupied in the heavy cyclotron mass pockets, and what is the occupation for each carrier pocket? A useful formula is:

$$n = \frac{\sqrt{2}}{\pi^2 \lambda^3} \sum_{l=0}^{l_F} \left(l'_F - l \right)^{1/2}$$

where $l'_F = E_F/(\hbar\omega_c^*) - 1/2$ and $\lambda = (\hbar/eB)^{1/2}$.

(c) What is the shift in Fermi level for part (a) relative to the Fermi level at zero magnetic field?

(d) Estimate the temperature range at which you would have to conduct your experiments to probe the effects calculated in parts (a), (b), and (c).

(e) Explain physically the dependence of the low temperature resistivity normal to the magnetic field for B in the range $0.9B_0 \leq B \leq 1.1B_0$. (No detailed calculation is expected.)

(f) If we prepare n-type Si (carrier concentration $10^{17}/\text{cm}^3$) in a quantum well of 5 nm thickness with quantum confinement along the (100) direction. At what magnetic field B_2 along (100) does the $n = 2$ Landau level pass through the Fermi level for this 2D electron gas? Explain physically the dependence of the low temperature resistivity normal to the magnetic field B, for B in the range $0.9B_2 \leq B \leq 1.1B_2$.

13.2 If a thin Si wire of 50 nm diameter with its axis oriented along (100) is prepared, estimate the magnetic field value above which boundary scattering by the wire boundary would be greatly reduced.

13.3 In graphene, what is the minimum magnetic field needed to observe magneto-oscillatory phenomena (i.e. $\omega_c\tau > 1$)? For graphene, we use the ficticious relativistic effective mass given by $m^* = E_F/v_F^2$. Assume v_F for graphene is 10^8cm/s and E_F=50meV. Estimate the scattering time, τ, using the Fermi velocity given here and an electron mean free path for the following two types of samples:

(a) Standard graphene–on–SiO_2 substrates, which typically have mean free paths of 100 nm.

(b) BN–encapsulated graphene samples, which typically have mean free paths longer than 1 μm.

13.4 Use the gauge $\mathbf{A} = -\mathbf{r} \times \mathbf{B}/2 = (-yB/2, xB/2)$, which is cylindrically symmetric. Show that the ground eigenstates of

$$\hat{H}_0 = \frac{1}{2}\left[\frac{1}{\imath}\frac{\partial}{\partial x} - \frac{y}{2}\right]^2 + \frac{1}{2}\left[\frac{1}{\imath}\frac{\partial}{\partial y} + \frac{x}{2}\right]^2 \tag{13.48}$$

(i.e. the lowest Laudau level status) can be written as

$$\psi(x, y) \propto z^m \exp[-|z|^2/4] \tag{13.49}$$

where $z = x - \imath y$ and m is the negative angular momentum about the B direction. [We shall later use the more customary $z = x + \imath y$, which is equivalent to taking B in the negative z direction and making some other sign changes].

13.5 A particle in a spherically symmetrical potential is known to be in an eigenstate of $\mathbf{L}^2$ and L_z with eigenvalues $\hbar^2\ell(\ell+1)$ and $m\hbar$, respectively.

(a) Prove that the expectation values between $|\ell m\rangle$ states satisfy

$$\langle L_x \rangle = \langle L_y \rangle = 0,$$

$$\langle L_x^2 \rangle = \langle L_y^2 \rangle = \frac{[\ell(\ell+1)\hbar^2 - m^2\hbar^2]}{2}.$$

Interpret this result semiclassically.

(b) Write explicit matrices for L_+, L_-, and L_x for $\ell = 2$.

13.6 The wave function of a particle subjected to a spherically symmetrical potential $V(r)$ is given by

$$\psi(\mathbf{r}) = (x + y + 3z) f(r)$$

(a) Is ψ an eigenfunction of $\mathbf{L}^2$? If so, what is the ℓ-value? If not, what are the possible values of ℓ that we may obtain when $\mathbf{L}^2$ is measured?

(b) What are the probabilities for the particle to be found in the various m_ℓ states?

13.7 Starting from the matrix elements for the position and momentum of the harmonic oscillator states,

$$\langle \ell | x | \ell' \rangle = \sqrt{\frac{\hbar}{2m_c^*\omega_c^*}} \left[\sqrt{\ell+1}\, \delta_{\ell',\ell+1} + \sqrt{\ell}\, \delta_{\ell',\ell-1} \right]$$

and

$$\langle \ell | p_x | \ell' \rangle = \sqrt{\frac{\hbar m_c^*\omega_c^*}{2}} \left[\sqrt{\ell+1}\, \delta_{\ell',\ell+1} + \sqrt{\ell}\, \delta_{\ell',\ell-1} \right]$$

(a) Find the corresponding matrix elements for $\langle \ell \mid x^2 \mid \ell' \rangle$ and $\langle \ell \mid p^2 \mid \ell' \rangle$.

(b) Show that the equipartition theorem applies to harmonic oscillator states: half the total energy goes into kinetic energy, and half into potential energy.

(c) Using the matrix elements in part (a), explain the degeneracy of the Landau levels in the limit $B \to 0$, as (ℓ, k_x) goes into (k_x, k_y).

13.8 Suppose that a magnetic field of 10 T is applied along a (100) direction to a silicon crystal doped n-type with $10^{18}/\text{cm}^3$ arsenic impurity atoms.

(a) What is the electron concentration for each of the 6 conduction band carrier pockets in zero magnetic field and in a field of 10 T? Note that carriers empty out of the carrier pockets that have large Landau level separations.

(b) At what magnetic field will the carrier pockets with the light cyclotron masses be completely emptied out?

(c) Is there a magnetic field direction for which there is no transfer of carriers between carrier pockets?

13.9 The effective mass of both electrons and holes in semiconducting materials can be measured using the cyclotron resonance technique. Discuss how these experiments are performed.

13.10 Is it physically reasonable to treat a bulk semiconductor in a strong magnetic field as a one dimensional system? Explain under which conditions this approximation is expected to be valid.

Suggested Readings

1. Ashcroft, Mermin, *Solid State Physics*. Chapter 14
2. C. Kittel, *Introduction to Solid State Physics*, 6th edn., pp. 239–249

Chapter 14
The Quantum Hall Effect (QHE)

14.1 Introduction to the Quantum Hall Effect

The observations of the quantum Hall effect (QHE), and The Fractional Quantum Hall Effect (FQHE), which is mentioned in Sect. 14.6, were made possible by advances in the preparation of high mobility materials with electrons and/or holes serving as physical realizations of a 2D electron gas. The MOSFET devices (see Sect. 12.2) and the modulation-doped heterostructures (see Sect. 12.3) that give rise to the formation of a 2D electron gas in a narrow interface region of typical samples. In this Chapter we present a simple view of the physics of the quantum Hall effect and the two-dimensional electron gas used to study the physics of the quantum Hall Effect.

The “Quantum Hall Effect” (QHE) is the step–like increase in the Hall resistance ρ_{xy} in units of h/e^2 as the magnetic field is increased (see Fig. 14.1). Each step in (ρ_{xy}) is accompanied by a vanishing of the magnetoresistance (i.e., $\rho_{xx} = 0$), as shown in Fig. 14.1. For an ordinary 3D electron gas, ρ_{xy} increases linearly with magnetic field B and the magnetoresistance ρ_{xx} increases as B^2. The quantum Hall effect is a strictly 2D phenomenon which can be observed in semiconductors containing a 2D electron gas region (e.g., in a modulation–doped superlattice as in Sect. 12.3). A second requirement for observation of the Quantum Hall Effect is a very high carrier mobility, so that no carrier scattering occurs until the carrier has completed many cyclotron orbits ($\omega_c \tau \gg 1$). A third prerequisite for the observation of the quantum Hall effect is that the Landau level separation of the magnetic levels is large compared with $k_B T$. Thus the QHE is normally observed at very high magnetic fields, very low temperatures and in very high mobility samples. Typical results for the Hall conductance and electrical conductivity as magnetic field magnitude are shown in Fig. 14.1 as model results for the magnetic field dependence of the Hall resistance and electrical resistance of carriers in Quantum Hall regime.

M. Dresselhaus et al., *Solid State Properties*, Graduate Texts in Physics,
https://doi.org/10.1007/978-3-662-55922-2_14

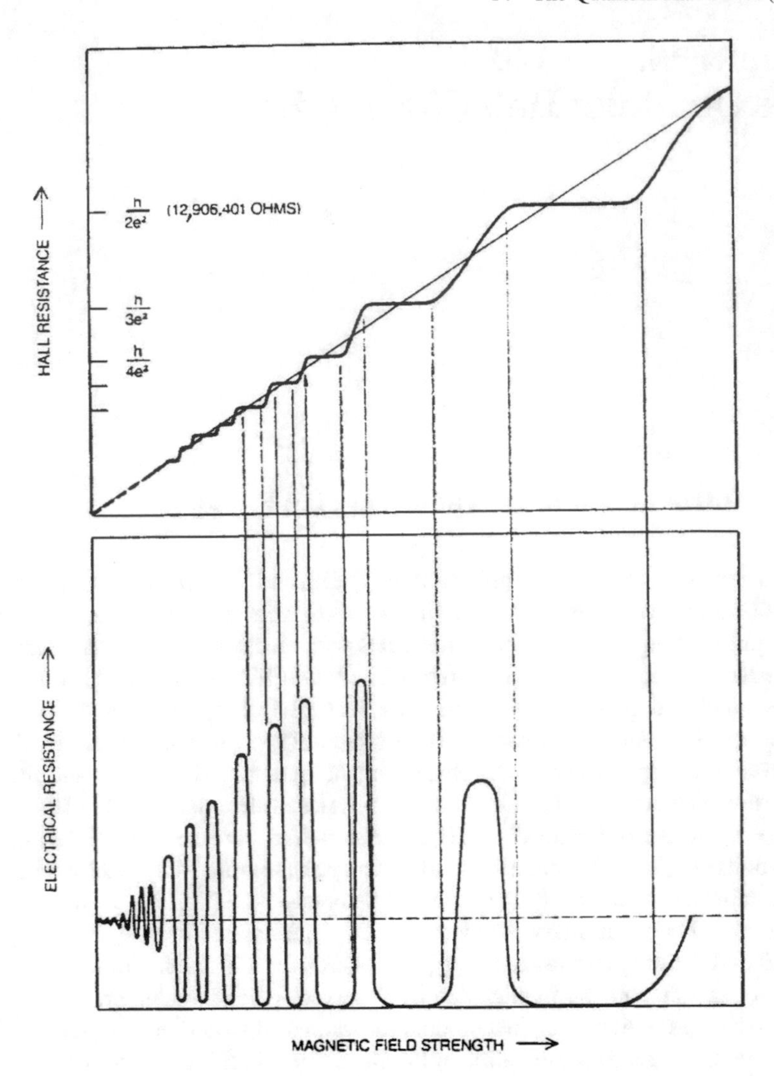

Fig. 14.1 The quantum Hall effect. As shown in the upper panel, the Hall resistance shows plateaux, with each plateau coinciding with the disappearance of the sample's electrical resistance. On the plateaux, the Hall resistance remains constant, while the magnetic field strength is varied. At each of these plateaux, the value of the Hall resistance is precisely equal to $h/(\ell e^2)$, where ℓ is an integer, while the magnetoresistance component vanishes $\rho_{xx} = 0$. (Note: plateaux is the preferred plural of plateau.)

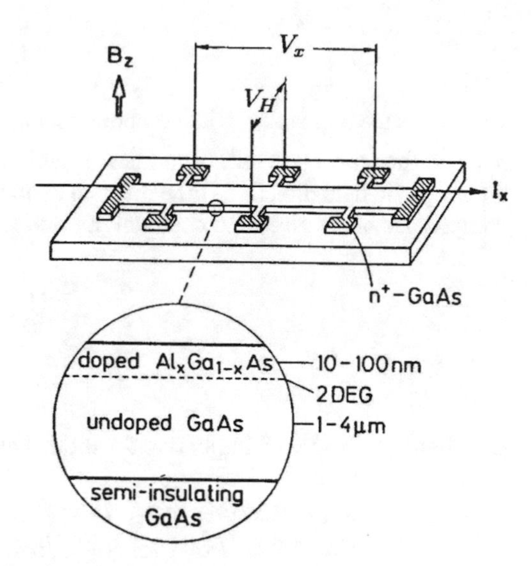

Fig. 14.2 Typical geometry of a sample used for Hall effect measurements. The formation of a 2D electron gas (2DEG) in a GaAs heterostructure is shown as on inset below in the enlargement of the cross section. The Hall voltage V_H and the voltage drop V_x are measured under the constant current condition $I_x =$ constant as a function of the magnetic field B_z perpendicular to the 2D electron gas

14.2 Basic Relations for 2D Hall Resistance

The conventional 3D Hall effect is usually measured in a long sample in which a fixed current I_x is flowing in the x–direction and a magnetic field B is applied in the z–direction. The Lorentz force on the electrons $e\mathbf{v} \times \mathbf{B}$ is compensated by the Hall electric field E_H in the y–direction to prevent the flow of current in the y–direction. The geometry for the Hall measurements is shown in Fig. 14.2. The two voltages V_x (driving voltage) in the x–direction and V_H (Hall voltage) in the y–direction are measured. The longitudinal (R_x) and Hall (R_H) resistances are defined in terms of the current flow I_x as:

$$\begin{aligned} R_x &= V_x / I_x \\ R_H &= V_H / I_x . \end{aligned} \tag{14.1}$$

In general, the conductivity tensor ($\overleftrightarrow{\sigma}$) and the resistivity ($\overleftrightarrow{\rho}$) tensor relate the current density ($\mathbf{j}$) and the electric field ($\mathbf{E}$) vectors, and the vector relations in 2D are written as:

$$\mathbf{j} = \begin{pmatrix} j_x \\ j_y \end{pmatrix} = \overleftrightarrow{\sigma} \cdot \mathbf{E} = \begin{pmatrix} \sigma_{xx} & \sigma_{xy} \\ \sigma_{yx} & \sigma_{yy} \end{pmatrix} \begin{pmatrix} E_x \\ E_y \end{pmatrix} \tag{14.2}$$

and in terms of the resistivity as

$$\mathbf{E} = \begin{pmatrix} E_x \\ E_y \end{pmatrix} = \overleftrightarrow{\rho} \cdot \mathbf{J} = \begin{pmatrix} \rho_{xx} & \rho_{xy} \\ \rho_{yx} & \rho_{yy} \end{pmatrix} \begin{pmatrix} j_x \\ j_y \end{pmatrix} \tag{14.3}$$

with the required relation between $\overleftrightarrow{\sigma}$ and $\overleftrightarrow{\rho}$ written in tensorial form:

$$\overleftrightarrow{\sigma} \cdot \overleftrightarrow{\rho} = \overleftrightarrow{1} \tag{14.4}$$

where $\overleftrightarrow{1}$ is the unit matrix with components (δ_{ij}). Since the off–diagonal xy components of the tensor $\overleftrightarrow{\rho}$ result from the magnetic field, they are odd under reversal of the magnetic field direction (time reversal symmetry), yielding the relation between components of the electrical conductivity and resistivity

$$\begin{aligned} \sigma_{xx} &= \sigma_{yy}, \\ \sigma_{yx} &= -\sigma_{xy}, \\ \rho_{xx} &= \rho_{yy}, \\ \rho_{yx} &= -\rho_{xy}. \end{aligned} \tag{14.5}$$

Equations 14.4 and 14.5 imply that for the 2D electron gas:

$$\begin{aligned} \rho_{xx} &= \sigma_{xx}/(\sigma_{xx}^2 + \sigma_{xy}^2), \qquad & \rho_{xy} &= -\sigma_{xy}/(\sigma_{xx}^2 + \sigma_{xy}^2) \\ \sigma_{xx} &= \rho_{xx}/(\rho_{xx}^2 + \rho_{xy}^2), \qquad & \sigma_{xy} &= -\rho_{xy}/(\rho_{xx}^2 + \rho_{xy}^2). \end{aligned} \tag{14.6}$$

An especially interesting implication of these formulae is that in a 2D system, when $\sigma_{xx} = 0$ but $\sigma_{xy} \neq 0$, then ρ_{xx} is also zero (and vice versa). This means that (as long as σ_{xy} is finite), the vanishing of the longitudinal conductivity implies that the longitudinal resistivity also vanishes. This is precisely the situation that occurs in the quantum Hall effect, and is fundamental to this phenomenon.

We now relate the resistance parameters that are measured (R_x and R_H) to the current density $\mathbf{j}$ and the electric fields $\mathbf{E}$. For a long device (as shown in Fig. 14.2), $j_y = 0$, so that R_H is related to the resistivity components ρ_{xx} and ρ_{xy} via:

$$\begin{aligned} R_x &= V_x/I_x = (L/W) \cdot (E_x/j_x)|_{j_y=0} = (L/W)\rho_{xx} \\ R_H &= V_H/I_x = (E_y/j_x)\big|_{j_y=0} = \rho_{xy} \end{aligned} \tag{14.7}$$

Note that the units of the resistivity in 2D is $\Omega/\square$, that is Ohms per square, and that R_H in 2D has the same units as ρ_{xy}.

In the presence of a DC magnetic field $\mathbf{B} = B\hat{z}$, and in the relaxation–time approximation, the classical equation of motion for the carriers is written as:

$$\frac{d\mathbf{v}}{dt} = \frac{e}{m^*}(\mathbf{E} + \mathbf{v} \times \mathbf{B}) - \mathbf{v}/\tau \tag{14.8}$$

where $\mathbf{v}$ denotes the drift velocity of the carriers, and the charge on the electron is taken as a negative number. Using the relation $\mathbf{j} = ne\mathbf{v}$, we can write

$$\begin{aligned} \sigma_0 E_x &= j_x - \omega_c \tau j_y \\ \sigma_0 E_y &= \omega_c \tau j_x + j_y \end{aligned} \tag{14.9}$$

where $\sigma_0 = ne^2\tau/m^*$ and $\omega_c = eB/m^*$. Finally, combining equations 14.7 and 14.9 with the condition $j_y = 0$, we can write:

$$E_y = \frac{\omega_c\tau}{\sigma_0} j_x \tag{14.10}$$

and the Hall resistance R_H becomes

$$R_H \equiv \frac{E_y}{j_x} = \frac{\omega_c\tau}{\sigma_0} = \frac{(eB/m^*)\tau}{(ne^2\tau/m^*)} = \frac{B}{ne} = \mathscr{R}B \tag{14.11}$$

where $\mathscr{R} = (1/ne)$ is called the Hall coefficient. We note here that R_H is proportional to the magnetic field. The result derived in (14.11) is valid for a classical system that ignores the quantization of the magnetic energy levels. This quantization effect becomes important in the limit $\omega_c\tau \gg 1$. The classical result for a 2D system is the same result as was previously obtained for the 3D system (see Sect. 10.2). Yet the experimental results for the 2D electron gas in a modulation–doped $GaAs/Ga_{1-x}Al_xAs$ interface (Fig. 14.1) exhibit the quantum Hall effect, where V_H (or ρ_{xy}) shows a series of flat plateaux as a function of magnetic field rather than a simple linear dependence in B (14.11). The reason for the steps in ρ_{xy} (or R_H) as a function of B is due to the density of states of the 2D electron gas in a magnetic field, as discussed below.

14.3 The 2D Electron Gas and the Quantum Hall Effect

We refer to the phenomenon shown in Fig. 14.1 as the quantum Hall effect because the values of R_H exhibit a plateau whenever

$$R_H = \frac{h}{\ell e^2} \qquad \ell = 1, 2, 3, \ldots \tag{14.12}$$

where ℓ is an integer. Figure 14.1 shows the results of Hall measurements on a modulation–doped $GaAs/Al_xGa_{1-x}As$ heterostructure. Here ρ_{xx} and ρ_{xy} are shown as a function of magnetic field for a heterostructure with a fixed density of carriers. These experiments (Fig. 14.3) are done at a low temperature (4.2 K), and the plateaux for ρ_{xy} can be observed very clearly, especially in the limit of small ℓ. (The plural of "plateau" is "plateaux".) The results shown in Fig. 14.1 indicate that R_H for the 2D electron gas is quantized. Detailed measurements show that R_H is given by (14.12) to an accuracy of better than 0.1 ppm (parts per million). This quantization is reported to be independent of the sample geometry, the temperature, the scattering mechanisms, or other parameters, including the physical system giving rise to the 2D electron gas. The accuracy of these results and their apparent independence of experimental parameters are very intriguing and, as we discuss below, are ultimately

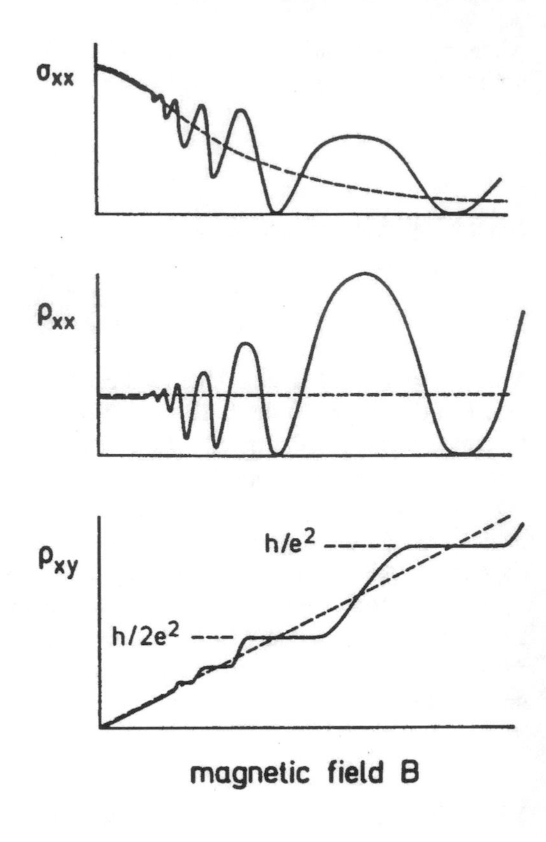

Fig. 14.3 Qualitative behavior for σ_{xx}, ρ_{xx} and ρ_{xy} of a two–dimensional electron gas with a fixed carrier density as a function of magnetic field B. The dotted lines represent the classical curves for a 3D electron gas. The effect of spin degeneracy is not included in these curves

due to a fundamental physical principle and therefore are very interesting to the field of metrology and NIST (National Institute of Science and technology) in the USA. A schematic diagram summarizing the general behavior of the 2D electron gas is given in Fig. 14.3 for ρ_{xy}, ρ_{xx} and σ_{xx} versus B, and also included in this diagram is the comparison with the behavior of a 3D electron gas.

Referring to the 2D conductivity $\overset{\leftrightarrow}{\sigma}$ and resistivity $\overset{\leftrightarrow}{\rho}$ tensors defined in equations 14.2 and 14.3, we can write $\overset{\leftrightarrow}{\rho}$ and $\overset{\leftrightarrow}{\sigma}$ in the region of the plateaux as

$$\overset{\leftrightarrow}{\rho}= \begin{pmatrix} 0 & -R_Q/i \\ R_Q/i & 0 \end{pmatrix} \tag{14.13}$$

and

$$\overset{\leftrightarrow}{\sigma}= \begin{pmatrix} 0 & i/R_Q \\ -i/R_Q & 0 \end{pmatrix} \tag{14.14}$$

where $R_Q = h/e^2$, and where $\rho_{xx} = \rho_{yy} = 0$ and $\sigma_{xx} = \sigma_{yy} = 0$. At these plateaux the power dissipation $P_{\rm diss}$ vanishes because

$$P_{\text{diss}} = \mathbf{j}{\cdot}\mathbf{E} = \mathbf{j}{\cdot} \overleftrightarrow{\rho} {\cdot}\mathbf{j} = \frac{1}{i}\begin{pmatrix} j_x & j_y \end{pmatrix}\begin{pmatrix} 0 & -R_Q \\ R_Q & 0 \end{pmatrix}\begin{pmatrix} j_x \\ j_y \end{pmatrix} = \frac{1}{i}\Big[-R_Q j_x j_y + R_Q j_y j_x\Big] = 0. \tag{14.15}$$

Thus at the plateaux we have no power dissipation and $\rho_{xy} = R_Q/e$ is independent of material, impurity level, sample geometry, while ρ_{xy} is just dependent on the fundamental constants h and e.

To explain the quantized Hall effect, let us first consider the carriers of the two dimensional electron gas to be free electrons at $T = 0$, but subjected to an applied magnetic field B normal to the plane of the 2D electron gas. The Landau quantization for B normal to the film surface gives completely quantized sub–band energies (for a simple band)

$$E_{n,\ell} = E_n + (\ell + 1/2)\hbar\omega_c \pm g^* \mu_B B = E_n + E_\ell \tag{14.16}$$

where $\omega_c \equiv eB/m^*$ is the cyclotron frequency and g^* is the effective g–factor as also for the 3D case, but now E_n pertains to the z–dependent bound state energy levels of the 2D electron gas. For the simplest case, the carrier density is arranged experimentally to be low so that only the lowest bound state ($n = 1$) is occupied. The number of states per unit area is found by noting that the energy is independent of the harmonic oscillator center.

Since the energy levels do not depend on the central position of the harmonic oscillator y_0, we can sum on all the k_x states to obtain the k_x degeneracy per unit area

$$g_{2D} = \frac{eB}{h}. \tag{14.17}$$

This degeneracy factor is here assumed to have the same value for each Landau level and is proportional to the magnetic field B and depends only on fundamental physical constants (i.e., e, h). In addition there is a degeneracy factor of 2 for the electron spin if the electron spin is not considered explicitly in writing the energy level equation, as is done here implicity.

Since there is no k_z dispersion for the 2D electron gas, the density of states in a magnetic field in two–dimensions consists of a series of singularities (δ-functions) as shown in Fig. 14.4b in contrast to the continuum of states in 3D, also shown in the figure. Multiplying (eB/h) by $\hbar\omega_c$ gives the number of 2D states in zero magnetic field that coalesce to form each Landau level. This number increases proportionally to the magnetic field as does also the Landau level separation.

If at a given magnetic field there are ℓ' filled Landau levels, the carrier concentration (neglecting spin) is given by $n_{2D} = \ell'(eB/2\pi\hbar)$, where n_{2D} is the carrier density associated with a given bound state $n = 1$. Thus the Hall resistance R_H in 2D becomes

$$R_H = \frac{B}{n_{2D}e} = \frac{h}{\ell' e^2} = \frac{R_Q}{\ell'} \tag{14.18}$$

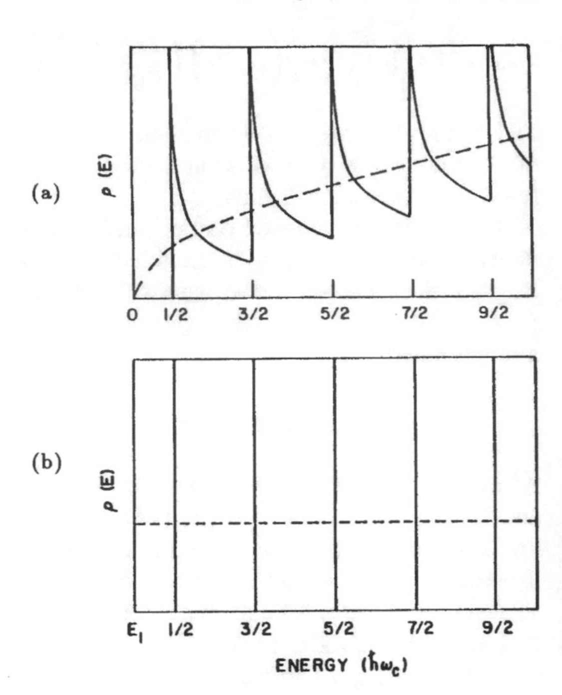

Fig. 14.4 Schematic density of states in a magnetic field and we also consider the case where the relaxation time for carriers $\rightarrow \infty$ for: **a** three–dimensions, where the energy is referred to the bottom of the band ($E = 0$) and **b** two–dimensions, where the energy is referred to the lowest bound state energy E_1. The energy is plotted in units of the cyclotron energy $\hbar\omega_c$. Dashed curves represent the density of states without a magnetic field. We note that in the 2D case the density of states in zero field is $m/\pi\hbar^2$, indicated by the dashed line. The filling per Landau level ℓ in a magnetic field is the degeneracy factor $g_{2D} = eB/2\pi\hbar$ or eB/h

for the ℓ' filled magnetic energy levels. We can then see that the unique property of the density of statess of a 2D electron gas in a magnetic field (see Fig. 14.4) leads to a Hall conductance at $T = 0$ that is quantized in multiples of e^2/h, (see Fig. 14.4).

Carrier filling in 2D is fundamentally different from that in 3D. As the Fermi level rises in 3D, because of the k_z degeneracy, all the Landau levels with subband extrema below E_F will become filled up. To the extent that the electron density is low enough so that only one bound state is occupied, each magnetic subband fills to the same number of carriers or carrier density at a given B field. In the region of the plateaux all Landau levels for $\ell \leq \ell'$ are filled and all Landau levels for $\ell > \ell'$ are empty so that for $k_B T \ll \hbar\omega_c$, very little carrier scattering can occur.

The electrons in the semiconductor heterostructure, however, are not free carriers: their behavior is influenced by the presence of the periodic ionic potential, impurities, and scattering phenomena (see Fig. 14.5). Therefore the simple explanation given above for a perfect crystal needs to be extended to account for these complicating effects. In Fig. 14.5 the two–dimensional density of states in a magnetic field is shown schematically both in the absence of disorder (Fig. 14.5a) and in the presence of disorder (Fig. 14.5b). Here we see that the δ–functions of Fig. 14.5a are now replaced by a continuous function $D(E)$ as shown in Fig. 14.5b. This figure also indicates schematically the magnetic field range over which electrical conduction occurs is

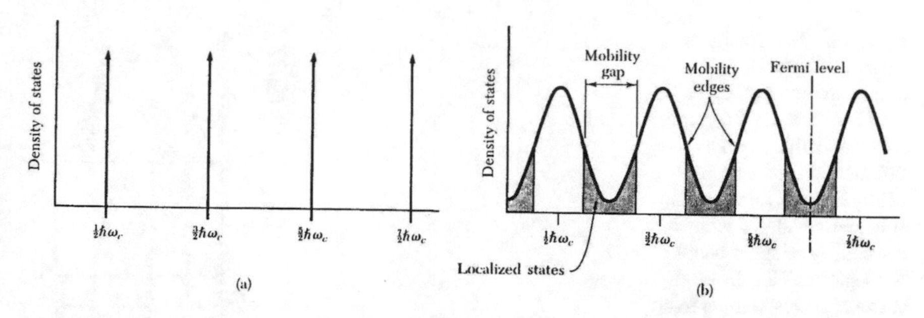

Fig. 14.5 Schematic representation of the 2D density of states in a magnetic field **a** without disorder and **b** with disorder. The shaded regions in (**b**) correspond to localized states

broadened. The figure further shows that in the tails of each Landau sub–band there exist regions of localized states (the shaded areas). The electrons associated with the mobility gap are in localized states that do not contribute to conduction. Much research has been done to show that the simple model described above accurately describes ρ_{xy} and ρ_{xx} for the 2D electron gas within the region of the plateaux in actual semiconductor devices.

Let us consider the diagram for the 2D density of states D(E) in a magnetic field shown in Fig. 14.5 for individual Landau levels. Suppose the magnetic field is just large enough so that the indicated 2D Landau level shown in the figure is completely filled and E_F lies at $\nu = 1$, where ν represents the fractional filling of the 2D Landau level for this magnetic energy level. Then as the magnetic field is further increased, the Fermi level falls. So long as the Fermi level remains within the region of the localized states, then $\sigma_{xx} = 0$. Thus when E_F lies in the shaded region, E_F is effectively in an energy gap where $\sigma_{xx} \equiv 0$ and the Hall conductance σ_{xy} remains on a plateau determined by $\ell' e^2/h$. As B increases further, E_F eventually reaches the unshaded region where σ_{xx} no longer vanishes and E_F passes through the mobile states (see Figs. 14.5 and 14.6), causing σ_{xy} to jump from $\ell' e^2/h$ to $(\ell' - 1)e^2/h$ as E_F passes through the mobile states. When the magnetic field is large enough for E_F to reach the localized states near $\nu = 0$, then σ_{xx} again vanishes and σ_{xy} now remains at the plateau $(\ell' - 1)e^2/h$ as the magnetic field continues to increase.

From these arguments we can conclude that the steps in ρ_{xy} and the zeros in ρ_{xx} are caused by the passage of a 2D Landau level through the Fermi level. When the effect of the electron spin is included, spin splitting of the Landau levels is expected to affect the quantum Hall effect measurements (Fig. 14.6). To see spin splittings effects the measurements must be made at sufficiently low temperatures (e.g., $T = 0.35$ K). Spin splitting effects of the $\ell = 1$ Landau level ($1 \downarrow$ and $1 \uparrow$) have been clearly seen experimentally. The observation of spin splitting in the Quantum Hall Effect thus requires high fields, low m_c^*, high mobility samples to achieve $\omega_c \tau \gg 1$ and low temperatures $k_B T \ll \hbar\omega_c$ to prevent thermal excitation between Landau levels.

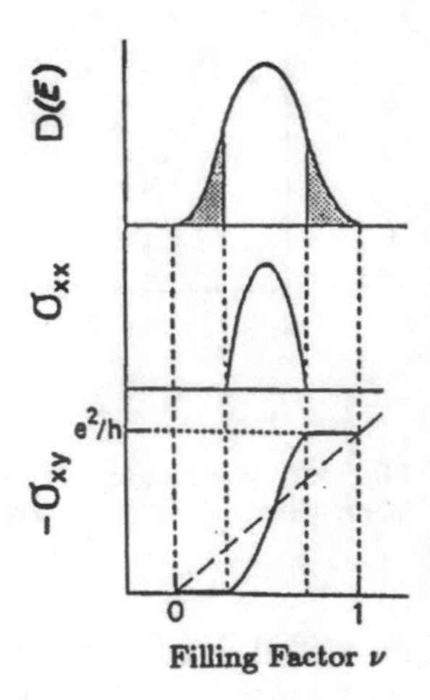

Fig. 14.6 The density of states $[D(E)]$, d.c. conductivity (σ_{xx}), and the Hall conductivity (σ_{xy}) are schematically shown as a function of the fractional filling factor ν for a Landau subband. Shaded regions in the density of states denote the regions of localized carriers, corresponding to an effective energy gap between magnetic subbands

14.4 Effect of Edge Channels and the Quantum Field Effect

In the simple explanation of the Quantum Hall effect, it is necessary to assume both localized states ($\sigma_{xx} = 0$) and extended states ($\sigma_{xx} \neq 0$). In taking into account the so-called edge channels, it is possible to explain more clearly why the quantization is so precise in the Quantum Hall Effect for real systems.

Referring to the derivation of the Landau levels for motion in a plane perpendicular to the magnetic field, we assume that only the lowest band state $n = 1$ is occupied and we neglect the interaction of the electron spin with the magnetic field. The wave function for an electron in the 2D electron gas can then be written as

$$\Psi_{2D}(x, y) = e^{ik_x x}\phi(y) \tag{14.19}$$

where $\phi(y)$ satisfies the harmonic oscillator equation

$$\left[\frac{p_y^2}{2m_c^*} + \frac{1}{2}m_c^*\omega_c^*(y - y_0)^2\right]\phi(y) = \left[\frac{p_y^2}{2m_c^*} + V(y)\right]\phi(y) = E_\ell\phi(y) \tag{14.20}$$

in which the harmonic oscillator center y_0 is given by

$$y_0 = \frac{\hbar k_x}{m_c^*\omega_c^*} = \lambda_B^2 k_x \tag{14.21}$$

and the harmonic oscillator energies are

$$E_\ell = (\ell + 1/2)\hbar\omega_c^*. \tag{14.22}$$

The characteristic magnetic length in (14.21)

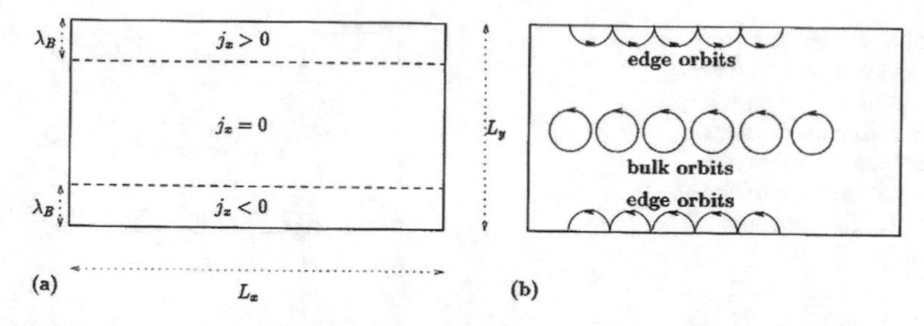

Fig. 14.7 Edge states and the orbits of carriers. **a** The edge regions shown here are defined by the characteristic length $\lambda_B = (\hbar/eB)^{1/2}$. Along each edge, **b** shows that all complete orbits give rise to a current j_x in the same direction, but the current direction is opposite for the two edges in (**b**). The bulk orbits, however, do not give rise to a current j_x

$$\lambda_B = \sqrt{\frac{\hbar}{eB}} = \frac{250\text{Å}}{\sqrt{B(\text{tesla})}} \tag{14.23}$$

relates to the real space orbit of the electron in a harmonic oscillator state (13.40) and λ_B depends only on the magnetic field, except for universal constants. Since the energy in (14.22) is independent of k_x, the electron velocity component x_x vanishes

$$v_x = \frac{1}{\hbar}\frac{\partial E}{\partial k_x} = 0 \tag{14.24}$$

and there is no net carrier current along $\hat{x}$.

The argument that the energy is independent of k_x, however, only applies to those harmonic oscillator centers y_0 that are interior to the sample. But if y_0 takes on a value close to the sample edge, i.e., $y_0 \simeq 0$ or $y_0 \simeq L_y$, then the electron is more influenced by the infinite potential barrier at the edge than the harmonic oscillator potential $V(y)$ associated with the magnetic field. Electrons in these edge orbits will be reflected at the edge potential barriers, and $V(y)$ is no longer strictly a harmonic oscillator potential. Since the potential $V(y)$ is perturbed, the energy will also be perturbed and the energy will then become dependent on k_x. Sincc thc harmonic oscillator orbit size is $\lambda_B\sqrt{\ell+1}$, the energy of the 2D electron gas depends on k_x only for a distance of approximately $\lambda_B\sqrt{\ell+1}$ from the sample edge.

The effect of the sample edges can be understood in terms of the edge orbits illustrated in Fig. 14.7.

All the harmonic oscillator orbits with y_0 values within λ_B of the edge will contribute to the current density j_x by the argument given in Fig. 14.7. The current I_ℓ contributed by the ℓth edge channel is

$$I_\ell = ev_{\ell,x}\left(\frac{dn}{dE_\ell}\right)\Delta\mu \tag{14.25}$$

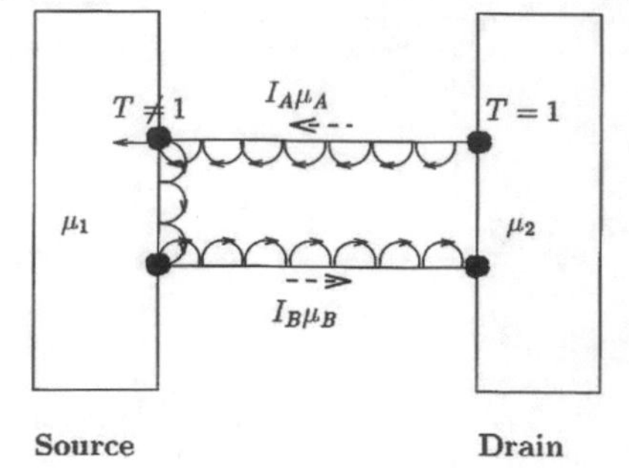

Fig. 14.8 Schematic diagram of current flow for edge channels. The dark circles denote the contacts between the edge channels and the electron reservoirs at electrochemical potentials μ_2 and μ_1

where (dn/dE_ℓ) is the 1D density of states and $\Delta\mu$ is the drop in chemical potential along the edge channel. We can then write

$$\frac{dn}{dE_\ell} = \frac{dn}{dk_x}\frac{dk_x}{dE_\ell} = \left(\frac{1}{2\pi}\right)\left(\frac{1}{\hbar v_{\ell,x}}\right) = \frac{1}{hv_{\ell,x}} \tag{14.26}$$

where $v_{\ell,x}$ is the velocity of the electrons in the x direction due to the carriers in channel ℓ. Substitution of (14.26) into (14.25) yields $I_\ell = (e/h)\Delta\mu$, which is independent of ℓ, so that the total current is obtained by summing over each of the edge channels to yield

$$I_x = \ell_c \frac{e}{h}\Delta\mu \tag{14.27}$$

where ℓ_c is the number of edge channels. If the conditions $\lambda_B \ll \ell_\phi$ and $\lambda_B \ll \ell_e$ are satisfied, where ℓ_ϕ and ℓ_e are, respectively, the inelastic and elastic scattering lengths, electrons are not likely to scatter across the sample (backscattering) because of the electron localization in the variable y ($\psi \sim \exp[-y^2/\lambda_B^2]$). The opposing directions of j_x along the two edges guarantees that the continuity equation is satisfied.

Let us now consider the electrochemical potential μ, which has a constant value along each edge channel, because of the absence of back scattering, as noted above. Two edge channels are shown in Fig. 14.7. From (14.27), we obtain the total current I_x in the upper and lower edges. The quantity $\Delta\mu$ in (14.25) and (14.27) denotes the potential drop between two points where the transmission coefficient T is unity ($T \equiv 1$). Thus for the upper edge channel in Fig. 14.8,

$$I_A = \ell_{cA}(e/h)(\mu_2 - \mu_A), \tag{14.28}$$

indicating that there is a reflection between the edge channel and the μ reservoir so that $T \neq 1$. For the lower channel

$$I_B = \ell_{cB}(e/h)(\mu_B - \mu_2). \tag{14.29}$$

where the number of edge channels for the two edges is the same, so that $\ell_{cA} = \ell_{cB} = \ell_c$. We thus obtain (Fig. 14.8):

$$I_x = I_A + I_B = \ell_c(e/h)(\mu_B - \mu_A). \tag{14.30}$$

Since the Hall voltage V_y is given by the difference in electrochemical potential in the y direction of the sample, we obtain

$$eV_y = \mu_B - \mu_A \tag{14.31}$$

so that

$$I_x = \ell_c(e^2/h)V_y. \tag{14.32}$$

The Hall resistance R_H then becomes

$$R_H = \frac{V_y}{I_x} = \frac{h}{e^2\ell_c} = \frac{R_Q}{\ell_c} \tag{14.33}$$

where $R_Q = h/e^2$ is the fundamental unit of resistance and ℓ_c is a quantum number denoting the number of edge channels. The edge channel picture thus provides another way to understand why the quantum Hall effect is associated with a fundamental constant of nature, $h/e^2 = R_Q$.

14.5 Precision of the Quantized Hall Effect and Applications

Because of the high precision with which the Hall resistance is quantized at integer fractions of h/e^2, we obtain

$$\ell' R_H = \frac{h}{e^2} = R_Q = 25,812.200\ \Omega \quad \ell' = 1, 2, 3, \ldots. \tag{14.34}$$

This quantity R_Q called the Klitzing (after the man Klaus von Klitzing, who discovered the Quantum Hall Effect experimentally) has since 1990 become the new IEEE resistance standard, and is known to an accuracy of $\sim 3 \times 10^{-8}$. When combined with the high precision with which the velocity of light is known, $c = 299,792,458 \pm 1.2$ m/s, the quantum Hall effect has become the primary technique for measuring the fine structure constant:

$$\alpha \equiv \frac{e^2}{\hbar}. \tag{14.35}$$

The fine structure constant must be known to high accuracy in tests of quantum electrodynamics (QED). The results for α from the QHE are not only of comparable accuracy to those obtained by other methods, but this determination of α is also independent of the QED theory. The QHE measurement thus acts as another independent verification of QED. It is interesting to note that the major source of uncertainty in the QHE result is the uncertainty in the calibration of the standard resistor used as a reference.

14.6 Fractional Quantum Hall Effect (FQHE)

When a two–dimensional electron gas is subjected to a sufficiently low temperature and an intense magnetic field ($B \| z$–axis), of magnitude greater than necessary to achieve the lowest quantum state in the quantum Hall effect, all electrons could be expected to remain in their lowest Landau level and spin state. In this limit, however, the possibility also exists, that the electrons will further order under the influence of their mutual interactions. Such ordering phenomena have been seen in $GaAs/Ga_{1-x}Al_xAs$ and other quantum well structures, where an apparent succession of correlated electron states has been found to have fractional occupations, ν, of their lowest Landau level. This ordering effect is called The Fractional Quantum Hall Effect (FQHE).

Just as for the quantum Hall effect discussed in Sect. 14.3, the fractional quantum Hall effect is characterized by minima in the electrical resistance and plateaux in the Hall resistance for current flow in the two–dimensional layers (x–direction). Whereas the integral quantum Hall effect occurs because of gaps in the density of mobile electron states at energies between the 2D Landau levels (see Fig. 14.4), the fractional quantization is interpreted in terms of new gaps in the spectrum of electron energy levels. These new energy levels appear predominantly at magnetic fields higher than the plateau for the $\ell = 0$ integral quantum Hall effect and are associated with electron-electron interactions.

The Fractional Quantum Hall Effect (FQHE) was first observed in the extreme quantum limit, for fractional filling factors ν

$$\nu = \frac{n_{2D} h}{eB} < 1, \tag{14.36}$$

where the 2D carrier density n_{2D} is given by

$$n_{2D} = \nu \left(\frac{eB}{h} \right). \tag{14.37}$$

This regime for the fractional quantum Hall Effect (FQHE) can be achieved experimentally at low carrier densities n_{2D}, high magnetic fields B, and very low temperatures T. The observations of the FQHE thus requires the Landau level spacing $\hbar\omega_c$

to exceed the zero field Fermi level

$$\hbar\omega_c = \frac{\hbar e B}{m_c^*} > E_F \tag{14.38}$$

where the Fermi level E_F for a single spin orientation is given by

$$E_F = \frac{2\pi n_{2D}\hbar^2}{m^*}. \tag{14.39}$$

This condition is equivalent to requiring the magnetic length or the cyclotron radius to be less than the inter–particle spacing $n_{2D}^{-1/2}$, where n_{2D} is the 2D electron density.

To observe electron ordering, it is desirable that electron–electron interactions be large and that electron–impurity interactions be small. This requires the minimization of both the uncertainty broadening of the electron levels and inhomogeneous broadening caused by potential fluctuations and electron scattering. Thus the observation of the fractional quantum Hall effect is linked to the availability of very high mobility samples containing a 2D electron gas in the lowest bound state level. The best samples for observing the FQHE have been the modulation–doped GaAs/$Al_xGa_{1-x}As$ interfaces, as shown in Fig. 14.9. In the fabrication of each device the n–doped regions have been confined to a single atomic layer (i.e., using δ–doping), far from the quantum well to achieve high carrier mobility.

The highest mobility materials that have been reported for modulation–doped MBE samples have been used for the observation of the FQHE. Measurements are made on photolithographically–defined Hall bridges using micro–ampere currents and Ohmic current and potential contacts. Experimental results for the resistivity ρ_{xx} and Hall resistance ρ_{xy} versus magnetic field in the fractional quantum Hall effect regime are shown in Fig. 14.10. Minima develop in the diagonal (in–plane) resistivity ρ_{xx} at magnetic fields corresponding both to integral Landau level filling and to certain fractional fillings of the Landau levels. The Hall resistivity ρ_{xy} develops plateaux at the same integral and fractional filling factors. The classical value of the Hall resistance ρ_{xy} for n carriers per unit area is $\rho_{xy} = B/ne$ and B/n is interpreted as the magnetic flux per carrier which is the flux quantum $\phi_0 = h/e$ divided by the Landau level filling ν, so that $B/n = h/e\nu$ is satisfied at filling factor ν.

The value of ρ_{xy} is $h/\nu e^2$ at filling factor ν. Plateaux were first measured at $\nu = 1/3$ (see Fig. 14.10) with ρ_{xy} equal to $h/\nu e^2$ to within one part in 10^4 in this early work.

In addition to quantization at quantum number 1/3, quantization has been observed at a number of other fractions $\nu = 2/3, 4/3, 5/3, 2/5, 3/5, 4/5, 2/7$ and others (see Fig. 14.11), suggesting that fractional quantization exists in multiple series, with each series based on the inverse of an odd integer. With the highest mobility materials, a fractional quantum Hall effect has recently been observed for an even integer denominator.

Only a certain specified set of fractions exhibit the fractional quantum Hall effect, corresponding to the relation

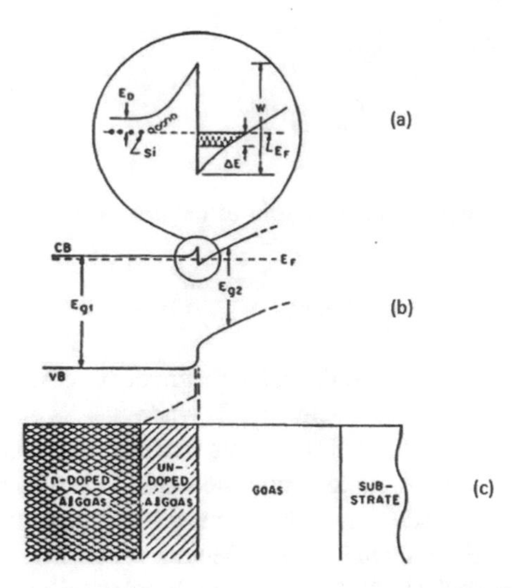

Fig. 14.9 **a** Schematic diagram of a modulation–doped n–type semiconductor GaAs/$Al_xGa_{1-x}As$ heterostructure and of its energy band structure. **b** CB and VB here refer to the conduction band (CB) and valence band (VB) edges; E_{g1} and E_{g2} are, respectively, the energy gaps of the $Al_xGa_{1-x}As$ and GaAs regions, while ΔE in (**a**) is the energy corresponding to the zero–magnetic–field filling of the lowest quantum subband of the two–dimensional electron gas, and E_F is the Fermi energy. W is the step height (band offset energy) between the GaAs conduction band and the $Al_xGa_{1-x}As$ conduction band at the interface. The two-dimensional electron gas in (**a**) lies in the GaAs region close to the undoped $Al_xGa_{1-x}As$ (see **c** showing the lowest diagram). The dopants used to introduce the n-type carriers are located in the region labelled n-doped AlGaAs

$$\nu = \frac{1}{p + \frac{\alpha_1}{p_1 + \frac{\alpha_2}{p_2 + \dots}}} \tag{14.40}$$

where the integers p is odd, p_i is even, and $\alpha_i = 0, \pm 1$. For example, $p = 3$, $p_i = 0$ and $\alpha_i = 0$ for all i yields a fractional filling factor of $1/3$, where the most intense fractional quantum Hall effect is observed. For $p = 3$, $p_1 = 1$ and $\alpha_1 = 1$ and all other coefficients taken to be zero gives $\nu = 2/3$. Equation 14.40 accounts for all the observed examples of the fractional quantum Hall effect except for the case of $\nu = 5/2$ mentioned above.

To explain the characteristics of the fractional quantum Hall effect, Laughlin proposed a many-electron wavefunction to account for the electron correlations responsible for the fractional quantum Hall effect:

$$\psi_m(z_1, z_2, z_3, \dots z_N) = \mathscr{C} \prod_{i<j}^{N} (z_i - z_j)^m \exp\left(-\frac{1}{4} \prod_{k}^{N} |z_k|^2 \right) \tag{14.41}$$

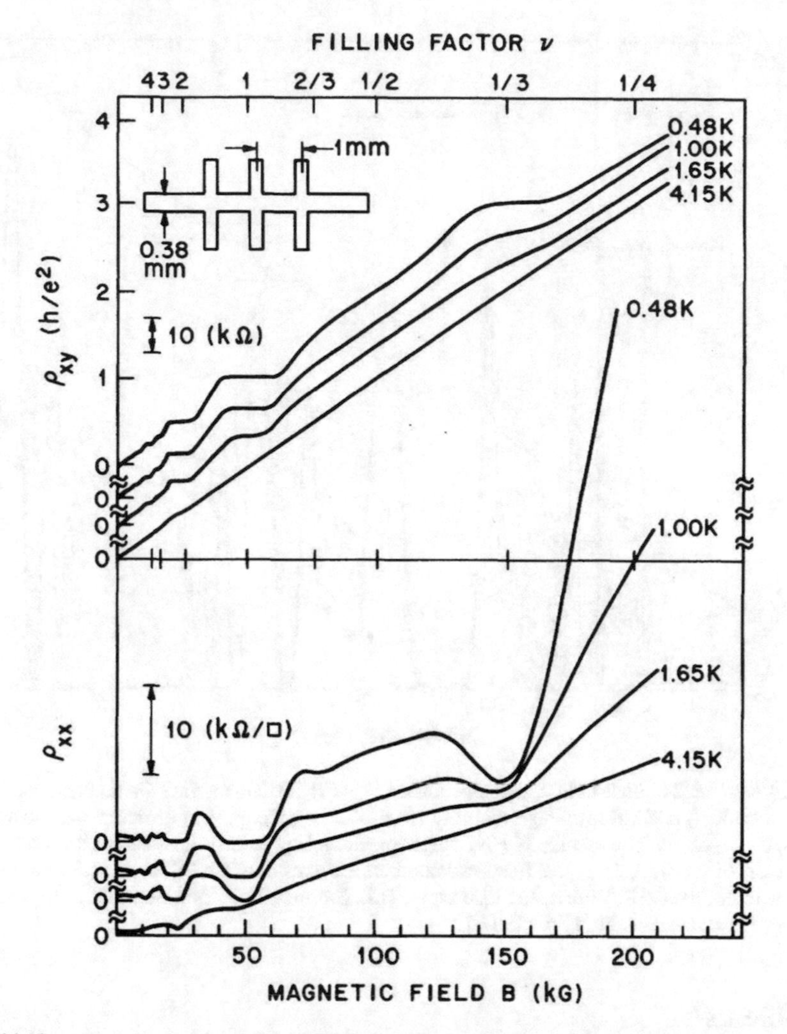

Fig. 14.10 **a** First observation of the FQHE in a GaAs/Al_xGa_{1-x}As modulation–doped heterostructure with an areal carrier density of $n = 1.23 \times 10^{11}$ electrons/cm^2 and an electron mobility of $\mu = 90{,}000\,\text{cm}^2$/Vs. **b** The Hall resistance ρ_{xy} assumes a plateau at fractional filling $\nu = 1/3$ indicating a fractional quantum number $\ell = 1/3$ (see top scale) shown at four low temperatures for both (**a**) and (**b**). The inset in (**b**) shows the geometry of the contacts with a center stage width of 0.38 mm and a separation of 1 mm between contacts. [D.C. Tsui, H.L. Störmer and A.C. Gossard, *Phys. Rev. Lett.* **48**, 1559 (1982)]

where $m = 1/\nu$ and ν is the filling factor. Research at the fundamental level is still on–going to gain further understanding of the fractional quantum Hall effect and related phenomena.

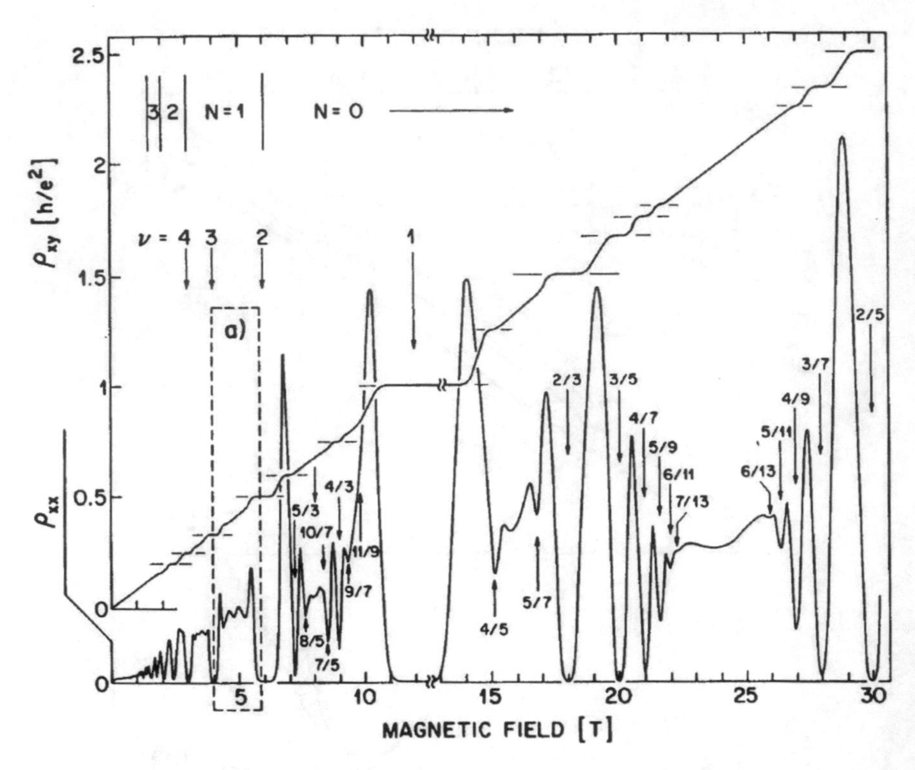

Fig. 14.11 Detailed high-field, low-temperature ($T \sim 0.1$ K) data on the FQHE (fractional quantum Hall effect) taken from a high mobility ($\mu \sim 1.3 \times 10^6\,\text{cm}^2/\text{V s}$) quantum well sample of $\text{GaAs}/\text{Ga}_{1-x}\text{Al}_x\text{As}$. The familiar IQHE (integer quantum Hall effect) characteristics appear at filling factors of $\nu = 1, 2, 3, \ldots$. All fractional numbers are a result of the FQHE. Fractions as high as 7/13 were observed. [R. Willett, J.P. Eisenstein, H.L. Störmer, D.C. Tsui, A.C. Gossard, and J.H. English *Phys. Rev. Lett.* **59**, 1776 (1987)]

Problems

14.1 (a) Assume that the conductivity tensor $\overleftrightarrow{\sigma}$ is an off–diagonal matrix,

$$\overleftrightarrow{\sigma} = \begin{bmatrix} 0 & -ie^2/h \\ ie^2/h & 0 \end{bmatrix}$$

Show that the measured Hall resistances R_H and longitudinal resistances R_x are independent of sample geometry.

(b) In order to observe the Quantum Hall Effect, why is it necessary for the electron gas be two dimensional?

14.2 In the integral quantum Hall effect, the electron density n of a 2D free electron gas is given by

$$n = i\frac{eB}{hc} \tag{14.42}$$

where i is an integer denoting the number of occupied magnetic energy levels, and the electron density n depends only on the magnetic field B.

(a) How do you reconcile this quantization of n with the 2D electron density measured in zero magnetic field?

(b) Suppose now that we have a 2D semiconductor with 4-fold symmetry having elliptical constant energy surfaces centered at the Brillouin zone boundary, where the E(**k**) relation for pocket #1 is written as

$$E(\mathbf{k}) = \frac{\hbar^2 k_x^2}{2m_{xx}} + \frac{\hbar^2 k_y^2}{2m_{yy}}, \tag{14.43}$$

in which $m_{xx} = m_0$ and $m_{yy} = 0.1m_0$. Starting with n_0 electrons/cm^2 at zero magnetic field, find the magnetic field B$_c$ (for $\mathbf{B} \parallel z$-axis) at which all the carriers in pocket #1 are transferred to carrier pocket #2.

(c) What is ρ_{xx} and ρ_{xy} for this value of magnetic field B$_c$?

(d) How do you reconcile the results in part (b) with the results in part (a)?

14.3 Suppose that you have a modulation doped (n–type) quantum well structure composed of layers of GaAs/Ga$_{1-x}$Al$_x$As such that the bulk carrier density in the GaAs is 10^{16}/cm^3 and the width of the quantum well is 80Å. (Use $m_e^* = 0.07m_0$, $E_{g1} = 1.42$ eV for GaAs, and $E_{g2} = 1.70\,eV$ for Ga$_{1-x}$Al$_x$As and $\Delta E_c = 3\Delta E_v$ for the band offsets). For simplicity in calculating the energy levels, use the energy eigenvalues of the infinite well.

(a) What is the quantum well widths range so that two bound states are contained in the quantum well at zero magnetic field. How many Landau levels are occupied at a field of 10 Tesla applied normal to the two dimensional electron gas? What is the fractional occupation of the last Landau level? The fractional occupation refers to the number of occupied states in the Landau level compared to the total number of states in the Landau level obtained from the degeneracy factor.

(b) Give design parameters for a quantum well that has only 1 bound state level, and this level has a filling factor or fractional occupation of 1/3.

Suggested Reading

R.E. Prange, S.M. Girvin, *The Quantum Hall Effect* (Springer, Berlin, 1987)

Part III
Optical Properties

Chapter 15
Review of Fundamental Relations for Optical Phenomena

15.1 Introductory Remarks on Optical Probes

The optical properties of solids provide an important tool for studying energy band structure, impurity levels, excitons, localized defects, lattice vibrations, and certain magnetic excitations. In such experiments, we measure some observable, such as reflectivity, transmission, absorption, ellipsometry or light scattering, and from these measurements we deduce the dielectric function $\varepsilon(\omega)$, the optical conductivity $\sigma(\omega)$, or the fundamental excitation frequencies. It is the frequency-dependent complex dielectric function $\varepsilon(\omega)$ or the complex conductivity $\sigma(\omega)$, which is directly related to the electronic energy band structure of solids.

The central question is the relationship between experimental observations and the electronic energy levels (energy bands) of the solid. In the infrared photon energy region, information on the phonon branches and on the electron-phonon interaction is obtained. These issues are the major concern of Part III of this book.

15.2 The Complex Dielectric Function and the Complex Optical Conductivity

Assuming no charge density in the absence of incident light, the complex dielectric function and complex optical conductivity are introduced through Maxwell's equations (in standard international (SI) units)

$$\nabla \times \mathbf{H} - \frac{\partial \mathbf{D}}{\partial t} = \mathbf{j} \tag{15.1}$$

$$\nabla \times \mathbf{E} + \frac{\partial \mathbf{B}}{\partial t} = 0 \tag{15.2}$$

M. Dresselhaus et al., *Solid State Properties*, Graduate Texts in Physics,
https://doi.org/10.1007/978-3-662-55922-2_15

$$\nabla \cdot \mathbf{D} = 0 \tag{15.3}$$

$$\nabla \cdot \mathbf{B} = 0 \tag{15.4}$$

in which **H**, **E** and **j** are the magnetic field, electric field and current density, while **D** and **B** are the electric and magnetic field in a medium. Modifying the magnitudes of the variables **E** and **H** thereby provides a probe of the properties of specific materials and their dependance on external variables, like temperature and pressure.

The constitutive equations relating **D**, **B** and **j** to **E** and **H** field are written as:

$$\mathbf{D} = \varepsilon \mathbf{E} \tag{15.5}$$

$$\mathbf{B} = \mu \mathbf{H} \tag{15.6}$$

$$\mathbf{j} = \sigma \mathbf{E} \tag{15.7}$$

Equations (15.5), (15.6) and (15.7) respectively, define the quantities ε, μ and σ from which the concepts of the complex dielectric function ε, the complex magnetic permeability μ, and the complex electrical conductivity σ, are defined and will be developed further in this chapter. When we discuss non–linear optics, these linear constitutive equations (15.5)–(15.7) must be generalized to include higher order terms in **EE** and **EEE**.

From Maxwell's equations and the constitutive equations, we obtain a wave equation for the variables **E** and **H** for electric and magnetic fields, respectively:

$$\nabla^2 \mathbf{E} = \varepsilon\mu \frac{\partial^2 \mathbf{E}}{\partial t^2} + \sigma\mu \frac{\partial \mathbf{E}}{\partial t} \tag{15.8}$$

and

$$\nabla^2 \mathbf{H} = \varepsilon\mu \frac{\partial^2 \mathbf{H}}{\partial t^2} + \sigma\mu \frac{\partial \mathbf{H}}{\partial t}. \tag{15.9}$$

For optical fields, we must look for a sinusoidal solution to (15.8) and (15.9)

$$\mathbf{E} = \mathbf{E}_0 e^{i(\mathbf{K}\cdot\mathbf{r} - \omega t)} \tag{15.10}$$

where **K** is a complex propagation constant, ω is the frequency of the light, and i is the imaginary unit. A solution similar to (15.10) is obtained for the **H** field. The real part of **K** can be identified as a wave vector, while the imaginary part of **K** accounts for the attenuation of the wave inside the solid material and corresponds to energy dissipation, thereby increasing the local temperature. Substitution of the plane wave solution (15.10) into the wave equation (15.8) yields the following relation for K:

$$K^2 = \varepsilon\mu\omega^2 + i\sigma\mu\omega. \tag{15.11}$$

If there were no losses in energy propagation, K would be equal to

$$K_0 = \omega\sqrt{\varepsilon\mu} \tag{15.12}$$

and would be a real number which has units of reciprocal length and K is thus identified as a wave vector. However, since there are losses in a conducting medium, we write

$$K = \omega\sqrt{\varepsilon_{\text{complex}}\mu} \tag{15.13}$$

where we have defined the complex dielectric function as

$$\varepsilon_{\text{complex}} = \varepsilon + \frac{i\sigma}{\omega} = \varepsilon_1 + i\varepsilon_2. \tag{15.14}$$

As shown in (15.14) it is customary to write ε_1 and ε_2 for the real and imaginary parts of $\varepsilon_{\text{complex}}$. From the definition in (15.14), it also follows that

$$\varepsilon_{\text{complex}} = \frac{i}{\omega}\left[\sigma + \frac{\varepsilon\omega}{i}\right] = \frac{i}{\omega}\sigma_{\text{complex}}, \tag{15.15}$$

where we define the complex conductivity σ_{complex} as:

$$\sigma_{\text{complex}} = \sigma + \frac{\varepsilon\omega}{i} = \sigma - i\varepsilon\omega \tag{15.16}$$

showing the phase relation between $\varepsilon_{\text{complex}}$ and σ_{complex}.

Now that we have defined the complex dielectric function $\varepsilon_{\text{complex}}$ and the complex conductivity σ_{complex}, these quantities are related to measurable materials properties, that are important for optical device applications:

1. observables such as the photon reflectivity and absorption which are measured in the laboratory;
2. properties of the solid material such as the carrier density, relaxation time, effective masses, energy band gaps, etc., which link materials properties to application opportunities.

15.2.1 Propagating Waves

Let us consider a wave propagating along the z direction. After substituting K in (15.10), the solution (15.11) to the wave equation (15.8) yields a plane wave

$$\mathbf{E}(z,t) = \mathbf{E}_0 e^{-i\omega t} e^{\left(i\omega z\sqrt{\varepsilon\mu}\sqrt{1+\frac{i\sigma}{\varepsilon\omega}}\right)}. \tag{15.17}$$

For the wave propagating in vacuum ($\varepsilon = \varepsilon_0$, $\mu = \mu_0$, $\sigma = 0$), (15.17) is reduced to a simple plane wave solution

$$\mathbf{E}(z,t) = \mathbf{E}_0 e^{i(Kz-\omega t)} \tag{15.18}$$

where $K = K_0 = \omega\sqrt{\varepsilon_0\mu_0}$ and $c = 1/\sqrt{\varepsilon_0\mu_0}$ is the speed of the eletromagnetic wave and the wave propagates in vacuum with light velocity c.

If the wave is propagating in a medium of finite electrical conductivity, the (15.17) can be written as

$$\mathbf{E}(z,t) = \mathbf{E}_0 e^{-\frac{\omega}{c}\tilde{k}z} e^{i(\frac{\omega}{c}\tilde{n}z-\omega t)} \tag{15.19}$$

where $\tilde{n}$ and $\tilde{k}$ are, respectively, the real and imaginary parts of the complex index of refraction defined as

$$\tilde{N}_{\text{complex}}(\omega) = \sqrt{\mu\varepsilon_{\text{complex}}} = \sqrt{\varepsilon\mu\left(1+\frac{i\sigma}{\varepsilon\omega}\right)} = \tilde{n}(\omega) + i\tilde{k}(\omega), \tag{15.20}$$

where $\tilde{k}$ is also called the extinction coefficient and has units of time/length. The *intensity* of the electric field, $|E|^2 = E_0^2 e^{-2\omega\tilde{k}z}$, decays as the wave propagates through the material. When the electric field intensity falls off to $1/e$ (where $e = 2.718$) of its value at the surface (let us set $z = 0$ at the surface), the wave will have traveled over a characteristic distance δ called the *electromagnetic skin depth* whose value is obtained from the condition

$$\frac{|E(0)|^2}{|E(\delta)|^2} = e = \frac{E_0^2}{E_0^2 e^{-2\frac{\omega}{c}\tilde{k}\delta}}, \tag{15.21}$$

so that

$$\delta = \frac{c}{2\omega\tilde{k}} = \frac{1}{\alpha_{\text{abs}}} \tag{15.22}$$

where $\alpha_{\text{abs}}(\omega)$ is the absorption coefficient for the solid at frequency ω, thereby defining both the electromagnetic skin depth δ and the absorption coefficient α_{abs}.

Since light is described by a transverse wave, there are two possible orthogonal directions for the **E** vector in a plane normal to the propagation direction and these directions determine the *polarization* of the light. For cubic materials, the index of refraction is the same along the two transverse directions. However, for anisotropic media, the index of refraction will be different for the two polarization directions.

15.3 Relation of the Complex Dielectric Function to Observables

In the previous Sect. 15.2.1, we introduced the complex index of refraction $\tilde{N}_{\text{complex}}$, which can be expressed in terms of the complex dielectric function $\varepsilon_{\text{complex}}$

$$\tilde{N}_{\text{complex}} = c\sqrt{\mu\varepsilon_{\text{complex}}} \tag{15.23}$$

where

$$K = \frac{\omega}{c}\tilde{N}_{\text{complex}} \tag{15.24}$$

Here, the wave vector K has the units of reciprocal length, and $\tilde{N}_{\text{complex}}$ is usually written in terms of its real and imaginary parts (see (15.20))

$$\tilde{N}_{\text{complex}} = \tilde{n} + i\tilde{k} = \tilde{N}_1 + i\tilde{N}_2. \tag{15.25}$$

The quantities $\tilde{n}$ and $\tilde{k}$ are collectively called the optical constants of the solid, where $\tilde{n}$ denotes the index of refraction and $\tilde{k}$ is the extinction coefficient. (We use the tilde over the optical constants $\tilde{n}$ and $\tilde{k}$ to distinguish them from the carrier density and wave vector which are denoted by n and k). The extinction coefficient $\tilde{k}$ vanishes for lossless materials. For non-magnetic materials, we can take $\mu = \mu_0$, and the approximation of a non-magnetic material will be made in writing the equations (15.26), (15.27) and (15.28) below. A generalization of $\mu > \mu_0$ yields a class of materials with interesting magnetic properties, and the special materials which allow μ and ε to have negative values are called metamaterials.

With this definition for $\tilde{N}_{\text{complex}}$ in (15.25), we can relate

$$\varepsilon_{\text{complex}} = \varepsilon_1 + i\varepsilon_2 = \varepsilon_o(\tilde{n} + i\tilde{k})^2 \tag{15.26}$$

yielding the important relations

$$\varepsilon_1 = \varepsilon_o(\tilde{n}^2 - \tilde{k}^2) \tag{15.27}$$

$$\varepsilon_2 = \varepsilon_o(2\tilde{n}\tilde{k}) \tag{15.28}$$

where we note that $\varepsilon_1, \varepsilon_2, \tilde{n}$ and $\tilde{k}$ are all frequency dependent, as well as being temperature and pressure dependent, etc.

Many important measurements of the optical properties of solids involve the simplest case of normal incidence reflectivity, which is illustrated in Fig. 15.1. Inside the solid, the electromagnetic wave will be attenuated relative to the incident value of E_0. We assume for the present discussion that the sample under consideration is thick enough so that reflections from the back surface can be neglected. We can then write the wave inside the solid for this one-dimensional propagation problem as

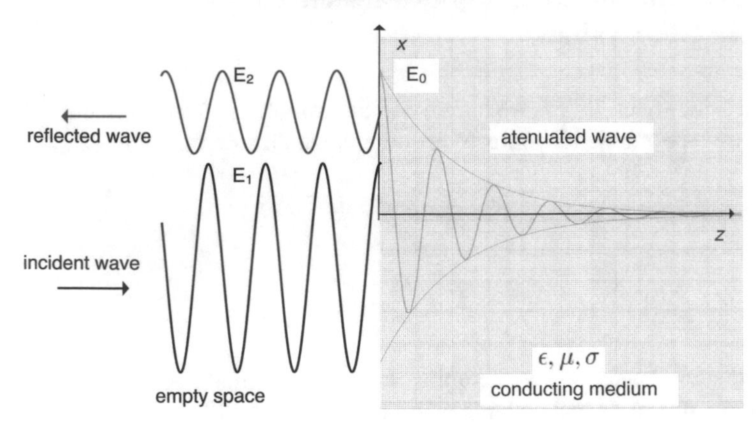

Fig. 15.1 Schematic diagram of normal incidence reflectivity

$$E_x = E_0 e^{i(Kz - \omega t)} \tag{15.29}$$

where x is the polarization direction and z is the direction of light propagation. Then the complex propagation constant for the light is given by $K = (\omega/c)\ \tilde{N}_{\text{complex}}$.

On the other hand, we know that in free space we have both an incident wave E_1 and a reflected wave E_2:

$$E_x = E_1 e^{i(\frac{\omega}{c}z - \omega t)} + E_2 e^{i(-\frac{\omega}{c}z - \omega t)} \tag{15.30}$$

in which c is the speed of light, z is the direction of the incident wave, $-z$ is that for the reflected wave, and x is a direction in the plane, parallel to the scattering surface. From (15.25) and (15.30), the continuity of E_x across the surface of the solid requires that

$$E_0 = E_1 + E_2. \tag{15.31}$$

where E_0 is the sum of the amplitude of the incident and reflected waves, which is the first relation between the field amplitudes. (For this discussion, anisotropy of the material in plane is neglected.) The second relation between E_0, E_1, and E_2 follows from the continuity condition for the tangential H_y field across the boundary of the solid, where the y direction is the in-plane direction perpendicular to the x direction, following the right hand rule. From Maxwell's equation (15.2) we have

$$\nabla \times \mathbf{E} = -\mu \frac{\partial \mathbf{H}}{\partial t} = i\mu\omega \mathbf{H} \tag{15.32}$$

which results in

$$\frac{\partial E_x}{\partial z} = i\mu\omega H_y. \tag{15.33}$$

The continuity condition on H_y thus yields a continuity relation for $\partial E_x/\partial z$ so that from (15.33)

$$E_0 K = E_1 \frac{\omega}{c} - E_2 \frac{\omega}{c} = E_0 \frac{\omega}{c} \tilde{N}_{\text{complex}} \tag{15.34}$$

or

$$E_1 - E_2 = E_0 \tilde{N}_{\text{complex}}, \tag{15.35}$$

which is the second equation between E, E_1 and E_2.

The normal incidence reflectivity $\mathscr{R}$ is then written as

$$\mathscr{R} = \left| \frac{E_2}{E_1} \right|^2 \tag{15.36}$$

which is most conveniently related to the reflection coefficient $\tilde{r}$ given by

$$\tilde{r} = \frac{E_2}{E_1}. \tag{15.37}$$

From (15.31) and (15.35), we have the results

$$E_2 = \frac{1}{2} E_0 (1 - \tilde{N}_{\text{complex}}) \tag{15.38}$$

$$E_1 = \frac{1}{2} E_0 (1 + \tilde{N}_{\text{complex}}) \tag{15.39}$$

so that the normal incidence reflectivity becomes

$$\mathscr{R} = \left| \frac{1 - \tilde{N}_{\text{complex}}}{1 + \tilde{N}_{\text{complex}}} \right|^2 = \frac{(1 - \tilde{n})^2 + \tilde{k}^2}{(1 + \tilde{n})^2 + \tilde{k}^2} \tag{15.40}$$

and the reflection coefficient for the wave itself is given by

$$\tilde{r} = \frac{1 - \tilde{n} - i\tilde{k}}{1 + \tilde{n} + i\tilde{k}} \tag{15.41}$$

where the reflectivity $\mathscr{R}$ is a number less than unity and $\tilde{r}$ has an amplitude of less than unity. We have now related one of the physical observables $\tilde{r}$ to the optical constants $\tilde{n}$ and $\tilde{k}$.

To relate these results to the power absorbed and the power transmitted at normal incidence, we utilize the following relation which expresses the idea that all the incident power is either reflected, absorbed, or transmitted

$$1 = \mathscr{R} + \mathscr{A} + \mathscr{T} \tag{15.42}$$

where $\mathscr{R}$, $\mathscr{A}$, and $\mathscr{T}$ are, respectively, the fraction of the power that is reflected, absorbed, and transmitted as illustrated in Fig. 15.1. At high temperatures, the most common observable is the emissivity, which is equal to the absorbed power for a black body or is equal to $1 - \mathscr{R}$ assuming $\mathscr{T}=0$. As an exercise, it is instructive to derive expressions for $\mathscr{R}$ and $\mathscr{T}$ when we have relaxed the restriction of no reflection from the back surface. Multiple reflections are encountered in optically thin films.

The discussion thus far has been directed toward relating the complex dielectric function or the complex conductivity to physical observables. If we know the optical constants, then we can find the reflectivity. We now want to ask the opposite question. Suppose we know the reflectivity, can we find the optical constants? Since there are two optical constants, $\tilde{n}$ and $\tilde{k}$, we need to make two independent measurements, such as the reflectivity at two different angles of incidence.

Nevertheless, even if we limit ourselves to normal incidence reflectivity measurements, we can still obtain both $\tilde{n}$ and $\tilde{k}$ provided that we make these reflectivity measurements for all frequencies. This is possible because, from a mathematical standpoint, the real and imaginary parts of a complex physical function are not independent. Because of causality, $\tilde{n}(\omega)$ and $\tilde{k}(\omega)$ are related to each other through the Kramers–Kronig relation, which we will discuss in Chap. 19. Since normal incidence measurements are easier to carry out in practice, it is quite possible to study the optical properties of solids with just normal incidence measurements, and then to do a Kramers–Kronig analysis of the reflectivity data to obtain the frequency–dependent dielectric functions $\varepsilon_1(\omega)$ and $\varepsilon_2(\omega)$ or the frequency–dependent optical constants $\tilde{n}(\omega)$ and $\tilde{k}(\omega)$.

In treating a solid, we will need to consider contributions to the optical properties from various electronic energy band processes, which can occur over a large frequency range. To begin with, there are **intraband processes** which correspond to the electronic conduction by free carriers, and hence are more important in electrically conducting materials, such as metals, semimetals, and degenerate semiconductors. These intraband processes can be understood in their simplest terms by the classical Drude theory (see Chap. 16), or in more detail by solving the classical Boltzmann equation or by using the quantum mechanical density matrix technique. In addition to the intraband (free carrier) processes, there are **interband processes** (see Chap. 17) which correspond to the absorption of electromagnetic radiation by an electron in an occupied state below the Fermi level, thereby inducing a transition to an unoccupied state in a higher band. This interband process is intrinsically a quantum mechanical process and must be discussed in terms of quantum mechanical concepts.

In practice, we consider in detail the contribution of only a few energy bands, namely the most important or dominant energy bands, to optical properties. In many cases we also restrict ourselves to the detailed consideration of only a portion of the Brillouin zone such as the band edges where the density of states is high and where strong interband transitions occur, especially in an energy range of particular interest to a given experiment. The **intraband** and **interband** contributions that are neglected in the present discussion are considered in Chap. 17 in an approximate way by introducing a core dielectric $\varepsilon_{\text{core}}$ constant, which is taken in the simplest approximation to be independent of frequency and external parameters, such as temperature, pressure, strain, etc.

15.4 Units for Frequency Measurements

The frequency of light is measured in several different units in the literature and here we list some of them which are particularly useful. The relation between the various units are: $1\,\text{eV} = 8065.5\,\text{cm}^{-1} = 2.418 \times 10^{14}\,\text{Hz} = 11{,}600\,\text{K}$. Also 1 eV corresponds to a wavelength of $1.2398\,\mu\text{m}$, and $1\,\text{cm}^{-1} = 0.12398\,\text{meV} = 3 \times 10^{10}\,\text{Hz}$.

Problems

15.1 Optical constants and attenuated field amplitudes.

(a) Starting from (15.11) show that the optical constants $\tilde{n}$ and $\tilde{k}$ (real ($\text{Re}(\tilde{N}_{\text{complex}})$) and imaginary ($\text{Im}(\tilde{N}_{\text{complex}})$) part of the complex propagation constant) are given by $\tilde{n} = c\sqrt{\frac{\varepsilon\mu}{2}}\left[1+\sqrt{1+\left(\frac{\sigma}{\varepsilon\omega}\right)^2}\right]^{1/2}$ and $\tilde{k} = c\sqrt{\frac{\varepsilon\mu}{2}}\left[-1+\sqrt{1+\left(\frac{\sigma}{\varepsilon\omega}\right)^2}\right]^{1/2}$.

(b) Consider a plane wave whose electric field **E** is polarized along x and propagating along z is given by $\mathbf{E}(z,t) = E_0 e^{-\frac{\omega}{c}\tilde{k}z} e^{i(\frac{\omega}{c}\tilde{n}z-\omega t)}\hat{\mathbf{i}}$. Use Maxwell's equations to demonstrate that the respective magnetic field is given by $\mathbf{H}(z,t) = \frac{1}{c}\frac{\tilde{N}}{\mu}E_0 e^{-\frac{\omega}{c}\tilde{k}z} e^{i(\frac{\omega}{c}\tilde{n}z-\omega t)}\hat{\mathbf{j}}$.

(c) Show that the ratio between the amplitudes of the fields is given by $\frac{H_0}{E_0} = \sqrt{\varepsilon/\mu\sqrt{1+\left(\frac{\sigma}{\varepsilon\omega}\right)^2}}$.

(d) Plot the fields $\mathbf{E}(z,t)$ and $\mathbf{H}(z,t)$ inside the solid material.

15.2 (a) Show that for a very good conductor the phase difference between the magnetic field **H** and the electric field **E** is $\pi/4$. (Hint: Use the result obtained in the previous problem and write the complex quantity $\tilde{N} = |\tilde{N}|e^{i\phi}$.)

(b) Verify the result obtained in (a) for a metal with a conductivity $\sigma \approx 10^7\,\Omega\text{m}^{-1}$.

15.3 (a) Show that the time averaged electromagnetic energy density of a plane wave in a conducting medium is given by $(K^2/2\mu\omega^2)E_0^2 e^{-2\frac{\omega}{c}\tilde{k}z}$.

(b) Show that the magnetic contribution to the electromagnetic energy density is given by $(\tilde{n}^2/c^2\mu)E_0^2 e^{-2\frac{\omega}{c}\tilde{k}z}$.

(c) Plot the result derived in (a) and (b) as a function of z and ω thereby showing that the magnetic contribution always dominates the energy in this situation.

15.4 (a) Show that for insulating materials, i.e., $\sigma \ll \omega\varepsilon$, the skin depth is independent of frequency and given by $\delta = \frac{c}{2\omega\tilde{k}} = \frac{1}{\sigma}\sqrt{\frac{\varepsilon}{\mu}}$.

(b) Estimate the skin depth in diamond and water, and give a numerical value in each case.
(c) What is the difference in absorption at a depth of 1000 m in the ocean and in an inland lake.

15.5 When a sample consists of more than one material and is non-homogeneous, the optical properties are modified in a non-trivial manner. One approach to approximate the optical response of a heterogeneous material in terms of its microstructure is by using an effective medium theory (EMT). EMT relates the dielectric function of a composite material with the dielectric function of the constituents materials, where the complex dielectric function, $\varepsilon_1 + i\varepsilon_2$, is that used in Maxwell's equations and is defined as

$$D = \varepsilon E = E + 4\pi P.$$

The dielectric function of a composite material can be easily solved for two situations given by (a) and (b) below.

(a) Show that if the internal boundaries are parallel to the applied electric field, the situation is analogous to capacitors in parallel, and the effective dielectric function is related to the dielectric function of the composites by the following equation:

$$\varepsilon_{composite} = \sum_j f_j \varepsilon_j$$

where ε_j and f_j are the complex dielectric function and volume fraction of each constituent material, j.

(b) Show that in the opposite limit, i.e., the boundaries are perpendicular to the applied electric field, the situation is analogous to capacitors in series and the composite's dielectric function is given by:

$$\varepsilon_{composite}^{-1} = \sum_j f_j \varepsilon_j^{-1}$$

These two cases define the absolute bounds to ε. The dielectric function of all composite materials lie on or within the region defined in the complex plane of ε.

15.6 (a) Derive a formula for the normal incidence reflectivity for a thin film of thickness t and the optical constants $\tilde{n}$ and $\tilde{k}$. Assume that $\tilde{n} \gg \tilde{k}$ and t is within a factor of 2 of the wavelength of light λ.
(b) Consider explicitly the case of light from a CO_2 laser ($\lambda = 10.6\,\mu$m) and a sample thickness $t = 5\,\mu$m and $t = 20\,\mu$m.
(c) Suppose that you have a superlattice of alternating thin films of dielectric constants ε_1 and ε_2, and thickness t_1 and t_2 respectively. Find the normal incidence reflectivity, neglecting optical losses (i.e., take the optical constants $k_1 = k_2 = 0$).

Suggested Readings

Bassani, Pastori-Parravicini, *Electronic States and Optical Transitions in Solids* (Pergamon Press, New York, 1975)
G. Bekefi, A.H. Barrett, *Electromagnetic Vibrations Waves and Radiation* (MIT Press, Cambridge, 1977)
J.D. Jackson, *Classical Electrodynamics* (Wiley, New York, 1975)
Yu, Cardona, *Fundamentals of Semiconductors* (Springer, Berlin, 1996)

Chapter 16
Drude Theory–Free Carrier Contribution to the Optical Properties

16.1 The Free Carrier Contribution

In this chapter, we relate the optical constants to the electronic properties of the solid material, by considering the real and imaginary parts of the dielectric function $\varepsilon = \varepsilon_1 + i\varepsilon_2$, the frequency dependence of ε_1 and ε_2, and considering both classical behavior, collective behavior, while providing background for studying the overall optical properties of solid state materials. One major contribution to the dielectric function that was introduced in Chap. 15 is through the "free carriers". Such free carrier contributions are very important in semiconductors and metals, and the main effect of free carriers can be understood in terms of a simple classical conductivity model, called the Drude model. This model is based on the classical equations of motion of an electron in an optical electric field, and gives the simplest theory for the so called optical constants. The classical equation of motion for the drift velocity $\boldsymbol{v}$ of the free carrier is given by

$$m\frac{d\boldsymbol{v}}{dt} + \frac{m\boldsymbol{v}}{\tau} = e\boldsymbol{E}_0 e^{-i\omega t} \tag{16.1}$$

where the relaxation time τ is introduced to provide a damping or dissipative term, $(m\boldsymbol{v}/\tau)$, and a sinusoidally time-dependent electric field $\boldsymbol{E}_0 e^{-i\omega t}$ provides the driving force. To respond to a sinusoidal applied field, the electrons undergo a sinusoidal motion which can be described as

$$\boldsymbol{v} = \boldsymbol{v}_0 e^{-i\omega t} \tag{16.2}$$

so that (16.1) becomes

$$(-mi\omega + \frac{m}{\tau})\boldsymbol{v}_0 = e\boldsymbol{E}_0 \tag{16.3}$$

and the amplitudes $\boldsymbol{v}_0$ and $\boldsymbol{E}_0$ are thereby related. The current density $\boldsymbol{j}$ is related to the drift velocity $\boldsymbol{v}_0$ and to the carrier density n by

M. Dresselhaus et al., *Solid State Properties*, Graduate Texts in Physics,
https://doi.org/10.1007/978-3-662-55922-2_16

$$\boldsymbol{j} = ne\boldsymbol{v}_0 = \sigma \boldsymbol{E}_0, \tag{16.4}$$

thereby introducing the electrical conductivity σ. Substitution for the drift velocity v_0 into (16.4) yields

$$\boldsymbol{v}_0 = \frac{e\boldsymbol{E}_0}{(m/\tau) - im\omega} \tag{16.5}$$

as well as the complex electrical conductivity

$$\sigma = \frac{ne^2\tau}{m(1 - i\omega\tau)}. \tag{16.6}$$

In writing σ in the Drude expression (16.6) for the free carrier conduction, we have suppressed the subscript in σ_{complex}, as is conventionally done in the literature. In what follows we will always write σ and ε to denote the complex conductivity and complex dielectric constant and suppress subscripts "complex" in order to simplify the notation. A more elegant derivation of the Drude expression can be made from the Boltzmann formulation, as is done in Part II under the heading of transport properties of solids.

In a real solid material, the same result as given above follows when the effective mass approximation can be used. Following the results for the dc conductivity obtained in Part II, an electric field applied in one direction can produce a force in another direction because of the anisotropy of the constant energy surfaces in actual solid materials. Because of the anisotropy of the effective mass tensor in solids, $\boldsymbol{j}$ and $\boldsymbol{E}$ are related by the tensorial relation,

$$j_\alpha = \sigma_{\alpha\beta} E_\beta \tag{16.7}$$

thereby defining the conductivity tensor $\sigma_{\alpha\beta}$ as a second rank tensor. For perfectly free electrons in an isotropic (or in a cubic) medium, the conductivity tensor is written as:

$$\overleftrightarrow{\sigma} = \begin{pmatrix} \sigma & 0 & 0 \\ 0 & \sigma & 0 \\ 0 & 0 & \sigma \end{pmatrix} \tag{16.8}$$

and we have our usual simple scalar expression $\boldsymbol{j} = \sigma \boldsymbol{E}$. However, in a solid material, $\sigma_{\alpha\beta}$ can have off-diagonal terms, because the effective mass tensors are related to the curvature of the energy bands $E(\boldsymbol{k})$ by the relation

$$\left(\frac{1}{m}\right)_{\alpha\beta} = \frac{1}{\hbar^2}\frac{\partial^2 E(\boldsymbol{k})}{\partial k_\alpha \partial k_\beta}. \tag{16.9}$$

The tensorial properties of the conductivity follow directly from the dependence of the conductivity on the reciprocal effective mass tensor.

As an example, semiconductors such as CdS and ZnO exhibit the wurtzite structure, which is a non-cubic structure. These semiconductors are *uniaxial* and contain an *optic axis* (which for the wurtzite structure is usually taken to be along the c-axis), along which the velocity of propagation of light is independent of the polarization direction. Along other directions of propagation, the velocity of light is different for the two polarization directions, giving rise to a phenomenon called *birefringence*, which is studied conveniently by optical techniques. Crystals with tetragonal or hexagonal symmetry are uniaxial. Crystals with lower symmetry can have two axes along which the light propagates at the same velocity for the two polarizations of light (but the actual velocities will be different from each other), and these crystals are therefore called *biaxial crystals*.

The constant energy surfaces for a large number of the common semiconductors have a small carrier concentration, so that these carrier pockets in reciprocal space can be described by ellipsoids as constant energy surfaces and the effective masses of the carriers in these simple ellipsoidal carrier pockets are given by an effective mass tensor $m_{\alpha\beta}$ which can be obtained from (16.9). It is a general result that for cubic materials (in the absence of externally applied stresses and magnetic fields), the conductivity for all electrons and all the holes is described by a single scalar quantity σ. To describe conduction processes in hexagonal materials, we need to introduce two constants: $\sigma_{\|}$ for conduction along the high symmetry axis and $\sigma_{\perp}$ for conduction in the basal plane perpendicular to the high symmetry axis. These results can be directly demonstrated by summing the contributions to the conductivity from all carrier pockets.

In narrow gap semiconductors the effective mass tensor, $m_{\alpha\beta}$, is itself a function of energy. If this is the case, the Drude formula is valid when $m_{\alpha\beta}$ is evaluated at the Fermi level and n is the total carrier density. Suppose now that the only conduction mechanism that we are treating in detail is the free carrier mechanism. Then we would consider all other contributions to σ in terms of the core dielectric constant $\varepsilon_{\text{core}}$ to obtain for the total complex dielectric function

$$\varepsilon(\omega) = \varepsilon_{\text{core}}(\omega) + i\sigma/\omega \tag{16.10}$$

where

$$\sigma(\omega) = \left(ne^2\tau/m^*\right)(1 - i\omega\tau)^{-1} \tag{16.11}$$

in which σ/ω denotes the imaginary part of the free carrier contribution to the complex dielectric function of (16.10). If there were no free carrier absorption, $\sigma = 0$ and $\varepsilon = \varepsilon_{\text{core}}$, and, $m^* \neq m$ so that in empty space $\varepsilon = \varepsilon_{\text{core}} = \varepsilon_{\text{o}}$. Substituting (16.11) into 16.10 gives

$$\varepsilon = \varepsilon_{\text{core}} + \frac{i}{\omega}\frac{ne^2\tau}{m(1 - i\omega\tau)} = (\varepsilon_1 + i\varepsilon_2) = \varepsilon_{\text{o}}(\tilde{n} + i\tilde{k})^2. \tag{16.12}$$

It is of interest to consider the expression in (16.12) in two limiting cases: low and high frequencies, as described in Sects. 16.2 and 16.3.

16.2 Low Frequency Response: $\omega\tau \ll 1$

In the low frequency regime ($\omega\tau \ll 1$), we obtain from (16.12)

$$\varepsilon \simeq \varepsilon_{\text{core}} + \frac{ine^2\tau}{m\omega}. \tag{16.13}$$

Since the free carrier term in (16.13) shows a $1/\omega$ dependence, as $\omega \to 0$, this term dominates in the low frequency limit. The core dielectric constant $\varepsilon_{\text{core}}$ in (16.13) is typically 16 for geranium, 12 for silicon and perhaps 100 or more, for narrow gap semiconductors like PbTe. It is also of interest to note that the core contribution $\varepsilon_{\text{core}}$ and free carrier contribution $ine^2\tau/m\omega$ are 90° out of phase in the complex plane.

To find the optical constants $\tilde{n}$ and $\tilde{k}$, we need to take the square root of ε in (16.12). Since we will see below that $\tilde{n}$ and $\tilde{k}$ are both large in the low frequency regime, we can for the moment ignore the core contribution $\varepsilon_{\text{core}}$ in (16.13) and we then obtain:

$$\sqrt{\varepsilon} \simeq \sqrt{\frac{ne^2\tau}{m\omega}}\sqrt{i} = \sqrt{\varepsilon_o}(\tilde{n} + i\tilde{k}) \tag{16.14}$$

and using the identity

$$\sqrt{i} = e^{\frac{\pi i}{4}} = \frac{1+i}{\sqrt{2}} \tag{16.15}$$

we obtain

$$\tilde{n} = \tilde{k} = \sqrt{\frac{ne^2\tau}{2\varepsilon_o m\omega}}. \tag{16.16}$$

We see that in the low frequency limit $\tilde{n} \approx \tilde{k}$, and that $\tilde{n}$ and $\tilde{k}$ are both large. Therefore the normal incidence reflectivity can be simply written as

$$\mathscr{R} = \frac{(\tilde{n}-1)^2) + \tilde{k}^2}{(\tilde{n}+1)^2) + \tilde{k}^2} \simeq \frac{\tilde{n}^2 + \tilde{k}^2 - 2\tilde{n}}{\tilde{n}^2 + \tilde{k}^2 + 2\tilde{n}} = 1 - \frac{4\tilde{n}}{\tilde{n}^2 + \tilde{k}^2} \simeq 1 - \frac{2}{\tilde{n}}. \tag{16.17}$$

Thus, the Drude theory shows that at low frequencies a material with a large concentration of free carriers (e.g., a metal) is almost a perfect reflector.

16.3 High Frequency Response: $\omega\tau \gg 1$

In the $\omega\tau \gg 1$ limit, (16.12) can be approximated by:

$$\varepsilon \simeq \varepsilon_{\text{core}} - \frac{ne^2}{m\omega^2}. \tag{16.18}$$

As the frequency becomes large, the $1/\omega^2$ dependence of the free carrier contribution guarantees that free carrier effects will become less important, and other processes will dominate. In practice, these other processes are the interband processes which in (16.18) were previously dealt with in a very simplified form through the core dielectric constant $\varepsilon_{\text{core}}$, but now become the dominant process. Using this new approximation in the high frequency limit, we can neglect the free carrier contribution in (16.18) to obtain

$$\sqrt{\varepsilon} \cong \sqrt{\varepsilon_{\text{core}}} \equiv \text{real number}. \tag{16.19}$$

Equation 16.19 implies that $\tilde{n} > 0$ and $\tilde{k} = 0$ in the limit of $\omega\tau \gg 1$, with

$$\mathscr{R} \to \frac{(\tilde{n}-1)^2}{(\tilde{n}+1)^2} \tag{16.20}$$

where $\tilde{n} = \sqrt{\varepsilon_{\text{core}}/\varepsilon_{\text{o}}}$. Thus, in the limit of very high frequencies, the Drude contribution is unimportant and the behavior of all materials is similar to that of a dielectric.

16.4 The Plasma Frequency

The plasma medium is defined as a neutral system in which some of the carriers are free to move under the action of eletromagentic fields. Metals or degenerate semiconductors are very good examples of a plasma medium because both have free carriers. Thus, at very low frequencies the optical properties of semiconductors exhibit a metal-like behavior, while at very high frequencies their optical properties are like those of insulators. A characteristic frequency at which the material changes from metallic behavior to a dielectric response is called the plasma frequency $\hat{\omega}_p$, which is defined as that frequency at which the real part of the dielectric function vanishes $\varepsilon_1(\hat{\omega}_p) = 0$. Thus, the plasma frequency defines the boundary region below which the transverse electromagnetic wave propagates in a given medium containing free carriers that attenuate the wave propagation, and a wave at higher frequency (above $\hat{\omega}_p$) that propagates without attenuation because the phase change of the wave is too fast to be affected by the pressure of free carriers. In summary, if $\omega > \hat{\omega}_p$ the wave propagates (insulating-like medium) and if $\omega < \hat{\omega}_p$ the waves are attenuated (conducting-like medium). According to the Drude theory (16.12), we have

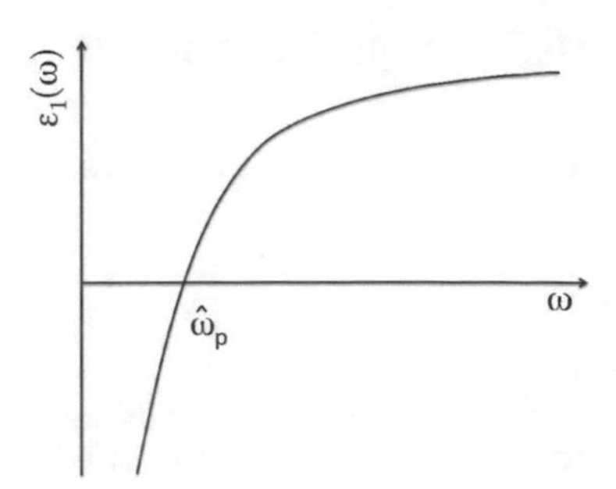

Fig. 16.1 The frequency dependence of $\varepsilon_1(\omega)$, showing the definition of the plasma frequency $\hat{\omega}_p$ by the relation $\varepsilon_1(\hat{\omega}_p) = 0$

$$\varepsilon = \varepsilon_1 + i\varepsilon_2 = \varepsilon_{\text{core}} + \frac{i}{\omega}\frac{ne^2\tau}{m(1 - i\omega\tau)} \cdot \left(\frac{1 + i\omega\tau}{1 + i\omega\tau}\right) \tag{16.21}$$

where we have written ε in a form which exhibits its real and imaginary parts explicitly. We can then write the real and imaginary parts $\varepsilon_1(\omega)$ and $\varepsilon_2(\omega)$ as:

$$\varepsilon_1(\omega) = \varepsilon_{\text{core}} - \frac{ne^2\tau^2}{m(1 + \omega^2\tau^2)} \qquad \varepsilon_2(\omega) = \frac{1}{\omega}\frac{ne^2\tau}{m(1 + \omega^2\tau^2)}. \tag{16.22}$$

The free carrier term makes a negative contribution to ε_1 which tends to cancel the core contribution, shown schematically in Fig. 16.1.

We see in Fig. 16.1 that $\varepsilon_1(\omega)$ vanishes at some frequency ($\hat{\omega}_p$) so that we can write

$$\varepsilon_1(\hat{\omega}_p) = 0 = \varepsilon_{\text{core}} - \frac{ne^2\tau^2}{m(1 + \hat{\omega}_p^2\tau^2)} \tag{16.23}$$

which yields

$$\hat{\omega}_p^2 = \frac{ne^2}{m\varepsilon_{\text{core}}} - \frac{1}{\tau^2} = \omega_p^2 - \frac{1}{\tau^2}\,. \tag{16.24}$$

Since the term $(-1/\tau^2)$ in (16.24) is usually small compared with ω_p^2, it is customary to neglect this term and to identify the plasma frequency with ω_p defined by

$$\omega_p^2 = \frac{ne^2}{m\varepsilon_{\text{core}}} \tag{16.25}$$

in which the screening of free carriers occurs through the core dielectric constant $\varepsilon_{\text{core}}$ of the medium. If $\varepsilon_{\text{core}}$ is too large, then $\varepsilon_1(\omega)$ never goes negative and there is no plasma frequency. The condition for the existence of a plasma frequency is

$$\varepsilon_{\text{core}} < \frac{ne^2\tau^2}{m}. \tag{16.26}$$

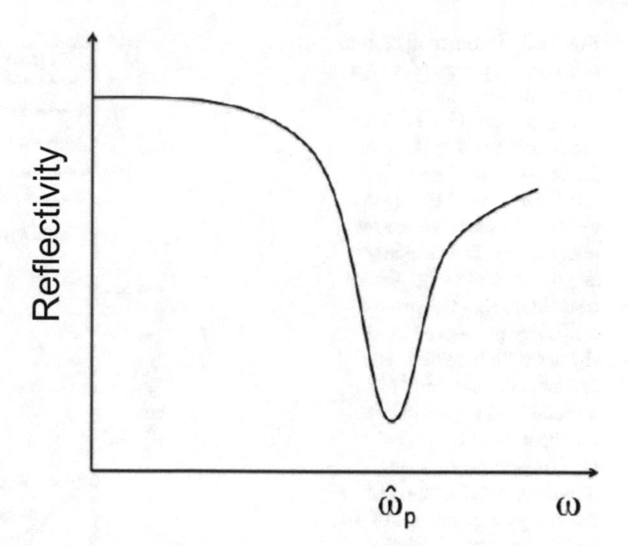

Fig. 16.2 Reflectivity $\mathscr{R}$ dependence on photon frequency for a metal or a degenerate semiconductor. The $\mathscr{R}$ goes to zero at the plasma frequency $\hat{\omega}_p$

The quantity ω_p in (16.25) is called the screened plasma frequency in the literature. Another quantity called the unscreened plasma frequency, which is obtained from (16.22) by setting $\varepsilon_{\text{core}} = \varepsilon_{\text{o}}$, is also used in the literature.

The dependence of the reflectivity spectra (versus wavelength) for various donor concentrations for heavily doped n-type InSb is shown in Fig. 16.3. The dependence of the plasma frequency on the carrier concentration is readily visible from the data in Fig. 16.3. This profile is also observed in metals.

At low frequencies, free carrier conduction dominates, and the reflectivity $\mathscr{R}$ is 100%. In the high frequency limit, we have

$$\mathscr{R} \sim \frac{(\tilde{n}-1)^2}{(\tilde{n}+1)^2} \tag{16.27}$$

Here, $\mathscr{R}$ is large if $\tilde{n} \gg 1$. In the vicinity of the plasma frequency, $\varepsilon_1(\hat{\omega}_p)$ is small by definition; furthermore, $\varepsilon_2(\omega_p)$ is also small, since from (16.22).

$$\varepsilon_2(\omega_p) = \frac{ne^2\tau}{m\omega_p(1+\omega_p^2\tau^2)} \tag{16.28}$$

and if $\omega_p\tau \gg 1$

$$\varepsilon_2(\omega_p) \cong \frac{\varepsilon_{\text{core}}}{\omega_p\tau} \tag{16.29}$$

so that $\varepsilon_2(\omega_p)$ is also small. With $\varepsilon_1(\omega_p) = 0$, we have from (16.27) $\tilde{n} \cong \tilde{k}$ and $\varepsilon_2(\omega_p) = 2\varepsilon_{\text{o}}\tilde{n}\tilde{k} \simeq 2\varepsilon_{\text{o}}\tilde{n}^2$. We thus see that $\tilde{n}$ tends to be small near ω_p and consequently $\mathscr{R}$ is also small (see Fig. 16.2). The steepness of the dip at the plasma

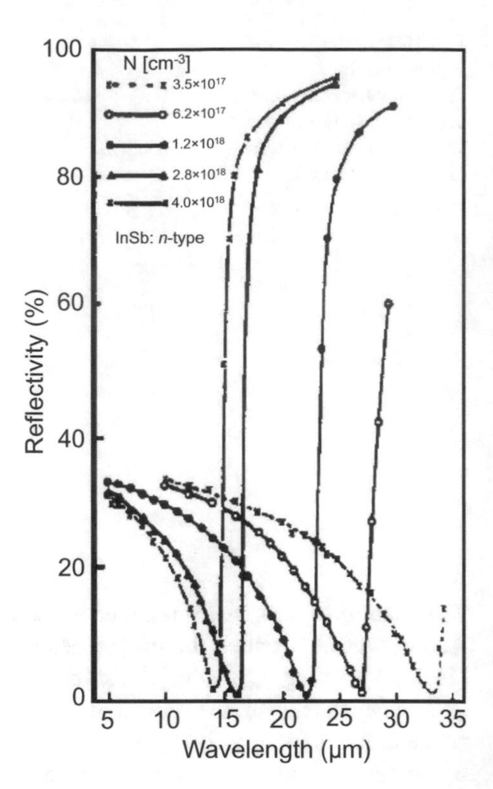

Fig. 16.3 Room temperature reflectivity spectra of n-type InSb with a carrier concentration (labeled N in the figure) varying between 3.5×10^{17} cm^{-3} and 4.0×10^{18} cm^{-3} (see inset upper left). Here we see two phenomena: The reflectivity $\mathscr{R}$ goes to zero at the plasma frequency, ω_p. The plasma frequency ω_p moves to shorter wavelengths λ as the carrier concentration N increases. The solid curves are theoretical fits to the experimental data points, including consideration of the energy dependence of m^* due to the strong interband coupling (called non-parabolic effects)

frequency in Fig. 16.2 is governed by the relaxation time τ; the longer the relaxation time τ, the sharper is the plasma structure near $\hat{\omega}_p$.

In metals, free carrier effects are almost always studied by optical reflectivity techniques because of the high optical absorption of metals at low frequency. For metals, the free carrier conductivity appears to be quite well described by the simple Drude theory. In studying free carrier effects in semiconductors, it is usually more accurate to use absorption techniques, which are discussed in Chap. 18. Because of the connection between the optical and the electrical properties of a solid through the conductivity tensor, transparent materials are expected to be poor electrical conductors, while highly reflecting materials are expected to be reasonably good electrical conductors. It is, however, possible for a material to have its plasma frequency just below visible frequencies, so that the material will be a good electrical conductor, and yet be transparent at visible frequencies. Because of the close connection between the optical and electrical properties, free carrier effects are sometimes exploited in the determination of the carrier density in instances where Hall effect measurements are difficult to make.

The contribution of holes to the optical conduction is of the same sign as for the electrons, since the conductivity depends on an even power of the charge ($\sigma \propto e^2$). In terms of the complex dielectric constant, we can write the contribution from electrons and holes as

$$\varepsilon = \varepsilon_{\text{core}} + \frac{i}{\omega}\left[\frac{n_e e^2 \tau_e}{m_e(1 - i\omega\tau_e)} + \frac{n_h e^2 \tau_h}{m_h(1 - i\omega\tau_h)}\right] \tag{16.30}$$

where the parameters n_e, τ_e, and m_e pertain to the electron carriers and n_h, τ_h, and m_h are for the holes. The plasma frequency is again found by setting $\varepsilon_1(\omega) = 0$. If there are multiple electron or hole carrier pockets, as is common for semiconductors, the contributions from each carrier type to the complex dielectric function ε is additive, using a formula similar to (16.30).

16.5 Plasmon Resonant Nanoparticles

The plasma frequency of most bulk metals lies in the ultraviolet wavelength range. When metal nanoparticles are made much smaller than the wavelength of light, this plasmon resonance (which occurs approximately when the real part of the dielectric function ε_1 goes to zero) is shifted into the visible region of the electromagnetic spectrum. Figure 16.4 shows an atomic force microscope (AFM) image and a UV-Vis absorption spectrum of 25 nm gold nanoparticles. The spectrum shows plasmon resonant absorption near $\lambda = 532$ nm. When irradiated at this plasmon resonant wavelength, immense plasmonic charge and intense electric fields occur at the surface of the nanoparticle, which can exceed the incident electric field intensity by more than three orders of magnitude. These immense plasmonic fields have been utilized for surface enhanced Raman spectroscopy (SERS), plasmon enhanced photocatalysis and dye sensitized solar cells. (See, for example, Kneipp, et al., Physical Review Letters, 78, 1667 (1997) and Hou, et al. Advanced Functional Materials, 23, 1612–1619 (2013)).

The plasmon resonance of metal nanostructures depends very strongly on the size, shape, and separation of metal nanostructures. Figure 16.5a shows an SEM image of a 5 nm gold film deposited by electron beam evaporation, which is not thick enough to form a continuous film, but instead forms an island-like structure. The plasmon absorption of this film, plotted in Fig. 16.5c, is broadened and redshifted because

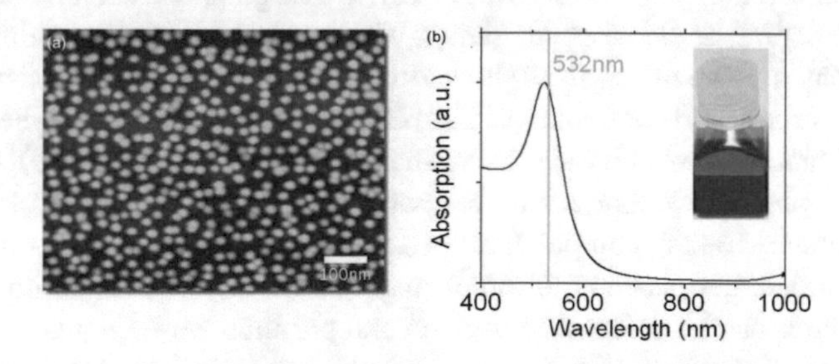

Fig. 16.4 **a** Atomic force microscope image of 25 nm diameter gold nanoparticles. **b** Absorption spectra of 25 nm gold nanoparticles showing plasmon resonant absorption near $\lambda = 532$ nm. Inset shows a photograph of 25 nm gold nanoparticles in solution

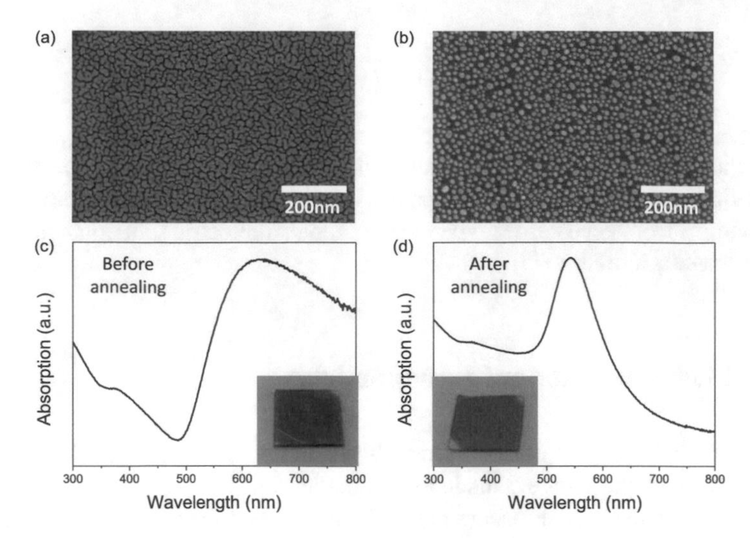

Fig. 16.5 **a**, **b** SEM images and **c**, **d** absorption spectra of a 5 nm Au thin film before (left) and after (right) thermal annealing in air at 600°C for half one hour. The insets show photographs of the Au nanoparticle film **c** before and **d** after annealing

of the large inhomogeneity in size, shape, and separation of these islands. After annealing, these islands form well separated spheres that are more uniform in size, shape, and separation, and the well-defined plasmon resonance near λ=532 nm is recovered.

16.6 Surface Plasmon Polaritons in Graphene

Due to its high carrier mobility and tunable carrier Fermi level, graphene has emerged as a promising materials for tunable terahertz to mid-infrared plasmonics. In contrast to conventional plasmonic materials, plasmons in graphene exhibit a rich array of unique features: (i) The Fermi level of carriers in graphene can be readily tuned by chemical doping or electrostatic gating, which can significantly modify the plasmonic responses of the material, as shown in Fig. 16.6. (ii) Graphene plasmons (GP) demonstrate strong field confinement. The plasmon wavelengths in graphene are typically 1 to 3 orders of magnitude smaller than the light wavelength. (iii) The charge carriers in graphene have long mean free paths, resulting in a relatively long optical relaxation time ($\sim$100 fs), compared to $\sim$10 fs in gold. This results in lower plasmon dissipation and longer plasmon lifetimes in graphene. (iv) The crystalline graphene lattice structures can be defect-free over several plasmon wavelengths.

By patterning graphene, for example, into ribbons or disks, localized plasmon modes can be excited by incident light. For incident light polarized perpendicular to the ribbon, distinctive absorption peaks originating from plasmon oscillations

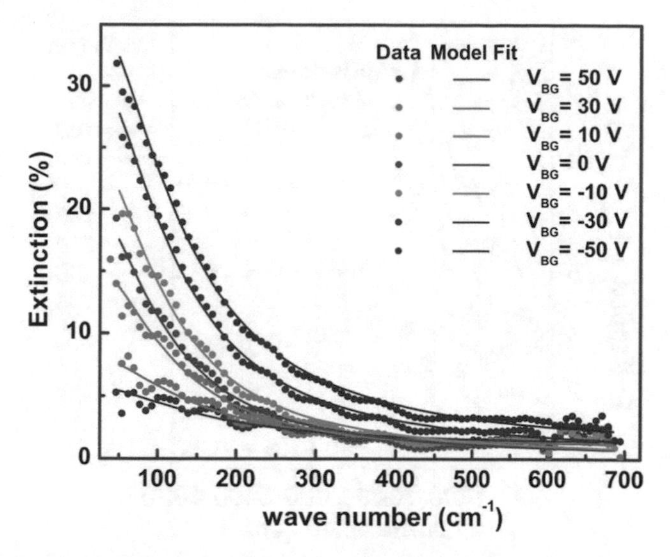

Fig. 16.6 Extinction spectrum of CVD graphene sample subject to electrical biasing by the back-gate applied through 285 nm SiO_2 dielectric. As the back-gate voltage (V_{BG}) varies from −50V to 50V, the extinction ratio is modulated by up to 30% at low frequency

dominate the optical response. The dependence of the graphene plasmon frequency on ribbon width and carrier density exhibits a characteristic power-law relation for the two-dimensional massless Dirac fermions. The plasmon frequency scales with $W^{1/2}$ and $n^{1/4}$, where W is the ribbon width and n is the carrier density, as predicted by the random phase approximation (RPA). In addition to the tunable surface plasmons in graphene ribbons, plasmon hybridization in coupled graphene ribbons has been observed and the splitting of GPs into bulk and edge modes under high magnetic fields has also been demonstrated.

The strong field confinement in GPs also facilitates the study of strong light-matter coupling or Vacuum Rabi Splitting (VRS). In graphene, the very small mode confinement compensates for the low quality factors (Q∼10) for graphene plasmons in mid-infrared regime. Large vacuum Rabi splitting and Purcell factors are predicted due to strong light matter interaction assisted by graphene plasmons. More interestingly, the tunable E_F of graphene provides an elegant approach to manipulate the coupling strength.

The plasmons in graphene can also couple strongly to polar phonons from the substrate, such as SiO_2 and atomically thin hexagonal boron nitride (h-BN). As shown in Fig. 16.7, the graphene plasmon resonance couples to the IR-active E_{1U} phonon mode at 1373 cm^{-1} in monolayer h-BN. The clear anti-crossing characteristics in the spectrum near 1373 cm^{-1} correspond to hybridized plasmon-phonon modes near the surface polar phonon frequency of monolayer h-BN. This phenomena, named the phonon induced transparency (PIT), is analogous to electromagnetically induced transparency (EIT), typically resulting from the destructive interference between a direct transition and an indirect transition pathway.

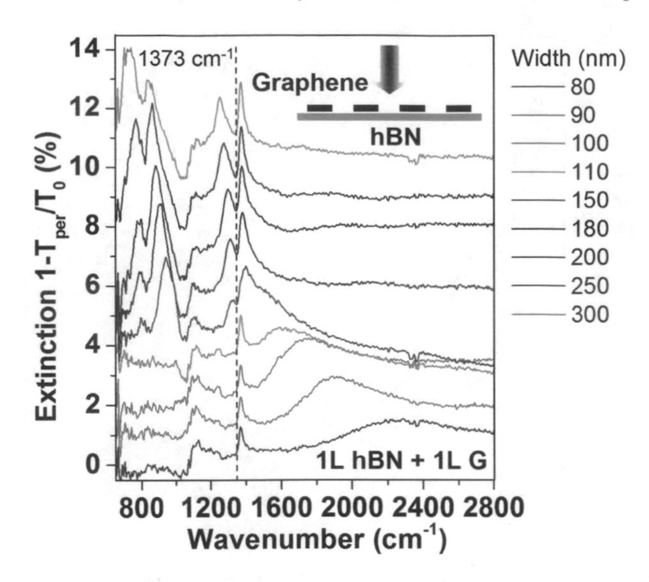

Fig. 16.7 The scaling of the Plasmon resonance and coupling with the E_{1u} phonon in single-layer h-BN. Extinction spectra are shown for graphene ribbon width varying from 80 nm to 300 nm

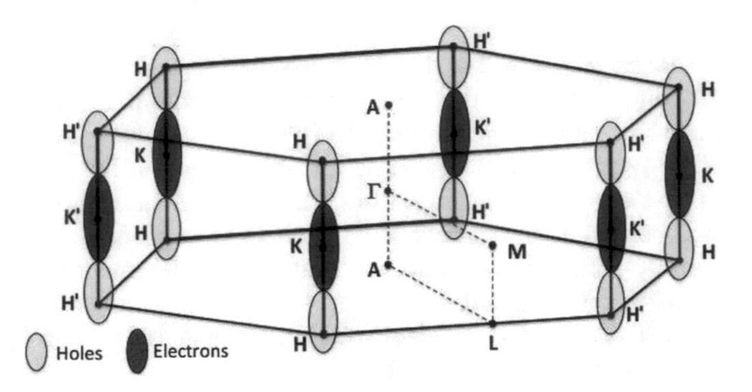

Fig. 16.8 Brillouin zone for graphite showing the high symmetry points and the location of both the electron and hole pockets

Furthermore, in Bernal-stacked bilayer graphene, the plasmons can also couple to the intrinsic IR-active phonons of bilayer graphene in a similar way. This EIT-like phenomenon in graphene plasmonic material with tunable resonance characteristics may open the door to a range of novel applications in room temperature nonlinear quantum optics, quantum information processing, and slow light.

In the following Chapter, we will treat another absorption process which is due to electron and/or hole interband transitions. In the above discussion, interband transitions were included in an extremely approximate way. That is, interband transitions were treated through a frequency independent core dielectric constant $\varepsilon_{\text{core}}$ (see (16.12)). In Chap. 17 we consider the frequency dependence of this important contribution to the optical properties.

Problems

16.1 Find the plasma frequency ω_p for a semimetal for light incident in an in-plane direction (such as the x-direction) for the two light polarization directions $\boldsymbol{E} \perp z$ and $\boldsymbol{E} \parallel z$. The 6 electron pockets are located about the K point at the center of the Brillouin zone edge HKH. Assume mass components for the electron carriers $m_t = 0.06\, m_0$ and $m_\ell = m_0$, where m_0 is the free electron mass, and $E_F = 50\,\text{meV}$ above the conduction band minima. As an intermediate step you will need to find E_F^h for holes.

16.2 (a) Assuming one free electron per Cu atom, calculate the plasma frequency for metallic Cu. Why then does Cu have a characteristic reddish color?

(b) ReO_3 is a reddish semi-transparent material to visible radiation, yet a good electrical conductor. How can this be explained?

(c) Sapphire is a good thermal conductor, yet a material that is optically transparent to visible light. How can this be explained?

(d) Doped SiO_2 is used as a transparent electrode. Explain how it is possible that this material can be used to make good electrical contacts and yet is transparent to visible light.

16.3 This problem involves the plasma frequency at room temperature for heavily doped p-type GaAs. Assume an acceptor impurity concentration $N_A = 10^{18}/\text{cm}^3$ (assume all acceptor levels are occupied by electrons), and use $m_{lh} = 0.07 m_0, m_{hh} = 0.6 m_0$, $m_e = 0.07\, m_0$, $E_{\text{gap}} = 1.4\,\text{eV}$, $\varepsilon_{\text{core}} = 15$ for the band parameters and for the core dielectric constant.

(a) Find the position of the Fermi level for this 3 dimensional sample.
(b) What is the approximate concentration of electrons? of holes?
(c) Find the plasma frequency for this semiconductor at $T = 300\,\text{K}$.
(d) Calculate the change in band gap between the bulk and the 5 nm quantum well.
(e) If the GaAs (with the same materials parameters) is made into a quantum well of 5 nm thickness, what is the expected change in the plasma frequency?
(f) What is the change in functional form of the onset of the absorption edge for the quantum well relative to the bulk?
(g) What is the effect of making a 5 nm quantum well on the exciton binding energy?

16.4 Suppose that you have an fcc semimetal with 2 atoms per unit cell. Suppose that the electrons are at the L points, $\pi/a(1, 1, 1)$, ($m_l^* = 0.3 m_0$ and $m_t^* = 0.1 m_0$) in the Brillouin zone and the holes are in a single carrier pocket at the Γ point ($k = 0$) with $m_h^* = 0.3 m_0$, and assume that the energy band overlap for this semimetal is 10 meV.

(a) Find the position of the Fermi level for the 3 dimensional semimetal at $T = 0$ K.
(b) Find the plasma frequency for this semimetal at $T = 0$ K (assume $\epsilon_{\text{core}} = 1.0$ and $\omega_p^2 \tau^2 \gg 1$).

(c) Suppose that the semimetal is now prepared as a thin layer (quantum well) between alkali halide insulating barriers with the (001) crystalline direction normal to the thin layer of the semimetal (layer thickness = 50 nm). Does the plasma frequency increase or decrease relative to part (b) and why?
(d) What is the change in the frequency dependence of the absorption coefficient for light as the thickness of the semimetal layer decreases and a semimetal-semiconductor transition occurs?
(e) How does the semimetal-semiconductor transition affect the optical reflectivity?, the transmission?, the photoconductivity? What photon energy would you use for these experiments?

16.5 (a) Find the plasma frequency for intrinsic Si at room temperature, assuming $\epsilon_{\text{core}} = 12$, and only thermally excited carriers are involved in the plasma.
(b) Using the results in (a), write an expression for the free carrier contribution to $\epsilon(\omega)$.
(c) Now suppose that we apply a magnetic field normal to the surface of the sample. Find the effect of this magnetic field on the plasma frequency using right and left incident circularly polarized light.

16.6 The frequency of the uniform plasmon mode of a sphere is determined by the depolarization field $\mathbf{E} = -4\pi\mathbf{P}/3$ of the sphere, where the polarization $\mathbf{P} = -ne\mathbf{r}$, with $\mathbf{r}$ as the average displacement of the electrons of concentration n.

(a) Show that the resonance frequency of the electron gas is given by

$$\omega_0^2 = 4\pi ne^2/3m.$$

Since all electrons participate in the oscillation, such an excitation is called a collective excitation or collective mode of the electron gas.
(b) Use similar methods to find the frequency of the uniform plasmon mode of a sphere placed in a constant uniform magnetic field $\mathbf{B}$. Let $\mathbf{B}$ be along the z axis. The solution should go to the cyclotron frequency $\omega_c = eB/mc$ in one limit and to $\omega_0 = (4\pi ne^2/3m)^{1/2}$ in another limit. Consider the motion in the $x - y$ plane.

16.7 (a) Suppose that a hexagonal material has n electrons/cm^3 in the conduction band which consists of ellipsoidal carrier pockets along each of the edges of the Brillouin zone. Find an expression for the plasma frequency for the polarization $\boldsymbol{E} \parallel \hat{x}, \hat{y}$ and $\boldsymbol{E} \parallel \hat{z}$. Take m_l and m_t as the effective mass components for the conduction band pockets.
(b) If a magnetic field is applied along the (001) direction, find the change in the plasma frequency as a function of magnetic field with the $\boldsymbol{E}$ field in the xy-plane. To determine ω_p, first find the magnetoconductivity. What is the effect of the magnetic field where $\boldsymbol{E} \parallel (001)$?

16.8 This problem considers the complex dielectric function for free carriers in a magnetic field.

(a) Find the complex dielectric constant due to free carriers in a magnetic field assuming the zero field dispersion relation $E(\boldsymbol{k}) = \hbar^2k^2/2m^*$. Consider the Poynting vector for the light along the external magnetic field $\boldsymbol{B}$ (neglect interband effects), and consider right and left circular polarization.
(b) Find the dependence of the plasma frequency on magnetic field $\boldsymbol{B}$.
(c) Sketch the optical reflectivity in a magnetic field relative to the zero field reflectivity for right and left circularly polarized radiation.

16.9 Consider a metal with a free electron concentration n and placed in a uniform magnetic field $\boldsymbol{B} = B\hat{k}$ in the z direction. The electric current density in the xy plane is related to the electric field by the relations:

$$j_x = \sigma_{xx}E_x + \sigma_{xy}E_y$$

$$j_y = \sigma_{yx}E_x + \sigma_{yy}E_y$$

Consider frequencies in which $\omega \gg \omega_c$, and $\omega \gg 1/\tau$, where ω_c is the cyclotron resonance frequency given by $\omega_c = eB/m$ and τ is the time between electron collisions.

(a) By solving the equations above, show that the components of the magnetoconductivity tensor are given by:

$$\sigma_{xx} = \sigma_{yy} = i\omega_p^2/\omega$$

$$\sigma_{yx} = -\sigma_{xy} = \omega_c\omega_p^2/\omega^2$$

where $\omega_p = \sqrt{ne^2/m}$ is the plasmon resonance frequency (in Sect. 16.4, we have derived the screened plasma frequency, which contains ϵ_{core}. In the present problem consider $\varepsilon_{\text{core}} = \varepsilon_{\text{o}}$).
(b) By using the tensorial equation $\boldsymbol{\epsilon} = \mathbf{1} + i/\omega\boldsymbol{\sigma}$ and considering an electromagnetic wave propagating with wavevector $\boldsymbol{k} = k\hat{k}$, show that the dispersion relation for this wave in the medium is given by

$$c^2k^2 = \omega^2 - \omega_p^2 \pm \omega_c\omega_p^2/\omega.$$

16.10 You may wish to use MATLAB to obtain numerical answers to this problem. Determine the absorption coefficient due to free-carrier absorption at a temperature of 300 K and a free-space wavelength of 0.9μm for a partially compensated GaAs sample with $N_d = 2 \times 10^{18}$ cm^{-3} and $N_a = 1 \times 10^{18}$ cm^{-3} for the donor and acceptor dopant concentrations. Where does the Fermi level lie? For this sample, assume that ionized impurity scattering is the dominant scattering mechanism at room temperature.

Suggested Readings

K.S. Novoselov et al., Proc. Natl. Acad. Sci. USA **102**, 10451 (2005)
M. Freitag, et al., Nat Commun. **4**, (2013)
L. Ju et al., Nat. Nanotechnol. **6**, 630 (2011)
H. Yan et al., Nat. Photonics **7**, 394 (2013)
V.W. Brar et al., Nano Lett. **13**, 2541 (2013)
F. Koppens et al., Nano Lett. **11**, 3370 (2011)
S.D. Sarma et al., Phys. Rev. Lett. **102**, 206412 (2009)
J. Christensen et al., ACS Nano. **6**, 431 (2011)
H. Yan et al., Nano. Lett. **12**, 3766 (2012)
Y. Jia et al., ACS Photonics **2**, 907–912 (2015)
H. Yan et al., Nano Lett. **14**, 4394 (2013)
T. Low et al., Phys. Rev. Lett. **112**, 116801 (2014)
H. Tanji-Suzuki et al., Science **333**, 1266 (2011)
M.F. Yanik et al., Phys. Rev. Lett. **93**, 233903 (2004)
L.V. Hau et al., Nature **397**, 594 (1999)
N.W Ashcroft, N.D Mermin, *Solid State Physics* (Holt, Rinehart and Winston, 1976)
C. Kittel, *Introduction to Solid State Physics*, 7th edn. (Wiley, 1996)
K. Kneipp et al., Phys. Rev. Lett. **78**, 1667 (1997)
W. Hou et al., Adv. Funct. Mater. **23**, 1612–1619 (2013)

Chapter 17
Interband Transitions

17.1 The Interband Transition Process

In a semiconductor at low frequencies, the principal electronic conduction mechanism is associated with free carriers. As the photon energy increases and becomes comparable to the energy gap, a new conduction process can occur due to optically excited carriers. A photon can excite an electron from an occupied state in the valence band to an unoccupied state in the conduction bands. This is called an **interband transition** and is represented schematically by the picture in Fig. 17.1. In this process the photon is absorbed, an excited electronic state is formed and a hole is left behind in the valence bands. This process is quantum mechanical in nature and we now discuss the main factors which rule these interband transitions.

1. We expect interband transitions to have a **threshold energy** at the energy gap. That is, we expect the frequency dependence of the real part of the conductivity $\sigma_1(\omega)$ due to an interband transition to exhibit a threshold as schematically shown in Fig. 17.1 for an allowed electronic transition.
2. The transitions are either **direct** (conserve crystal momentum $\mathbf{k}$: $E_v(\mathbf{k}) \rightarrow E_c(\mathbf{k})$) or **indirect** (a phonon is involved because the $\mathbf{k}$ vectors for the valence and conduction bands differ by the phonon wave vector $\mathbf{q}$). Conservation of crystal momentum yields $\mathbf{k}_{\text{valence}} = \mathbf{k}_{\text{conduction}} \pm \mathbf{q}_{\text{phonon}}$. In discussing the direct transitions, one might wonder about conservation of crystal momentum with regard to the photon. The reason we need not be concerned with the momentum of the photon is that it is very small in comparison to Brillouin zone dimensions. For a typical optical wavelength of 6000 Å, the wave vector for the photon $k_{\text{photon}} = 2\pi/\lambda \sim 10^5\,\text{cm}^{-1}$, while a typical dimension across the Brillouin zone (zone boundary ZB vector) is $k_{\text{ZB}} = 2\pi/a \sim 10^8\,\text{cm}^{-1}$. Here a is the lattice parameter which has order of $1\,\mathring{A}$. Thus, typical direct optical interband processes excite an electron from a valence to a conduction band without a significant change in the wave vector.
3. The transitions depend on the coupling between the valence and conduction bands and this is measured by the magnitude of the momentum matrix elements $|\langle v|\mathbf{p}|c\rangle|^2$ which couple the valence band state v and the conduction band state c.

M. Dresselhaus et al., *Solid State Properties*, Graduate Texts in Physics,
https://doi.org/10.1007/978-3-662-55922-2_17

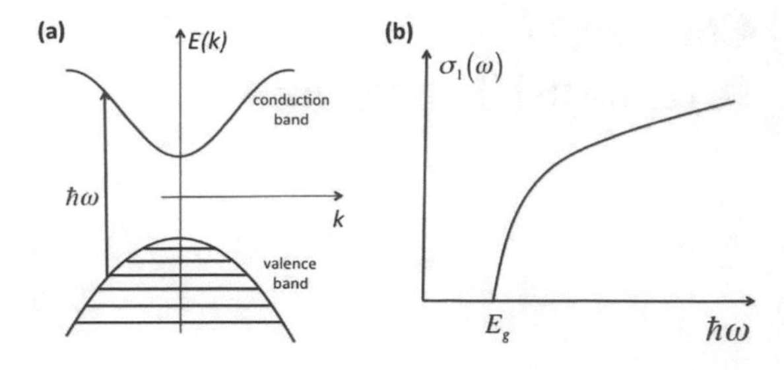

Fig. 17.1 **a** Schematic diagram of an allowed direct interband transition. **b** Real part of the conductivity for an allowed optical transition

This dependence results from Fermi's "Golden Rule" (see Appendix B) and from the discussion of the perturbation interaction Hamiltonian $\mathscr{H}'$ for the electromagnetic field with electrons in the solid (discussed in Sect. 17.2). Selection rules due to state symmetry can cause certain transitions to be forbidden.

4. Because of the Pauli Exclusion Principle, an interband transition occurs from an occupied state below the Fermi level to an unoccupied state above the Fermi level.
5. Since the optical properties are found by an integration over k space, the joint density of states (discussed in Sect. 17.4) is important. Photons of a particular energy are more effective in producing an interband transition if the energy separation between the 2 bands is nearly constant over many **k** values. In that case, there are many initial and final states which can be coupled by the same photon energy. This is perhaps easier to see if we allow a photon to have a small band width. That band width will be effective over many **k** values if $E_c(\mathbf{k}) - E_v(\mathbf{k})$ does not vary rapidly with **k**. Thus, we expect the interband transitions to be most important for **k** values near band extrema. That is, in Fig. 17.1a we see that states around $\mathbf{k} = 0$ make the largest contribution per unit bandwidth of the optical source. It is also for this reason that optical measurements are so important in studying energy band structure; the optical structure emphasizes band extrema and therefore provides special information about the energy bands at specific points in the Brillouin zone. Because of the dependence of the density of states and the joint density of states on the dimensionality of the system, the optical properties will be very sensitive to the dimensionality of a sample.

Although we will not derive the expression for the interband contribution to the conductivity, we will write it down here to show how all the physical ideas that were discussed above enter into the conductivity equation. We now write the conductivity tensor relating the interband current density j_α in the direction α which flows upon application of an electric field E_β in direction β

$$j_\alpha = \sigma_{\alpha\beta} E_\beta \tag{17.1}$$

as

$$\sigma_{\alpha\beta} = -\frac{e^2}{m^2} \sum_{i,j} \frac{[f(E_i) - f(E_j)]}{E_i - E_j} \frac{\langle i|p_\alpha|j\rangle\langle j|p_\beta|i\rangle}{[-i\omega + 1/\tau + (i/\hbar)(E_i - E_j)]} \tag{17.2}$$

in which the sum in (17.2) is over all valence and conduction band states labeled by i and j, respectively. Structure in the optical conductivity arises through a singularity in the resonant denominator of (17.2) $[-i\omega + 1/\tau + (i/\hbar)(E_i - E_j)]$ discussed above under properties (1) and (5). The appearance of the Fermi functions $f(E_i) - f(E_j)$ follows from the Pauli principle in property (4). The dependence of the conductivity on the momentum matrix elements accounts for the tensorial properties of $\sigma_{\alpha\beta}$ (interband) and relates to properties (2) and (3).

In semiconductors, interband transitions usually occur at frequencies above which free carrier contributions are important. If we now want to consider the total complex dielectric constant, we would write

$$\varepsilon = \varepsilon_{\text{core}} + \frac{i}{\omega}\left[\sigma_{\text{Drude}} + \sigma_{\text{interband}}\right]. \tag{17.3}$$

The term $\varepsilon_{\text{core}}$ contains the contributions from all processes that are not considered explicitly in (17.3); this would include both intraband and interband transitions that are not treated explicitly. We have now dealt with the two most important processes (intraband and interband transitions) involved in studies of the electronic properties of solid state materials. If we think about the optical properties for various classes of materials according to their conducting behavior, it is clear that major differences will be found from one class of materials to another as we briefly highlight below.

1. Insulators: In these materials the band gap is sufficiently large so that at room temperature, essentially no carriers are thermally excited across the band gap. This means that there is no free carrier absorption and that interband transitions only become important at relatively high photon energies (above the visible). Thus, insulators frequently are optically transparent in the visible frequency range.
2. Semiconductors: In these materials the band gap is small enough so that appreciable thermal excitation of carriers occurs at room temperature. Thus there is often appreciable free carrier absorption at room temperature either through thermal excitation or doping and the interband transitions occur in the infrared and visible energy ranges.
3. Metals: In these materials free carrier absorption is extremely important. Typical plasma frequencies are $\hbar\omega_p \cong 10$ eV which occur far out in the ultraviolet energy range. In the case of metals, interband transitions typically occur at frequencies where free carrier effects are still important.
4. Semimetals: Similar to metals, the optical properties of these materials exhibit only a weak temperature dependence with carrier densities almost independent of temperature. Although the carrier densities are low, the high carrier mobilities nevertheless guarantee a large contribution of the free carriers to the optical conductivity.

17.2 Hamiltonian for a Charge in an Electromagnetic Field

In this section we discuss how the optical field is inserted into the Hamiltonian in the form $\mathbf{p} \to \mathbf{p} - e\mathbf{A}$. Let us consider the classical equation of motion for a charge in an electromagnetic field:

$$\frac{d}{dt}(m\mathbf{v}) = e\,[\mathbf{E} + (\mathbf{v} \times \mathbf{H})] = e\left[-\nabla\phi - \frac{\partial \mathbf{A}}{\partial t} + \mathbf{v} \times (\nabla \times \mathbf{A})\right] \tag{17.4}$$

where ϕ and $\mathbf{A}$ are, respectively, the scalar and vector potentials, and $\mathbf{E}$ and $\mathbf{B}$ are the electric and magnetic fields given by

$$\mathbf{E} = -\nabla\phi - \partial\mathbf{A}/\partial t \tag{17.5}$$

$$\mathbf{B} = \nabla \times \mathbf{A}.$$

Using standard vector identities, the equation of motion (17.4) becomes

$$\frac{d}{dt}(m\mathbf{v} + e\mathbf{A}) = \nabla(-e\phi) + e\nabla(\mathbf{A} \cdot \mathbf{v}) \tag{17.6}$$

where $[\nabla(\mathbf{A} \cdot \mathbf{v})]_j$ denotes $v_i \partial A_i/\partial x_j$. We have used the Einstein summation convention that repeated indices are summed. We have also used the vector relation $\mathbf{a} \times (\mathbf{b} \times \mathbf{c}) = \mathbf{b}(\mathbf{a} \cdot \mathbf{c}) - \mathbf{c}(\mathbf{a} \cdot \mathbf{b})$ in (17.4)

$$\frac{dA}{dt} = \frac{\partial \mathbf{A}}{\partial t} + (\mathbf{v} \cdot \nabla)\mathbf{A} \tag{17.7}$$

and

$$[\mathbf{v} \times (\nabla \times \mathbf{A})]_i = v_j \frac{\partial A_j}{\partial x_i} - v_j \frac{\partial A_i}{\partial x_j}. \tag{17.8}$$

If we write the Hamiltonian as

$$\mathscr{H} = \frac{1}{2m}(\mathbf{p} - e\mathbf{A})^2 + e\phi \tag{17.9}$$

and then use Hamilton's equations

$$\mathbf{v} = \frac{\partial \mathscr{H}}{\partial \mathbf{p}} = \frac{1}{m}(\mathbf{p} - e\mathbf{A}) \tag{17.10}$$

$$\dot{\mathbf{p}} = -\nabla\mathscr{H} = -e\nabla\phi + e\mathbf{v} \cdot \nabla\mathbf{A} \tag{17.11}$$

we can show that (17.4) and (17.6) are satisfied, thereby verifying that (17.9) is the proper form of the Hamiltonian in the presence of an electromagnetic field, which has the same form as the Hamiltonian without an optical field except that $\mathbf{p} \rightarrow \mathbf{p} - e\mathbf{A}$. The same transcription is used when light is applied to a solid and it is then called the Luttinger transcription. The Luttinger transcription is used in the effective mass approximation where the periodic potential is replaced by the introduction of $\mathbf{k} \rightarrow -(1/i)\nabla$ and $m \rightarrow m^*$.

The reason why interband transitions depend on the momentum matrix element can be understood from perturbation theory. At any instance of time, the Hamiltonian for an electron in a solid in the presence of an optical field is

$$\mathcal{H} = \frac{(\mathbf{p} - e\mathbf{A})^2}{2m} + V(\mathbf{r}) = \frac{p^2}{2m} + V(\mathbf{r}) - \frac{e}{m}\mathbf{A} \cdot \mathbf{p} + \frac{e^2 A^2}{2m} \tag{17.12}$$

in which $\mathbf{A}$ is the vector potential due to the optical fields and $V(\mathbf{r})$ is the periodic potential. We also used the fact that $\mathbf{A} \cdot \mathbf{p} = \mathbf{p} \cdot \mathbf{A}$. This is valid because we adopt a Coulomb gauge ($\nabla \cdot \mathbf{A} = 0$) which implies that operators $\mathbf{p}$ and $\mathbf{A}$ commute. Thus, the one-electron Hamiltonian without optical fields is

$$\mathcal{H}_0 = \frac{p^2}{2m} + V(\mathbf{r}) \tag{17.13}$$

and the optical perturbation terms are

$$\mathcal{H}' = -\frac{e}{m}\mathbf{A} \cdot \mathbf{p} + \frac{e^2 A^2}{2m}. \tag{17.14}$$

Optical fields are generally very weak (unless generated by powerful lasers) and we usually consider only the term linear in $\mathbf{A}$ which is called the linear response regime. The form of the Hamiltonian in the presence of an electromagnetic field is derived in this section. The momentum matrix elements $\langle v|\mathbf{p}|c\rangle$ which determine the strength of optical transitions also govern the magnitudes of the effective mass components

The coupling of the valence and conduction bands through the optical fields (17.9), depends on the matrix element for the coupling to the electromagnetic field perturbation

$$\mathcal{H}' \cong -\frac{e}{m}\mathbf{p} \cdot \mathbf{A}. \tag{17.15}$$

With regard to the spatial dependence of the vector potential we can write

$$\mathbf{A} = \mathbf{A}_0 \exp[i(\mathbf{K} \cdot \mathbf{r} - \omega t)] \tag{17.16}$$

where for a loss-less medium the propagation wavevector $K = \tilde{n}\omega = 2\pi c\tilde{n}/\lambda$ is a slowly varying function of $\mathbf{r}$ since K is much smaller than typical wave vectors in solids. Here $\tilde{n}$, ω, and λ are, respectively, the real part of the index of refraction, the optical frequency, and the wavelength of light.

17.3 Relation Between Momentum Matrix Elements and the Effective Mass

Because of the relation between the momentum matrix element $\langle v|\mathbf{p}|c\rangle$, which governs the electromagnetic interaction with electrons and solids, and the band curvature $(\partial^2 E/\partial k_\alpha \partial k_\beta)$, the energy band diagrams provide important information on the strength of optical transitions. Correspondingly, knowledge of the optical properties can be used to infer experimental information about $E(\mathbf{k})$.

We now derive the relation between the momentum matrix element coupling the valence and conduction bands $\langle v|\mathbf{p}|c\rangle$ and the band curvature $(\partial^2 E/\partial k_\alpha \partial k_\beta)$. We start with Schrödinger's equation in a periodic potential $V(\mathbf{r})$ having the Bloch solutions

$$\psi_{n\mathbf{k}}(\mathbf{r}) = e^{i\mathbf{k}\cdot\mathbf{r}} u_{n\mathbf{k}}(\mathbf{r}), \tag{17.17}$$

$$\mathcal{H}\psi_{n\mathbf{k}}(\mathbf{r}) = E_n(\mathbf{k})\psi_{n\mathbf{k}}(\mathbf{r}) = \left[\frac{p^2}{2m} + V(\mathbf{r})\right] e^{i\mathbf{k}\cdot\mathbf{r}} u_{n\mathbf{k}}(\mathbf{r}) = E_n(\mathbf{k}) e^{i\mathbf{k}\cdot\mathbf{r}} u_{n\mathbf{k}}(\mathbf{r}). \tag{17.18}$$

Since $\mathbf{p}$ is an operator $(\hbar/i)\nabla$, we can write

$$\mathbf{p} e^{i\mathbf{k}\cdot\mathbf{r}} u_{n\mathbf{k}}(\mathbf{r}) = e^{i\mathbf{k}\cdot\mathbf{r}} (\mathbf{p} + \hbar\mathbf{k}) u_{n\mathbf{k}}(\mathbf{r}). \tag{17.19}$$

Therefore the differential equation for $u_{n\mathbf{k}}(\mathbf{r})$ becomes

$$\left[\frac{p^2}{2m} + V(\mathbf{r}) + \frac{\hbar\mathbf{k}\cdot\mathbf{p}}{m} + \frac{\hbar^2 k^2}{2m}\right] u_{n\mathbf{k}}(\mathbf{r}) = E_n(\mathbf{k}) u_{n\mathbf{k}}(\mathbf{r}) \tag{17.20}$$

giving the following differential equation for the periodic function $u_{nk}(\mathbf{r}) = u_{nk}(\mathbf{r} + \mathbf{R}_m)$

$$\left[\frac{p^2}{2m} + V(\mathbf{r}) + \frac{\hbar\mathbf{k}\cdot\mathbf{p}}{m}\right] u_{n\mathbf{k}}(\mathbf{r}) = \left[E_n(\mathbf{k}) - \frac{\hbar^2 k^2}{2m}\right] u_{n\mathbf{k}}(\mathbf{r}) \tag{17.21}$$

which we write as follows to put (17.21) in the canonical form for application of the perturbation theory formulae

$$\mathcal{H}_0 = \frac{p^2}{2m} + V(\mathbf{r}) \tag{17.22}$$

$$\mathcal{H}' = \frac{\hbar\mathbf{k}\cdot\mathbf{p}}{m} \tag{17.23}$$

$$\mathcal{E}_n(\mathbf{k}) = E_n(\mathbf{k}) - \frac{\hbar^2 k^2}{2m} \tag{17.24}$$

to yield

$$[\mathscr{H}_0 + \mathscr{H}']u_{n\mathbf{k}}(\mathbf{r}) = \mathscr{E}_n(\mathbf{k})u_{n\mathbf{k}}(\mathbf{r}). \tag{17.25}$$

Assume that we know the solution to (17.25) about a special point $\mathbf{k}_0$ in the Brillouin zone which could be a band extremum, such as $\mathbf{k}_0 = 0$. Then the perturbation formulae (17.22)–(17.25) allow us to find the energy and wave function for states near $\mathbf{k}_0$. For simplicity, we carry out the expansion about the center of the Brillouin zone $\mathbf{k} = 0$, which is the most important case in practice; the extension of this argument to an energy extremum at arbitrary $\mathbf{k}_0$ is immediate. Perturbation theory to second order then gives:

$$\mathscr{E}_n(\mathbf{k}) = E_n(0) + (u_{n,0}|\mathscr{H}'|u_{n,0}) + \sum_{n' \neq n} \frac{(u_{n,0}|\mathscr{H}'|u_{n',0})(u_{n',0}|\mathscr{H}'|u_{n,0})}{E_n(0) - E_{n'}(0)}. \tag{17.26}$$

The first order term $(u_{n,0}|\mathscr{H}'|u_{n,0})$ in (17.26) normally vanishes about an extremum because of inversion symmetry, with $\mathscr{H}'$ being odd under inversion and the two wavefunctions $u_{nk}(\mathbf{r})$ both being even or both being odd. Since

$$\mathscr{H}' = \frac{\hbar \mathbf{k} \cdot \mathbf{p}}{m} \tag{17.27}$$

the matrix element is then written as

$$(u_{n,0}|\mathscr{H}'|u_{n',0}) = \frac{\hbar}{m}\mathbf{k} \cdot (u_{n,0}|\mathbf{p}|u_{n',0}). \tag{17.28}$$

We now apply (17.26) to optical transitions, for the simplest case of a two band model. Here we assume that:

1. bands n and n' (valence (v) and conduction (c) bands) are close to each other and far from other bands
2. interband transitions occur between these two bands are separated by an energy gap E_g.

We note that the perturbation theory is written in terms of the energy $\mathscr{E}_n(k)$

$$\mathscr{E}_n(k) = E_n(\mathbf{k}) - \frac{\hbar^2 k^2}{2m}. \tag{17.29}$$

Assuming that the first order term in perturbation theory (17.26) can be neglected by parity (even and oddness) arguments, we obtain for $\mathscr{E}_n(k)$ about $\mathbf{k}$= 0

$$\mathscr{E}_n(\mathbf{k}) = E_n(0) + \frac{\hbar^2}{m^2} k_\alpha k_\beta \frac{|(v|p_\alpha|c)(c|p_\beta|v)|}{E_g} \tag{17.30}$$

or in terms of the energy eigenvalues of Schrödinger's equation (17.18)

$$E_n(\mathbf{k}) = E_n(0) + \frac{\hbar^2 k^2}{2m} + \frac{\hbar^2}{m^2} k_\alpha k_\beta \frac{|(v|p_\alpha|c)(c|p_\beta|v)|}{E_g}. \tag{17.31}$$

We define the effective mass tensor by the relation

$$E_n(\mathbf{k}) = E_n(0) + \frac{\hbar^2}{2} \sum_{\alpha,\beta} k_\alpha k_\beta \left(\frac{1}{m^*}\right)_{\alpha\beta} \tag{17.32}$$

so that

$$\left(\frac{1}{m^*}\right)_{\alpha\beta} = \frac{\delta_{\alpha\beta}}{m} + \frac{2}{m^2} \frac{|(v|p_\alpha|c)(c|p_\beta|v)|}{E_g} \tag{17.33}$$

where $\delta_{\alpha\beta}$ is the unit matrix. This discussion shows that the non-vanishing momentum matrix element is responsible for the inequality between the free electron m and the effective mass m^* in the solid. With regard to the optical properties of solids we note that the same momentum matrix element that governs the effective mass formula (17.33) also governs the electromagnetic interaction given by (17.15). Thus small effective masses tend to give rise to strong coupling between valence and conduction bands and large values for $|(v|p|c)|^2$. On the other hand, small effective masses lead to a small density of states because of the $m^{*3/2}$ dependence of the density of states as will be discussed in details in the following sections.

17.4 The Joint Density of States

The detailed calculation of the contribution to the frequency dependent complex dielectric function $\varepsilon(\omega) = \varepsilon_1(\omega) + i\varepsilon_2(\omega)$ due to interband transitions is rather difficult. It is therefore instructive to obtain, by use of the Fermi Golden Rule (17.34), an approximate solution for the contribution of interband transitions to the complex dielectric function $\varepsilon(\omega)$ and what is called the joint density of states $\rho_{cv}(\hbar\omega)$, which couples the valence band to the conduction band by a photon of energy $\hbar\omega$. By this calculation, we obtain the probability per unit time $W_\mathbf{k}$ that a photon of energy $\hbar\omega$ makes a transition at a given $\mathbf{k}$ point in the Brillouin zone:

$$W_\mathbf{k} \cong \frac{2\pi}{\hbar} |\langle v|\mathscr{H}'|c\rangle|^2 \delta[E_c(\mathbf{k}) - E_v(\mathbf{k}) - \hbar\omega] \tag{17.34}$$

in which the matrix element for the electromagnetic perturbation $\mathscr{H}'$ is taken between the valence and conduction band Bloch states at wave vector $\mathbf{k}$, and the δ-function $\delta[E_c - E_v - \hbar\omega]$, which expresses energy conservation, and is also evaluated at the same wave vector $\mathbf{k}$. In writing (17.34), we exploit the fact that the wave vector for the light is small compared to the Brillouin zone dimensions. Because the electronic states in the Brillouin zone are quasi–continuous functions of $\mathbf{k}$, to obtain the lineshape for an interband transition, we must integrate $\omega_{\mathbf{k}}$ over all wave vectors $\mathbf{k}$ in the Brillouin zone. By recognizing that both the perturbation matrix elements $\langle v|\mathcal{H}'|c\rangle$ and the joint density of states ρ_{cv} are $\mathbf{k}$-dependent, we obtain upon integration of (17.34) over $\mathbf{k}$ space the total transition probability per unit time

$$W = \frac{2\pi}{\hbar}\int |\langle v|\mathcal{H}'|c\rangle|^2 \frac{2}{8\pi^3}\delta(E_c(\mathbf{k}) - E_v(\mathbf{k}) - \hbar\omega)\, d^3k \tag{17.35}$$

for a 3D system. For 2D and 1D systems, we replace $[d^3k/(2\pi)^3]$ by $[d^2k/(2\pi)^2]$ and $[dk/(2\pi)]$, respectively. The perturbation Hamiltonian for the electromagnetic interaction in (17.34) is simply

$$\mathcal{H}' = -\frac{e\mathbf{A}\cdot\mathbf{p}}{m} \tag{17.36}$$

where the time dependence of the vector potential $\mathbf{A}$ has already been taken into account in formulating time dependent perturbation theory and in the use of the Fermi Golden Rule (see Appendix B), so that the vector potential $\mathbf{A}$ in (17.36) is a vector with only a spatial dependence. In taking matrix elements of the perturbation Hamiltonian $\mathcal{H}'$ in (17.36), we need then only consider matrix elements of the momentum operator $\rho_{cv}(\hbar\omega)$ that connect the valence and conduction bands. In practical cases, it is often not necessary to evaluate these matrix elements explicitly because it is precisely these momentum matrix elements that determine the experimentally measured effective masses (see Sect. 17.3). If we assume for simplicity that $|\langle v|\mathcal{H}'|c\rangle|^2$ is independent of $\mathbf{k}$, then the remaining term to be considered in calculating the integral in (17.35) is the joint density of states between the valence and conduction bands $\rho_{cv}(\hbar\omega)$. For a 3D system, we thus define $\rho_{cv}(\hbar\omega)$ as

$$\rho_{cv}(\hbar\omega) \equiv \frac{2}{8\pi^3}\int \delta[E_c(\mathbf{k}) - E_v(\mathbf{k}) - \hbar\omega]\, d^3k \tag{17.37}$$

and $\rho_{cv}(\hbar\omega)$ is the number of states per unit volume per unit energy range which occurs with an energy difference between the conduction and valence bands equal to the incident photon energy. As explained above, $\rho_{cv}(\hbar\omega)$ can be evaluated in a corresponding manner for 2D and 1D systems.

17.5 Connecting Optical Properties and the Joint Density of States

We would now like to look at this joint density of states (17.37) in more detail to see why the optical properties of solids give unique information about the electronic energy band structure. The main point is that optical measurements preferentially

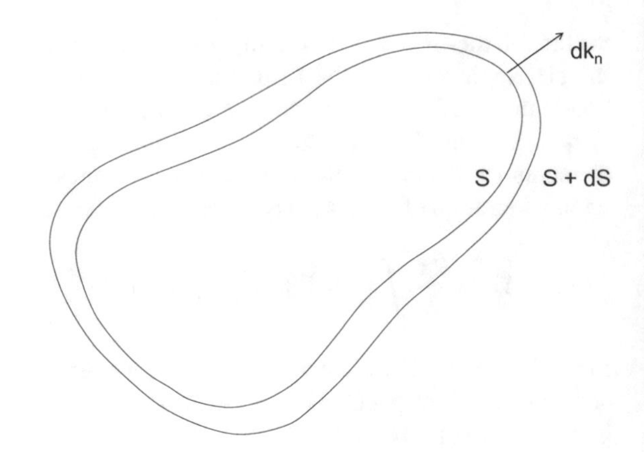

Fig. 17.2 Adjacent constant energy difference surfaces in reciprocal space, S and $S + dS$, where the energy difference is between the valence and conduction bands $d(E_c - E_v)$, and dk_n is the normal to the surface in k-space between the constant energy difference

provide information about the electronic energy bands at *particular* **k** *points* in the Brillouin zone, usually points of high symmetry and near energy band extrema. This can be understood by casting $\rho_{cv}(\hbar\omega)$ from (17.37) into a more transparent form. We start with the definition of the joint density of states given in (17.37). It is convenient to convert this integral over **k**-space to an integral over energy. This is done by introducing a constant energy surface S in k-space such that the energy difference $E_c - E_v = \hbar\omega$ is the incident photon energy. Then we can introduce the constant energy surfaces S and $S + dS$ in reciprocal space (see Fig. 17.2) as corresponding to a constant energy difference between the conduction and valence bands at each **k** point in the Brillouin zone so that

$$d^3k = dS\, dk_n \tag{17.38}$$

where dk_n is an element of a wave vector normal to S, as shown in Fig. 17.2, and follows each segment of the energy contour of Fig. 17.2.

By definition of the gradient, we have $|\nabla_k E| dk_n = dE$ so that for constant energy surfaces with an energy difference $E_c - E_v$ we write:

$$|\nabla_k(E_c - E_v)| dk_n = d(E_c - E_v). \tag{17.39}$$

Therefore

$$d^3k = dk_n dS = dS\left[\frac{d(E_c - E_v)}{|\nabla_k(E_c - E_v)|}\right] \tag{17.40}$$

so that the joint density of states becomes:

$$\rho_{cv}(\hbar\omega) = \frac{2}{8\pi^3}\int\int\int \frac{dS\, d(E_c - E_v)\delta(E_c - E_v - \hbar\omega)}{|\nabla_k(E_c - E_v)|}. \tag{17.41}$$

We now carry out the above integral over $d(E_c - E_v)$ to obtain an integral over the surface indicated in Fig. 17.2

$$\rho_{cv}(\hbar\omega) = \frac{2}{8\pi^3} \int\int \frac{dS}{|\nabla_k(E_c - E_v)|_{E_c - E_v = \hbar\omega}}. \tag{17.42}$$

Of special interest are those points in the Brillouin zone where $(E_c - E_v)$ is stationary so that $\nabla_k(E_c - E_v)$ becomes very small. At such points, called *joint critical points*, the denominator of the integrand in (17.42) vanishes and especially large contributions can be made to $\rho_{cv}(\hbar\omega)$. This interpretation of (17.42) can also be understood on the basis of physical considerations. Around critical points, the photon energy $\hbar\omega = (E_c - E_v)$ is highly effective in inducing electronic transitions over a relatively larger region of the Brillouin zone than would be the case for transitions occurring around non-critical points. The relatively large contributions to the transition probability for critical points gives rise to "structure" observed in the frequency dependence of the optical properties of solids. Critical points generally occur at high symmetry points in the Brillouin zone, though this is not necessarily the case, and it is important to look for large contributions to $\rho_{cv}(\hbar\omega)$ coming from each of these situations, both in the case of bulk materials and in nanostructures.

As an illustration, let us consider the energy bands of the semiconductor germanium (see Fig. 17.3). Here we see that both the valence and conduction bands have extrema at the Γ point, $\mathbf{k} = 0$, although the lowest conduction band minimum is located at the L point. For the band extrema at $\mathbf{k} = 0$, the condition $[E_c(k=0) - E_v(k=0)] = \hbar\omega$ gives rise to critical points in the joint density of states. Notice also that around the L points, extrema occur in both the valence and conduction bands, and a critical point therefore results at the L point in

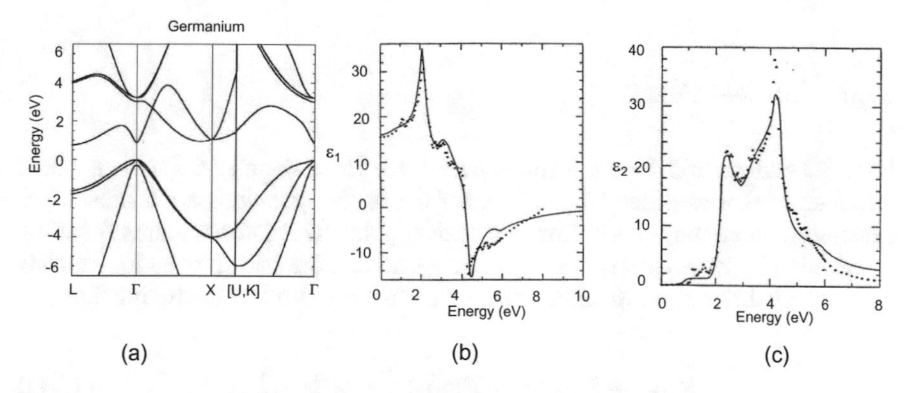

Fig. 17.3 **a** $E(\mathbf{k})$ for a few high symmetry directions in germanium, neglecting the spin-orbit interaction. The frequency dependence of the real **b** (ε_1) and **c** imaginary (ε_2) parts of the dielectric function for germanium $\varepsilon = \varepsilon_1 + i\varepsilon_2$. The solid curves are obtained from an analysis of experimental normal-incidence reflectivity data, while the dots are calculated from an energy band model. In particular $\varepsilon_2(\omega)$ for germanium provides an excellent example for illustrating the 4 kinds of critical points: M_0, M_1, M_2 and M_3 (see Sect. 17.6)

germanium. Since the energy difference $[E_c - E_v]$ has a relatively small gradient as we move away from the L point, this critical point participates more effectively in the interband transitions that are observed in the actual optical spectra. In fact, for germanium, Fig. 17.3 shows that there are large regions along the (100) and (111) axes where the energy separation between valence and conduction bands $(E_c - E_v)$ is roughly constant. These large regions in k-space make very large contributions to the complex dielectric function $\varepsilon(\omega)$. We can see these features directly by looking at the frequency dependence of the real (ε_1) and imaginary (ε_2) parts of the dielectric function for germanium (see Fig. 17.3b, c). Here we see that at low photon energies (below ~2 eV), where the interband transitions from the $\Gamma_{25'}$ valence band to the $\Gamma_{2'}$ conduction band dominate, the contributions to the real and imaginary parts of the dielectric function are small. On the other hand, the contributions from the large regions of the Brillouin zone along the (100) and (111) axes between 2 and 5 eV are very much more important, as is seen in Fig. 17.3 for both $\varepsilon_1(\omega)$ and $\varepsilon_2(\omega)$.

In describing these strong contributions to the dielectric function of germanium, we say that the valence and conduction bands track each other, and in this way they produce a large contribution to the joint density of states over large regions of the Brillouin zone for certain energy differences between the conduction and valence bands. A similar situation occurs in silicon and in common III-V semiconductors. The diagram in Fig. 17.3 shows that beyond ~5 eV there is no longer any significant tracking of the valence and conduction bands of germanium. Consequently, the magnitudes of $\varepsilon_1(\omega)$ and $\varepsilon_2(\omega)$ fall sharply for $\hbar\omega$ beyond ~5 eV. The absolute magnitudes of ε_1 and ε_2 for germanium and other semiconductors crystallizing in the diamond or zincblende structure are relatively large. We will see shortly, when we discuss the Kramers–Kronig relations in Chap. 19, that these large magnitudes of ε_1 and ε_2 are responsible for the large value of $\varepsilon_1(\omega \to 0)$ in these materials. For germanium $\varepsilon_1(0)$ is 16 from Fig. 17.3b.

17.6 Critical Points

For a 3D system, critical points (often called Van Hove singularities) are classified into four categories depending on whether the band separations are increasing or decreasing as we move away from the critical point. This information is found by expanding $[E_c(\mathbf{k}) - E_v(\mathbf{k})]$ in a Taylor series around the critical point $\mathbf{k_0}$ which is at an energy difference extremum, where we can write as a first approximation:

$$E_c(\mathbf{k}) - E_v(\mathbf{k}) = E_g(\mathbf{k_0}) + \sum_{i=1}^{3} a_i(k_i - k_{0i})^2 \tag{17.43}$$

in which the energy gap at the expansion point $\mathbf{k_0}$ is written as $E_g(\mathbf{k_0})$ and as we move away from $\mathbf{k_0}$ in three dimensions, the sum is taken over the three directions x, y, and z. The coefficients a_i in (17.43) are related to the second derivative of the

Type of Singularity	Number of Negative a_i's	Joint Density of States
M_0 Minimum	0	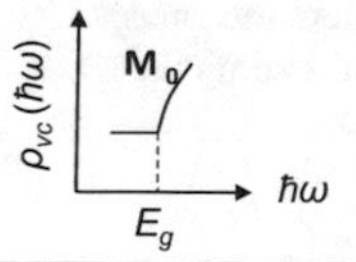
M_1 Saddle Point	1	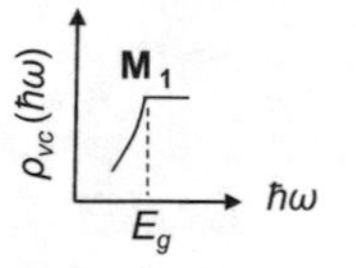
M_2 Saddle Point	2	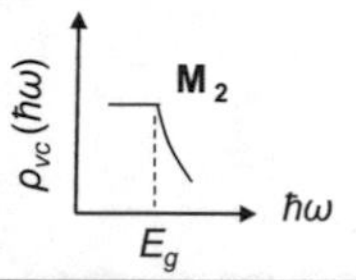
M_3 Maximum	3	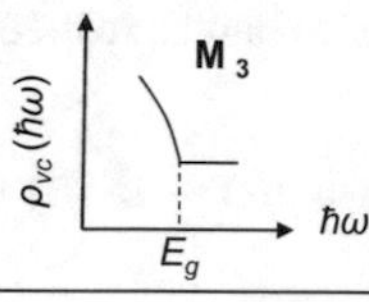

Fig. 17.4 Summary of the joint density of states for a 3D system near each of the distinct types of critical points: M_0, M_1, M_2 and M_3

interband energy difference by $2a_i = (\partial^2/\partial k_i^2)[E_c(\mathbf{k}) - E_v(\mathbf{k})]$. The classification of the critical points in a 3D system shown in Fig. 17.4 is made according to how many a_i coefficients in (17.43) are *negative*. The shapes given for the joint density of states curves of Fig. 17.4 are obtained, as is here illustrated, for the case of an M_0 singularity for a 3D system. In the case of 2D and 1D systems, there are 3 and 2 types of critical points, respectively, using the same definition of the coefficients a_i to define the type of critical point.

As an example, we will calculate $\rho_{cv}(\hbar\omega)$ for an M_0 singularity in a 3D system, assuming simple parabolic bands for simplicity (see Fig. 17.5). Here,

$$E_c(\mathbf{k}) = \frac{E_g}{2} + \frac{\hbar^2 k^2}{2m_c} \tag{17.44}$$

and

$$E_v(\mathbf{k}) = \frac{-E_g}{2} - \frac{\hbar^2 k^2}{2m_v} \tag{17.45}$$

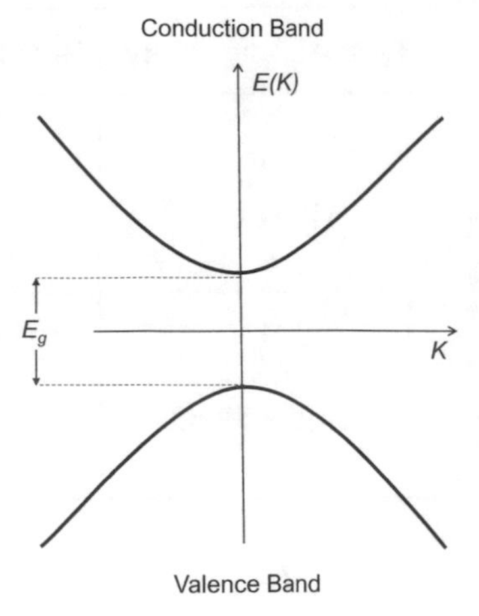

Fig. 17.5 Electronic energy bands associated with an M_0 critical point for a 3D system, where the energy gap E_g of the semiconductor is indicated

where E_g is the energy gap, and m_c and m_v are effective masses for the conduction and valence bands, respectively, and m_v is taken as a positive number. We thus obtain

$$E_c(\mathbf{k}) - E_v(\mathbf{k}) = E_g + \frac{\hbar^2 k^2}{2}\left(\frac{1}{m_c} + \frac{1}{m_v}\right) = E_g + \frac{\hbar^2 k^2}{2m_r} \tag{17.46}$$

where we define the reduced effective mass m_r through the relation

$$\frac{1}{m_r} = \frac{1}{m_c} + \frac{1}{m_v}. \tag{17.47}$$

Taking the gradient of the energy difference $E_c - E_v$ as a function of k yields

$$\nabla_k(E_c - E_v) = \frac{\hbar^2 \mathbf{k}}{m_r} \tag{17.48}$$

so that the joint density of states becomes

$$\rho_{cv}(\hbar\omega) = \frac{2}{8\pi^3}\int \frac{dS}{|\nabla_k(E_c - E_v)|}_{E_c - E_v = \hbar\omega} \tag{17.49}$$

or

$$\rho_{cv}(\hbar\omega) = \frac{2}{8\pi^3}\left[\frac{4\pi}{\hbar^2}\left(\frac{k^2 m_r}{k}\right)\right]_{E_c - E_v = \hbar\omega} = \left[\frac{m_r}{\pi^2\hbar^2}k\right]_{E_c - E_v = \hbar\omega}. \tag{17.50}$$

We evaluate k in (17.50) from the condition

$$E_c - E_v = \hbar\omega = E_g + \frac{\hbar^2 k^2}{2m_r} \tag{17.51}$$

or

$$k = \left[\frac{2m_r}{\hbar^2} (\hbar\omega - E_g) \right]^{1/2} \tag{17.52}$$

so that

$$\begin{aligned} \rho_{cv}(\hbar\omega) &= \frac{1}{2\pi^2} \left[\frac{2m_r}{\hbar^2} \right]^{3/2} \sqrt{\hbar\omega - E_g} \quad && \hbar\omega > E_g \\ &= 0 && \hbar\omega < E_g \end{aligned} \tag{17.53}$$

as shown in Fig. 17.4 for an M_0 critical point. The expression for $\rho_{cv}(\hbar\omega)$ in (17.53) is not singular for a 3D system but rather represents a discontinuity in the slope at $\hbar\omega = E_g$, which corresponds to a *threshold* for the absorption process, as discussed in Sect. 17.1

On the other hand, the situation is quite different for the joint density of states corresponding to an M_0 critical point for a 3D system in a magnetic field. At a critical point in k-space, the joint density of states in a magnetic field *does* show singularities and the density of states in a magnetic field can approach infinity. These singularities in the magnetic field dependence of the joint density of states make it possible to carry out resonance experiments in solids, despite the quasi–continuum of the energy levels in the electronic dispersion relations $E(\mathbf{k})$ in zero magnetic field.

We note that we can also have M_0-type critical points for electronic energy bands that look like Fig. 17.6a or like Fig. 17.6b. It is clear that the energy difference $E_c - E_v$ in Fig. 17.6b varies more slowly, as a function of k around the critical point than it does in Fig. 17.6a. Thus, bands that tend to "track" each other in k-space, as in Fig. 17.6b have an exceptionally high joint density of states and contribute strongly to the optical properties. Above we gave examples of electronic energy bands with very high values for $\varepsilon_1(\omega)$ and $\varepsilon_2(\omega)$ due to bands that track each other as are

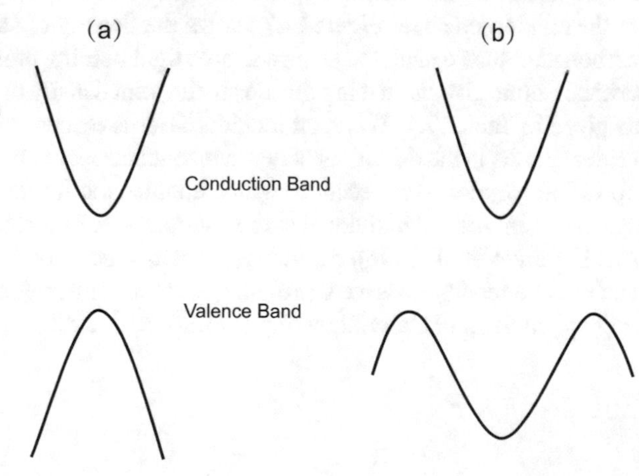

Fig. 17.6 Two cases of band extrema which are associated with M_0 critical points. **a** Conduction band minimum and a valence band maximum and **b** Both conduction and valence bands showing an M_0 type minimum. The dispersion relation shown in (**b**) for the valence band is sometimes called a "camel-back" structure

Table 17.1 Functional forms for the joint density of states $\rho_{vc}(\hbar\omega)$ for various types of critical points M_0, M_1, M_2 and M_3 below and above the energy gap E_g for 3D, 2D, and 1D systems, where C denotes a constant, not dependent on energy

	Type	$\hbar\omega < E_g$	$\hbar\omega > E_g$
3D	M_0	0	$(\hbar\omega - E_g)^{1/2}$
	M_1	$C - (E_g - \hbar\omega)^{1/2}$	C
	M_2	C	$C - (\hbar\omega - E_g)^{1/2}$
	M_3	$(E_g - \hbar\omega)^{1/2}$	0
2D	M_0	0	C
	M_1	$-\ln(E_g - \hbar\omega)$	$-\ln(\hbar\omega - E_g)$
	M_2	C	0
1D	M_0	0	$(\hbar\omega - E_g)^{-1/2}$
	M_0	$(E_g - \hbar\omega)^{-1/2}$	0

found in common semiconductors like germanium along the Λ (111) direction (see Figs. 17.3b, c).

In addition to the M_0 critical points, we have M_1, M_2, and M_3 critical points in 3D systems. The functional forms for the joint density of states for $\hbar\omega < E_g$ and $\hbar\omega > E_g$ are given in Table 17.1. This table also gives the corresponding expressions for the electronic density of states for 2D and 1D systems. From Table 17.1, we see that in 2D, the M_0 and M_2 critical points correspond to discontinuities in the joint density of states at E_g, while the M_1 singularity corresponds to a saddle point logarithmic divergence. In the case of the 1D system, both the M_0 and M_1 critical points are singular, as discussed below.

17.7 Critical Points in Low Dimensional Materials

Because of the presence of critical points, large enhancement effects in 1D and 2D materials enable the observation of spectra from a very small amount of matter in the condensed state. Figure 17.7 shows the density of states of a semiconducting carbon nanotube calculated using a simple tight binding model. In this 1D system, the critical points give rise to singularities in the joint density of states, of the form $E^{-1/2}$, as given in Table 17.1. When an incident laser is resonant with a transition between critical points in the density of states, a resonance occurs in the Raman scattering and optical absorption cross section, which enables the Raman and photoluminescence spectra of an isolated individual carbon nanotube to be measured (see Jorio et al. 2001 and Lefebvre et al. 2004). Similarly, Raman spectra of individual graphene layers can also be readily observed, providing detailed information about layer thickness, doping, strain, and material quality (Ferrari et al. 2006).

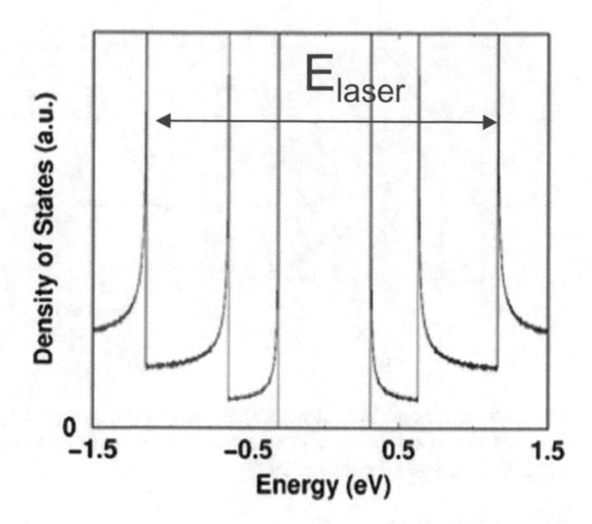

Fig. 17.7 Calculated energy dependence of the density of states of a semiconducting carbon nanotube. These calculations use a tight binding model

Problems

17.1 This problem is intended to make you familiar with time-dependent perturbation theory. Suppose we have a two level system and that initially the system is in the ground state E_1. Suppose that we apply a perturbation $\mathcal{H}'$ arising from an electric field with time variation $e^{i\omega t}$ which is switched on at $t = 0$ and switched off at $t = t_1$.

(a) Find an expression for the probability that the system in state E_1 at time t_1 and at time $2t_1$. Consider the case when the system is far from resonance and also when near resonance ($\omega \simeq \omega_2 - \omega_1$).
(b) Suppose that $\mathcal{H}'$ is associated with a very high power laser. What happens in part (a) if $\mathcal{H}'$ is very large? What is the temperature of the system under these condition?

17.2 This problem considers the joint density of states for a 1D electron gas.

(a) Find an expression for the joint density of states for a 1D electron gas (e.g., a quantum wire of 10 nm diameter) for each of the pertinent types of singularities (M_0, ...).
(b) How many types of such singularities are there in 1D?
(c) What is the selection rule for allowed interband transitions in (a)?

17.3 Verify the form of the critical points given in Table 17.1.

17.4 Label the critical points in the real (ε_1) and imaginary (ε_2) parts of the dielectric function for silicon in Fig. 17.8.

17.5 Label the critical points in the real (ε_1) and imaginary (ε_2) parts of the dielectric function for GaAs in Fig. 17.9.

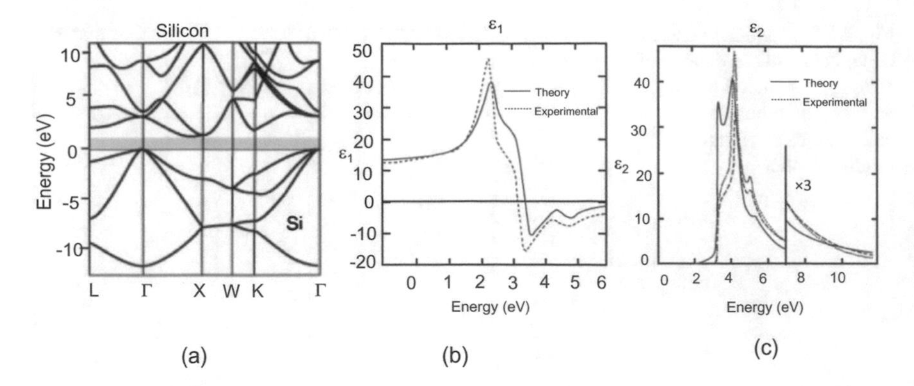

Fig. 17.8 Electronic band structure and the real and imaginary parts of the complex dielectric function ($\varepsilon_1(\omega)$ and $\varepsilon_2(\omega)$) for silicon, taken from Yu and Cardona, Fundamentals of Semiconductors, pp. 251–258, Springer, Berlin, 1999

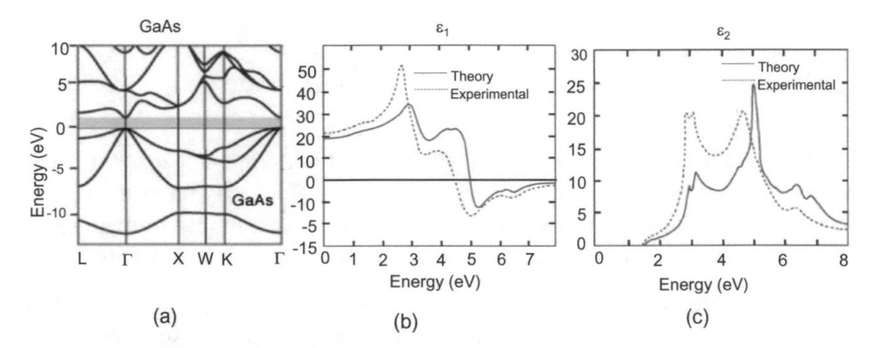

Fig. 17.9 Electronic band structure and real and imaginary parts of the complex dielectric function ($\varepsilon_1(\omega)$ and $\varepsilon_2(\omega)$) for GaAs, taken from Yu and Cardona, Fundamentals of Semiconductors, pp. 251–258, Springer, Berlin, 1999

17.6 Show that (17.34) can be derived from time-dependent perturbation theory, as outlined in Appendix B.

17.7 Consider an M_0 critical point for direct interband transitions in a semiconductor and assume that a parabolic expansion of the energy bands about their band extrema is valid to very large energy values ($E \to \infty$).

(a) Using the imaginary part of the dielectric function $\varepsilon_2(\omega)$ for an M_0

$$\varepsilon_2(\omega) = A(x-1)^{1/2} \text{ for } \quad x > 1$$

$$\varepsilon_2(\omega) = 0 \qquad\qquad \text{for } \quad x < 1$$

where $x = \hbar\omega/E_g$ and

$$A = \frac{2e^2(2\mu)^{3/2}}{m^2\omega^2\hbar^3}|p_{cv}|^2 E_g^{1/2}$$

obtain the corresponding $\varepsilon_1(\omega)$ assuming a constant momentum matrix element $|p_{cv}|^2$, independent of frequency.

(b) Using the analytic expressions in part (a), use MATLAB to plot both the imaginary and real parts of the dielectric function $\varepsilon(\omega) = \varepsilon_1(\omega) + i\varepsilon_2(\omega)$.

17.8 CdSe is a direct gap semiconductor with $E_g = 1.6$ eV. The matrix element for direct transitions between the valence and conduction bands is 6.1×10^{-20} g cm s^{-1} and the effective mass for electrons is 0.08 m_0 and for holes $0.45m_0$ where m_0 is the free electron mass. The refractive index in the vicinity of the energy gap E_g is approximately 3.

(a) Calculate the derivative of the normal-incidence reflectivity with respect to photon energy dR/dE near the energy gap E_g, assuming a one electron band-to-band transition and light propagation along the c-axis (Hint: Prove that near the lowest energy gap, the derivative dR/dE is proportional to $d\varepsilon_1/dE$, and then use the results of Problem 19.6).

(b) Plot the derivative function and sketch by free hand how this curve would be modified by Coulomb interactions between the electron and the hole (i.e. excitonic effects).

17.9 The joint density of states for two simple parabolic bands with 3D dispersion relations E(k) = $\pm\ E_g/2 \pm \hbar^2k^2/(2m^*)$ about an M_0 singularity is

$$\rho_{vc} = \frac{1}{2\pi^2}\left[\frac{2m_r}{\hbar^2}\right]^{3/2}\sqrt{\hbar\omega - E_g} \qquad (17.54)$$

(a) Find the corresponding joint density of states for two 2D parabolic bands, for which the above dispersion relation E(k) holds in the xy-plane, but there is no dispersion along the z-direction.

(b) From the result in (a) does the effect of the confinement of the carriers in a quantum well increase or decrease the optical absorption intensity per unit area? Why is your answer physically reasonable?

17.10 (a) Calculate and sketch the joint density of states $\rho_{vc}(\hbar\omega)$ as a function of $\hbar\omega$ for an M_3 singularity in a 3D, 2D, and 1D crystal.

(b) Consider explicitly the spatial dependence of the vector potential

$$\mathbf{A} = \mathbf{A}_0 \exp[i(\mathbf{k}\cdot\mathbf{r} - \omega t)],$$

calculate the matrix element for the electromagnetic interaction between Bloch states including the effect of the spaticial dependence of the vector potential

$$\langle n'\mathbf{k}'|(e/mc)(\mathbf{p}\cdot\mathbf{A})|n\mathbf{k}\rangle.$$

On the basis of your result in (a) show that the conservation of crystal momentum **k** is a good approximation for optical transitions.

Suggested Readings

F. Bassani, G. Pastori-Parravicini, *Electronic States and Optical Transitions in Solids* (Pergamon Press, Oxford, 1975)

M. Fox, *Optical Properties of Solids* (Oxford University Press, Oxford, 2001)

W. Jones, N.H. March, *Theoretical Solid State Physics* (Wiley, New York, 1973), pp. 806–814

A. Jorio, M.S. Dresselhaus, R. Saito, G. Dresselhaus, *Raman Spectroscopy in Graphene Related Systems* (Wiley, New York, 2011)

O. Madelung, *Introduction to Solid State Theory* (Springer, Berlin, 2012), pp. 262–271

Y. Yu, M. Cardona, *Fundamentals of Semiconductors* (Springer, Berlin, 2013), pp. 251–258

References

A.C. Ferrari, J.C. Meyer, V. Scardaci, C. Casiraghi, M. Lazzeri, F. Mauri, S. Piscanec, D. Jiang, K.S. Novoselov, S. Roth, A.K. Geim, Raman spectrum of graphene and graphene layers. Phys. Rev. Lett. **97**, 187401 (2006)

A. Jorio, J.H. Hafner, C.M. Lieber, M. Hunter, T. McClure, G. Dresselhaus, M.S. Dresselhaus, Structural (n, m) determination of isolated single-wall carbon nanotubes by resonant Raman scattering. Phys. Rev. Lett. **86**, 1118 (2001)

J. Lefebvre, J.M. Fraser, P. Finnie, Y. Homma, Photoluminescence from an individual single-walled carbon nanotube. Phys. Rev. B **69**, 075403 (2004)

Chapter 18
Absorption of Light in Solids

18.1 The Absorption Coefficient

Measurement of the absorption of light is one of the most important techniques for optical measurements in solid state materials. In the optical absorption measurements, we are concerned with the light intensity $I(z)$ after the light beam traverses of a thickness z of material as compared with the incident light intensity I_0, thereby defining the absorption coefficient $\alpha_{\mathrm{abs}}(\omega)$ as:

$$I(z) = I_0 e^{-\alpha_{\mathrm{abs}}(\omega) z} \tag{18.1}$$

where the frequency dependence of the absorption coefficient is shown schematically in Fig. 18.1.

Since the intensity $I(z)$ depends on the square of the field variables, it immediately follows that

$$\alpha_{\mathrm{abs}}(\omega) = 2\omega\tilde{k}(\omega) \tag{18.2}$$

where the factor of 2 results from the definition of $\alpha_{\mathrm{abs}}(\omega)$ in terms of the light intensity, which is proportional to the *square* of the optical fields. This expression tells us that the absorption coefficient is proportional to $\tilde{k}(\omega)$, the imaginary part of the complex index of refraction (i.e., extinction coefficient), which is usually associated with power loss from the incident optical beam. We note that (18.2) applies to free carrier absorption in semiconductors in the limit $\omega\tau \gg 1$, and $\omega \gg \omega_p$, where ω_p is the plasma frequency for the free carriers.

We will now show that the frequency dependence of the absorption coefficient for a bulk system (3D) is quite different for the various physical processes that occur in the optical properties of solids as follow:

M. Dresselhaus et al., *Solid State Properties*, Graduate Texts in Physics,
https://doi.org/10.1007/978-3-662-55922-2_18

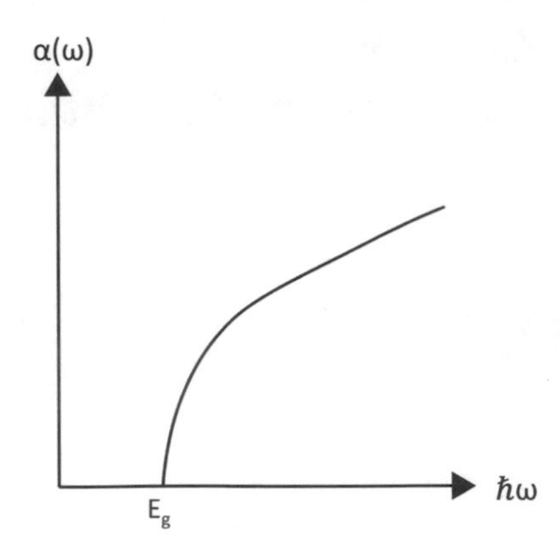

Fig. 18.1 Frequency dependence of the absorption coefficient near a threshold for probing the interband transitions in solid state materials such a band gap energy for direct optical transition or an indirect optical transition

1. **Free carrier absorption**
   a. Typical semiconductor $\alpha_{\text{abs}}(\omega) \sim \omega^{-2}$
   b. Metals at low frequencies $\alpha_{\text{abs}}(\omega) \sim \omega^{\frac{1}{2}}$
2. **Direct interband transitions**
   a. Form of absorption coefficient $\alpha_{\text{abs}}(\omega) \sim \frac{(\hbar\omega - E_g)^{\frac{1}{2}}}{\hbar\omega}$
   b. Conservation of crystal momentum
   c. Relation between m^* and the momentum matrix element
   d. Form of $\alpha_{\text{abs}}(\omega)$ for direct symmetry forbidden transitions $\sim \frac{(\hbar\omega - E_g)^{\frac{3}{2}}}{\hbar\omega}$
3. **Indirect interband transitions**
   a. Form of absorption coefficient $\alpha_{\text{abs}}(\omega) \sim (\hbar\omega - E_g \pm \hbar\omega_q)^2$
   b. Phonon absorption and emission processes

The summary given above is for 3D systems. In the case of 2D and 1D systems, the functional dependence is sensitive to the dimensionality of the system for each process.

18.2 Free Carrier Absorption in Semiconductors

For free carrier absorption we use the relation for the complex dielectric function $\varepsilon(\omega) = \varepsilon_1(\omega) + i\varepsilon_2(\omega)$ given by

$$\varepsilon(\omega) = \varepsilon_0 + \frac{i\sigma(\omega)}{\omega} \tag{18.3}$$

where ε_0 is the core dielectric constant in the optical frequency range above the lattice mode frequencies and ε_0 is assumed to be independent of ω, and this approximation can be made for many semiconductors. For example, for chemical elements with low atomic numbers like silicon, the introduction of suitable common dopants can be handled within this simple model. The electronic polarizability is related to the frequency-dependent electrical conductivity $\sigma(\omega)$ by the frequency-dependent Drude term

$$\sigma(\omega) = \frac{ne^2\tau}{m^*(1 - i\omega\tau)}. \tag{18.4}$$

The plasma frequency ω_p is then given by the vanishing of $\varepsilon_1(\omega)$, that is $\varepsilon_1(\omega_p) = 0$ or

$$\omega_p^2 = \frac{ne^2}{m^*\varepsilon_0}. \tag{18.5}$$

For semiconductors, the core dielectric constant ε_0 is typically a large number ($\varepsilon_0 \gg 1$) and the contribution due to the free carriers is small at infrared and visible frequencies. For metals, the free carrier absorption is dominant over the entire optical frequency range.

For semiconductors, the typical frequency range of interest is that above the optical phonon frequencies, and for these frequencies it is generally true that $\omega\tau \gg 1$. The generic expression for $\varepsilon(\omega)$ follows from expanding (18.3):

$$\varepsilon(\omega) = \varepsilon_0 + \frac{ine^2\tau(1 + i\omega\tau)}{m^*\omega[1 + \omega^2\tau^2]} = \varepsilon_0 + \frac{i\varepsilon_0\omega_p^2\tau(1 + i\omega\tau)}{\omega[1 + \omega^2\tau^2]} \tag{18.6}$$

which for $\omega\tau \gg 1$ becomes

$$\varepsilon(\omega) \simeq \varepsilon_0 + \frac{i\varepsilon_0\omega_p^2}{\omega^3\tau} - \frac{\varepsilon_0\omega_p^2}{\omega^2}. \tag{18.7}$$

In the range of interest for optical measurements in typical semiconductors, the relation $\omega \gg \omega_p$ is generally satisfied. It is then convenient to express the complex dielectric function $\varepsilon(\omega)$ in terms of the optical constants $\tilde{n}(\omega)$ and $\tilde{k}(\omega)$ according to the definition $\varepsilon(\omega) = [\tilde{n}(\omega) + i\tilde{k}(\omega)]^2$, where $\tilde{n}(\omega)$ is the optical index of refraction and $\tilde{k}(\omega)$ is the optical extinction coefficient. We can then write for the real part of the dielectric function as:

$$\varepsilon_1(\omega) \equiv \tilde{n}^2(\omega) - \tilde{k}^2(\omega) \approx \varepsilon_0 \tag{18.8}$$

where the index of refraction $\tilde{n}(\omega)$ is large and the extinction coefficient $\tilde{k}(\omega)$ is small, and light penetrates through a thin film (few nm thick) without significant absorption. For the imaginary part of the dielectric function, we have

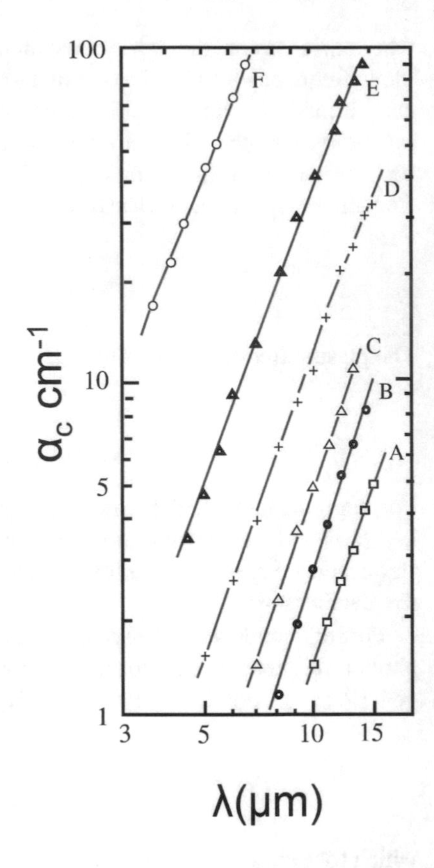

Fig. 18.2 Free carrier absorption (plotted as log(α) versus. log(λ)) in n-type InAs at room temperature for six different carrier concentrations (in units of 10^{17} cm^{-3}) A: 0.28; B: 0.85; C: 1.4; D: 2.5; E: 7.8; and F: 39.0. Reproduced with permission from Springer-Verlag, Yu and Cardona, Fundamentals of Semiconductors: Physics and Materials Properties, 1996

$$\varepsilon_2(\omega) \equiv 2\tilde{n}(\omega)\tilde{k}(\omega) \approx 2\sqrt{\varepsilon_0}\,\tilde{k}(\omega) = \frac{\varepsilon_0\omega_p^2}{\omega^3\tau} \tag{18.9}$$

which is small, since $\omega_p \ll \omega$. Thus, the absorption coefficient can be written as:

$$\alpha_{\text{abs}}(\omega) = 2\omega\tilde{k}(\omega) \simeq 2\omega\frac{\varepsilon_0\omega_p^2}{2\sqrt{\varepsilon_0}\omega^3\tau} = \frac{\sqrt{\varepsilon_0}\omega_p^2}{\omega^2\tau} \tag{18.10}$$

and thus $\alpha_{\text{abs}}(\omega)$ is proportional to $1/\omega^2$ or to λ^2 for free carrier absorption in semiconductors for the case where $\omega\tau \gg 1$ and $\omega \gg \omega_p$. Figure 18.2 shows a plot of the optical absorption coefficient for InAs versus wavelength on a log-log plot for various carrier densities, showing that $\alpha_{\text{abs}}(\omega) \sim \lambda^p$, where the exponent, p, is between 2 and 3 for a wide range of donor concentrations.

18.3 Free Carrier Absorption in Metals

The typical limits of validity for metals are somewhat different than for semiconductors. In particular we consider here the case where $\omega\tau \ll 1$, $\omega \ll \omega_p$, $|\varepsilon_0| \ll \sigma/\omega$, so that $\tilde{n} \simeq \tilde{k}$ is obtained from (18.11). Thus we obtain

$$\varepsilon(\omega) \simeq \frac{i\sigma}{\omega} \simeq \frac{ine^2\tau}{\omega m^*} \simeq i\varepsilon_2(\omega) \equiv 2i\tilde{n}\tilde{k} \simeq 2i\tilde{k}^2. \tag{18.11}$$

It is worth mentioning that n is used as the carrier density, while $\tilde{n}$ is the real part of the index of refraction. This gives us the extinction coefficient $\tilde{k}(\omega)$ as

$$\tilde{k}(\omega) = \sqrt{\frac{ne^2\tau}{2m^*\omega}} \tag{18.12}$$

and the absorption coefficient becomes:

$$\alpha_{\text{abs}}(\omega) = 2\omega\tilde{k}(\omega) = \sqrt{\frac{2\pi\omega ne^2\tau}{m^*}} \tag{18.13}$$

For this limit $\alpha_{\text{abs}}(\omega)$ is proportional to $\sqrt{\omega}$. Usually, the convenient observable for metals is the reflectivity. In the limit appropriate for metals, $\tilde{n} = \tilde{k}$, and both $\tilde{n}$ and $\tilde{k}$ are large. We thus have

$$\mathscr{R} = \frac{(\tilde{n}-1)^2 + \tilde{k}^2}{(\tilde{n}+1)^2 + \tilde{k}^2} = \frac{\tilde{n}^2 - 2\tilde{n} + 1 + \tilde{k}^2}{\tilde{n}^2 + 2\tilde{n} + 1 + \tilde{k}^2} = 1 - \frac{4\tilde{n}}{\tilde{n}^2 + \tilde{k}^2 + 2\tilde{n} + 1} \tag{18.14}$$

$$\mathscr{R} \approx 1 - \frac{4\tilde{n}}{\tilde{n}^2 + \tilde{k}^2} \approx 1 - \frac{2}{\tilde{n}}. \tag{18.15}$$

However, from (18.12) and the condition $\tilde{n} \approx \tilde{k} \gg 1$, we obtain

$$\tilde{n}(\omega) \simeq \sqrt{\frac{me^2\tau}{2m^*\omega}} \tag{18.16}$$

so that the reflectivity shows a frequency dependence

$$\mathscr{R}(\omega) \simeq 1 - 2\sqrt{\frac{2m^*\omega}{ne^2\tau}}. \tag{18.17}$$

Equation (18.17) is known as the Hagen–Rubens relation which holds well for most metals in the infrared region of the spectrum and also applies to degenerately doped semiconductors below the plasma frequency where interband transitions are not so important.

18.4 Direct Interband Transitions

To calculate the absorption due to direct interband transitions we again start with the definition for the absorption coefficient $\alpha_{\text{abs}}(\omega)$, which is defined as the power removed from the incident beam per unit volume per unit incident flux of electromagnetic energy:

$$\alpha_{\text{abs}}(\omega) = \frac{(\hbar\omega) \times \text{number of transitions/unit volume/unit time}}{\text{incident electromagnetic flux}}. \tag{18.18}$$

The incident electromagnetic flux appearing in the denominator of (18.18) is calculated from the Poynting vector

$$\mathscr{S} = \frac{1}{2} Re(\mathbf{E}^* \times \mathbf{H}). \tag{18.19}$$

By adopting the Coulomb Gauge $\nabla \cdot \mathbf{A} = 0$, it is convenient to relate the field variables ($\mathbf{E}, \mathbf{H}, \mathbf{B}$) to $\mathbf{A}$, the vector potential:

$$\mathbf{E} = -\frac{\partial \mathbf{A}}{\partial t} = i\omega\mathbf{A} \tag{18.20}$$

$$\mu\mathbf{H} = \mathbf{B} = \nabla \times \mathbf{A}. \tag{18.21}$$

In non-magnetic materials, we can take the permeability μ to be μ_0. In taking the cross product $\nabla \times \mathbf{A}$, we assume a plane wave of the form

$$\mathbf{A} = \mathbf{A}_0 e^{i(\mathbf{K}\cdot\mathbf{r}-\omega t)} \tag{18.22}$$

where the propagation constant for the light is denoted by the wave vector $\mathbf{K}$. We thus obtain the Poynting vector in terms of the vector potential, $\mathbf{A}$

$$\mathscr{S} = \frac{1}{2} Re\Big[- i\omega\mathbf{A}^* \times (i\mathbf{K} \times \mathbf{A})\Big] \tag{18.23}$$

or

$$\mathscr{S} = \frac{\omega}{2} Re\Big[(\mathbf{A}^* \cdot \mathbf{A})\mathbf{K} - (\mathbf{A}^* \cdot \mathbf{K})\mathbf{A}\Big]. \tag{18.24}$$

Utilizing the fact that for a transverse plane wave $\mathbf{A}^* \cdot \mathbf{K} = 0$, we obtain from (18.24)

$$\mathscr{S} = \frac{\omega\tilde{n}\omega}{2} |A|^2 \hat{K} \tag{18.25}$$

where $\tilde{n}$ denotes the real part of the complex index of refraction and $\hat{K}$ is a unit vector along the Poynting vector. This quantity $|\mathscr{S}|$ in (18.25) becomes the denominator in

(18.18), which is the expression defining the absorption coefficient. The transition probability/unit time/unit volume is calculated from the "Fermi Golden Rule"

$$W = \frac{2\pi}{\hbar} \mid \mathscr{H}'_{vc} \mid^2 \rho_{cv}(\hbar\omega). \tag{18.26}$$

If we wish to consider the absorption process at finite temperature, we also need to include the Fermi functions to represent the occupation of the states at finite temperature

$$f(E_v)[1 - f(E_c)] - f(E_c)[1 - f(E_v)] \tag{18.27}$$

in which the first group of terms represents the absorption process, which further depends on the valence band (v) being nearly full and the conduction band (c) being nearly empty. The second group of terms represents the emission process which proceeds if there are occupied conduction states and unoccupied valence states. Clearly, the Fermi functions in (18.27) simply reduce to $[f(E_v) - f(E_c)]$. The matrix elements $|\mathscr{H}'_{vc}|^2$ in (18.26) can be written in terms of the electromagnetic interaction Hamiltonian

$$\mathscr{H}'_{vc} = \langle v|\mathscr{H}'_{em}|c\rangle = -\left(\frac{e}{m}\right)\langle v|\mathbf{A}(\mathbf{r}, t) \cdot \mathbf{p}|c\rangle. \tag{18.28}$$

We show in Sect. 18.5 that the matrix element $\langle v|\mathbf{A}(\mathbf{r}, t) \cdot \mathbf{p}|c\rangle$ coupling the valence and conduction bands for the electromagnetic interaction is diagonal in wave vector $\mathbf{k}$ since the wave vector for light $\mathbf{K}$ is small relative to Brillouin zone dimensions. As a result, the spatial dependence of the vector potential can be ignored. Thus the square of the matrix elements coupling the valence and conduction bands becomes

$$|\mathscr{H}'_{vc}|^2 = \left(\frac{e}{m}\right)^2 |A|^2 |\langle v|p|c\rangle|^2, \tag{18.29}$$

where $|\langle v|p|c\rangle|^2$ couples states with the same electron wave vector k in the valence and conduction bands. Since $|\langle v|p|c\rangle|^2$ is slowly varying with k in comparison to $\rho_{cv}(\hbar\omega)$, it is convenient to neglect the k dependence of $|\langle v|p|c\rangle|^2$. Thus for direct interband transitions, we obtain the following expression for the absorption coefficient

$$\alpha_{\text{abs}}(\omega) = \frac{(\hbar\omega)[\frac{2\pi}{\hbar}(\frac{e}{m})^2|A|^2|\langle v|p|c\rangle|^2\rho_{cv}(\hbar\omega)][f(E_v) - f(E_c)]}{|A|^2\omega\tilde{n}\omega/2} \tag{18.30}$$

or

$$\alpha_{\text{abs}}(\omega) = \frac{4\pi^2 e^2}{m^2\tilde{n}\omega}|\langle v|p|c\rangle|^2\rho_{cv}(\hbar\omega)[f(E_v) - f(E_c)] \tag{18.31}$$

where $\tilde{n}$ in (18.30) and (18.31) denotes the index of refraction.

To get a physical idea about the functional forms of the quantities in (18.31), we will consider a rather simplified picture of two simple parabolic bands with an electron making an allowed optical transition from the valence band to the conduction band via a non-vanishing momentum matrix element coupling them. Writing the joint density of states from (17.37) for the case of an M_0 critical point (as occurs near $k = 0$ for many semiconductors)

$$\rho_{cv}(\hbar\omega) = \frac{1}{2\pi^2}\left(\frac{2m_r}{\hbar^2}\right)^{3/2}\sqrt{\hbar\omega - E_g} \tag{18.32}$$

where m_r is the reduced mass for the valence and conduction bands, we can estimate the absorption coefficient $\alpha_{\text{abs}}(\omega)$. At very low temperature, a semiconductor has an essentially filled valence band and an empty conduction band; that is $f(E_v) = 1$ and $f(E_c) = 0$. We can estimate $|\langle v|p|c\rangle|^2$ from the effective mass sum-rule (17.33)

$$|\langle v|p|c\rangle|^2 \simeq \frac{m_0 E_g}{2}\frac{m_0}{m^*} \tag{18.33}$$

where m_0 is the free electron mass. After substitution of (18.32) and (18.33) into (18.31), we obtain the following frequency dependence of the absorption coefficient fordirect allowed interband transitions:

$$\alpha_{\text{abs}}(\omega) \propto \frac{1}{\omega}\sqrt{\hbar\omega - E_g} \tag{18.34}$$

so that the direct optically–allowed interband transitions are characterized by a threshold at the energy gap E_g, as shown in Fig. 18.1. We thus see a very different frequency dependence of $\alpha_{\text{abs}}(\omega)$ for the various physical processes.

It is sometimes convenient to relate the optical absorption coefficient to the imaginary part of the dielectric function

$$\varepsilon_2(\omega) = \frac{\tilde{n}}{\omega}\alpha_{\text{abs}}(\omega) \tag{18.35}$$

which from (18.31) becomes

$$\varepsilon_2(\omega) = \left(\frac{e}{m\omega}\right)^2 |\langle v|p|c\rangle|^2 \rho_{cv}(\hbar\omega)[f(E_v) - f(E_c)]. \tag{18.36}$$

If we introduce the dimensionless quantity f_{vc}, which is usually called the oscillator strength and is defined by

$$f_{vc} = \frac{2|\langle v|p|c\rangle|^2}{m[E_c(k) - E_v(k)]} = \frac{2|\langle v|p|c\rangle|^2}{m\hbar\omega}, \tag{18.37}$$

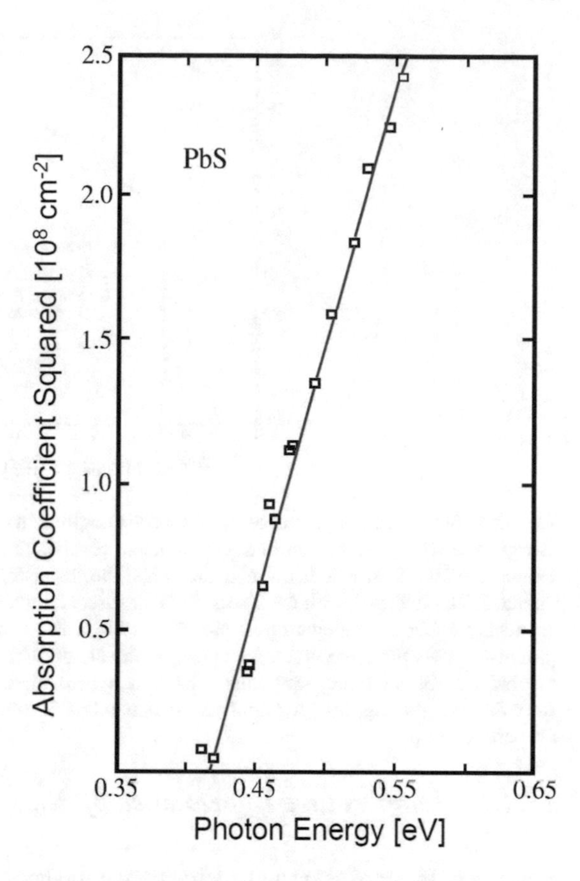

Fig. 18.3 Plot of the square of the absorption coefficient of PbS as a function of photon energy showing the linear dependence of $[\alpha_{\rm abs}(\omega)]^2$ on $\hbar\omega$. The intercept with the x-axis defines the direct energy gap. Reproduced from Springer-Verlag, Yu and Cardona, Fundamentals of Semiconductors: Physics and Materials Properties, 1996

we obtain the following result for $\varepsilon_2(\omega)$ at $T = 0$ in terms of the oscillator strength f_{vc} and the density of states

$$\varepsilon_2(\omega) = \left(\frac{2e^2\hbar}{m\omega}\right) f_{vc}\rho_{cv}(\hbar\omega). \tag{18.38}$$

We further discuss how $\varepsilon_1(\omega)$ for interband transitions is obtained from $\varepsilon_2(\omega)$ in Sect. 19.2 using the Kramers–Kronig relation.

To illustrate the fit between these simple models and the behavior of the absorption coefficient near the fundamental absorption edge, we show in Fig. 18.3, a plot of $[\alpha_{\rm abs}]^2$ versus. $\hbar\omega$ for PbS, with the intercept of $[\alpha_{\rm abs}]^2$ on the photon energy axis giving the direct energy band gap. By plotting $\alpha_{\rm abs}(\omega)$ on a log scale versus. $\hbar\omega$, a more accurate value for the energy gap can also be obtained, as shown in Fig. 18.4 for InSb.

The derivation of the functional form for the absorption coefficient for direct forbidden transitions proceeds as in the derivation of (18.31), except that $|\langle v|p|c\rangle|^2$ is now dependent on k^2 so that $\alpha_{\rm abs}(\omega)$ shows a $(\hbar\omega - E_g)^{3/2}$ threshold dependence for direct forbidden interband transitions, where "forbidden" means forbidden by symmetry to the lowest order of approximation.

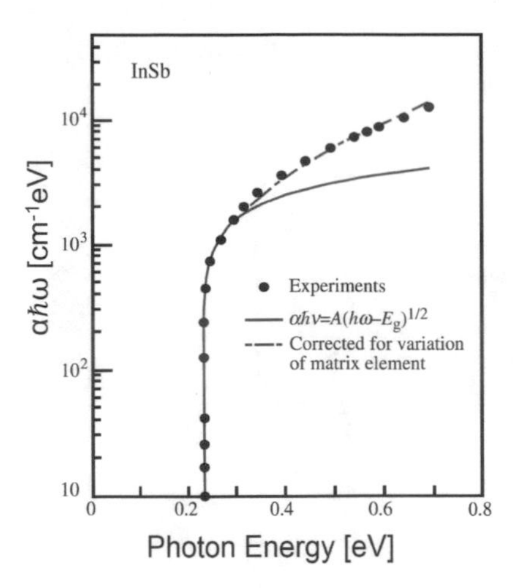

Fig. 18.4 Semilogarithmic plot of the absorption coefficient of InSb at 5 K as a function of photon energy. The filled circles represent experimental results. The curves have been calculated using various models. Best results are obtained when the dependence of the matrix elements on k are included. The intercept with the x-axis gives the direct bandgap of InSb, which can be found more accurately using a semilogarithmic plot than using a linear plot as in Fig. 18.3. The logarithmic plot also shows the large increases in α_{abs} at the absorption edge of a direct gap semiconductor and indicate the photon energy range where the approximations used here are valid. Reproduced from Springer-Verlag, Yu and Cardona, Fundamentals of Semiconductors: Physics and Materials Properties, 1996

18.4.1 Temperature Dependence of E_g

Because of the expansion and contraction of the lattice with temperature, the various band parameters, particularly the energy gap is expected to be temperature dependent. Although calculations are available to predict and account for the T dependence of the band gap at the fundamental absorption edge (threshold), $E_g(T)$ is best found by empirical fits to experimental data. Below, we give expressions for such fits which are useful for research purposes for several common semiconductors.

For group IV and III–V compound semiconductors, $E_g(T)$ decreases with increasing T, as shown above, but for IV–VI compounds, $E_g(T)$ increases with increasing T.

18.4.2 Dependence of the Absorption Edge on Fermi Energy

For lightly doped semiconductors, E_F lies in the bandgap and the absorption edge occurs at E_g, neglecting excitonic effects which are discussed in Chap. 20. However, for heavily doped semiconductors, E_F lies within the valence or conduction bands and the threshold for optical absorption is shifted. This shift in the absorption edge is often referred to as the Burstein shift, and is illustrated in Fig. 18.5, where it is shown that the threshold for absorption occurs when

Table 18.1 Temperature dependence of gap energy E_g

$E_g(T) = 1.165 - 2.84 \times 10^{-4}T$ (eV)	Si
$E_g(T) = 0.742 - 3.90 \times 10^{-4}T$ (eV)	Ge
$E_g(T) = 1.522 - \frac{5.8\times10^{-4}T^2}{T+300}$ (eV)	GaAs
$E_g(T) = 2.338 - \frac{6.2\times10^{-4}T^2}{T+460}$ (eV)	GaP
$E_g(T) = 263 + \sqrt{400 + (0.506T)^2}$ (meV)	PbS
$E_g(T) = 125 + \sqrt{400 + (0.506T)^2}$ (meV)	PbSe
$E_g(T) = 171.5 + \sqrt{164 + [0.44(T + 20)]^2}$ (meV)	PbTe

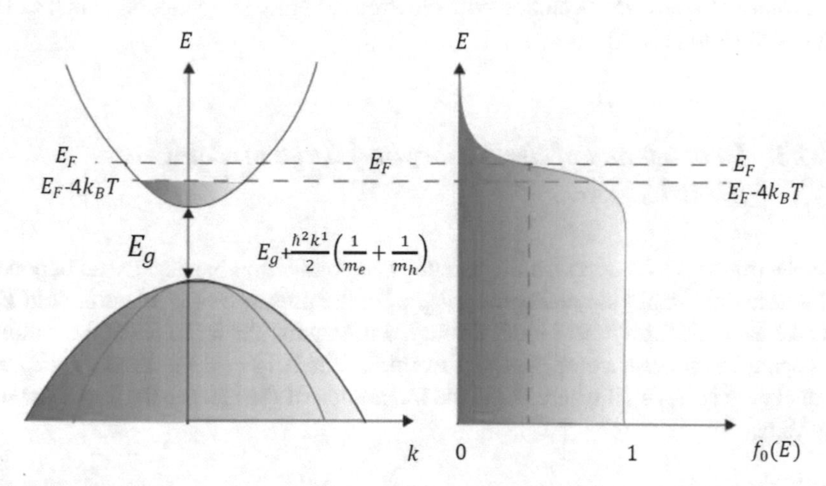

Fig. 18.5 Diagram showing how the fundamental absorption edge of an n-type semiconductor is shifted to higher energy by heavy doping. The wave vector for the Burstein shift k_{BS} is defined in (18.39) and involves the reduced mass of both the electrons and holes

$$\hbar\omega = E_g + \frac{\hbar^2 k_{\text{BS}}^2}{2}\left(\frac{1}{m_e^*} + \frac{1}{m_h^*}\right) = E_g + \frac{\hbar^2 k_{\text{BS}}^2}{2m_r^*} \tag{18.39}$$

in which m_r^* is the reduced mass, $(1/m_r^*) = (1/m_e^*) + (1/m_h^*)$, and k_{BS} is the wave vector at the Fermi level corresponding to the Burstein shift defined in (18.39).

Referring to (18.27) where we introduce the probability that the initial state is occupied and the final state is unoccupied, we find that since doping affects the position of the Fermi level, the Fermi functions will depend on carrier concentration for heavily doped semiconductors. In particular the quantity $(1 - f_0)$ denoting the availability of final states will be sensitively affected by the Burstein shift. If we then write

$$\frac{\hbar^2 k_{\mathrm{BS}}^2}{2m_e^*} = E - E_c \tag{18.40}$$

where E_c is the energy at the bottom of the conduction band, then the probability that the final state is empty is

$$1 - f_0 = \frac{1}{1 + \exp[(E_F - E)/k_B T]} = \frac{1}{1 + \exp\left[\frac{E_F - E}{k_B T} - \frac{(\hbar\omega - E_g)m_h^*}{(m_e^* + m_h^*)k_B T}\right]} \tag{18.41}$$

and (18.41) should be used for finding the probability of final states in evaluating $f(E_c)$ in (18.31). Referring to Fig. 18.5, we see that transitions to the conduction band can start at $E_F - 4k_B T$ given by (18.39). The Fermi level is at E_F and some states above E_F are also occupied with electrons at finite temperature, which can be calculated from (18.31).

18.4.3 Dependence of the Absorption Edge on Applied Electric Field

The electron wave functions in the valence and conduction bands have an exponentially decaying amplitude in the energy gap. In the presence of an electric field **E**, a valence band electron must tunnel through a triangular barrier to reach the conduction band. In the absence of photon absorption, the height of the barrier is E_g and its thickness is $E_g/e|\mathbf{E}|$ where $|\mathbf{E}|$ is the magnitude of the electric field, as shown in Fig. 18.6a.

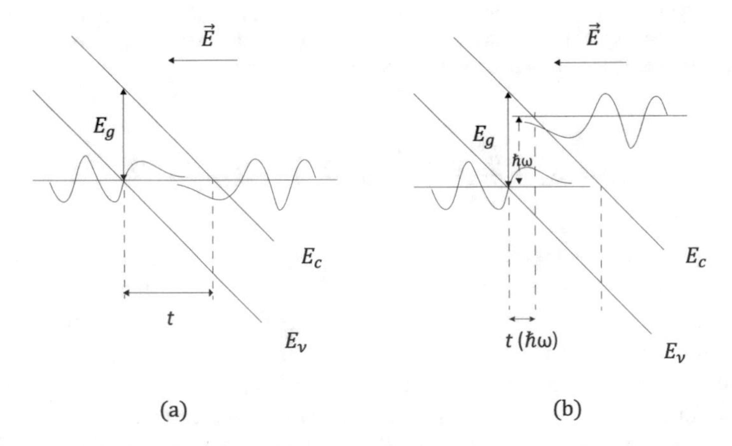

Fig. 18.6 Energy band diagram in an electric field showing the wavefunction overlap **a** without and **b** with the absorption of a photon of energy $\hbar\omega$

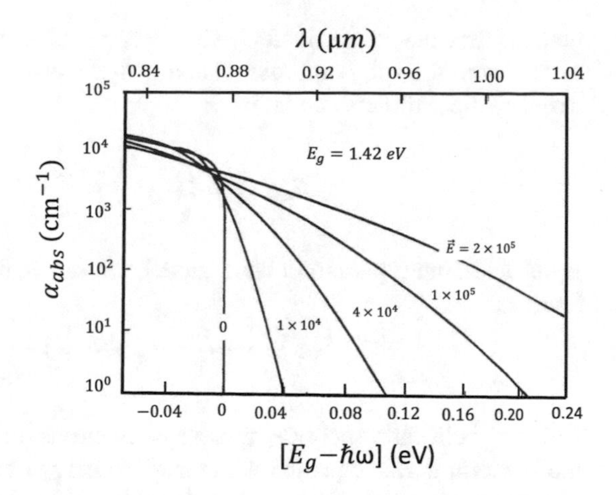

Fig. 18.7 Electric field and photon energy dependence of the band-to-band absorption for GaAs, which is a direct band gap semiconductor

The effect of the photon, as shown in Fig. 18.6b, is to lower the energy barrier thickness to

$$t(\hbar\omega) = \frac{E_g - \hbar\omega}{e|\mathbf{E}|} \tag{18.42}$$

so that the tunneling probability is enhanced by photon absorption. Figure 18.7 shows the absorption edge being effectively lowered by the presence of the electric field, and the effect of the electric field on α_{abs} is particularly pronounced just below the zero field band gap. The effect of an electric field on the fundamental absorption edge is called the Franz–Keldysh effect.

18.4.4 Dependence of the Absorption Edge on Applied Magnetic Field

When a strong magnetic field is applied to a solid, the electrons will move in orbits with a characteristic frequency called the cyclotron frequency ω_c, and these energy levels are quantized with a harmonic oscillator–like energy spectrum given by

$$E_n = \left(n + \frac{1}{2}\right)\hbar\omega_c \tag{18.43}$$

where $\omega_c = eB/m^*$, m^* is the effective mass, and $n = 0, 1, 2, 3, \ldots$ giving rise to quantized subbands called "Landau levels". If we consider a typical semiconductor submitted to a strong magnetic field applied along the z direction, the motion of electrons (holes) in the conduction (valence) bands is confined into quantized orbits in the $x - y$ plane while the motion along the z-axis does not depend on magnetic field. The electron wavector is continuous along the z direction, but k_x and k_y are quantized by the magnetic field. The energy levels for electron motion will be a sum of energy in the sub-bands associated with the orbits plus the free energy along the magnetic field direction, that is

$$E_n(k_z) = \left(n + \frac{1}{2}\right)\frac{e\hbar B}{m^*} + \frac{\hbar^2 k_z^2}{2m^*}. \tag{18.44}$$

By considering a parabolic band model, we can write the energy for electrons and holes as

$$E_n^{e,h}(k_z) = \pm\frac{E_g}{2} \pm \left(n + \frac{1}{2}\right)\frac{e\hbar B}{m^*_{e,h}} \pm \frac{\hbar^2 k_z^2}{2m^*_{e,h}} \tag{18.45}$$

where E_g is the gap energy, + represents electrons (e) and − represents holes (h), and the last term in this equation is the kinetic energy of an electron along the direction of the magnetic field. The optical direct transition energy $\hbar\omega$ between the valence and conduction bands can be written as

$$\hbar\omega = E_n^+(K_z) - E_n^-(K_z) = E_g + (n + \frac{1}{2})\frac{e\hbar B}{\mu} + \frac{\hbar^2 k_z^2}{2\mu} \tag{18.46}$$

where μ is the reduced mass defined as $\frac{1}{\mu} = \frac{1}{m_e} + \frac{1}{m_h}$. From this equation it is clear that the effect of a magnetic field is to blueshift the absorption energy threshold ($n = 0$) by the amount $\frac{1}{2}\frac{e\hbar B}{\mu}$. This phenomena is illustrated in Fig. 18.8, which shows the room temperature optical absorption spectrum of germanium as a function of magnetic field. The oscillations in the spectrum, which are due to transitions between Landau levels, and the separation between the peaks can be used to determine the effective masses of electrons and holes.

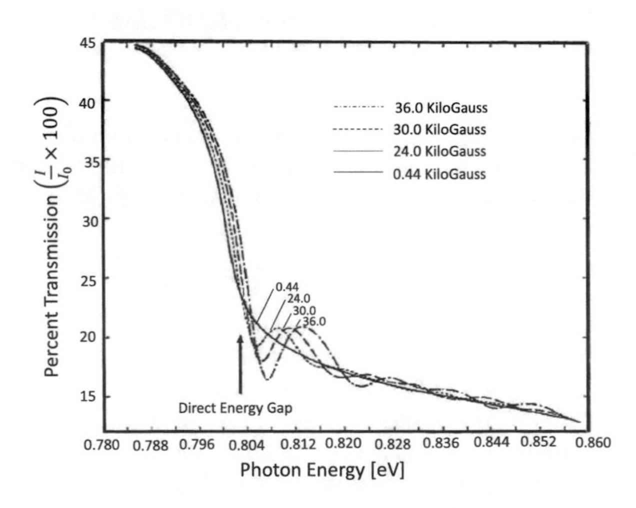

Fig. 18.8 Optical transmission spectrum of Ge as a function of magnetic field intensity, measured in kiloGauss

18.5 Conservation of Crystal Momentum in Direct Optical Transitions

For clarity we now show why the momentum matrix elements coupling two Bloch states for a perfect crystal are diagonal in $\mathbf{k}$ and conserve crystal momentum. It is this property of the momentum matrix elements that is responsible for direct interband transitions in the presence of a magnetic field.

We write the momentum matrix elements coupling two bands (for example, the valence and conduction bands) as

$$\langle n'\mathbf{k}'|\mathbf{p}|n,\mathbf{k}\rangle = \int d^3r \left[e^{-i\mathbf{k}'\cdot\mathbf{r}} u^*_{n'k'}(\mathbf{r}) \left(\frac{\hbar}{i}\nabla\right) e^{i\mathbf{k}\cdot\mathbf{r}} u_{nk}(\mathbf{r}) \right]. \tag{18.47}$$

Operating with ∇ for the $\mathbf{p}$ vector on the product function of the Bloch state yields

$$\langle n'\mathbf{k}'|\mathbf{p}|n,\mathbf{k}\rangle = \int d^3r e^{-i\mathbf{k}'\cdot\mathbf{r}} u^*_{n'k'}(\mathbf{r}) e^{i\mathbf{k}\cdot\mathbf{r}} \left(\hbar\mathbf{k} + \frac{\hbar}{i}\nabla\right) u_{nk}(\mathbf{r}). \tag{18.48}$$

Now the term in $\hbar\mathbf{k}$ can be integrated immediately to give $\hbar\mathbf{k}\delta_{nn'}\delta(\mathbf{k}-\mathbf{k}')$ and is thus diagonal in both the band index and crystal momentum. This term therefore does not give rise to interband transitions. The remaining term in (18.48) is the function $u^*_{n'k'}(\mathbf{r})\nabla u_{nk}(\mathbf{r})$ which is periodic under the translation $\mathbf{r} \to \mathbf{r} + \mathbf{R_n}$, where $\mathbf{R_n}$ is any lattice vector. But any spatially periodic function can be Fourier expanded

$$\sum_m F_m e^{i\mathbf{G_m}\cdot\mathbf{r}} = \frac{\hbar}{i} u^*_{n'k'}(\mathbf{r})\nabla u_{nk}(\mathbf{r}) \tag{18.49}$$

in terms of the reciprocal lattice vectors $\mathbf{G_m}$. We thus obtain for the integral in (18.48) factors of the form

$$\int d^3r e^{i(\mathbf{k}-\mathbf{k}')\cdot\mathbf{r}} F_m e^{i\mathbf{G_m}\cdot\mathbf{r}} \tag{18.50}$$

which vanish unless

$$\mathbf{k} - \mathbf{k}' + \mathbf{G_m} = 0. \tag{18.51}$$

Since $\mathbf{k}-\mathbf{k}'$ must be within the first Brillouin zone, $\mathbf{k}$ and $\mathbf{k}'$ can only differ by the reciprocal lattice vector $\mathbf{G}_m \equiv 0$. Thus, (18.50) vanishes unless $\mathbf{k} = \mathbf{k}'$ and we have demonstrated that because of the periodicity of the crystal lattice, the momentum matrix elements coupling two bands can only do so at the same value of crystal momentum $\mathbf{k}$. Since the probability for optical transitions involves the same momentum matrix elements as occur in the determination of the effective mass in the transport properties, study of the optical properties of a solid also bears an important relation to the transport properties of that material. If the finite wave vector of the light is included, then the spatial dependence of the vector potential must also be included, as a correction. In some cases the interband transition is not allowed at the high symmetry point for symmetry reasons. As we then move away from the high symmetry point, transitions can occur as the wave function is expanded in a Taylor expansion about the high symmetry point.

18.6 Indirect Interband Transitions

In making indirect transitions, the semiconductor can either emit or absorb a phonon of energy $\hbar\omega_q$

$$\hbar\omega = E_f - E_i \pm \hbar\omega_q \tag{18.52}$$

in which E_f and E_i are, respectively, the energies of the final and initial electron states and the $\pm$ signs refer to phonon emission (+ sign) or absorption (– sign).

To review indirect interband transitions in a semiconductor, we derive below an expression for the absorption coefficient for the situation where a phonon is absorbed in the indirect process, as shown schematically in Fig. 18.9. Similar arguments can then be applied to the case where a phonon is emitted. The conservation of energy principle is applied to the total process, consisting of the direct optical transition and the absorption of a phonon $\hbar\omega_q$, yielding

$$\hbar\omega = E_g - \hbar\omega_q + \frac{\hbar^2(\mathbf{k}_n - \mathbf{k}_c)^2}{2m_n} + \frac{\hbar^2 k_p^2}{2m_p} \tag{18.53}$$

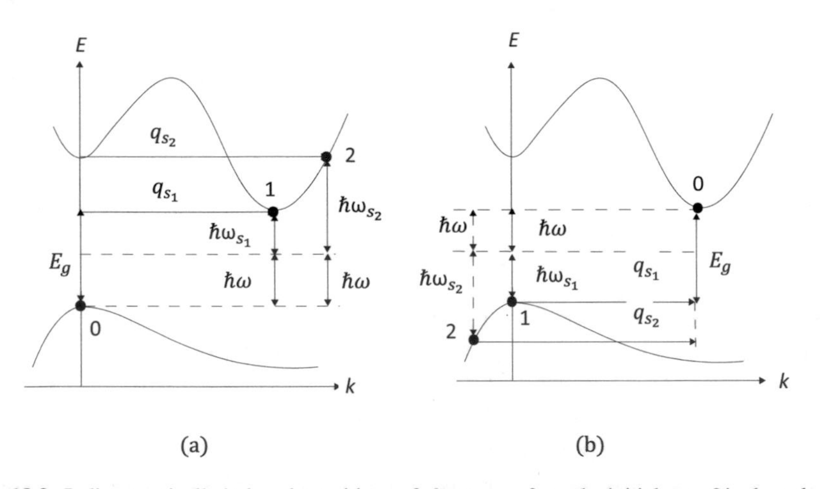

Fig. 18.9 Indirect optically induced transitions of electrons **a** from the initial state 0 in the valence band to final states 1 and 2 in the conduction band, and **b** from initial states 1 and 2 in the valence band to the final state 0 in the conduction band. In both (**a**) and (**b**) a phonon labeled by $(\hbar\omega_s, q_s)$ is absorbed in the indirect transition process

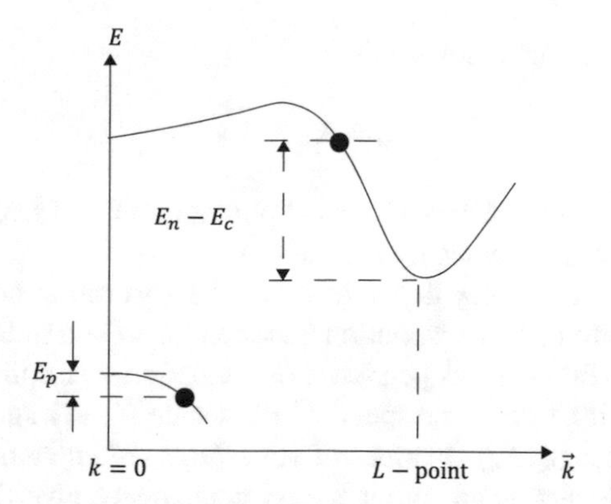

Fig. 18.10 Schematic diagram of an indirect transition showing the notation used in the text. E_c is the energy of the L-point conduction band at wave vector k_c, while E_c is the thermal energy gap. The valence band maximum E_v is taken as the zero of energy. E_n and $(\mathbf{k}_n - \mathbf{k}_c)$, respectively, denote the energy and momentum of an excited electron, while E_p and $\mathbf{k}_p$, respectively, denote the corresponding parameters for the holes near $\mathbf{k} = 0$. It is customary to place the zero of energy at the valence band maximum

in which the notation in (18.53) is defined in Fig. 18.10, and E_g is the thermal gap or energy difference between the conduction band minimum (e.g., at the L-point) and the valence band maximum at the Γ-point of the Brillouin zone. The negative sign in front of the phonon energy $\hbar\omega_q$ in (18.53) corresponds to the phonon absorption process. In (18.53), the term $\hbar(\mathbf{k}_n - \mathbf{k}_c)$ denotes the difference between the crystal momentum $\hbar\mathbf{k}_n$ of an excited electron in the L–point conduction band and the crystal momentum $\hbar\mathbf{k}_c$ at the L–point conduction band minimum. Thus, the kinetic energy of the excited electron with crystal momentum $\hbar\mathbf{k}_n$ is

$$E_n - E_c = \frac{\hbar^2(\mathbf{k}_n - \mathbf{k}_c)^2}{2m_n} \tag{18.54}$$

where E_n is the energy above the conduction band minimum E_c, and m_n in (18.54) is the effective mass of an electron near the conduction band minimum.

Since the valence band extremum is at $\mathbf{k} = 0$, then $\hbar\mathbf{k}_p$ is the crystal momentum for the hole that is created when the electron is excited, corresponding to the kinetic energy of the p-type hole

$$E_p = \frac{\hbar^2 k_p^2}{2m_p}. \tag{18.55}$$

The sign convention that is used in this discussion is to take E_p as a positive number and the zero of energy is taken at the valence band maximum (see Fig. 18.10). In terms of these sign conventions, conservation of energy yields

$$\hbar\omega = E_g - \hbar\omega_q + (E_n - E_c) + E_p \tag{18.56}$$

and conservation of momentum requires

$$\mathbf{q} = \mathbf{k}_n - \mathbf{k}_p \tag{18.57}$$

where $\mathbf{q}$ is the wave vector for the absorbed phonon. In Fig. 18.9, the phonon energy and the wave vector are denoted by $\hbar\omega_{s_i}$ and q_{s_i}.

We now find the frequency dependence of the absorption edge for the indirect transitions in order to make a distinction between direct and indirect transitions just from looking at the frequency dependence of the optical absorption data. Let us then consider the transition from some specific initial state E_p to a specific final state E_n. The density of states $\rho_c(E_n)$ (number of states/unit volume/unit energy range) for the final state conduction band has an energy dependence given by

$$\rho_c(E_n) \propto (E_n - E_c)^{1/2}. \tag{18.58}$$

Using the conservation of energy relation in (18.56), the dependence of $\rho_c(E_n)$ on E_p can be expressed through the relation

$$\rho_c(E_n) \propto (\hbar\omega - E_g - E_p + \hbar\omega_q)^{1/2}. \tag{18.59}$$

Thus we see that transitions to a state E_n take place from a range of initial states, since E_p, can vary between $E_p = 0$, where all of the kinetic energy is given to the electron, and in the opposite limit where $E_n - E_c = 0$ and all of the kinetic energy is given to the hole. Let the energy δ denote the range of possible valence band energies between these limits

$$\delta = \hbar\omega - E_g + \hbar\omega_q. \tag{18.60}$$

The density of initial states for the valence band has an energy dependence given by

$$\rho_v(E_p) \propto E_p^{1/2} \tag{18.61}$$

where we are using the convention $E_v \equiv 0$ for defining the zero of energy, so that E_p vanishes at the top of the valence band. Thus the effective density of states for the phonon absorption process is found by summing over all E_p values which conserve energy,

$$\rho(\hbar\omega) \propto \int_0^\delta \rho_c(E_n)\rho_v(E_p)dE_p \propto \int_0^\delta \sqrt{\delta - E_p}\,\sqrt{E_p}\,dE_p. \tag{18.62}$$

The integral in (18.62) can be carried out through integration by parts, utilizing the notation $u = E_p$, and $v = \delta - E_p$, and writing the limits of the integration in terms of the variable E_p

$$\int_0^\delta \sqrt{uv}\,du = \frac{\delta - 2v}{4}\sqrt{uv}\,\Bigg|_0^\delta + \frac{\delta^2}{4}\tan^{-1}\sqrt{\frac{u}{\delta - u}}\,\Bigg|_0^\delta = \frac{\delta^2\pi}{8}. \tag{18.63}$$

Substitution in (18.60) for δ in (18.62) and (18.63) results in

$$\rho(\hbar\omega) \propto \frac{\pi}{8}(\hbar\omega - E_g + \hbar\omega_q)^2 \tag{18.64}$$

which gives the frequency dependence for the indirect interband transitions involving *phonon absorption*. Also, the probability for the absorption of a phonon is proportional to the Bose–Einstein factor

$$n(\hbar\omega_q) = \frac{1}{\exp(\hbar\omega_q/k_B T) - 1} \tag{18.65}$$

so that the absorption coefficient for indirect transitions in which a phonon is absorbed becomes

$$\alpha_{\text{abs}}(\omega) = \mathscr{C}_a \frac{(\hbar\omega - E_g + \hbar\omega_q)^2}{\exp(\hbar\omega_q/k_B T) - 1} \tag{18.66}$$

where $\mathscr{C}_a$ is a constant for the phonon absorption process.

To find the absorption coefficient for the indirect absorption process that involves the emission of a phonon, we must find the effective density of states for the emission process. The derivation in this case is very similar to that given above for phonon absorption, except that the energy conservation condition now involves the phonon energy with the opposite sign. Furthermore, the probability of emission of a phonon is proportional to $[n(\hbar\omega_q) + 1]$, which is given by

$$[n(\hbar\omega_q) + 1] = 1 + [e^{\hbar\omega_q/k_B T} - 1]^{-1} = \frac{1}{1 - e^{-\hbar\omega_q/k_B T}} \tag{18.67}$$

so that the absorption constant for phonon emission becomes

$$\alpha_{\text{ems}}(\omega) = \mathscr{C}_e \frac{(\hbar\omega - E_g - \hbar\omega_q)^2}{1 - \exp(-\hbar\omega_q/k_B T)} \tag{18.68}$$

where $\mathscr{C}_e$ is a constant for the phonon emission process.

At low temperatures, the phonon emission process dominates because there are so few phonons available for the absorption process. Furthermore, as a function of the photon energy, different thresholds are obtained for the absorption and emission processes. In the absorption process, absorption starts when $\hbar\omega = E_g - \hbar\omega_q$ (see Fig. 18.11), while for the emission process, the optical absorption starts when $\hbar\omega = E_g + \hbar\omega_q$. So if we plot $\sqrt{\alpha_{\text{abs}}(\omega)}$ versus. $\hbar\omega$, as is shown in Fig. 18.11, then $\sqrt{\alpha_{\text{abs}}(\omega)}$ in the low photon energy range will go as $\sqrt{\alpha_{\text{abs}}(\omega)} \propto (\hbar\omega - E_g + \hbar\omega_q)$, while $\sqrt{\alpha_{\text{ems}}(\omega)}$ will be proportional to $(\hbar\omega - E_g - \hbar\omega_q)$ for the emission process. Experimentally, a superposition of the absorption and emission processes will be observed and the two terms will have a different temperature dependence.

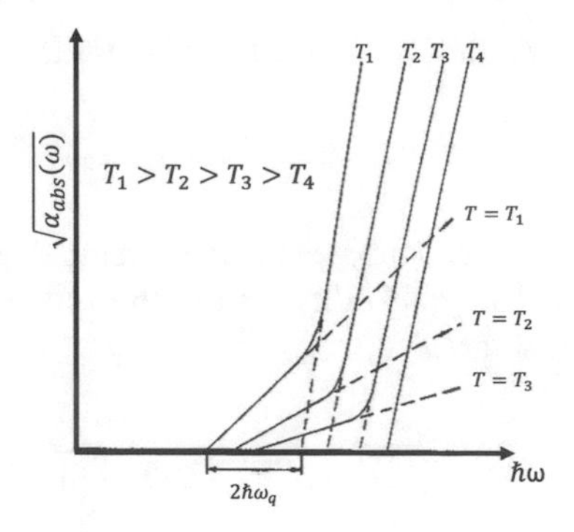

Fig. 18.11 Schematic diagram showing the frequency dependence of the square root of the absorption coefficient for indirect interband transitions near the thresholds for the phonon emission and absorption processes. The curves are for four different temperatures. At the lowest temperature (T_4) the phonon emission process dominates, while at the highest temperature (T_1) the phonon absorption process is most important at low photon energies. The magnitude of twice the phonon energy is indicated on the x axis (see Yu and Cardona, Fundamentals of Semiconductors: Physics and Materials Properties)

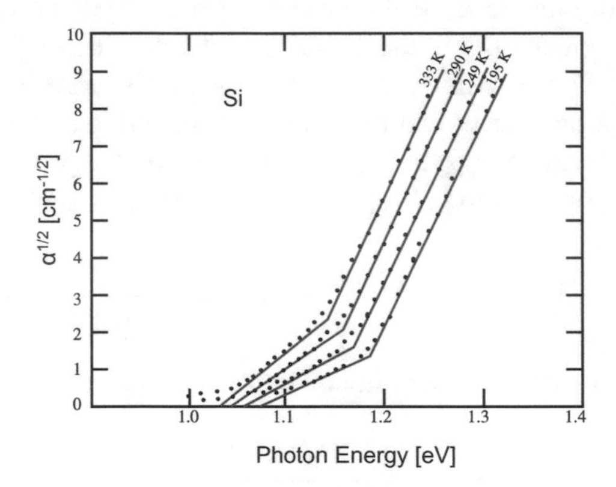

Fig. 18.12 Plots of the square root of the absorption coefficients of Si versus photon energy at several temperatures. The two segments of a straight line drawn through the experimental points represent the two contributions associated with phonon absorption and emission. (From Macfarlane, et al., Phys. Rev. **111**, 1249 (1958))

Some experimental data illustrating indirect interband transitions are given in Fig. 18.12. The shift of the curves in Fig. 18.12 as a function of photon energy is due to the temperature dependence of the indirect gap in silicon. In Fig. 18.12 it is easy to separate out the lower energy absorption contribution which is associated with the phonon absorption process (compare Figs. 18.11 and 18.12). At higher energies it is also easy to separate out the phonon emission contribution. By carrying out measurements at several different temperatures it is therefore possible to obtain a more accurate value for $\hbar\omega_q$. Figure 18.12 shows that the phonon absorption process becomes more favorable as the temperature is raised, while the emission process is less sensitive to temperature. The physical reason behind this is that for the absorption process to occur in the first place, phonons of the appropriate wave vector **q** must be available. In Ge, the phonon assisted process requires phonons of wave vector **q** extending from Γ to L, while for Si we need a phonon **q**–vector extending from Γ to Δ_{min} (where Δ_{min} corresponds to the Δ point conduction band minimum). Since lattice vibrations are thermally excited, there are few available phonons at low temperatures, but more are available at high temperatures. On the other hand, phonon emission does not depend upon the availability of phonons since the emission process itself generates phonons; for this reason the phonon emission process is relatively insensitive to temperature. Since silicon is a relatively hard material (with a Debye temperature of $\theta_D = 658\,\text{K}$), there will only be a few large wavevector phonons excited at room temperature. Therefore the phonon emission process will dominate in the optical absorption for photon energies where such emission is energetically possible. These arguments account for the different slopes observed for the phonon absorption and emission contributions to the absorption coefficient of Fig. 18.12.

Another complication that arises in real materials is that there are several types of phonons present for a given **q**-vector, i.e., there are acoustic and optical branches, and for each branch there are longitudinal and transverse modes. An example of the analysis of optical absorption data to obtain the frequencies of the various phonons at $q = 0$ is given in Fig. 18.13, where $\alpha_{\text{abs}}^{1/2}$ versus $\hbar\omega$ is plotted for the indirect gap semiconductor GaP, from which it is possible to measure $\hbar\omega_q$ for various LO, LA, TO and TA phonons. Today such optical data are seldom taken, because it is now customary to use inelastic neutron diffraction data to plot out the entire phonon dispersion curve for each of the phonon branches. When the phonon frequencies

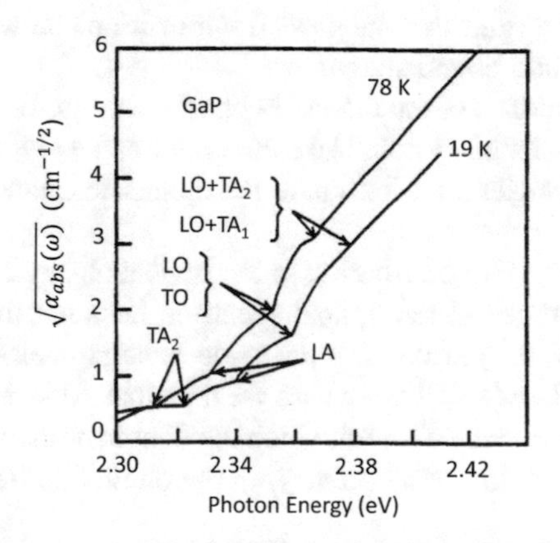

Fig. 18.13 Plots of the square root of the absorption coefficients of GaP versus photon energy at two different temperatures. The labels denote the various absorption thresholds associated with the emission of various phonon modes. The observation of these phonon modes is made possible by the enhanced absorption associated with excitons at the absorption threshold. The apparent shift in the phonon frequencies is mostly due to the variation of the bandgap energy with temperature (see Figs. 18.11 and 18.12). Reproduced from Springer-Verlag, Yu and Cardona, Fundamentals of Semiconductors: Physics and Materials Properties with permission

are high, electron energy loss spectroscopy can be helpful in obtaining $\omega_q(q)$ for the various phonon branches as is discussed in Chap. 6. In the case of graphite, it has been shown how resonance Raman spectroscopy can be used to obtain important information about the phonon dispersion relations for a material with lower symmetry (see Saito et al., Phys. Rev. Lett., **88**, 027401 (2002).).

Problems

18.1 Consider the Burstein shift phenomena for heavily doped semiconductors. Determine the energy at which appreciable interband (band-to-band) absorption begins at 300 K for a p-type GaAs sample $E_g = 1.42\,\text{eV}$ with a hole concentration of $1 \times 10^{20}\,\text{cm}^{-3}$. Take into account both the light- and heavy-hole bands, which are degenerate at the Γ point in the Brillouin zone ($m_{lh} = 0.074m_0$, $m_{hh} = 0.62m_0$).

18.2 Suppose that you prepare a quantum well structure by molecular beam epitaxy (MBE) from GaAs ($E_g = 1.42\,\text{eV}$) and $\text{Al}_x\text{Ga}_{1-x}\text{As}$ ($E_g = 1.80\,\text{eV}$), where $x = 0.3$ so that $\text{Al}_x\text{Ga}_{1-x}\text{As}$ is a direct gap semiconductor. Assume that the band off–set of the conduction band is three times greater than in the valence band ($m_e = 0.067m_0$).

(a) Assuming a width of the quantum well of $L_z = 15\,\text{nm}$, find the photon energies at which optical absorption can take place due to optical transitions between the highest heavy hole and light hole bound states to the lowest conduction band bound state? Use the approximation of an infinite rectangular well in obtaining the energy levels in the bound state.
(b) Are there selection rules that suppress the transitions between selected valence and conduction band bound states?
(c) What is the dependence of the threshold photon energy for optical transitions on L_z for a lightly n-doped system? Use the notation $n = n_0$ and $\tau = \tau_0$.
(d) What is the free carrier contribution to the dielectric function at room temperature?
(e) What is the free carrier contribution to the optical absorption coefficient?
(f) What is the difference in the optical spectrum between the superlattice where $\text{Al}_x\text{Ga}_{1-x}\text{As}$ ($x = 0.3$) is used as the wide bandgap semiconductor [part (a)] and the case of AlAs ($x = 1.0$), where we note that AlAs is an indirect bandgap semiconductor ($E_g = 2.2\,\text{eV}$), for which the X point conduction band minimum is 0.23 eV below the lowest Γ point ($k = 0$) conduction band.

18.3 In many physical cases, the momentum matrix element coupling the highest lying valence band and the lowest lying conduction band vanishes by symmetry at the extremal point $\mathbf{k}_0$. Thus, optical transitions at $\mathbf{k}_0$ are "forbidden". However, in these cases the momentum matrix element is non–zero as we move away from $\mathbf{k}_0$ by an arbitrary amount. This gives rise to "forbidden direct interband transitions" which have a different frequency dependence for the optical absorption coefficient than their "allowed counterparts".

(a) By making a Taylor expansion of the wave function $\psi_{n\mathbf{k}}(\mathbf{r})$ about the band extremum $\mathbf{k}_0$, where $\mathbf{k} = \mathbf{k}_0 + \boldsymbol{\kappa}$, find the dependence of the matrix element $\langle v|\mathbf{p}|c\rangle$ on the magnitude of $\boldsymbol{\kappa}$, assuming that the matrix element vanishes by symmetry at $\mathbf{k}_0$.
(b) Using the result from part (a), find the frequency dependence of the optical absorption coefficient for the case of a forbidden direct interband transition around an M_0 type critical point.
(c) Compare the frequency dependence of the optical absorption coefficient in (b) to that for direct allowed interband transitions and for indirect optical transitions.
(d) What is the frequency dependence of the optical absorption coefficient for a two-dimensional electron gas for allowed and forbidden interband transitions?

18.4 Temperature dependence and isotopic shift of the bandgaps:

(a) Show that the temperature (T) dependence of an interband gap energy E_g can be written as

$$E_g(T) - E_g(0) = A\left(\frac{2}{\exp[\hbar\Omega/(k_B T)] - 1} + 1\right),$$

where A is a temperature-independent constant, k_B is the Boltzmann constant, and $\hbar\Omega$ represents an average phonon energy. Hint: the term inside the parenthesis in the above equation represents the ensemble-averaged square of the phonon displacement.
(b) Show that $\Delta E_g(T) = E_g(T) - E_g(0)$ becomes linear in T in the limit of $k_B T \gg \hbar\Omega$.
(c) For small T, $\Delta E_g(T)$ can also be written as

$$\Delta E_g(T) = \left(\frac{\partial E_g}{\partial V}\right)_T \left(\frac{dV}{dT}\right)_P \Delta T + \left(\frac{\partial E_g}{\partial T}\right)_V \Delta T$$

where the first term describes the change in E_g caused by thermal expansion. Its sign can be positive or negative. The second term is the result of electron-phonon interaction. Its sign is usually negative. Estimate the contribution of these effects to $E_g(0)$ by extrapolating $E_g(T)$ to $T = 0$ using its linear dependence at large T. The resultant energy is known as the *renormalization* of the bandgap at $T = 0$ by electron-phonon interaction. Determine this energy for the E_0 gap of Ge from Fig. 18.14.
(d) The result in part (c) can be used to estimate the dependence of the bandgap on isotopic mass. Since the bonding between atoms is not affected by the isotopic mass, the average phonon energy $\hbar\Omega$ in solids with two identical atoms per unit cell, like Ge, can be assumed to depend on atomic mass M as $M^{-1/2}$. Calculate the difference in the E_0 bandgap energies between the following isotopes: ^{70}Ge, ^{74}Ge, and ^{76}Ge.

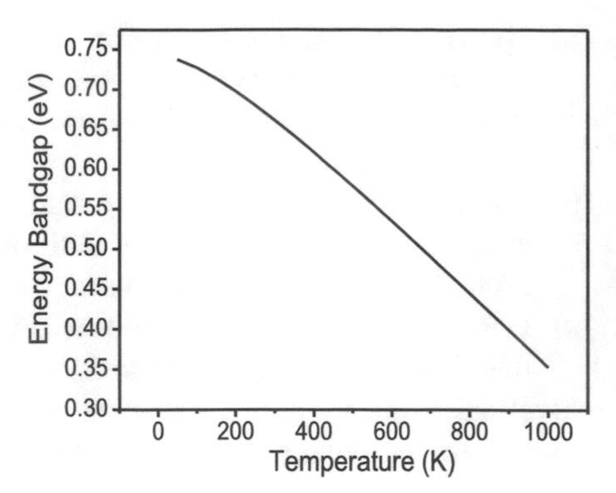

Fig. 18.14 Ge energy bandgap versus temperature

18.5 For a two level system with energy levels E_1 and E_2 (where $E_1 < E_2$), assume that before time $t = 0$ when a light wave of frequency ω and intensity I_0 is applied, the system is in the ground state E_1.

(a) Find the transition probability for transitions to the state E_2 as a function of time t. Consider the response of the system as a signal of frequency ω is tuned over the resonant frequency ω_R allowing measurement of the linewidth in energy at $\hbar\omega_R = E_2 - E_1$.
(b) Suppose that the system is in state E_2 at time t_0 when the light wave is switched off, find an expression for the probability that state E_2 is still occupied after a time $(t_f - t_0)$.
(c) Sketch the occupation of states E_1 and E_2 over the time interval $0 \leq t \leq t_f$ and indicate the change in behavior occurring at time t_0.

18.6 For GaAs nanowires, it is very difficult to determine the free carrier concentration using traditional Hall effect measurements. As an alternative, photoluminescence measurements can be used to estimate the carrier density based on the Burstein shift.

(a) For an n-type GaAs nanowire, a blueshift of 20 meV is observed in the photoluminescence emission relative to an undoped nanowire. Based on this information, estimate the Fermi level and carrier density of these nanowires.
(b) How would your answer change if this shift was observed in p-doped GaAs nanowires? Explain your answer.

Suggested Readings

F. Bassani, G. Pastori-Parravicini, *Electronic States and Optical Transitions in Solids* (Pergamon Press, New York, 1975). Chapter 5

R. Saito, A. Jorio, A.G. Souza Filho, G. Dresselhaus, M.S. Dresselhaus, M.A. Pimenta, Probing phonon dispersion relations of graphite by double resonance Raman scattering. Phys. Rev. Lett. **88**, 027401 (2002)

C.M. Wolfe, N. Holonyak, G.E. Stillman, *Physical Properties of Semiconductors* (Prentice Hall, Englewood Cliffs, 1989). Chapter 7

P.Y. Yu, M. Cardona, *Fundamentals of Semiconductors* (Springer, Berlin, 1996). Chapter 6

J.M. Ziman, *Principles of the Theory of Solids* (Cambridge University Press, Cambridge, 1972). Chapter 8

Chapter 19
Optical Properties of Solids over a Wide Frequency Range

19.1 Kramers–Kronig Relations

Measurement of the absorption coefficient (Chap. 18) gives the imaginary part of the complex index of refraction, while the reflectivity is sensitive to a more complicated combination of $\varepsilon_1(\omega)$ and $\varepsilon_2(\omega)$. Thus, from measurements such as the frequency dependent absorption $\alpha_{\text{abs}}(\omega)$, we often have insufficient information to determine $\varepsilon_1(\omega)$ and $\varepsilon_2(\omega)$ independently. However, if we know either $\varepsilon_1(\omega)$ or $\varepsilon_2(\omega)$ over a wide frequency range, then $\varepsilon_2(\omega)$ or $\varepsilon_1(\omega)$ can be determined from the Kramers–Kronig relations given by

$$\varepsilon_1(\omega) - 1 = \frac{2}{\pi}\mathscr{P}\int_0^\infty \frac{\omega'\varepsilon_2(\omega')}{\omega'^2 - \omega^2}d\omega' \tag{19.1}$$

and

$$\varepsilon_2(\omega) = -\frac{2}{\pi}\mathscr{P}\int_0^\infty \frac{\omega'\varepsilon_1(\omega')}{\omega'^2 - \omega^2}d\omega' \tag{19.2}$$

in which $\mathscr{P}$ denotes the principal value of the integrals. The Kramers–Kronig relations are very general mathematical relations and are based on causality, linear response theory and the boundedness of physical observables.

Since the Kramers–Kronig relations relate $\varepsilon_1(\omega)$ and $\varepsilon_2(\omega)$ to each other, if either of these functions is known as a function of ω, the other is completely determined. Because of the form of these relations (19.1) and (19.2), it is clear that the main contribution to $\varepsilon_1(\omega)$ comes from the behavior of $\varepsilon_2(\omega')$ near $\omega' \approx \omega$ due to the resonant denominator in these two equations. What this means physically is that to obtain $\varepsilon_1(\omega)$, we really should know $\varepsilon_2(\omega')$ for all ω', but it is more important to know $\varepsilon_2(\omega')$ in the frequency range in the vicinity of ω than elsewhere. This property is greatly exploited in the analysis of optical reflectivity data, where measurements are generally available over a finite range of ω' values. Some kind of extrapolation procedure must then be used for those frequencies ω' that are experimentally unavailable.

M. Dresselhaus et al., *Solid State Properties*, Graduate Texts in Physics,
https://doi.org/10.1007/978-3-662-55922-2_19

We now give a derivation of the Kramers–Kronig relations, after some introductory material.

This theorem is generally familiar to electrical engineers in the context of causality. If a system is linear and obeys causality (i.e., there is no output before the input is applied), then the real and imaginary parts of the system function are related by a Hilbert transform. Let us now apply this causality concept to the polarization in a solid resulting from the application of an optical electric field. We have the constitutive equation which defines the polarization of a solid material:

$$\varepsilon \mathbf{E} = \mathbf{D} = \mathbf{E} + 4\pi \mathbf{P} \tag{19.3}$$

so that

$$\mathbf{P} = \frac{\varepsilon - 1}{4\pi} \mathbf{E} \equiv \alpha(\omega) \mathbf{E} \tag{19.4}$$

where $\alpha(\omega)$ defines the polarizability, and $\mathbf{P}$ is the polarization/unit volume or the response of the solid to an applied field $\mathbf{E}$. The polarizability $\alpha(\omega)$ is the system function

$$\alpha(\omega) = \alpha_r(\omega) + i\alpha_i(\omega) \tag{19.5}$$

in which we have explicitly written the real and imaginary parts $\alpha_r(\omega)$ and $\alpha_i(\omega)$, respectively. Let $E(t) = E_0\delta(t)$ be an impulse optical field at $t = 0$. Then from the definition of a δ-function, we have:

$$E(t) = E_0\delta(t) = \frac{E_0}{\pi} \int_{0-}^{\infty} \cos \omega t d\omega. \tag{19.6}$$

The response to this impulse field yields an in-phase term proportional to $\alpha_r(\omega)$ and an out-of-phase term proportional to $\alpha_i(\omega)$, where the polarization vector is given by

$$\mathbf{P}(t) = \frac{E_0}{\pi} \int_{0-}^{\infty} \left[\alpha_r(\omega) \cos \omega t + \alpha_i(\omega) \sin \omega t \right] d\omega, \tag{19.7}$$

in which $\alpha(\omega)$ is written for the complex polarizability (see (19.5)). Since $\mathbf{P}(t)$ obeys causality and is bounded, we find that the integral of $\alpha(\omega)e^{-i\omega t}$ is well behaved along the contour C' in Fig. 19.1a as $R \to \infty$ and no contribution to the integral is made along the contour C' in the upper half plane. Furthermore, the causality condition that $\mathbf{P}(t)$ vanishes for $t < 0$ requires that $\alpha(\omega)$ have no poles in the upper half plane shown in Fig. 19.1a.

To find an explicit expression for $\alpha(\omega)$ we must generate a pole on the real axis. Then we can isolate the behavior of $\alpha(\omega)$ at some point ω_0 by taking the principal value of the integral. We do this with the help of Cauchy's theorem. Since $\alpha(\omega)$ has no poles in the upper half-plane, the function $[\alpha(\omega)/(\omega - \omega_0)]$ will have a single pole along the axis at $\omega = \omega_0$ (see Fig. 19.1b). If we run our contour just above the

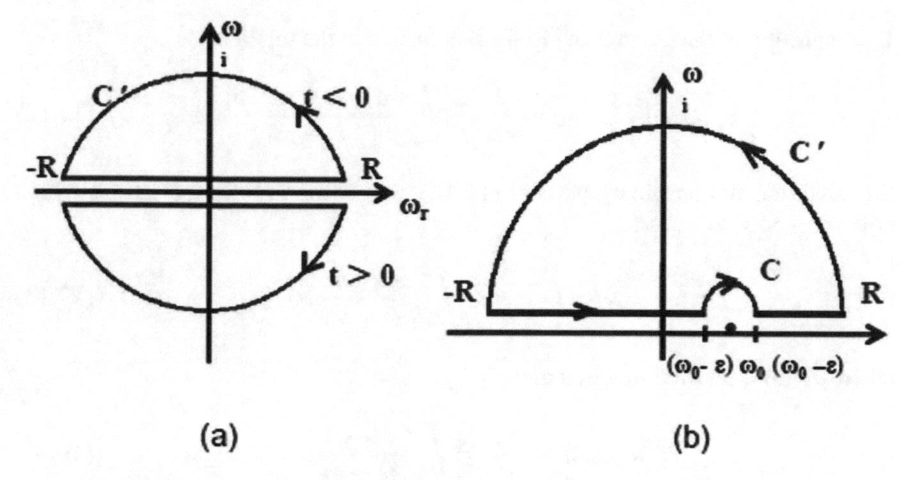

Fig. 19.1 **a** Contours used in evaluating the complex polarizability integral of (19.7). **b** Contour used to evaluate (19.9)

real axis, there are no poles in the upper-half plane and the integral around the closed contour vanishes:

$$\oint \frac{\alpha(\omega)}{\omega - \omega_0} d\omega = 0. \tag{19.8}$$

Let us now consider the integral taken over the various portions of this closed contour:

$$\int_{C'} \frac{\alpha(\omega)}{\omega - \omega_0} d\omega + \int_{-R}^{\omega_0 - \varepsilon} \frac{\alpha(\omega)}{\omega - \omega_0} d\omega + \int_{C} \frac{\alpha(\omega)}{\omega - \omega_0} d\omega + \int_{\omega_0 + \varepsilon}^{R} \frac{\alpha(\omega)}{\omega - \omega_0} d\omega = 0. \tag{19.9}$$

The contribution over the contour C' vanishes since $\alpha(\omega)$ remains bounded, while $[1/(\omega - \omega_0)] \to 0$ as $R \to \infty$ (see Fig. 19.1b). Along the contour C, we use Cauchy's theorem to obtain

$$\lim_{\varepsilon \to 0} \int_{C} \frac{\alpha(\omega)}{\omega - \omega_0} d\omega = -\pi i \alpha(\omega_0) \tag{19.10}$$

in which $\alpha(\omega_0)$ is the residue of $\alpha(\omega)$ at $\omega = \omega_0$ and the minus sign is written because the contour C is taken clockwise. We further define the principal part $\mathscr{P}$ of the integral in the limit $R \to \infty$ and $\varepsilon \to 0$ as

$$\lim_{\substack{R \to \infty \\ \varepsilon \to 0}} \int_{-R}^{\omega_0 - \varepsilon} \frac{\alpha(\omega)}{\omega - \omega_0} d\omega + \int_{\omega_0 + \varepsilon}^{R} \frac{\alpha(\omega)}{\omega - \omega_0} d\omega \to \mathscr{P} \int_{-\infty}^{\infty} \frac{\alpha(\omega)}{\omega - \omega_0} d\omega. \tag{19.11}$$

The vanishing of the integral in (19.8) thus results in the relation

$$\alpha_r(\omega_0) + i\alpha_i(\omega_0) = \frac{1}{\pi i}\mathscr{P}\int_{-\infty}^{\infty}\frac{\alpha_r(\omega) + i\alpha_i(\omega)}{\omega - \omega_0}d\omega. \tag{19.12}$$

Equating real and imaginary parts of (19.12), we get the following relations which hold for $-\infty < \omega < \infty$:

$$\alpha_r(\omega_0) = \frac{1}{\pi}\mathscr{P}\int_{-\infty}^{\infty}\frac{\alpha_i(\omega)}{\omega - \omega_0}d\omega \tag{19.13}$$

where $\alpha_r(\omega)$ is an even function and

$$\alpha_i(\omega_0) = \frac{-1}{\pi}\mathscr{P}\int_{-\infty}^{\infty}\frac{\alpha_r(\omega)}{\omega - \omega_0}d\omega \tag{19.14}$$

where $\alpha_i(\omega)$ is an odd function of ω.

We would like to write these relations in terms of integrals over positive frequencies. We can do this by utilizing the even- and oddness of $\alpha_r(\omega)$ and $\alpha_i(\omega)$. If we now multiply the integrand by $(\omega + \omega_0)/(\omega + \omega_0)$ and make use of the even- and oddness of the integrands, we get:

$$\alpha_r(\omega_0) = \frac{1}{\pi}\mathscr{P}\int_{-\infty}^{\infty}\frac{\alpha_i(\omega)(\omega + \omega_0)}{\omega^2 - \omega_0^2}d\omega = \frac{2}{\pi}\mathscr{P}\int_{0}^{\infty}\frac{\omega\alpha_i(\omega)d\omega}{\omega^2 - \omega_0^2} \tag{19.15}$$

$$\alpha_i(\omega_0) = \frac{-1}{\pi}\mathscr{P}\int_{-\infty}^{\infty}\frac{\alpha_r(\omega)(\omega + \omega_0)}{\omega^2 - \omega_0^2}d\omega = -\frac{2}{\pi}\mathscr{P}\int_{0}^{\infty}\frac{\omega_0\alpha_r(\omega)d\omega}{\omega^2 - \omega_0^2}. \tag{19.16}$$

We have now obtained the Kramers–Kronig relations. To avoid explicit use of the principal value of a function, we can subtract out the singularity at ω_0, by writing

$$\alpha_r(\omega_0) + i\alpha_i(\omega_0) = \frac{1}{\pi i}\int_{-\infty}^{\infty}\left(\frac{\alpha(\omega) - \alpha(\omega_0)}{\omega - \omega_0}\right)\left(\frac{\omega + \omega_0}{\omega + \omega_0}\right)d\omega. \tag{19.17}$$

Using the evenness and oddness of $\alpha_r(\omega)$ and $\alpha_i(\omega)$ we then obtain

$$\alpha_r(\omega_0) = \frac{2}{\pi}\int_{0}^{\infty}\frac{\omega\alpha_i(\omega) - \omega_0\alpha_i(\omega_0)}{\omega^2 - \omega_0^2}d\omega \tag{19.18}$$

and

$$\alpha_i(\omega_0) = -\frac{2}{\pi}\int_{0}^{\infty}\frac{\omega_0\alpha_r(\omega) - \omega_0\alpha_r(\omega_0)}{\omega^2 - \omega_0^2}d\omega. \tag{19.19}$$

To obtain the Kramers–Kronig relations for the dielectric function itself, we just substitute

$$\varepsilon(\omega) = 1 + 4\pi\alpha(\omega) = \varepsilon_1(\omega) + i\varepsilon_2(\omega) \tag{19.20}$$

into (19.18) and (19.19) to obtain

$$\varepsilon_1(\omega_0) - 1 = \frac{2}{\pi}\int_0^\infty \frac{\omega'\varepsilon_2(\omega') - \omega_0\varepsilon_2(\omega_0)}{\omega'^2 - \omega_0^2}d\omega' \tag{19.21}$$

and

$$\varepsilon_2(\omega_0) = \frac{-2}{\pi}\int_0^\infty \frac{\omega_0\varepsilon_1(\omega') - \omega_0\varepsilon_1(\omega_0)}{\omega'^2 - \omega_0^2}d\omega'. \tag{19.22}$$

The Kramers–Kronig relations (19.21) and (19.22) are very general and depend, as we have seen, on the assumptions of causality, linearity and boundedness. From this point of view, the real and imaginary parts of a "physical" quantity Q can be related by making the identification

$$Q_{\text{real}} \rightarrow \alpha_r \tag{19.23}$$

$$Q_{\text{imaginary}} \rightarrow \alpha_i. \tag{19.24}$$

Thus, we can identify $\varepsilon_1(\omega) - 1$ with $\alpha_r(\omega)$, and $\varepsilon_2(\omega)$ with $\alpha_i(\omega)$. The reason, of course, why the identification $\alpha_r(\omega)$ is made with $[\varepsilon_1(\omega) - 1]$ rather than with $\varepsilon_1(\omega)$ is that if $\varepsilon_2(\omega) \equiv 0$ for all ω, we want $\varepsilon_1(\omega) \equiv 1$ for all ω (the dielectric constant for free space).

Thus, if we are interested in constructing a Kramers–Kronig relation for the optical constants, then we again want to make the following identification for the optical constants $(\tilde{n} + i\tilde{k})$

$$[\tilde{n}(\omega) - 1] \rightarrow \alpha_r(\omega) \tag{19.25}$$

$$\tilde{k}(\omega) \rightarrow \alpha_i(\omega). \tag{19.26}$$

From (19.21) and (19.22), we can obtain the Kramers–Kronig relations for the optical constants $\tilde{n}(\omega)$ and $\tilde{k}(\omega)$

$$\tilde{n}(\omega) - 1 = \frac{2}{\pi}\int_0^\infty \frac{\omega'\tilde{k}(\omega') - \omega\tilde{k}(\omega)}{\omega'^2 - \omega^2}d\omega' \tag{19.27}$$

and

$$\tilde{k}(\omega) = -\frac{2}{\pi}\int_0^\infty \frac{\omega\tilde{n}(\omega') - \omega\tilde{n}(\omega)}{\omega'^2 - \omega^2}d\omega' \tag{19.28}$$

where we utilize the definition relating the complex dielectric function $\varepsilon(\omega)$ to the optical constants $\tilde{n}(\omega)$ and $\tilde{k}(\omega)$, where $\varepsilon(\omega) = [\tilde{n}(\omega) + i\tilde{k}(\omega)]^2$.

It is useful to relate the optical constants to the reflection coefficient $r(\omega)\exp[i\theta(\omega)]$ defined by

$$r(\omega)\exp[i\theta(\omega)] = \frac{\tilde{n}(\omega) - 1 + i\tilde{k}(\omega)}{\tilde{n}(\omega) + 1 + i\tilde{k}(\omega)} \tag{19.29}$$

in which the conjugate variables are $\ln r(\omega)$ and $\theta(\omega)$, while the reflectivity is given by $\mathscr{R}(\omega) = r^2(\omega)$. From (19.29), we can then write

$$\tilde{n}(\omega) = \frac{1 - r^2(\omega)}{1 + r^2(\omega) - 2r(\omega)\cos\theta(\omega)} \tag{19.30}$$

$$\tilde{k}(\omega) = \frac{2r(\omega)\sin\theta(\omega)}{1 + r^2(\omega) - 2r(\omega)\cos\theta(\omega)} \tag{19.31}$$

so that once $r(\omega)$ and $\theta(\omega)$ are found, the frequency dependent optical constants $\tilde{n}(\omega)$ and $\tilde{k}(\omega)$ are determined. In practice, $r(\omega)$ and $\theta(\omega)$ are found from the reflectivity $\mathscr{R}(\omega)$, which is measured over a wide frequency range and is modeled outside the measured range by use of the following relations. Using the Kramers–Kronig relation, we can also consider the conjugate variables $\ln r(\omega)$ and $\theta(\omega)$, from which $\theta(\omega)$ is found:

$$\ln r(\omega) = \frac{2}{\pi}\int_0^\infty \frac{\omega'\theta(\omega') - \omega\theta(\omega)}{\omega'^2 - \omega^2} d\omega' \tag{19.32}$$

$$\theta(\omega) = -\frac{2\omega}{\pi}\int_0^\infty \frac{\ln r(\omega') - \ln r(\omega)}{\omega'^2 - \omega^2} d\omega'. \tag{19.33}$$

where $\ln \mathscr{R}(\omega) = 2\ln r(\omega)$.

From a knowledge of the frequency-dependent reflectivity $\mathscr{R}(\omega)$, for example, the reflection coefficient $r(\omega)$ and the phase of the reflectivity coefficient $\theta(\omega)$ can be found. We can then find the frequency dependence of the optical constants $\tilde{n}(\omega)$ and $\tilde{k}(\omega)$, which in turn yield the frequency dependent dielectric functions $\varepsilon_1(\omega)$ and $\varepsilon_2(\omega)$. Starting with the experimental data for the reflectivity $\mathscr{R}(\omega)$ for germanium in Fig. 19.2a, the Kramers–Kronig relations are used to obtain results for $\varepsilon_1(\omega)$ and $\varepsilon_2(\omega)$ for germanium, as shown in Fig. 19.2b.

The Kramers–Kronig relations for the conjugate variables $\varepsilon_1(\omega)$ and $\varepsilon_2(\omega)$; $\tilde{n}(\omega)$ and $\tilde{k}(\omega)$; and $\ln r(\omega)$ and $\theta(\omega)$ are all widely used in quantitative studies of the optical properties of specific materials, as for example germanium, as shown in Fig. 19.2. For pedagogic purposes, germanium is simpler because germanium is a direct bandgap semiconductor for which the bandgap occurs at $k = 0$, the Γ point is in the Brillouin zone.

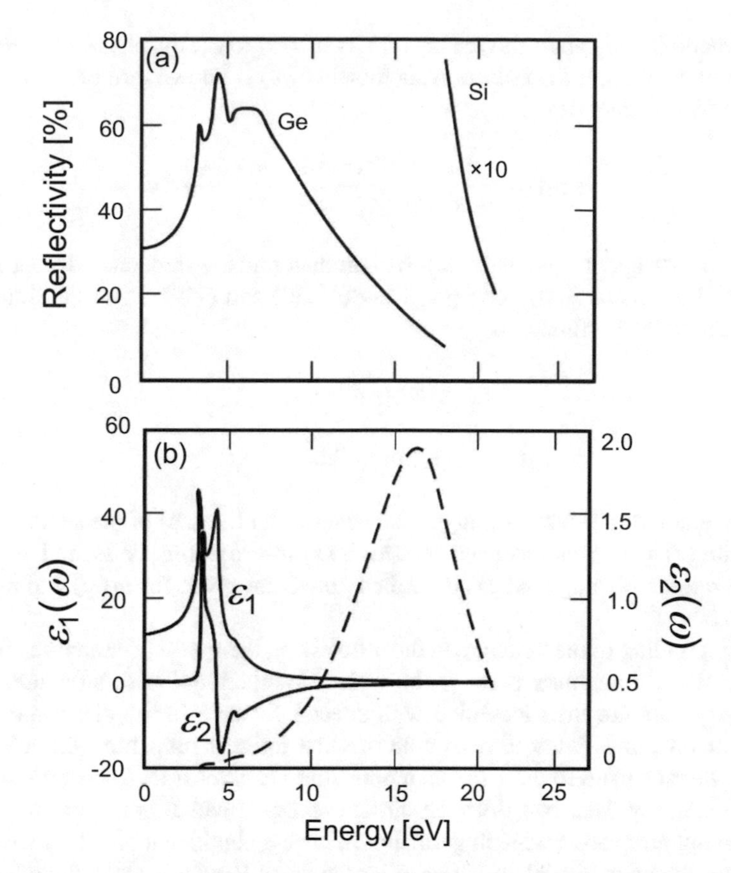

Fig. 19.2 **a** Frequency dependence of the reflectivity of Ge over a wide frequency range. **b** Plot of the real $[\varepsilon_1(\omega)]$ and imaginary $[\varepsilon_2(\omega)]$ parts of the dielectric functions for Ge obtained by a Kramers–Kronig analysis of the reflectivity data in part (**a**)

19.2 Optical Properties and Band Structure

If we are interested in studying the optical properties near the band edge, such as the onset of indirect transitions or of the lowest direct interband transitions, then we should carry out absorption measurements (Chap. 18) to determine the absorption coefficient $\alpha_{\text{abs}}(\omega)$ and thus identify the type of process that is dominant (indirect, direct, allowed, forbidden, etc.) at the band edge. However, if we are interested in the optical properties of a semiconductor over a wide frequency range, then we want to treat all energy bands and interband transitions within a few eV from the Fermi level on an equal footing. Away from the band edge, the optical absorption coefficients become too high in energy for conventional absorption techniques to be useful, and reflectivity measurements are made instead. Experimentally, it is most convenient to carry out reflectivity measurements at normal incidence. From these measurements, the Kramers–Kronig analysis (see Sect. 19.1) is used to get the phase angle $\theta(\omega)$ for some frequency ω_0, if the reflection coefficient $r(\omega)$ is known throughout the entire range of photon energies

$$\theta(\omega_0) = -\frac{2\omega_0}{\pi} \int_0^{\infty} \frac{\ln r(\omega) - \ln r(\omega_0)}{\omega^2 - \omega_0^2} d\omega. \tag{19.34}$$

From a knowledge of $r(\omega)$ and $\theta(\omega)$, we can then find the frequency dependence of the optical constants $\tilde{n}(\omega)$ and $\tilde{k}(\omega)$ using (19.30) and (19.31) and the frequency-dependent dielectric functions

$$\varepsilon_1(\omega) = \tilde{n}^2 - \tilde{k}^2 \tag{19.35}$$

$$\varepsilon_2(\omega) = 2\tilde{n}\tilde{k}. \tag{19.36}$$

As an example of such an analysis, let us consider the case of the semiconductor germanium. The normal incidence reflectivity is shown in Fig. 19.2a, and the results of the Kramers–Kronig analysis described above are given for $\varepsilon_1(\omega)$ and $\varepsilon_2(\omega)$ in Fig. 19.2b.

Corresponding to the structure in the reflectivity, there will be structure observed in the real and imaginary parts of the dielectric function. These structures in the reflectivity data are then identified with special features in the electronic energy band structure. It is interesting to note that the indirect transition (0.66 eV) from the $\Gamma_{25'}$ valence band to the L_1 conduction band has almost no impact on the measured reflectivity data. Nor does the direct band gap, which is responsible for the fundamental absorption edge in germanium, have a significant effect on the reflectivity data shown in Fig. 19.2a. These effects are small on the scale of the reflectivity structures shown in Fig. 19.2a and must be looked for with great care in a narrow frequency range where structure in the absorption data is found for these particular structures. The big contribution to the dielectric constant comes from interband transitions $L_{3'} \rightarrow L_1$ for which the joint density of states is large over large volumes of the Brillouin zone. The sharp rise in $\varepsilon_2(\omega)$ at 2.1 eV is associated with the $L_{3'} \rightarrow L_1$ transition. For higher photon energies, large volumes of the Brillouin zone contribute until a photon energy of about 5 eV is reached. Above this photon energy, we cannot find bands that track each other closely enough to give interband transitions with intensities of large enough magnitude to make a significant contribution to observables in the visible optical frequency range.

19.3 Modulated Reflectivity Experiments

If we wish to study the *critical point* contributions to the optical reflectivity in more detail, it is useful to carry out modulated reflectivity measurements. If, for example, a small periodic perturbation is applied to a sample then there will be a change in

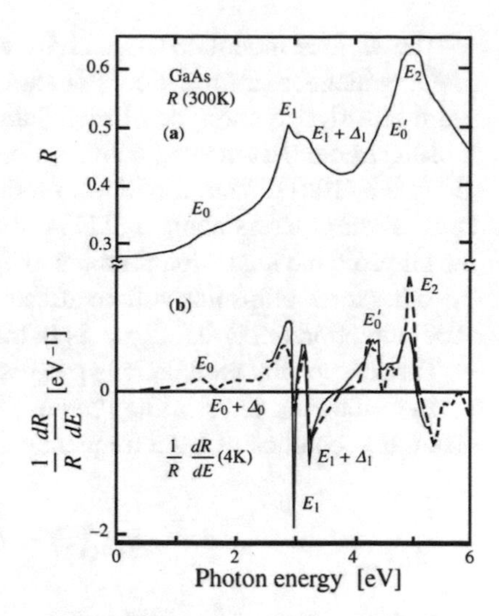

Fig. 19.3 Reflectance and frequency modulated reflectance spectra for GaAs. **a** Room temperature reflectance spectrum and **b** the wavelength modulated spectrum $(1/R)(dR/dE)$ at 4 K (the solid curve is experimental and the broken curve is calculated using a pseudopotential band structure model. (Adapted from Yu and Cardona)

the reflectivity at the frequency of that perturbation. The frequency dependence of this change in reflectivity is small (parts in 10^3 or 10^4) but it is measurable. As an example, we show in Fig. 19.3, results for the reflectivity $R(\omega)$ and for the wavelength modulated reflectivity $(1/R)(dR/dE)$ of GaAs, where the energy variable E is related to the probing photon frequency ω by $E = \hbar\omega$. Structure at E_0 would be identified with the direct band gap, while the structure at $E_0 + \Delta_0$ corresponds to a transition from the split-off valence band at $\mathbf{k} = 0$ which arises through the spin-orbit interaction. It is interesting to note that the modulated reflectivity $(1/R)(dR/dE)$ in Fig. 19.3b provides a more sensitive probe of the value of E_0 and Δ_0 than the reflectivity itself in Fig. 19.3a.

In the vicinity of a critical point, the denominator in the joint density of states is small, so that a small change in photon energy can produce a significant change in the joint density of states. Hence, modulation spectroscopy techniques emphasize critical points. There are a number of parameters that can be varied in these modulation spectroscopy experiments, including the following Table 19.1:

Table 19.1 List of parameters and corresponding modulated reflectivity measurements

Electric field – electro-reflectance
Wavelength – wavelength modulation
Stress – piezo-reflectance
Light intensity – photo-reflectance
Temperature – thermo-reflectance
Fermi level – gate modulation

The various modulated reflectivity experiments are complementary rather than yielding identical information. For example, certain structures in the reflectance are more sensitive to one type of modulation technique than to another. If we wish to look at structure associated with the L point (111 direction) transitions, then a stress along the (100) direction will not produce as important a symmetry change as the application of stress along a (111) direction; with a stress along a (111) direction, the ellipsoid having its longitudinal axis along (111) will be affected one way while the other three ellipsoids will be affected in another way. However, stress along the (100) direction treats all ellipsoids in the same way.

The reason why modulation spectroscopy emphasizes critical points can be seen by the following physical argument. For a direct interband transition, the optical absorption coefficient has a frequency dependence

$$\alpha_{abs}(\omega) = C\frac{\sqrt{(\hbar\omega - E_g)}}{\hbar\omega}. \tag{19.37}$$

Therefore, a plot of $\alpha_{abs}(\omega)$ vs. $\hbar\omega$ exhibits a threshold [Fig. 19.4a], but shows no singularity in the frequency plot. However, when we take the frequency derivative of (19.37)

$$\frac{\partial\alpha_{abs}(\omega)}{\partial\omega} = \frac{C}{2\omega}\left(\hbar\omega - E_g\right)^{-1/2} - \frac{C}{\hbar\omega^2}\left(\hbar\omega - E_g\right)^{1/2} \tag{19.38}$$

a sharp structure is obtained in the modulated reflectivity due to the singularity in the first term of (19.37) at $\hbar\omega = E_g$ [see Figs. 19.4a and 19.4b].

If we modulate the incident light with any arbitrary parameter x, then

$$\frac{\partial\alpha_{abs}}{\partial x} = \frac{\partial\alpha_{abs}}{\partial\omega}\frac{\partial\omega}{\partial x}. \tag{19.39}$$

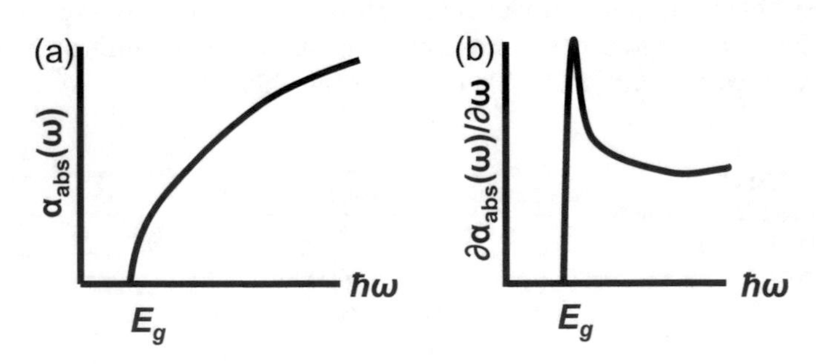

Fig. 19.4 Physical picture of **(a)** the frequency dependence of the optical absorption coefficient showing a threshold for interband transitions at the band gap. **(b)** The derivative of **(a)** with respect to frequency, which is the quantity measured in the modulated reflectivity experiments, shows a sharp singularity associated with the threshold energy

Additional structure in the reflectivity is expected as x is varied, and this behavior is shown schematically in Fig. 19.4b.

Thus all modulation parameters can be expected to produce singularities in the optical absorption. For some variables such as stress, the modulated signal is sensitive to both the magnitude and the *direction* of the stress relative to the crystal axes. For thermomodulation, the spectrum is sensitive to the magnitude of the thermal pulses, but the response is independent of crystalline direction in a highly symmetric crystal structure. Thermomodulation is, however, especially sensitive to transitions from and to the Fermi level.

Thus, the various modulation techniques can be used in optical studies to obtain additional information about symmetry, which can then be used for more reliable identification of the physical mechanism most important in causing a particular structure to appear in the optical properties. The modulation technique specifically emphasizes interband transitions associated with particular points in the Brillouin zone. The identification of where in the Brillouin zone a particular transition is occurring is one of the most important and difficult problems in optical studies of solids. It is often not the case that we have reliable band models available to us when we start to do optical studies. For this reason, studying the symmetry of actual samples provides a very powerful tool for studying their optical properties.

The high sensitivity of modulation spectroscopy provides valuable information about the band structure that would be difficult to obtain otherwise, and some examples of actual materials are cited below. One example of the use of modulation spectroscopy is to determine the temperature dependence of the bandgap of a semiconductor through the temperature modulation technique, as shown in Fig. 19.5 for the direct Γ point gap in Ge. This measurement takes advantage of the high resolution of modulation spectroscopy and is especially useful for measurements at elevated temperatures.

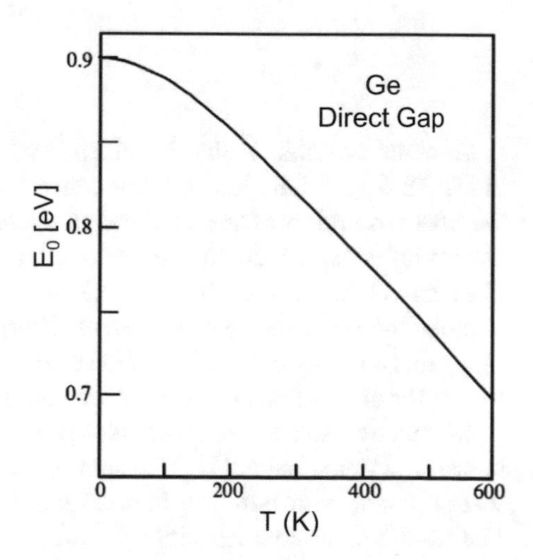

Fig. 19.5 Temperature dependence of the direct band gap (E_0) of Ge using the thermal modulation measurement technique many years ago. (J.S. Kline, F.H. Pollak and M.Cardona, Helv. Phys. Acta., 41, 968 (1968))

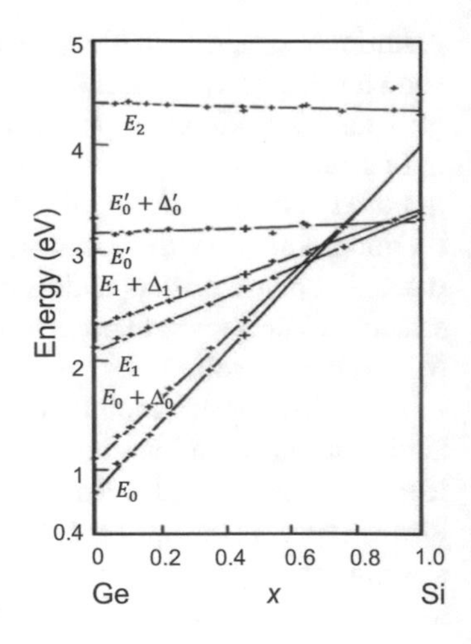

Fig. 19.6 Dependence of the energies of the E_0, $E_0 + \Delta_0$, E_1, $E_1 + \Delta_1$, E_0', and E_2 electro-reflectance peaks on x in the amount Si added to Ge in the $Ge_{1-x}Si_x$ alloy system at room temperature (C. Parks, Phys. Rev.B, 49, 14244 (1994))

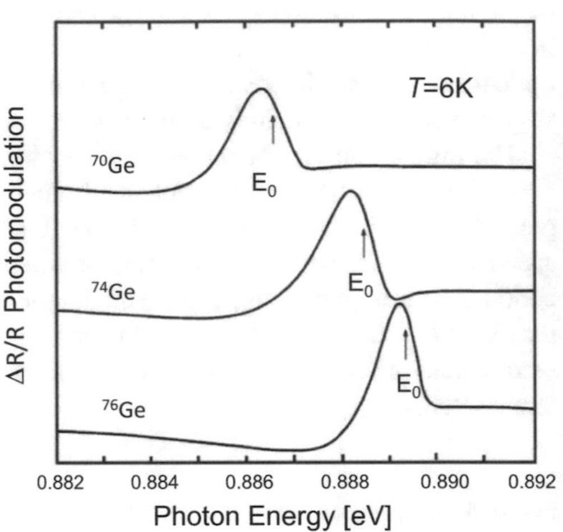

Fig. 19.7 Photo-modulated reflectivity spectra of Ge showing the E_0 direct gap at $k = 0$ for single crystals of nearly isotopically pure ^{70}Ge, ^{74}Ge, and ^{76}Ge, at $T = 6$ K. Note the remarkable dependence of the direct bandgap E_0 on the isotopic composition

Another example is the dependence of the various band separations identified in Fig. 19.6 as a function of alloy concentration x in $Ge_{1-x}Si_x$ alloys. Here again the high resolution of the modulation spectroscopy is utilized, but now using the electro-reflectance technique to get much more detailed information.

A third example of the use of a related experimental technique is the isotope effect to probe the dependence of the direct absorption edge of Ge, as shown in Fig. 19.7. Using this technique the effect of the electron-phonon interaction contribution to the shift in the absorption edge can be separated from the purely electronic contribution.

Modulation spectroscopy has also been applied to studying interband transitions in metals. For example, Fig. 19.8 shows modulated spectroscopy results from a gold surface taken with both the thermal modulation and piezoreflectance techniques. The results show that transitions involving states at the Fermi level (either initial

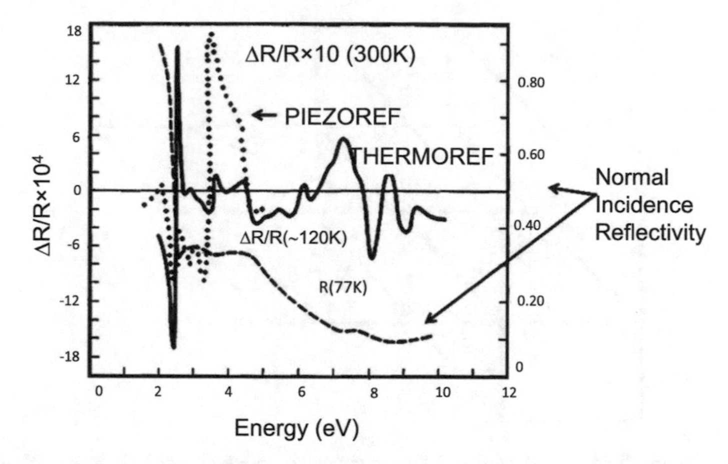

Fig. 19.8 Thermo-reflectance and normal incidence reflectivity spectra of gold near liquid nitrogen temperature (from W.J. Scouler, Phys. Rev. Letters 18, 445 (1967)) together with the room temperature piezoreflectance spectrum (M. Garfunkel, J.J. Tiemann, and W.E. Engeler, Phys. Rev. 148, 698 (1966))

or final states) are more sensitively seen using thermal modulation because small temperature variations affect the Fermi tail of the distribution function strongly. Thus, thermo-reflectance measurements on the noble metals give a great deal of well–resolved structure, as illustrated in Fig. 19.8, where the electro-reflectance and piezoreflectance measurements are compared. In this figure, we see that in gold the piezoreflectance is much more sensitive than are the ordinary reflectivity measurements near 4 eV, but the thermoreflectance technique is most powerful for transitions made to states near the Fermi level.

19.4 Ellipsometry and Measurement of the Optical Constants

Ellipsometry has become a standard method for measuring the complex dielectric function or the complex optical constants $\tilde{N} = \tilde{n} + i\tilde{k}$ of a material. Since two quantities are measured in an ellipsometry measurement, $\tilde{n}$ and $\tilde{k}$ can both be determined at a single frequency. The ellipsometry measurements are usually made over a range of frequencies, especially for frequencies well above the fundamental absorption edge where semiconductors become highly absorbing. At these higher frequencies very thin samples would be needed if the method of interference fringes were used to determine $\tilde{n}$, which is a very simple method for measuring the wavelength in a non-absorbing medium. One drawback of the ellipsometry technique is the high sensitivity of the technique to the quality and cleanliness of the surface. Ellipsometry is limited by precision considerations to measurements on samples with absorption

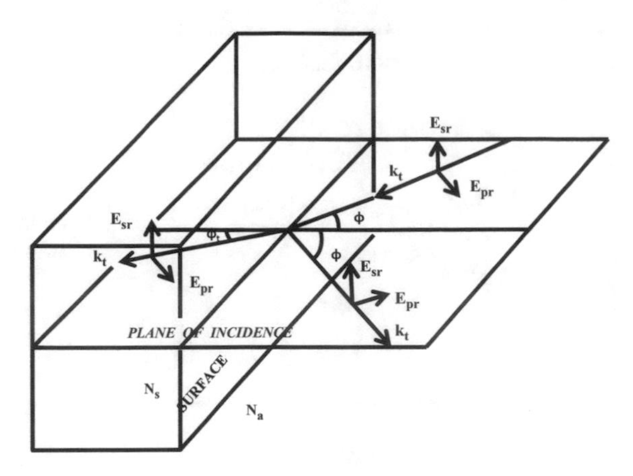

Fig. 19.9 Electric field vectors resolved into p and s components, for light that is incident (i), reflected (r), and transmitted (t) at an interface between these two media each with complex indices of refraction (in the medium) $\tilde{N}_a$ and $\tilde{N}_s$ at the surfaces. The propagation vectors are labeled by wave vectors for the incident, reflected and transmitted light components $\mathbf{k}_i$, $\mathbf{k}_r$, and $\mathbf{k}_t$

coefficients $\alpha_{\text{abs}} > 1 - 10\ \text{cm}^{-1}$. Ellipsometers can be made to operate in the near infrared, visible and near ultraviolet frequency regimes, and data acquisition can be made fast enough to do real time monitoring of $\varepsilon(\omega)$.

In the ellipsometry method, the reflected light with polarizations "p" (parallel) and "s" (perpendicular) to the plane of incidence [see Fig. 19.9] is measured as a function of the angle of incidence ϕ and the light frequency ω. The corresponding reflectances $R_s = |r_s|^2$ and $R_p = |r_p|^2$ are related to the complex dielectric function $\varepsilon(\omega) = \varepsilon_1(\omega) + i\varepsilon_2(\omega) = (\tilde{n} + i\tilde{k})^2$ by the Fresnel equations which can be derived from the boundary conditions on the fields at the interface between two surfaces with complex dielectric functions ε_a and ε_s as shown in Fig. 19.9. From the figure, we see that the complex reflection coefficients for polarizations s and p are

$$r_s = \frac{E_{sr}}{E_{si}} = \frac{\tilde{N}_a \cos\phi - \tilde{N}_s \cos\phi_t}{\tilde{N}_a \cos\phi + \tilde{N}_s \cos\phi_t} \tag{19.40}$$

and

$$r_p = \frac{E_{pr}}{E_{pi}} = \frac{\varepsilon_s \tilde{N}_a \cos\phi - \varepsilon_a \tilde{N}_s \cos\phi_t}{\varepsilon_s \tilde{N}_a \cos\phi + \varepsilon_a \tilde{N}_s \cos\phi_t} \tag{19.41}$$

in which

$$\tilde{N}_s \cos\phi_t = (\varepsilon_s - \varepsilon_a \sin^2\phi)^{1/2} \tag{19.42}$$

and r_s and r_p are the respective reflection coefficients, ε_s and $\tilde{N}_s$, respectively, denote the complex dielectric function and the complex index of refraction within the medium, while ε_a and $\tilde{N}_a$ are the corresponding quantities outside the medium (which is usually vacuum or air). When linearly polarized light, that is neither s- nor p-polarized, is incident on a medium at an oblique angle of incidence ϕ, the reflected light will be elliptically polarized. The ratio (σ_r) of the complex reflectivity coefficients $r_p/r_s \equiv \sigma_r$ is then a complex variable which is measured experimentally in terms of its phase (or the phase shift relative to the linearly polarized incident light) and its magnitude, which is the ratio of the axes of the polarization ellipse of the reflected light [see Fig. 19.9]. These are the two measurements that are made in ellipsometry. The complex dielectric function of the medium $\varepsilon_s(\omega) = \varepsilon_1(\omega) + i\varepsilon_2(\omega)$ can then be determined from the angle ϕ, the complex reflectivity coefficient ratio σ_r, and the dielectric function ε_a of the ambient environment using the relation

$$\varepsilon_s = \varepsilon_a \sin^2\phi + \varepsilon_a \sin^2\phi \tan^2\phi \left(\frac{1-\sigma_r}{1+\sigma_r}\right)^2, \tag{19.43}$$

and in a vacuum environment $\varepsilon_a = 1$.

The experimental set-up for ellipsometry measurements is shown in Fig. 19.10. Light from a tunable light source is passed through a monochromator to select a frequency ω and the light is then polarized linearly along the direction of the applied electric field $\mathbf{E}$ to yield the I_s and I_p incident light intensities. After reflection, the light is elliptically polarized along $\mathbf{E}(t)$ as a result of the phase shifts that E_{pr} and E_{sr} have each experienced. The compensator introduces a phase shift $-\theta$ which cancels the $+\theta$ phase shift induced by the reflection at the sample surface, so that the light becomes linearly polarized again as it enters the analyzer. If the light is polarized at an angle of $\pi/2$ with respect to the analyzer setting, then no light reaches the detector. Thus at every angle of incidence and every frequency, the dielectric function $\varepsilon(\omega, \phi)$ is determined by (19.43) from measurement of the magnitude and phase of the

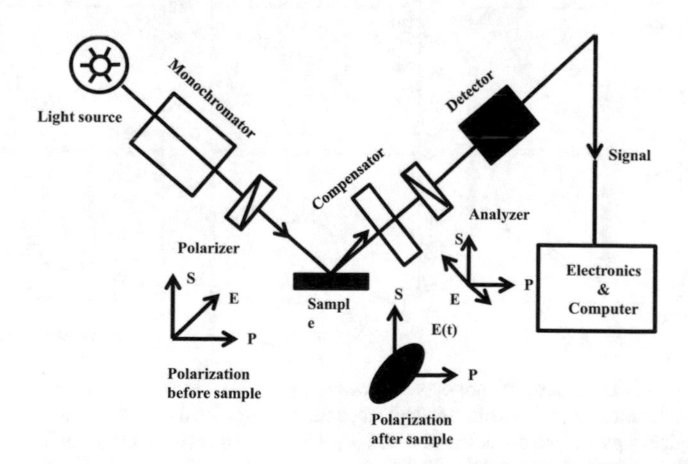

Fig. 19.10 Schematic diagram of an ellipsometer, where P and S denote polarizations parallel and perpendicular to the plane of incidence, respectively

reflection coefficient σ_r. The determination of the optical constants can be made by using the normal incidence reflectivity data taken over a wide frequency range and using the Kramers–Kronig analysis, as discussed in Sect. 19.2, to determine the optical constants $\tilde{n}(\omega)$ and $\tilde{k}(\omega)$.

19.5 Kramers–Kronig Relations in 2D Materials

In Fig. 19.11, the measurements by Li et al. are shown for the complex dielectric functions $\varepsilon(E)$ obtained from four different transition metal dichalcogenide monolayers. The experimental reflectance spectra are shown together with the $\varepsilon_1(\omega)$ and $\varepsilon_2(\omega)$ results obtained by a Kramers–Kronig analysis. Here, the photo-reflectance spectra are measured over a wide frequency range (from 1.5–3 eV). Results for the photo-reflectance of some different transition metal dichalcogenides MoS_2, $MoSe_2$, WS_2, and WSe_2 are also shown (see Li et al. (2014)).

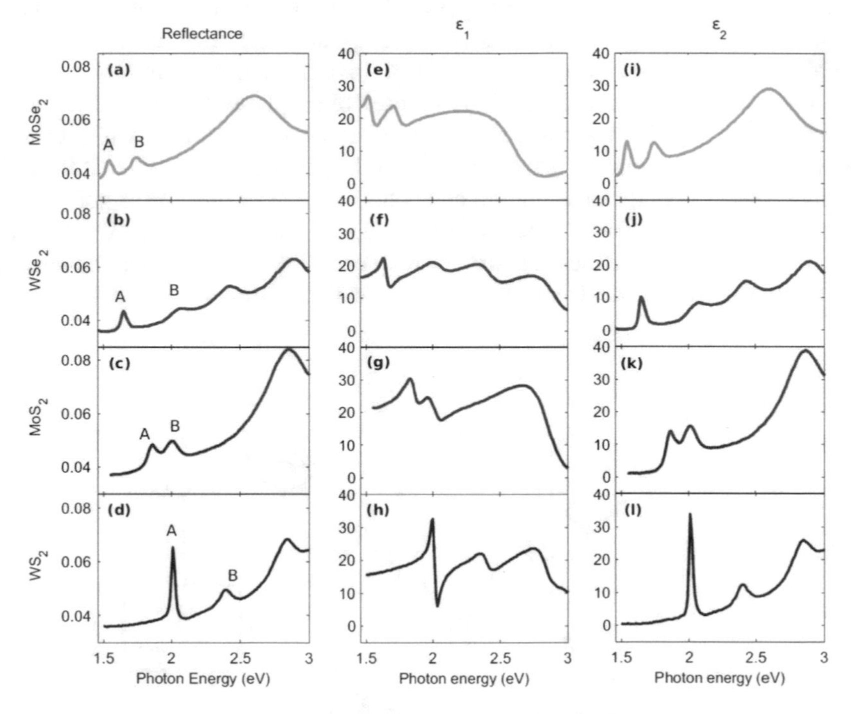

Fig. 19.11 Optical response of monolayers of $MoSe_2$, WSe_2, MoS_2, and WS_2 exfoliated on a fused silica substrate: (a–d) Measured reflectance spectra. (e-h) Real part of the dielectric function, ε_1. (i-l) Imaginary part of the dielectric function, ε_2. The peaks labeled A and B in (a)(d) correspond to excitons from the two spin-orbit split transitions at the K point of the Brillouin zone. (Taken from Li et al. (2014))

19.6 Summary

The Kramers–Kronig relations are bidirectional mathematical relations, connecting the real and imaginary parts of the dielectric function. These relations are often used to calculate the real part from the imaginary part (or vice versa) of the dielectric function $\varepsilon(\omega)$. Both the real and imaginary parts of the dielectric function can be obtained from a single experimentally measured spectrum (e.g., absorption or reflection) by use of the Kramers–Kronig relations. These are general relations, applicable to many other complex functions in physical and mathematical systems, and are also known under the names of the Sokhotski-Plemelj theorem and the Hilbert transform.

Problems

19.1 Suppose that we model the interband transitions in Ge as a step function $\varepsilon_2(\omega) = \varepsilon_l$ for $E_{min} < \hbar\omega < E_{max}$ (see diagram) and $\varepsilon_2(\omega) = 0$ otherwise, as shown in Fig. 19.12.

(a) Use the Kramers–Kronig relation to find an expression for $\varepsilon_1(\omega)$ for all ω. Take $\varepsilon_1 = 1$ in the limit $\omega = \infty$ and express your answer in terms of E_{max}, E_{min}, and ε_l.
(b) For which photon energies does $\varepsilon_1(\omega)$ exhibit structure? Is your answer physically reasonable and why?
(c) Obtain an explicit expression for $\varepsilon_1(0)$ at zero frequency, and use this result to explain why narrow gap semiconductors tend to have large dielectric constants at $\omega = 0$.

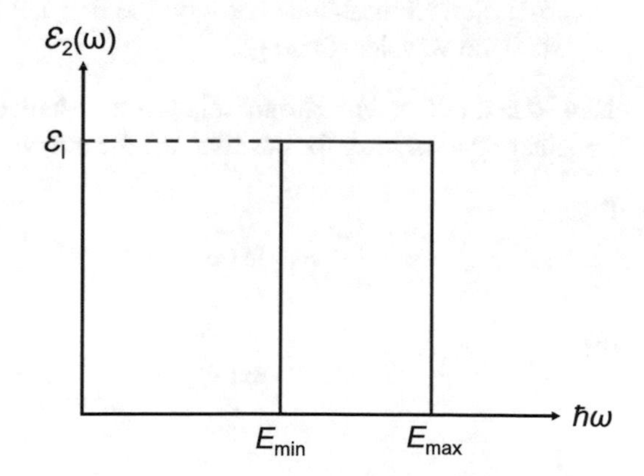

Fig. 19.12 The sample step function model used in problem 19.1 for the interband transitions in Ge

(d) Use the Kramers–Kronig relations to show the sum rule

$$\frac{ne^2}{m} = \frac{1}{2\pi^2}\int_0^\infty \varepsilon_2(\omega)\omega d\omega$$

where n is the total carrier density of the semiconductor at a temperature T.

19.2 Using the Kramers–Kronig relation

$$\varepsilon_1(\omega_0) - 1 = \frac{2}{\pi}\int_0^\infty \frac{\omega'\varepsilon_2(\omega') - \omega_0\varepsilon_2(\omega_0)}{\omega'^2 - \omega_0^2}d\omega'$$

explain why $\varepsilon_1(0)$ at $\omega_0 = 0$ is so large for Si [$\varepsilon_0(\text{Si}) = 12$] relative to glass for which $\varepsilon_0 < 3$.

19.3 According to Johnson and Christy's paper on the Optical Constants of Noble Metals (see Johnson et al. (1972): 4370), the equation for the dielectric permittivity in the Drude free electron theory is given by

$$\varepsilon(\omega) = 1 - \frac{\omega_p^2}{\omega(\omega + \frac{1}{\tau})}$$

where $\varepsilon(\omega)$ is the dielectric function at frequency (ω). Here, ω_p is the plasma frequency, and $\gamma = \frac{1}{\tau}$ is the collision frequency. Use values for gold $m^* = 0.99m_0$, and $\tau = 9.3 \times 10^{-15}s$.

(a) Plot the real and imaginary parts of the dielectric function (ε_1 and ε_2) for Au over the wavelength range from 400-1000 nm.
(b) Using the Kramers–Kronig relation, calculate ε_1 from ε_2 and then calculate ε_2 from ε_1. Plot these together with the original functions from (a) over the 400–1000 nm wavelength range.

19.4 Use the Kramers–Kronig relation to calculate the real part of $\varepsilon(\omega)$, given the imaginary part of $\varepsilon(\omega)$ for positive (ω) for the two cases:

(a)

$$\frac{\varepsilon_2(\omega)}{\varepsilon_0} = \lambda[\theta(\omega - \omega_1) - \theta(\omega - \omega_2)], \omega_2 > \omega_1 > 0$$

(b)

$$\frac{\varepsilon_2(\omega)}{\varepsilon_0} = \frac{\lambda\gamma\omega}{(\omega_0^2 - \omega^2)^2 + \gamma^2\omega^2}$$

19.5 Show that if a linear response function, such as the linear electric susceptibility $\chi(\omega)$ or the dielectric function $(\varepsilon(\omega) - 1)$, satisfies the following two conditions: (1) it is analytic in the upper half of the complex ω-plane and (2) it approaches zero sufficiently fast as ω approaches infinity, it satisfies the Kramers–Kronig relation.

19.6 An electromagnetic wave travels inside a dielectric material with a complex index of refraction $\tilde{N} = \tilde{n} + i\tilde{k} = c/c_1$, and is reflected at the plane boundary (xy-plane). If the **E** vector of the incident wave is in the x-direction, the reflection coefficient for the **E** field is

$$r = \frac{\tilde{n}\cos\theta_1 - \cos\theta_2}{\tilde{n}\cos\theta_1 + \cos\theta_2}$$

and if the **H** vector is in the x-direction, the reflection coefficient is

$$r' = \frac{\cos\theta_1 - \tilde{n}\cos\theta_2}{\cos\theta_1 + \tilde{n}\cos\theta_2}$$

(a) Use boundary conditions at the interface to derive the above equation for r.
(b) Suppose that θ_1 is made large enough so that total internal reflection occurs. With the aid of the above equations, find the phase angles ϕ and ϕ' of the reflected waves E_r and E'_r in each of the two cases, defining the phase angle of the incident wave to be zero at the boundary. (Note that $\cos\theta$ is imaginary under conditions of total internal reflection.) Show that

$$\tan\left(\frac{\phi' - \phi}{2}\right) = \frac{\cos\theta_1\sqrt{\sin^2\theta_1 - (1/\tilde{n}^2)}}{\sin^2\theta_1}.$$

(c) For a linearly polarized incident wave, it is possible for the reflected wave to be circularly polarized. If this is possible, what must be the polarization direction of the incident wave? Write an equation that determines the required angle of incidence?
(d) Determine the smallest value of the index of refraction for which (b) is possible, and find the corresponding angle of incidence.

19.7 (a) Suppose that we apply a magnetic field along the (001) direction normal to the surface of a sample. Find the dependence of ε_1 and ε_2 on the magnetic field. (Hint: Use of right and left circularly polarized fields will be helpful with this problem.)
(b) Find the dependence of the plasma frequency on magnetic field using right and left incident circularly polarized light. Sketch the result of the magnetic field on the optical reflectivity for right and left circular polarized light.

Suggested Readings

Yu, Cardona, *Fundamentals of Semiconductors* (Springer, Berlin, 1996). Sects. 6.1.3 and 6.6
Jones, March, *Theoretical Solid State Physics* (1973), pp. 787–793
Jackson, *Classical Electrodynamics* (1999), pp. 306–312
Peter B. Johnson, R.-W. Christy, Optical constants of the noble metals. Phys. Rev. B **6**(12), 4370 (1972)

Reference

Y.L. Li, A. Chernikov, X. Zhang, A. Rigosi, H.M. Hill, A.M. van der Zande, D.A. Chenet, E.M. Shih, J. Hone, T.F. Heinz, Measurement of the optical dielectric function of monolayer transition-metal dichalcogenides: MoS_2, $MoSe_2$, WS_2, and WSe_2. Phys. Rev. B **90**, 205422 (2014)

Chapter 20
Impurities and Excitons

20.1 Impurity Level Spectroscopy

Selected impurities are frequently introduced into semiconductors to make them n–type or p–type. The introduction of impurities into a crystal lattice not only shifts the Fermi level, but also results in a perturbation to the periodic potential, giving rise to bound impurity levels which often occur in the band gap of the semiconductor.

Impurities and defects in semiconductors can be classified according to whether they result in a minor or major perturbation to the periodic potential. Any disturbance to the periodic potential results in energy levels differing from the energy levels of the perfect crystal. However, when these levels occur within the energy band gap of a semiconductor or of an insulator, they are most readily identified, and these are the levels which give rise to well-defined optical spectra. Impurity levels are classified into two categories:

1. Shallow levels
2. Deep levels

Shallow (deep) levels cause a minor (major) perturbation to the periodic potential of the crystal lattice. Whereas shallow levels can be treated in pertubation theory, deep levels require more in-depth calculations. Impurities are also classified according to whether they give rise to electron carriers (donors) or hole carriers (acceptors). We will now discuss the main aspects of optical spectra for impurities in crystals.

M. Dresselhaus et al., *Solid State Properties*, Graduate Texts in Physics,
https://doi.org/10.1007/978-3-662-55922-2_20

20.2 Shallow Impurity Levels

An example of a shallow impurity level in a semiconductor is a hydrogenic donor level in a semiconductor like Si, Ge or a III–V compound. Let us briefly review the origin of shallow donor levels in n–type semiconductors, where the electronic conduction is predominantly made by electron carriers.

Suppose that we add donor impurities such as arsenic, which has 5 valence electrons, to germanium which has 4 valence electrons. Each germanium atom in the perfect crystal makes 4 bonds to its tetrahedrally placed neighbors. For the arsenic impurity in the germanium lattice, four of the valence electrons will participate in the tetrahedral bonding to the germanium neighbors, but the fifth electron will be attracted back to the arsenic impurity site because the arsenic ion on the site has a positive charge. Within the effective mass approximation, this interaction is described by the Coulomb perturbation Hamiltonian,

$$\mathscr{H}'(r) = -\frac{e^2}{\varepsilon r} \tag{20.1}$$

where ε is the static dielectric constant, which is $16\varepsilon_0$ for germanium and $12\varepsilon_0$ for silicon. This Coulomb interaction is screened by the static dielectric constant of the semiconductor. The approximation of taking ε to be independent of distance is, however, not valid for values of r comparable to lattice dimensions, as discussed below.

In simple terms, $\mathscr{H}'$ given by (20.1) is the same as for the hydrogen atom except that the charge is now $e/\sqrt{\varepsilon}$ and the mass, which enters the kinetic energy term, is the effective mass m^* of the charge carriers. Since the levels in the hydrogen atom are given by the Bohr energy levels E_n^{hydrogen}

$$E_n^{\text{hydrogen}} = -\frac{m_0 e^4}{2\hbar^2 \varepsilon_0^2 n^2} \tag{20.2}$$

then the energy levels in the hydrogenic impurity problem are, to a first approximation, given by hydrogenic levels E_n^{impurity}

$$E_n^{\text{impurity}} = -\frac{m^* e^4}{2\hbar^2 \varepsilon^2 n^2}. \tag{20.3}$$

The impurity levels are shown schematically in Fig. 20.1, where the donor levels are seen to lie in the gap below the conduction band minimum.

For the hydrogen atom, the ground state energy $E_1^{\text{hydrogen}} = -13.6\,\text{eV}$, but for germanium $E_1^{\text{impurity}} \sim 6 \times 10^{-3}\,\text{eV}$ for the lowest impurity level, where we have used a value of $m^* = 0.12 m_0$ representing an average of the effective mass over the entire conduction band pocket. From measurements such as the optical absorption spectra, we find that the thermal energy gap (which is the energy difference between

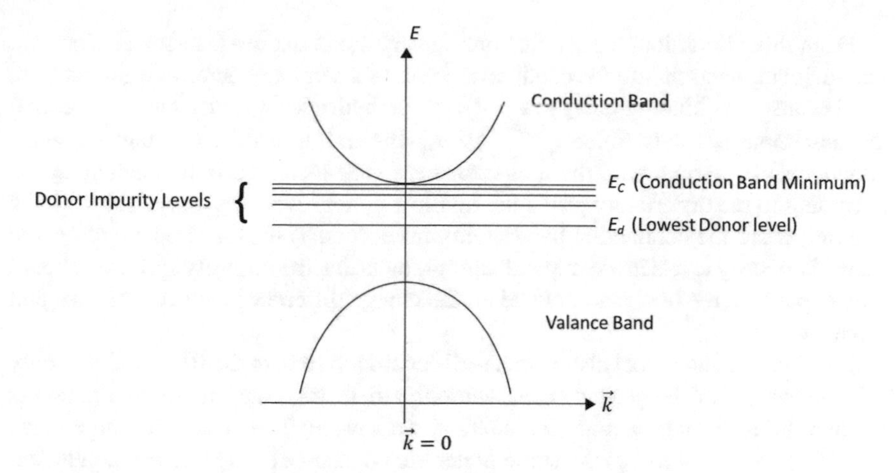

Fig. 20.1 Hydrogenic impurity donor levels in a typical semiconductor

the L point lowest conduction band and the Γ point highest valence band) is 0.66 eV at room temperature. But the donor level manifold is only 6×10^{-3} eV wide (ranging from the $E_1 = E_d$ level to the ionization limit E_c at the conduction band edge) so that these impurity levels are very close to the bottom of the conduction band, as shown in Fig. 20.1.

Another quantity of interest in this connection is the "orbital radius" of the impurity. Unless the orbital radius is greater than a few crystal lattice dimensions, it is not meaningful to use a dielectric constant independent of **r** in constructing the perturbation Hamiltonian, since the dielectric constant used for $\mathscr{H}'_{\text{impurity}}$ is conceptually meaningful only for a continuum medium. It is, however, of interest to calculate the hydrogenic Bohr radius using the usual recipe for the hydrogenic atom

$$r_n^{\text{hydrogen}} = \frac{n^2 \hbar^2 \varepsilon_0}{m_0 e^2} \tag{20.4}$$

where $\hbar = 1.054 \times 10^{-34}$ J·s, the mass of the free electron is $m_0 = 9.109 \times 10^{-31}$ kg, and the charge on the electron is $e = 1.602 \times 10^{-19}$ C. The value for the Bohr radius in the hydrogen atom is $r_1^{\text{hydrogen}} = 0.5$ Å. For the screened hydrogenic states corresponding to the impurity in a crystaline semiconductor or insulator, we have

$$r_n^{\text{impurity}} = \frac{n^2 \hbar^2 \varepsilon}{m^* e^2} \tag{20.5}$$

which is larger than the hydrogen Bohr radius by a factor $\varepsilon_0 m_0 / \varepsilon m^*$. Using typical values of these quantities for germanium, we get a ground state radius $r_1^{\text{impurity}} \sim 70$ Å. Thus, the electron travels over many lattice sites in germanium before being scattered, and for this reason the dielectric constant approximation used in (20.1) is valid.

From this discussion, we see that only a very small energy is needed to ionize a bound donor electron into the conduction band of a semiconductor like germanium, and because this binding energy is small, these hydrogenic donor levels are called shallow impurity levels. Since $r_n^{\text{impurity}} \gg a$, where a is the lattice constant in a semiconductor like germanium, these electrons are well localized in momentum space according to the uncertainty principle. Shallow donor levels are associated with the k–point where the conduction band minimum occurs. Thus, the simple hydrogenic view of impurity levels in a semiconductor predicts that the impurity spectrum should only depend on the host material and on the charge difference between the host and impurity.

This hydrogenic model also works well for silicon (where the Bohr radius is only $\approx$20 Å) except for the ground state, where the dielectric constant approximation is not as valid as for germanium. For small r, we have $\varepsilon(r) \to \varepsilon_0$ and for large r, we have $\varepsilon(r) \to \varepsilon$, where ε is the static dielectric constant of the bulk semiconductor. Thus, a spatial dependence for $\varepsilon(r)$ needs to be assumed and this spatial dependence can be incorporated into a variational calculation. The inclusion of screening effects by the introduction of a spatial dependence to the dielectric function $\varepsilon(r)$, is called the "central cell correction" and this correction has to be used for calculating the ground state of shallow impurities in Si.

The impurity spectra are studied most directly by infrared absorption spectra and optical transmission measurements. As an example of such spectra, Fig. 20.2

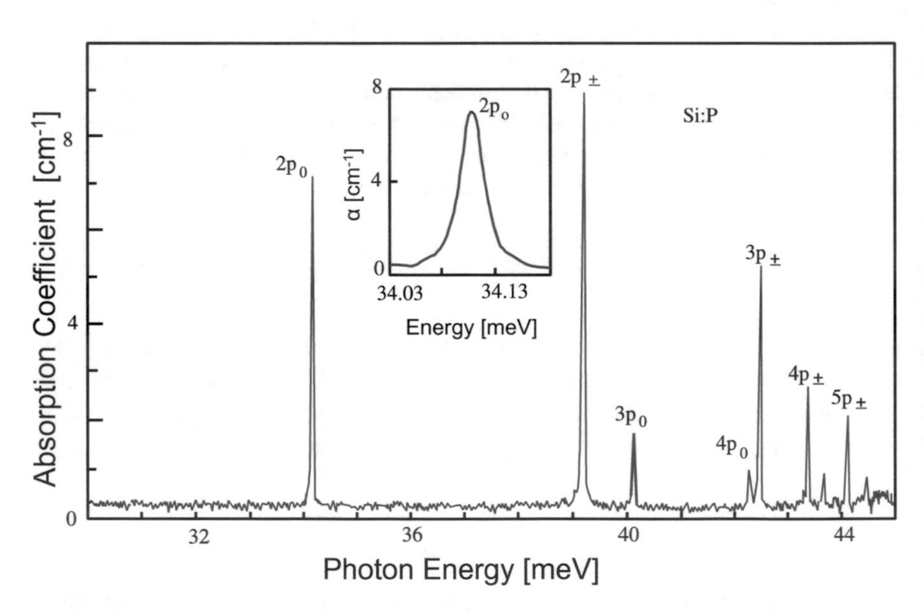

Fig. 20.2 Absorption spectrum of phosphorus donors in Si for a sample at liquid helium temperature containing $\sim 1.2 \times 10^{14}$ cm^{-3} phosphorus impurities. The inset shows the $2p_0$ line on an expanded horizontal scale. (Taken from Yu and Cardona, Fundamentals of Semiconductors, Springer Verlag (1996).)

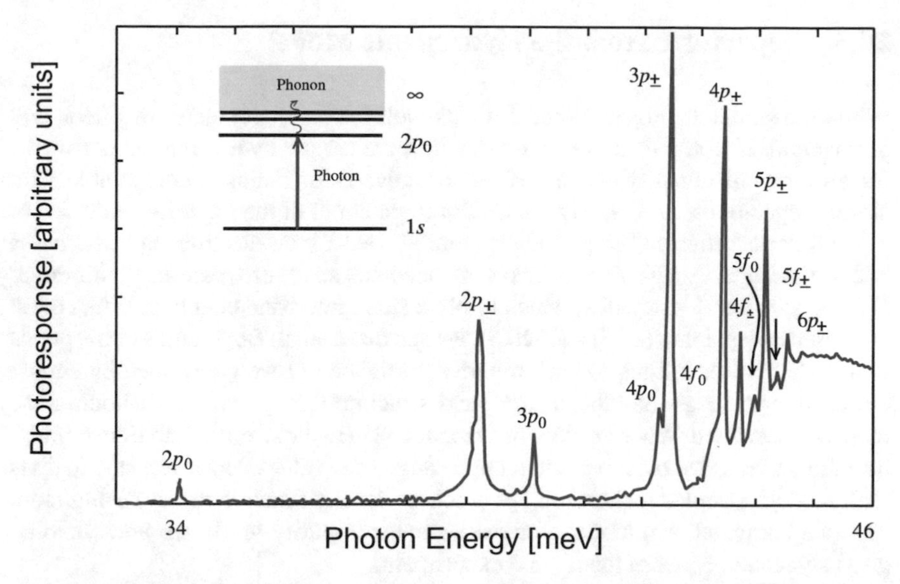

Fig. 20.3 Photo-thermal ionization spectrum of phosphorus-doped Si measured by modulation spectroscopy, which is a particularly useful technique for resolving the higher lying impurity levels. The inset shows schematically the photo-thermal ionization process for a donor impurity. (Taken from Yu and Cardona, Fundamentals of Semiconductors, Springer Verlag (1996).)

shows the absorption spectrum of phosphorus impurities in Si. Note that the photon energies used in these measurements are small so that far infrared frequencies must be employed. The ground state donor level is a $1s$ state and allowed transitions are made to a variety of p–states. Since the constant energy surface is ellipsoidal, the $2p$ levels break up into a $2p(m_l = 0)$ level and a $2p(m_l = \pm 1)$ level which is doubly degenerate (see Fig. 20.2). Transitions from the $1s$ level to both kinds of p levels occur, and account for the sharp features in the spectrum shown in Fig. 20.2. The sensitivity of the spectra is somewhat improved using modulated spectroscopy techniques as shown in Fig. 20.3, where transitions to higher quantum states ($n = 6$) and to higher angular momentum states (f levels where $\ell = 3$) can be resolved, noting that electric dipole transitions always occur between states of opposite parity. For both Figs. 20.2 and 20.3, the initial state is the $1s$ impurity ground state. Analysis of such spectra gives the location in energy of the donor impurity levels, including the location of the ground state donor level, which is more difficult to calculate because of the central cell correction, but can be nicely measured experimentally.

In absorption measurements, the impurity level transitions are observed as peaks. On the other hand, impurity spectra can also be taken using transmission techniques, where the impurity level transitions appear as minima in the transmission spectra.

20.3 Departures from the Hydrogenic Model

While the simple hydrogenic model works well for the donor states in silicon and germanium, it would be naive to assume that the simple hydrogenic model works for all kinds of impurity centers. If the effective Bohr radius is comparable with atomic separations, then clearly the Coulomb potential of the impurity center is not a small perturbation to the periodic potential seen by an electron. Specific cases where the impurity effective Bohr radius becomes small are materials with either (1) a large m^* or (2) a small ε, which imply a small interband coupling. When these conditions are put into (20.3) and (20.5), we see that a small Bohr radius corresponds to a large E_n value. Thus "deep" impurity levels are not well described by simple effective mass theory and the energy band structure throughout the Brillouin zone must be considered. When an electron is localized in real space, a suitable description in momentum space must include a large range of **k** values. However, calculations with modern calculational techniques (such as density function theory) using more accurate computational models for the acceptor impurity levels are now yielding good agreement between theory and experiment.

When the impurity concentration becomes so large that the Bohr orbits for neighboring impurity sites start to overlap, the impurity levels start to broaden, and eventually impurity bands are formed. These impurity bands tend to be only half filled because of the Coulomb repulsion which inhibits placement of both a spin up and a spin down carrier in the same impurity level. When these impurity bands lie close to a conduction or valence band extremum, the coalescence of these impurity levels with band states produces in a smearing out of the threshold of the fundamental absorption edge as observed in absorption measurements. When the impurity band broadening becomes sufficiently large that the electron wavefunction extends to adjacent sites, metallic conduction can occur. The onset of metallic conduction is called the Mott metal–insulator transition.

20.4 Vacancies, Color Centers and Interstitials

Closely related to the impurity problem is the vacancy problem. When a compound semiconductor crystallizes, the melt usually is slightly off stoichiometry with respect to the concentration of anions and cations. As an example, suppose that we prepare PbTe with Pb and Te concentrations in the melt that are stoichiometric to 0.01% and that some of the more volatile Te at growth temperature is lost in vapor phase evaporation. This means that there will be a slight excess of one of the atomic species relative to the other. This deficiency of one of the atomic species shows up in the crystal lattice as an atomic vacancy. Such a vacancy represents a strong local perturbation of the crystal potential which again cannot be modeled in terms of hydrogenic impurity models. Such vacancy centers further tend to attract impurity atoms to form vacancy-impurity complexes. Furthermore, an excess of one stoichiometric type could also

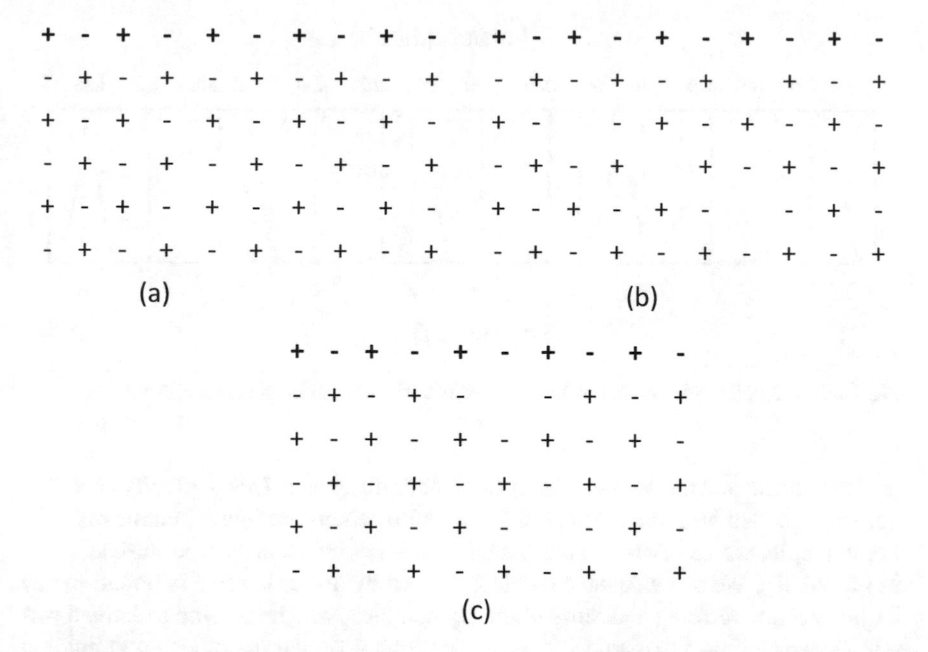

Fig. 20.4 Schematic representation of various possible arrangements of both vacancies and interstitials: **a** a perfect ionic crystal, **b** an ionic crystal with positive and negative ion vacancies, and **c** an ionic crystal with positive and negative ion vacancies and interstitials

form interstitials. Both of these defects are difficult to model theoretically because their spatial localization requires participation of energy states throughout the Brillouin zone. Defect centers generally give rise to energy states within the band gap of semiconductors and insulators. Such defect centers are often studied by optical techniques because their presence strongly modifies the optical properties.

In Fig. 20.4 various types of point defects are shown. Figure 20.4a illustrates a perfect ionic crystal while Fig. 20.4b shows an ionic crystal with vacancies. This particular collection of vacancies is of the Schottky type (equal numbers of positive and negative ion vacancies). Schottky point defects also include neutral vacancies. Finally, Fig. 20.4c shows both vacancies and interstitials. When a + (–) ion vacancy is near a + (–) ion interstitial, this defect configuration is called a Frenkel-type point defect.

One important defect in ionic insulating crystals is the *F-center* ("Farbe" or color center). We see in Fig. 20.4b that the negative ion vacancy acts like a $+ve$ charge (absence of a $-ve$ charge). This effective $+ve$ charge tends to bind to an electron. The binding of an electron to a $-ve$ ion vacancy is called an *F-center*. These F-centers give rise to absorption bands in the visible spectrum. Without F-centers, these crystals are usually clear and transparent. The F-center absorption band causes crystals with defects to appear colored, having the color of the transmitted light. When the crystals are heated to high enough temperatures, these defects can be

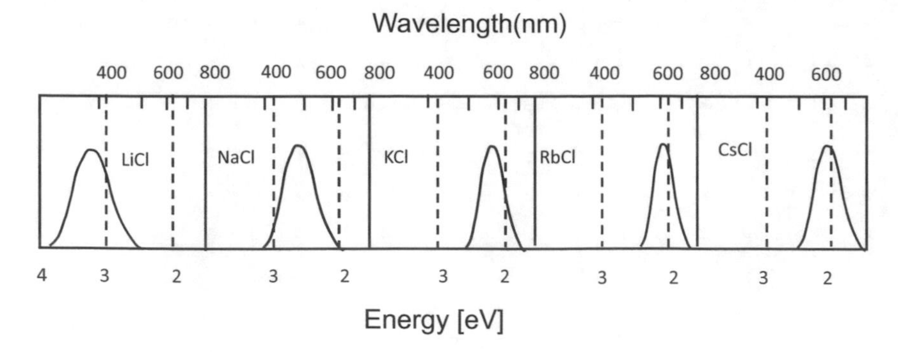

Fig. 20.5 Examples of F–center absorption lines in various alkali halide ionic crystals

made to anneal and the colored absorption bands disappear. This photophysical phenomena is called *bleaching*. Many different color centers are found in ionic crystals. For example, we can have a hole bound to a $+ve$ ion vacancy (see dashed circle in Fig. 20.4b). We can also have a defect formed by a vacancy that is bound to any impurity atom, forming a vacancy-impurity complex, which can bind a charged carrier. Or we can have two adjacent vacancies (one $+ve$ and the other $-ve$) binding an electron to a hole. Further generalizations are also found. These defect centers are collectively called *color centers* and each color center has its own characteristic absorption band. In Fig. 20.5, we see an example of absorption bands due to F-centers in several insulating alkali halides. In all cases the absorption bands are very broad, in contrast with the sharp impurity lines which are observed in the far infrared for shallow impurity level transitions in semiconductors (see Figs. 20.2 and 20.3). In the case of the vacancy defect there is a considerable lattice distortion around each vacancy site as the neighboring atoms rearrange their electronic bonding arrangements.

20.4.1 Schottky Defects

We will now use simple statistical mechanical arguments to estimate the concentration of Schottky defects. Let E_s be the energy required to take an atom from a lattice site inside the crystal to the surface. If n is the number of vacancies, the change in internal energy resulting from vacancy generation is $U = nE_s$. Now the number of ways that n vacancy sites can be picked from N lattice sites is $N!/[(N-n)!n!]$, so that the formation of vacancies results in an increase in entropy of

$$S = k_B \ln \frac{N!}{(N-n)!n!} \tag{20.6}$$

and a change in free energy of

$$F = U - TS = nE_s - k_B T \ln \frac{N!}{(N-n)!n!}. \tag{20.7}$$

Using Stirling's approximation for $\ln x!$ when x is large, we write

$$\ln x! \cong x \ln x - x. \tag{20.8}$$

Equilibrium is achieved when $(\partial F/\partial n) = 0$, so that at equilibrium we have

$$E_s = k_B T \left[\frac{\partial}{\partial n} \ln \frac{N!}{(N-n)!n!} \right] = k_B T \ln \frac{N-n}{n} \tag{20.9}$$

from which we write

$$\frac{n}{N-n} = \exp\left[-\frac{E_s}{k_B T} \right] \tag{20.10}$$

or

$$\frac{n}{N} \sim \exp\left[-\frac{E_s}{k_B T} \right] \tag{20.11}$$

since $n \ll N$. The vacancy density is small because for $E_s \sim 1\,\text{eV}$ and $T \sim 300\,\text{K}$, we get $(n/N) \sim e^{-40} \sim 10^{-17}$.

In the case of vacancy pair formation in an ionic crystal (a Schottky defect), the number of ways to make n separated pairs is $[N!/(N-n)!n!]^2$, so that for Schottky vacancy pair formation, we have

$$\frac{n_p}{N} \sim \exp\left[-\frac{E_p}{2k_B T} \right] \tag{20.12}$$

where n_p is the pair vacancy density and E_p is the energy required for pair formation.

These arguments can readily be extended to the formation of Frenkel defects (see Problem section) and it can be shown that if N' is the density of possible interstitial sites, then the density of occupied interstitial sites is

$$n_i \simeq (NN')^{\frac{1}{2}} \exp\left[-\frac{E_i}{2k_B T} \right] \tag{20.13}$$

where E_i is the energy to remove an atom from a lattice site to form an interstitial defect site.

20.5 The Concept and Spectroscopy of Excitons

An exciton denotes a system of an electron and a hole bound together by their Coulomb interaction. When a photon excites an electron into the conduction band, a hole is left behind in the valence band, the electron, having a negative charge will be attracted to this hole and may (provided the energy is large enough) bind to the positively charged hole forming a neutral quasi-particle called an *exciton*. Depending on the binding energy, the radius of the excitons in real space can extend from several unit cells (free excitons) to the same order as the size of the unit cell (bound excitons) as shown in Fig. 20.6. The free excitons are present in semiconductors while bound excitons are typical of insulators and molecular crystals. Using the simplest possible approximation, the exciton levels have been treated as a hydrogenic system where a charge $-e$ is bonded to a positive charge. Thus, because of the attractive Coulomb potential, the exciton binding energy is attractive and represents a lower energy state than the band states. Excitons are important in the optical spectra of bulk semiconductors especially at low temperatures, and their exciton levels are important for device applications since light emitting diodes and semiconductor lasers operation often involve excitons. However, because of the confinement of carriers in low dimensional systems, exciton effects become much more important in the case of quantum wells, superlattices, carbon nanotubes, transition metal dichalcogenides, and devices based on deliberately nanostructured materials. The topic of excitons in low dimensional semiconductor systems (quantum wells and carbon nanotubes) is discussed in Sect. 20.5.4.

We will now use the effective mass approximation to find the Wannier–Mott exciton spectrum near an interband threshold. Let us here assume that the exciton was created by a photon with energy slightly less than the direct energy gap E_g. The Schrödinger equation for the two-body exciton packet wave function Φ is written in the effective mass approximation as:

$$\left[\frac{p_e^2}{2m_e^*} + \frac{p_h^2}{2m_h^*} - \frac{e^2}{\varepsilon|\mathbf{r}_e - \mathbf{r}_h|}\right]\Phi = E\Phi \tag{20.14}$$

thereby including the Coulomb binding energy $\varepsilon(|\mathbf{r}_e - \mathbf{r}_h|)$ of the electron–hole pair. For simplicity, we assume that the dielectric constant ε is independent of $\mathbf{r}_e$ and $\mathbf{r}_h$ and corresponds to a large spatial extension of the exciton in a semiconductor. We introduce new coordinates for the spatial separation $\mathbf{r}$ between the electron and hole

$$\mathbf{r} = \mathbf{r}_e - \mathbf{r}_h \tag{20.15}$$

and for the center of mass coordinate $\boldsymbol{\rho}$ given by

$$\boldsymbol{\rho} = \frac{m_e^*\mathbf{r}_e + m_h^*\mathbf{r}_h}{m_e^* + m_h^*}. \tag{20.16}$$

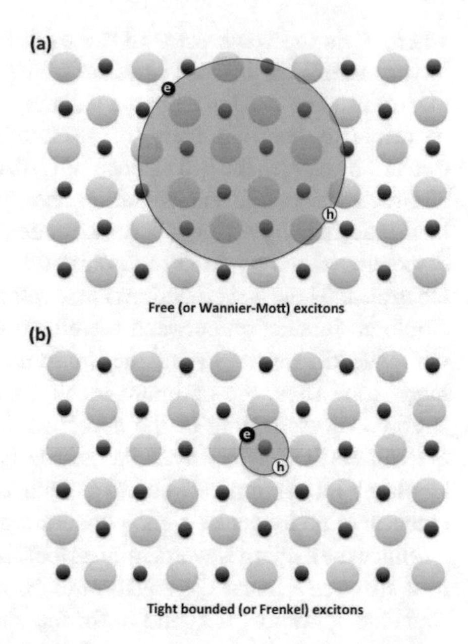

Fig. 20.6 Schematic diagrams for free (Wannier–Mott) and bound (Frenkel) excitons. e and h stand for an electron and a hole, respectively

We now separate the Schrödinger equation (20.14) into an equation for the relative motion of the electron and hole in the exciton wave packet $F(\mathbf{r})$ and an equation of motion for the center of mass $G(\boldsymbol{\rho})$

$$\Phi(\mathbf{r}_e, \mathbf{r}_h) = F(\mathbf{r})G(\boldsymbol{\rho}). \tag{20.17}$$

Thus (20.14) becomes

$$\left[\frac{p_\rho^2}{2(m_e^* + m_h^*)} + \frac{p_r^2}{2m_r^*} - \frac{e^2}{\varepsilon_0 r}\right] F(\mathbf{r})G(\boldsymbol{\rho}) = EF(\mathbf{r})G(\boldsymbol{\rho}) \tag{20.18}$$

where the reduced effective mass is given by

$$\frac{1}{m_r^*} = \frac{1}{m_e^*} + \frac{1}{m_h^*} \tag{20.19}$$

to obtain an eigenvalue equation for $G(\boldsymbol{\rho})$

$$\left[\frac{p_\rho^2}{2(m_e^* + m_h^*)}\right] G(\boldsymbol{\rho}) = \Lambda G(\boldsymbol{\rho}) \tag{20.20}$$

which is of the free particle form and has eigenvalues

$$\Lambda(K) = \frac{\hbar^2 K^2}{2(m_e^* + m_h^*)} \tag{20.21}$$

where K is the wavevector of the exciton. This single particle picture of carriers is very useful for discussing exciton dispersion. In a semiconducting material, an electron can be excited from the valence to the conduction energy band, by gaining more than the band gap energy of the material. The energy difference E_{ii} for an optical transition between the i-th valence and i-th conduction bands in a one-electron picture is directly related to the excitation energy. An excitonic picture, however, can not be represented by a single particle model, and we can not generally use the energy dispersion relations directly to obtain the excitation energy for the exciton. If the electron and hole wavefunctions are localized in the same spatial region, the attractive Coulomb interaction between the electron and hole increases the binding energy, while the kinetic energy and the Coulomb repulsion between the electrons becomes large, too. Thus, the optimum localized distance determines the exciton binding energy. The screening of the attractive Coulomb interaction by other conduction electrons is the reason why excitons usually are not important in metals. The repulsive Coulomb interaction between two electrons causes the wavevector k for an excited electron to no longer be a good quantum number.

Since the exciton wavefunction is localized in real space, the exciton wavefunction in k space is a linear combination of Bloch wavefunctions with different k states. Thus the definition of k_c and k_v for the electron and the hole might not be so clear. However, since the exciton wavefunction is localized in k space too (the Fourier transform of a Gaussian in real space is a Gaussian in k space), we can define k_c or k_v as the central position of the corresponding wavefunctions in k space.

When we consider an optical transition in a crystal, we expect a vertical transition, $k_c = k_v$, [Fig. 20.7a] where k_c and k_v are, respectively, the wavevectors of the electron and hole. The wavevector of the center of mass for the exciton is defined by $K = (k_c - k_v)/2$, while the relative coordinate is defined by $k = k_c + k_v$, in which we note that the hole (created by exciting an electron) has the opposite sign for its wavevector and effective mass as compared to the electron. The exciton has an energy dispersion as a function of K which represents the translational motion of an exciton.

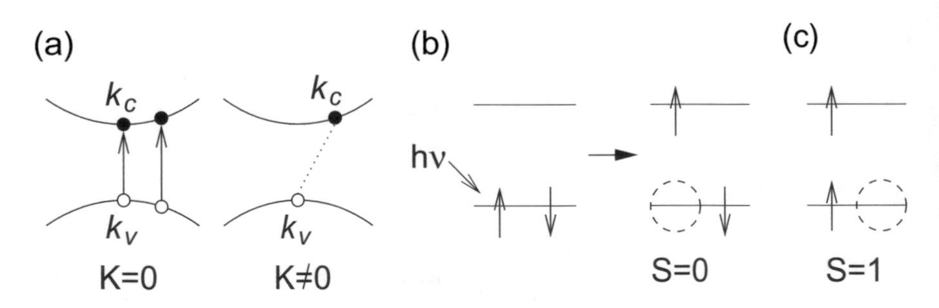

Fig. 20.7 **a** A singlet exciton formed at exciton wavevector $K = 0$ in a crystal where $k_c = k_v$ (left), at either the band extremum or away from the band extremum. If $k_c \neq k_v$, $K \neq 0$, giving rise to a dark exciton (right) (see text). **b** When a photon is absorbed by an electron with spin $\uparrow$ (left), we get a singlet exciton ($S = 0$, right). If the spin of the electron is $\uparrow$, we define the spin of the hole that is left behind as $\downarrow$. **c** A triplet exciton ($S = 1$), which is a dark exciton (Jorio, 2011)

Thus only the $K = 0$ exciton can recombine by emitting a photon. Correspondingly, a $K \neq 0$ exciton cannot recombine directly to emit a photon and therefore is a dark exciton. Recombination emission for $K \neq 0$ is, however, possible by a phonon assisted process which we call an indirect transition.

The Exciton Spin

When we discuss the interaction between an electron and a hole, the definition of the total spin for an exciton is a bit different from the conventional idea of two electrons in a molecule or crystal. A hole is a different "particle" from an electron, but, nevertheless, an exchange interaction between the electron and the hole exists, just like for two electrons in a hydrogen molecule.

When an electron absorbs a photon, an electron, for example with spin $\uparrow$ is excited to an excited state as shown in Fig. 20.7b, leaving behind a hole at the energy level that the electron with up-spin had previously been. This hole has not only a wavevector of $-k$ and an effective mass of $-m^*$, but also is defined to be in a spin down hole state. The exciton thus obtained [Fig. 20.7b] is called a spin singlet, with $S = 0$, since the definition of S for the two-level model shown here is in terms of the two actual electrons that are present, and in this sense the definition for the two actual electrons and for the $S = 0$ exciton are identical. This is valid because the mediated electric dipole transition does not change the total spin of the ground state, which is $S = 0$. It should also be mentioned that Fig. 20.7b does not represent an $S = 0$ eigenstate. To make an eigenstate, we must take the antisymmetric combination of the state shown in Fig. 20.7b with an electron $\downarrow$ and hole $\uparrow$. In contrast, a triplet exciton ($S = 1$) can be represented by two electrons, one in the ground state and the other in an excited state to give a total spin of $S = 1$ [Fig. 20.7c]. For the triplet state in Fig. 20.7c, we define the hole to have a spin $\uparrow$ and the resulting state shown here is an eigenstate ($m_s = 1$) for $S = 1$. We further note that a triplet exciton can not be recombined by emitting a photon because of the Pauli principle. We call such an exciton "a dark exciton". The spin conversion by a magnetic field could flip a spin and lead to the recombination of the triplet exciton. An exchange interaction between a hole and an electron works only for $S = 0$ [see Fig. 20.7b] and thus the $S = 1$ state in Fig. 20.7c has a lower energy than the $S = 0$ state and the exchange interaction for the $S = 0$ exciton can be understood as the difference in the interaction energy between two electrons [one at the position of the excited electron and the other at the position of the hole left behind as in Fig. 20.7b] and the energy of the $S = 1$ exciton which has no exchange energy [Fig. 20.7c]. It should be noted that for the more familiar case of just two electrons, the exchange interaction works for the $S = 1$ case and therefore the $S = 1$ state lies lower in energy than the $S = 0$ state.

The free particle solutions for the center of mass problem of (20.21) show that the exciton can move freely as a unit through the crystal. The momentum of the center of mass for a direct band gap exciton is small because of the small amount of momentum imparted to the excitation by the light.

The Schrödinger equation in the coordinate system of relative motion can be written as

$$\left[\frac{p_r^2}{2m_r^*} - \frac{e^2}{\varepsilon_0 r}\right] F(\mathbf{r}) = E_n F(\mathbf{r}), \tag{20.22}$$

where (20.22) has the functional form of the Schrödinger equation for a hydrogen atom with eigenvalues E_n for quantum numbers n (where $n = 1, 2, \ldots$) given by

$$E_n = -\frac{m_r^* e^4}{2\hbar^2 \varepsilon^2 n^2}, \tag{20.23}$$

and the total energy for the exciton levels is then

$$E = \Lambda(K) + E_n. \tag{20.24}$$

The energy levels of (20.23) look like the donor impurity spectrum, but instead of the effective mass of the conduction band m_e^* we now have the reduced effective mass m_r^* given by (20.19). Since m_r^* has a smaller magnitude than m_e^* as seen in (20.19), we conclude that the exciton binding energy is less than the impurity ionization energy for a particular solid.

20.5.1 Exciton Effects in Bulk Materials

An example of a spectrum showing exciton effects is presented in Fig. 20.8. The points are experimental and the solid curves are a fit of the data points to the imaginary part of dielectric function $\varepsilon_2(\omega)$ for excitons given by

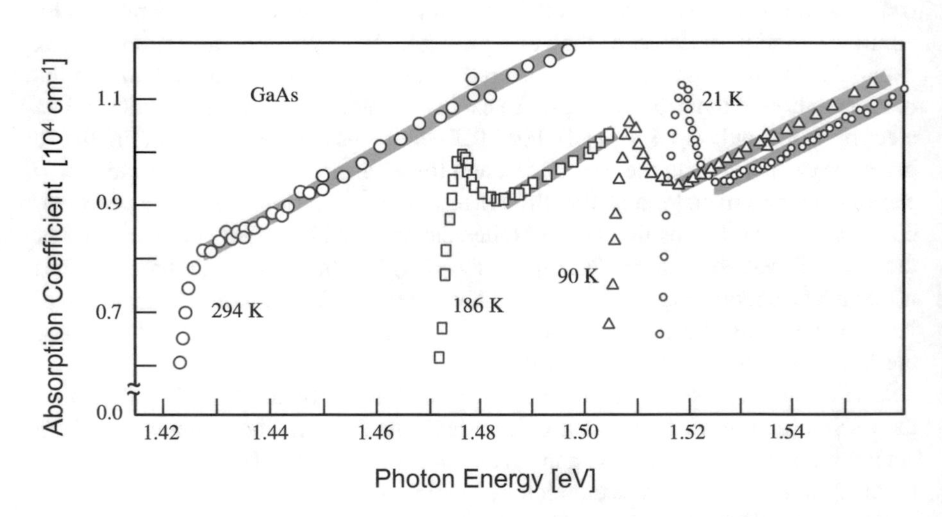

Fig. 20.8 Excitonic absorption spectra in GaAs near its bandgap for several sample temperatures. The lines drawn through the 21, 90 and 294 K data points represent fits with theory. (Taken from Yu and Cardona, Fundamentals of Semiconductors, Springer Verlag (1996).)

Table 20.1 Exciton binding energy (E_1) and Bohr radius (r_1) in some direct bandgap semiconductors with the zinc-blende structure (taken from Yu and Cardona, Fundamentals of Semiconductors, Springer Verlag (1996))

Semiconductor	E_1 (meV)	E_1 (theory) (meV)	r_1 (Å)
GaAs	4.9	4.4	112
InP	5.1	5.14	113
CdTe	11	10.71	12.2
ZnTe	13	11.21	11.5
ZnSe	19.9	22.87	10.7
ZnS	29	38.02	10.22

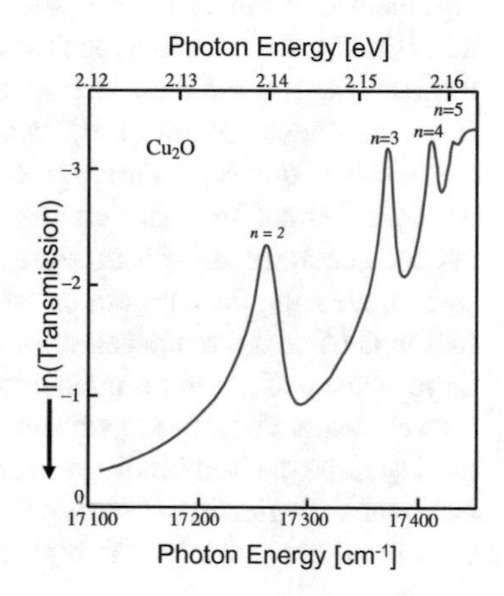

Fig. 20.9 Low temperature absorption spectrum of Cu_2O (plotted as the log of the optical transmission) showing the excitonic p series associated with its "dipole-forbidden" band edge in Cu_2O, with the photon energy given in both eV and cm^{-1}. (Taken from Yu and Cardona, Fundamentals of Semiconductors, Springer Verlag (1996).)

$$\varepsilon_2(\omega) = \frac{8\pi |\langle v|p|c\rangle|^2 m_r^{*3}}{3\omega^2\varepsilon^3} \sum_{n=1}^{\infty} \frac{1}{n^3}\delta(\omega - \omega_n), \tag{20.25}$$

where the sum is over all the exciton bound states. From Table 20.1, we see that the binding energy for excitons for GaAs is 4.9 meV and the effective Bohr radius is 112 Å, which is very much larger than the lattice spacing in GaAs.The various exciton lines contributing to the exciton absorption profiles in Fig. 20.8 are unresolved even for the data shown for the lowest temperature of 21 K. A material for which the higher exciton energy levels ($n = 2, 3, \ldots$) of the Rydberg series are resolved is Cu_2O as can be seen in Fig. 20.9. The observation of these higher states is attributed to the forbidden nature of the coupling of the valence and conduction bands, giving rise to a strict selection rule that only allows coupling to exciton states with p symmetry. Since the observation of transitions for $n \geq 2$ requires p exciton states, the $n = 1$ exciton is forbidden in the Cu_2O spectrum, and the exciton lines start at $n = 2$. The transitions are sharp, and well resolved exciton lines up to $n = 5$ can be identified in Fig. 20.9.

The exciton spectrum appears to be quite similar to the impurity spectrum of shallow impurity states. These two types of spectra are distinguished through their respective dependences on impurity concentration. Suppose that we start with a very pure sample (10^{14} impurities/cm^3) and then dope the sample lightly (to 10^{16} impurities/cm^3). If the spectrum is due to donor impurity levels, the intensity of the spectral lines would tend to increase and perhaps the spectral linewidth would broaden somewhat. If, on the other hand, the spectrum is associated with an exciton, the spectrum would be attenuated because of screening effects associated with the charged impurities. Exciton states in 3D semiconductors are generally observed in very pure samples and at very low temperatures. The criterion is that the average Bohr orbit of the exciton is less than the distance between impurities. For the sake of this argument, consider an excitonic radius of $\sim$100 Å. If an impurity ion is located within this effective Bohr radius, then the electron–hole Coulomb interaction is screened by the impurity ion and the sharp spectrum associated with the excitons will disappear. A carrier concentration of 10^{16}/cm^3 corresponds to finding an impurity ion within every 100 Å from some lattice point. Thus the electron–hole coupling can be screened out by a charged impurity concentration as low as 10^{16}/cm^3. Low temperatures are needed generally to yield an energy separation of the exciton levels that is larger than $k_B T$. Increasing the temperature shifts the absorption edge and broadens the exciton line in GaAs. At a temperature of 20 K we have $k_B T \simeq 1.7$ meV which is nearly as large as the exciton binding energy of 4.9 meV found in Table 20.1, explaining why no well resolved exciton spectrum for higher quantum states is observed. For the case of Cu_2O, the exciton binding energy of the ground state ($1s$), were it to exist, would be 97 meV, neglecting central cell corrections. The large exciton binding energy in Cu_2O also helps resolve the higher quantum exciton states.

20.5.2 Classification of Excitons

The exciton model discussed above is appropriate for a *free* exciton and a *direct* exciton (see Fig. 20.6a). For the *direct* exciton, the initial excitation is accomplished in a **k**-conserving process without the intervention of phonons. The condition for forming the exciton without involving phonons is that the group velocity of both the electron and hole is the same, which implies that absolute values of the $\partial E(k)/\partial k$ slopes are equal. Therefore, the excitonic effects are expected to be very strong close to the band gap transition. In materials like silicon and germanium, the thermal band gap corresponds to an indirect energy gap. For these materials, the exciton is formed by an indirect phonon-assisted process and the exciton is consequently called an *indirect* exciton.

Indirect excitons can be formed either with the emission or absorption of a phonon. Since excitons are more important at low temperatures, the emission process is much more likely than the absorption process. Because of the large difference in crystal momentum $\hbar\mathbf{k}$ between the valence band extremum and the lowest conduction band minimum in these indirect gap semiconductors, the exciton may acquire a large center

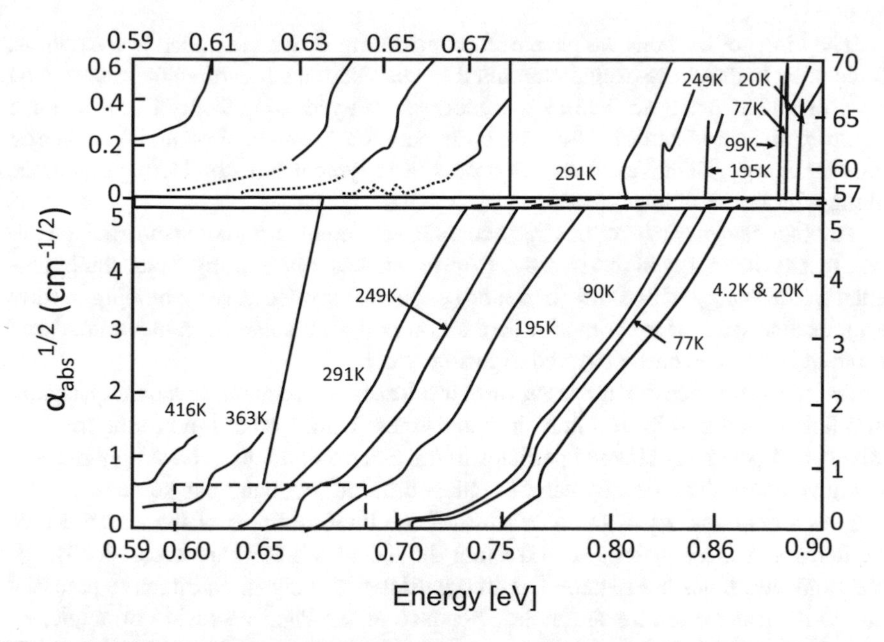

Fig. 20.10 Plot of the square root of the absorption coefficient vs. $\hbar\omega$ for Ge (because Ge is an indirect band gap semiconductor) for various temperatures, showing the effect of excitons. Features associated with both indirect and direct excitons are found. The upper left panel shows the detailed behavior at the onset of the indirect bandgap absorption, where the absorption is low and the upper right panel shows direct exciton phenomena where the absorption is high

of mass momentum corresponding to the momentum of the absorbed or emitted phonon $\hbar\mathbf{q}$. For the indirect exciton, a large range of crystal momentum $\hbar\mathbf{k}$ values are possible and hence the exciton levels spread out into bands as shown in the lower dashed rectangle of Fig. 20.10. This portion of the figure also appears in more detail in the upper left-hand corner. In Fig. 20.10 we also show in the upper right-hand corner the direct exciton associated with the Γ point conduction band for various temperatures. The shift in the absorption edge is associated with the decrease in band gap with increasing temperature. In Fig. 20.10, the individual exciton lines are not resolved, since a lower temperature would be needed for that.

Addition of impurities to suppress the exciton formation does not help with the identification of bandgaps in semiconductors since the presence of impurities broadens the band edges. It is for this reason that energy gaps are best found from optical data in the presence of a magnetic field.

For small distances from the impurity site or for small electron-hole separations, the effective mass approximation must be modified to consider central cell corrections explicitly. For example, central cell corrections are very important in Cu_2O so that the binding energy attributed to the $1s$ state is 133 meV, whereas the binding energy deduced from the Rydberg series shown in Fig. 20.9 indicates a binding energy of 97 meV.

The kinds of excitons we have been considering above are called *free* excitons. In contrast to these, are excitations called *bound* excitons. It is often the case that an electron and hole may achieve a lower energy state by locating themselves near some impurity site, in which case the exciton is called a *bound* exciton and has a larger binding energy. Bound excitons are observed in typical semiconducting materials, along with free excitons.

Another category of excitons that occurs in semiconductors is the molecular exciton. Just as the energy of two hydrogen atoms decreases in forming molecular hydrogen H_2, the energy of two free (or bound) excitons may decrease on binding to form a molecular state. More complicated exciton complexes can be contemplated and some of these have been observed experimentally.

As the exciton density increases, further interactions occur and eventually a quantum fluid called the electron-hole drop is formed. Unlike other fluids, both the negatively and positively charged particles in the electron-hole fluid have light masses. A high electron-hole density can be achieved in indirect band gap semiconductors such as silicon and germanium because of the long lifetimes of the electron-hole excitations in these materials. In treating the electron-hole drops theoretically, the electrons and holes are regarded as free particles moving in an effective potential due to the other electrons and holes. Because of the Pauli exclusion principle, no two electrons (or holes) can have the same set of quantum numbers. For this reason, like particles tend to repel each other spatially, but unlike particles do not experience this repulsion. In this discussion, two electrons are like particles, but one electron and one hole are unlike particles. Thus electron-hole pairs are formed and these pairs can be bound to each other to form an electron-hole drop. These electron-hole drops have been studied in the emission or luminescence spectra. Results for the luminescence spectra of Ge and Si at very low temperatures ($T \leq 2\,\text{K}$) are shown in Fig. 20.11. Luminescence spectra for germanium provide experimental evidence for electron-hole drops for electron-hole concentrations exceeding $10^{17}/\text{cm}^3$.

In insulators (as for example alkali halides), excitons are particularly important, but here they tend to be well localized in space because the effective masses of any carriers that are well localized tend to be large (see Fig. 20.6b). These localized excitons, called *Frenkel* excitons, are much more strongly bound and must be considered on the basis of a much more complicated exciton theory. It is only for the excitons which extend over many lattice sites, the *Wannier* excitons, that effective mass theory can be used. And even here many-body effects must be considered to solve the problem with any degree of accuracy – already an electron bound to a hole is a two-body problem so that one-electron effective mass theory is generally not completely valid.

In studying the optical absorption of the direct gap, the presence of excitons complicates the determination of the direct energy gap, particularly in alkali halides where the exciton binding energy is large. Referring to Fig. 20.12a, both Γ-point and L-point excitons are identified in the alkali halide ionic crystal KBr. The correspondence of the optical structure with the $E(\mathbf{k})$ diagram is shown by comparison of Fig. 20.12a and b. Here it is seen that the Frenkel exciton lines dominate the spectrum at the absorption edge, and we also see huge shifts in energy between the exciton

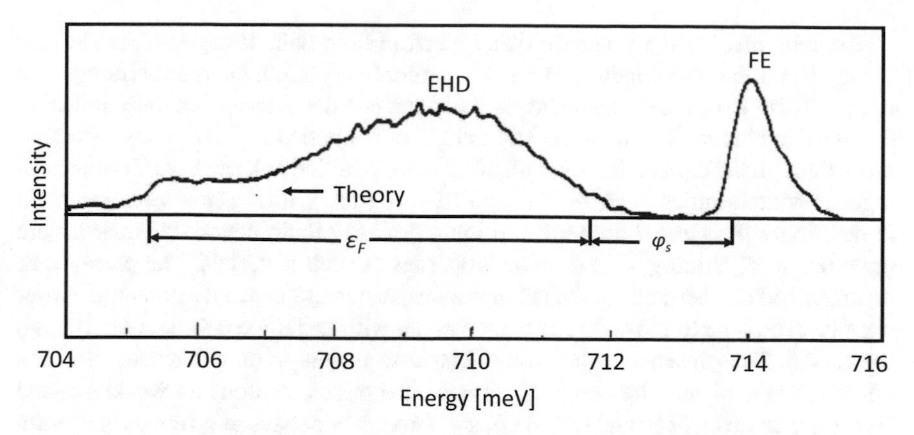

Fig. 20.11 Recombination radiation (or photoluminescence spectrum) of free electrons (FE) and of electron-hole drops (EHD) in Ge at low temperature 3.04 K. The Fermi energy in the electron-hole drop is ϵ_F and the cohesive energy of the electron-hole drop with respect to a free exciton is $\phi_s = 1.8$ meV. The critical concentration and temperature for forming an electron-hole drop in Ge are, respectively, $2.6 \times 10^{17}/\text{cm}^3$ and 6.7 K, and for this reason electron-hole drop experiments are done at low temperature

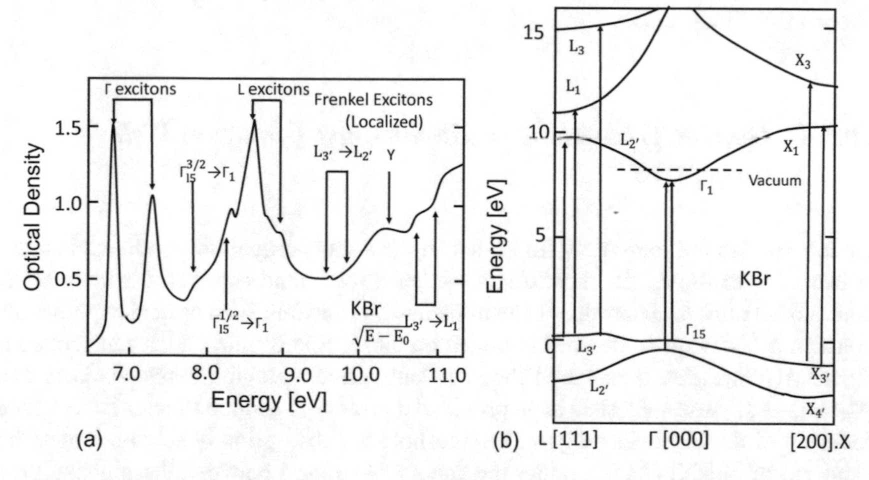

Fig. 20.12 **a** A spectrum of the optical density of KBr showing Frenkel excitons. The optical density is defined as $\log(1/\mathscr{T})$, where $\mathscr{T}$ is the optical transmission. **b** The energy bands of KBr, as inferred from tight-binding calculations of the valence bands and from the assignments of interband edges in optical experiments. The valence band in KBr is a Br $5p$ derived band. The conduction band would be dominated by the K $4s$ band with a higher lying band possibly a K $4p$ band. We note that the optical spectrum on the left is dominated by exciton effects and that direct band edge contributions are much less important. We further note that the exciton binding energy is on the order of an electron volt

lines and the direct absorption edge. These figures show the dominance of strongly bound, localized Frenkel excitons in the spectra of alkali halides.

Excitons involve the presence of an electron-hole pair. If instead, an electron is introduced into the conduction band of an ionic crystal, a charge rearrangement occurs. This charge rearrangement partially screens the electron, thereby reducing its effective charge. When an electric field is now applied and the charge starts to move through the crystal, it moves together with its lattice polarization. The electron together with its lattice polarization is called a *polaron*. While excitons are important in describing the *optical* properties of ionic or partly ionic materials, polarons are important in describing the *transport* properties of such materials. The presence of polarons leads to thermally activated mobilities, which implies that a potential barrier must be overcome to move an electron together with its lattice polarization through the crystal. The presence of polaron effects also results in an enhancement in the effective mass of the electron. Just as one categorizes excitons as weakly bound (Wannier) or strongly bound (Frenkel), a polaron may behave as a free particle with a relatively weak enhancement of the effective mass (a *large* polaron) or may be in a bound state with a finite excitation energy (a *small* polaron), depending on the strength of the electron-phonon coupling. Large polarons are typically seen in weakly ionic semiconductors, and small polarons in strongly ionic, large-gap materials. Direct evidence for large polarons in semiconductors has come from optical experiments in a magnetic field in the region where the cyclotron frequency ω_c is close to the optical phonon frequency ω_{LO}.

20.5.3 Optical Transitions in 2D Systems: Quantum Well Structures

Optical studies are extremely important in the study of quantum wells and superlattices. For example, the most direct evidence for bound states in quantum wells comes from optical absorption measurements. To illustrate such optical experiments consider a GaAs quantum well bounded on either side by the wider gap semiconductor $Al_xGa_{1-x}As$. Because of the excellent lattice matching between GaAs and $Al_xGa_{1-x}As$, these materials have provided a prototype semiconductor superlattice for study of the 2D electron gas. The threshold for absorption is now no longer the band gap of bulk GaAs but rather the energy separation between the highest lying bound state of the valence band and the lowest bound state of the conduction band. Since the valence band of GaAs is degenerate at $\mathbf{k} = 0$ and consists of light and heavy holes, there will be two $n = 1$ levels in the valence band for this 2D system. Since $E_n \propto 1/m^*$ for the quantum well $n = 1$ bound state level, the heavy hole subband extremum will be closer in energy to the band edge than that of the light hole as shown on the left side of Fig. 20.13. Also the density of states for the heavy hole subband will be greater than that of the light hole subband by a factor of $2m_{hh}{}^*/m_{lh}{}^*$. The optical absorption will thus show two peaks near the optical threshold as illustrated in the diagram in Fig. 20.13, with the lower energy peak associated with the heavy hole transition and the higher energy peak is for the light hole transition. These data are for

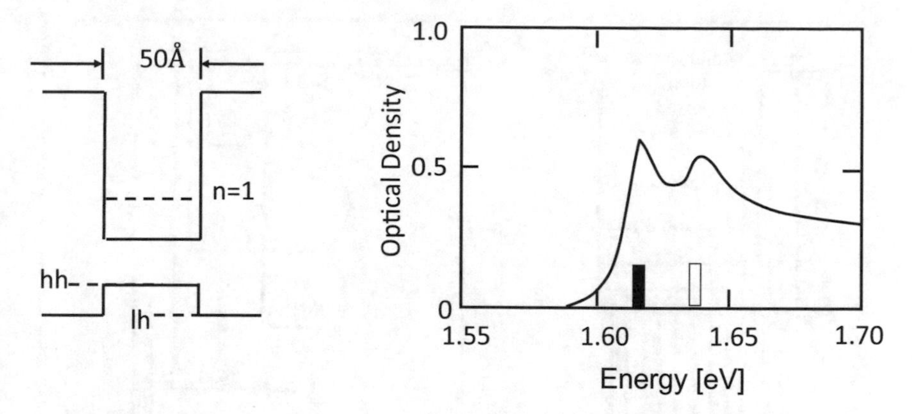

Fig. 20.13 Optical excitations in a quantum well (50 Å quantum well width) where the valence band has light holes (lh) and heavy holes (hh) (as in GaAs). The optical density, defined as $\log(1/\mathscr{T})$ where $\mathscr{T}$ is the transmission, shows peaks associated with each of these optical transitions

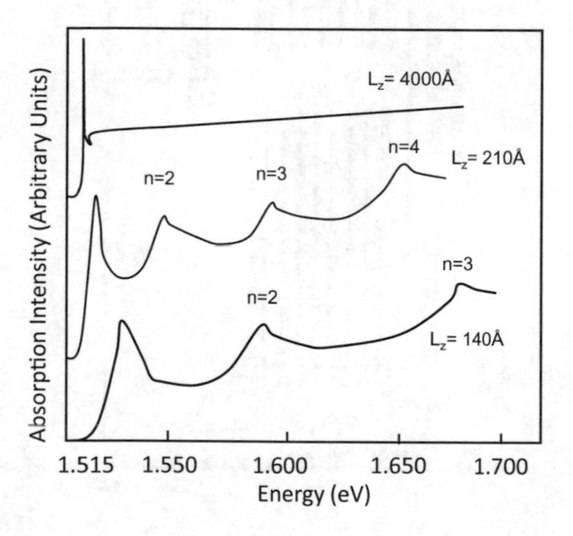

Fig. 20.14 Frequency dependence of the absorption for GaAs/$Al_{0.2}Ga_{0.8}As$ heterostructure superlattices of different thicknesses at optical frequencies. The exciton edge can be seen most clearly for interband transitions to the lowest conduction subband ($n = 1$), and a thick film (4000 Å), while the higher order transitions are best seen for thinner films

a sample with a quantum well width of 50 Å, which is small enough to contain a single bound state ($n = 1$), making use of the relation $E_n = \hbar^2\pi^2 n^2/(2m^* L_z)$, where L_z is the quantum well width. Since the optical absorption from a single quantum well is very weak, the experiment is usually performed in superlattice structures containing a periodic array of many equivalent quantum wells. In forming the superlattice structure, it is important that the barrier between the quantum wells is not too small in extent, because for small spatial separations between quantum wells and low band offsets at the interfaces, the eigenfunctions in adjacent wells become coupled and we no longer have a simple 2D electron gas in the quantum well.

For wider quantum wells containing several bound states (see Figs. 20.14 and 20.15), a series of absorption peaks are found for the various bound states, and the

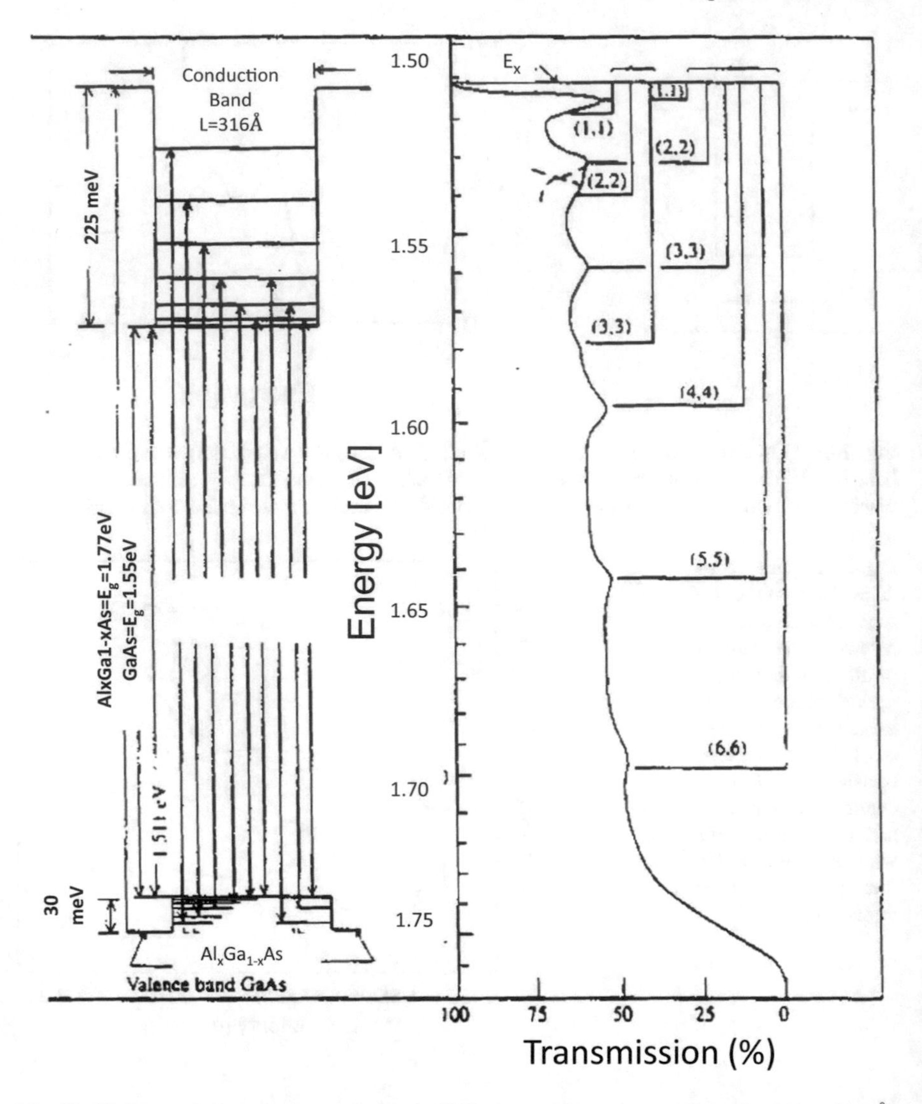

Fig. 20.15 Transmission spectrum of a GaAs/AlGaAs multi-quantum well (well width = 316 Å) measured as a function of photon energy at low temperature (right panel). The peaks labeled (n, n) have been identified with optical transitions from the nth heavy hole (hh) and light hole (lh) subbands to the nth conduction subband as shown by arrows in the band diagram in the left panel, where we see the valence band levels confined to a 30 meV range and those for the conduction band to a 225 meV range. Conduction (valence) band levels at higher (lower) energies are considered to be in continuum states. The values of the band offsets used in the analysis are given in the diagram, but these are not the most recent values

interband transitions follow the selection rule $\Delta n = 0$. This selection rule follows because of the orthogonality of the wave functions for different states n and n'. Thus to get a large n matrix element for coupling the valence and conduction band states, n' and n must be equal. As the width of the quantum well increases, the spectral features associated with transitions to the bound states become smaller in intensity and more closely spaced and eventually cannot be resolved. For the thickest films, the quantum levels are too close to each other to be resolved and only the bulk exciton peak is seen. For the 210 Å quantum well shown in Fig. 20.14, transitions for all 4 bound states within the quantum well are observed. In addition, excitonic behavior is observed on the $n = 1$ peak. For the 140 Å quantum well, the transitions are broader, and effects due to the light and heavy hole levels can be seen through the distorted lineshape (see Fig. 20.14). To observe transitions to higher bound states, the spectra in Fig. 20.15 are taken for a quantum well width of 316 Å, for which transitions up to (6, 6) are resolved. For such wide quantum wells, the contributions from the light holes are only seen clearly when a transition for a light hole state is not close to a heavy hole transition because of the lower density of states for the light holes (see Fig. 20.15).

Exciton effects are significantly more pronounced in quantum well structures than in bulk semiconductors, as can be understood from the following considerations. When the width d_1 of the quantum well is less than the diameter of the exciton Bohr orbit, the electron–hole separation will be limited by the quantum well width rather than by the larger Bohr radius, thereby significantly increasing the Coulomb binding energy and the intensity of the exciton peaks. Thus, small quantum well widths enhance exciton effects. Normally sharp exciton peaks in bulk GaAs are observed only at low temperature ($T \ll 77$ K); but in quantum well structures, excitons can be observed at room temperature, as shown in Fig. 20.14, which should be compared with Fig. 20.8 for 3D bulk GaAs.

The reason why the exciton line intensities are so much stronger in the quantum well structures is due to the reduction in the radius of the effective real space Bohr orbits, thereby allowing more k band–states to contribute to the optical transition. This argument is analogous to arguments made to explain why the exciton intensities for the alkali halides are huge [see Fig. 20.12a]. In the alkali halides the excitons have very small real space Bohr orbits so that large regions of k space can contribute to the exciton excitation intensity.

In the case of the quantum well structures, two exciton peaks are observed because the bound states for heavy and light holes have different energies, in contrast to the case of bulk GaAs where the $j = 3/2$ valence band states are degenerate at $k = 0$. This property was already noted in connection with Fig. 20.13 for the bound state energies. Because of the large phonon density available at room temperature, the ionization time for excitons is only 3×10^{-13} s. Also the presence of the electron–hole plasma strongly modifies the optical constants, so that the optical constants are strongly dependent on the light intensity, thereby giving rise to non–linear effects that are not easily observed in 3D semiconductors.

Because of the small binding energy of these exciton states in a semiconductor like GaAs, modest electric fields have a relatively large effect on the photon energy of the exciton peaks and on the optical constants. Application of an electric field perpendicular to the layers of the superlattice confines the electron and hole wave

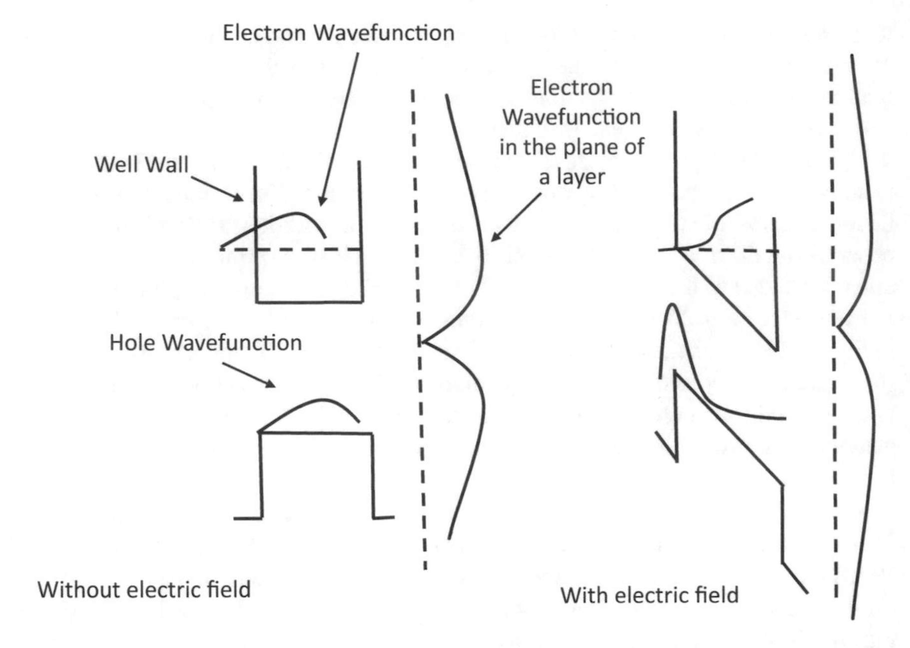

Fig. 20.16 Excitonic wave functions in a GaAs quantum well without (left) and with (right) an applied electric field. Because of the triangular potentials that are created by the electric field in the z–direction, the quantum well retains the electron and hole in a bound state at electric fields much higher than would be possible in a bulk classical ionization field

functions at opposite ends of the quantum well, as shown in Fig. 20.16. Because of this spatial separation, the excitons become relatively long lived and now recombine on a time scale of 10^{-9} s. Also because of the quantum confinement, it is possible to apply much higher (50 times) electric fields than is possible for an ionization field in a bulk semiconductor, thereby producing very large Stark red shifts of the exciton peaks, as shown in Fig. 20.17. This perturbation by the electric field on the exciton levels in a quantum confined structure is called the *quantum confined Stark effect*. This effect is not observed in bulk semiconductors. The large electric field–induced change in the optical absorption that is seen in Fig. 20.17 has been exploited for device applications.

The following mechanism is proposed to explain the quantum confined Stark effect when the electric field is applied perpendicular to the layers. This electric field pulls the electrons and holes toward opposite sides of the layers as shown in Fig. 20.16 resulting in an overall net reduction in the attractive energy of the electron–hole pair and a corresponding Stark (electric field induced) shift in the exciton absorption features. Two separate reasons explain the strong exciton peaks in quantum well structures. Firstly the walls of the quantum wells impede the electron and hole from tunneling out of the wells. Secondly, because the quantum wells are narrow (e.g., $\sim$ 100 Å) compared to the three–dimensional (3D) exciton size (e.g., $\sim$ 200 Å), the electron–hole interaction, although slightly weakened by the separation of electron

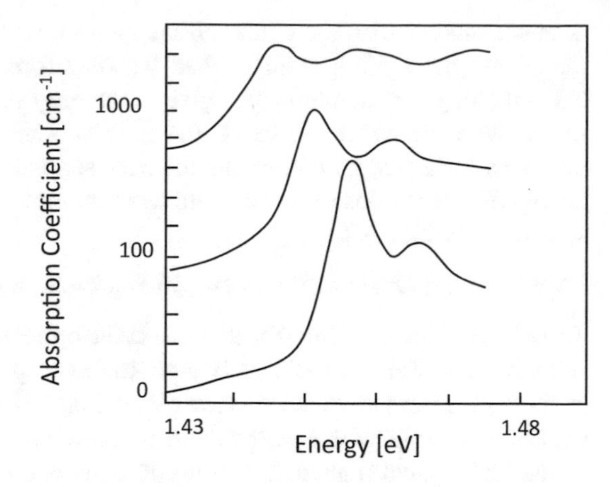

Fig. 20.17 The absorption spectra in $GaAs/Ga_{1-x}Al_xAs$ heterostructures for various values of applied electric field illustrating the large changes in optical properties produced by the quantum confined Stark effect. The electric fields normal to the layer planes are: **a** 10^4 V/cm, **b** 5×10^4 V/cm, and **c** 7.5×10^4 V/cm

and hole, is still strong, and well defined excitonic states can still exist. Thus, exciton resonances can remain to much higher fields than would be possible in the absence of this confinement, and large absorption shifts can be seen experimentally without excessive broadening.

20.5.4 Excitons in 0D and 1D Systems: Fullerene C_{60} and Carbon Nanotubes

The exciton binding energy commonly observed in semiconductors is of the order of $\sim$10 meV presenting discrete levels below the single particle excitation spectra as we discussed above. Due to this, optical absorption of exciton levels is usually observed only at low temperatures. However, in one-dimensional system such as a single wall carbon nanotube, the exciton binding energy can be as large as 1 eV so that exciton effects can be observed at room temperature. Therefore, excitons are key ingredients for explaining optical processes in carbon nanotubes and fullerenes, such as optical absorption, photoluminescence, and resonance Raman spectroscopy.

Both C_{60} and carbon nanotube are π conjugated sp^2 carbon systems and their exciton properties show many similarities. These systems have about the same diameter and their electronic density of states exhibit very narrow singularities. In C_{60}, the wavefunction for the lowest exciton energy level is not homogeneous because the electron and hole have their own molecular orbitals with different symmetries. On the other hand, the wavefunction for the lowest exciton energy in carbon nanotubes is homogeneous around the circumferential direction and is localized only along the tube axis direction. The range of the Coulomb interaction, U, is larger than the tube diameter and smaller than the length of a SWNT whereas for C_{60} the effective range of U is different. The energy band width of the highest occupied molecular orbital (HOMO) and the lowest unoccupied molecular orbital (LUMO) band in C_{60}

is much smaller than the Coulomb interaction, while in the case of the nanotube, the energy band width is larger than U. Therefore, in a SWNT the motion of the exciton along the nanotube axis gives an energy dispersion for the exciton, while in C_{60} the excitons are localized. The exciton binding energy in fullerenes is about 0.5 eV as measured by comparing the optical absorption energy (1.55 eV) with the energy difference observed by photo electron emission and inverse photo emission spectroscopy (2.3 eV).

Experimental Observation of Bright Excitons in Carbon Nanotubes

The experimental evidence for excitons in the optical transitions in carbon nanotubes initially came from non-linear two-photon absorption experiments by using very high-power laser pulses (see Dukovic et al. 2005). In such experiments, SWNTs are excited in a two-photon absorption process at energies (E_{laser}), somewhat above half of the first optical transition energy for semiconducting SWNTs ($E_{\text{laser}} \geq E_{11}^S/2$),

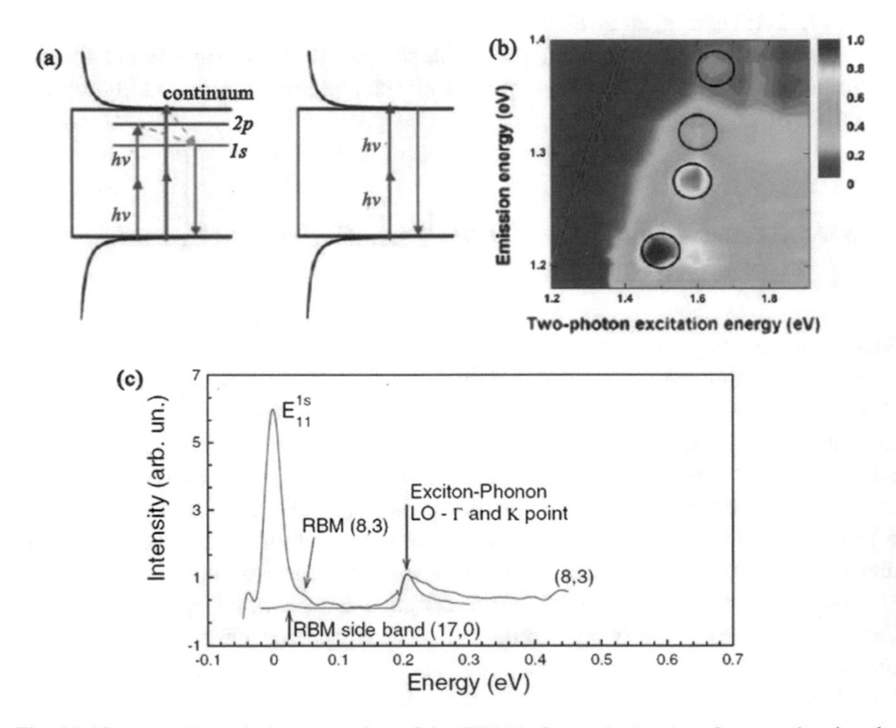

Fig. 20.18 **a** A schematic representation of the SWNT electronic density of states, showing the absorption of two photons and the emission of one photon with (left) and without excitons (right) is shown in the upper panel. **b** A 2D contour plot of the two-photon excitation spectra of SWNTs. The black circles label the higher two-photon absorption energy E_{TPA} and the lower-lying one-photon emission energy E_{em} for different (n, m) nanotubes. Emission always occurs at $E_{\text{em}} = E_{11}^S < E_{\text{TPA}}$. **c** Photoluminescence excitation spectrum taken at 1.30 eV showing the bright exciton close to the band edge, and the phonon side band approximately 200 meV higher in energy is identified with an (8, 3) SWNT

and subsequent light emission is observed around E^S_{11}. If the optically induced transitions E^S_{11} were related to free electron-hole pairs making band-to-band transitions, then the absorption of the two photons would occur at exactly $[E_{\text{laser}} = E^S_{11}/2]$, as schematically shown in the upper panel in Fig. 20.18a. However, if the E^S_{11} optical level were related to the creation of an exciton, then the energy for the absorption of two photons would be observed at an excited exciton state with an energy higher than E^S_{11}. Such a difference in energy between the absorption and emission processes was observed experimentally [see Fig. 20.18a], thus giving strong support for the presence of excitons in optical transitions of carbon nanotubes.

Further evidences for the excitonic nature of optical transitions in carbon nanotubes was provided by the experimental observation of exciton-phonon sidebands (see Plentz et al. 2005). Figure 20.18b shows the photoluminescence spectrum profile for the (8, 3) SWNT excited with 1.30 eV radiation where well defined resonances are found about 200 meV above each $E_{11}(A_2$ symmetry) feature. Asymmetric line shapes of the photoluminescence profile in Fig. 20.18b and the small upshift of the frequency of the phonon side peak relative to the G-band frequency $\hbar\omega_G$ are observed. The observed phonon side-band is assigned to a resonance identified with the absorption of light to a bound exciton-phonon state and the main contribution to this exciton-phonon bound state, shown in Fig. 20.18c, is attributed to the optical LO phonon at the **K** and Γ points of the graphene Brillouin zone. In such a situation, a significant fraction of the spectral weight should be transferred from the exciton peak to the exciton-phonon complex, and this transferred spectral weight is predicted to have a diameter dependence which was observed experimentally by Plentz et. al.

20.5.5 Excitons and Trions in Transition Metal Dichalcogenides

In 2010, Mak et al. 2010 showed that, by mechanically exfoliating the transition metal dichalcogenide MoS_2, an atomically thin layer of this material undergoes an indirect-to-direct bandgap transition. Similar observations were made in WSe_2 and $MoSe_2$. The tight confinement of the charge carriers in these two-dimensional materials results in extremely high exciton binding energies 400meV, which are stable at room temperature. As a result, the photophysics of these materials is quite different from most known bulk materials. For electrostatically gated MoS_2, charged excitons called *trions* are formed. Figure 20.19 shows the electrostatic gating/doping dependence of the photoluminescence of monolayer $MoSe_2$. Near zero doping, we observe mostly neutral (X^o) and impurity-trapped (X^I) excitons. With large electron (hole) doping, negatively (positively) charged excitons (X^+, X^-) dominate the spectrum. These charged and uncharged excitons are illustrated schematically in Fig. 20.19b.

20.5.6 Excitons in Transition Metal Dichalcogenide Heterojuctions

Another interesting aspect of exciton photophysics of layered transition metal dichalcogenides (TMDCs) occurs at the interface between two different TMDCs. Rivera et al. showed that interlayer excitons in monolayer $MoSe_2 - WSe_2$ heterostructures have strongly redshifted emission and substantially longer exciton lifetimes than either of the two constituent materials alone. Figure 20.20a illustrates the basic heterostructure configuration, in which monolayer WSe_2 is deposited on top of monolayer $MoSe_2$. Figure 20.20c shows the PL spectra taken from the WSe_2, $MoSe_2$, and heterostructure regions. For WSe_2 and $MoSe_2$, the exciton peaks all lie above 1.6 eV. In the heterostructure region, however, a large peak emerges below 1.4 eV due to an interlayer exciton. Figure 20.20b shows an energy band diagram of this interlayer exciton, in which the hole resides primarily in the WSe_2 while the electron resides in the $MoSe_2$ layer.

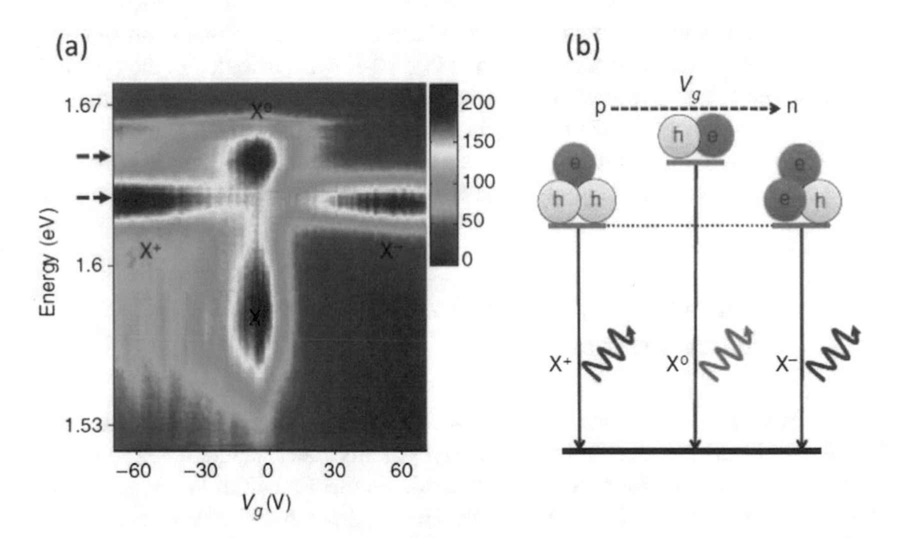

Fig. 20.19 Electrostatic control of exciton charge. **a** $MoSe_2$ PL (colour scale in counts) is plotted as a function of back-gate voltage. Near zero doping, we observe mostly neutral and impurity-trapped excitons. With large electron (hole) doping, negatively (positively) charged excitons dominate the spectrum. **b** Illustration of the gate-dependent trion and exciton quasi-particles and transitions. (Taken from Ross et al. 2013)

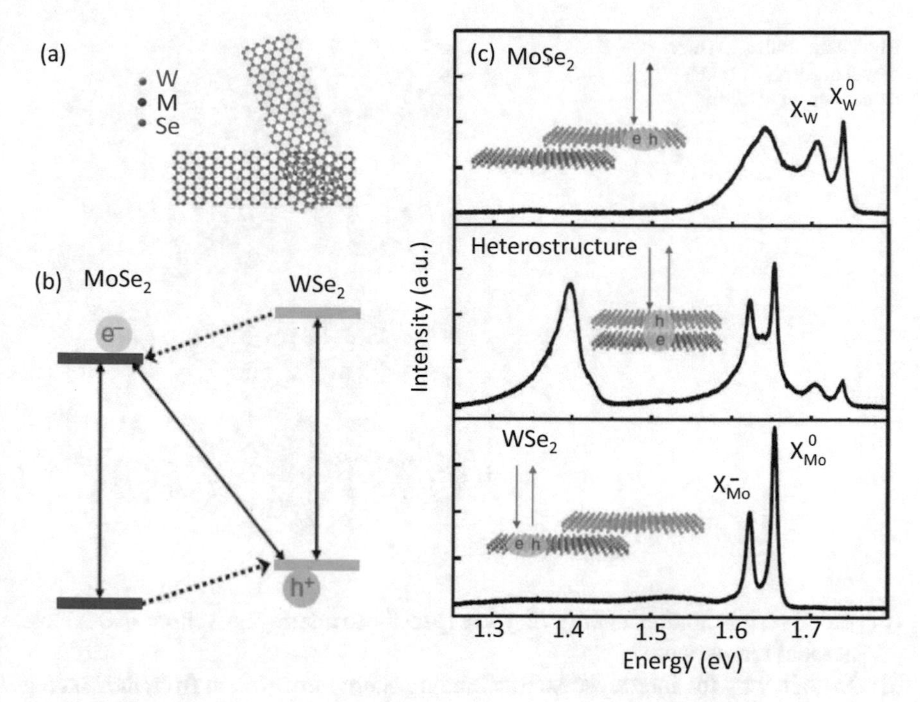

Fig. 20.20 Intralayer and interlayer excitons of a monolayer $MoSe_2 - WSe_2$ vertical heterostructure. **a** Schematic and **b** energy band diagram of $MoSe_2 - WSe_2$ heterostructure. **c** Photoluminescence spectra of individual monolayers and the heterostructure at 20 K under 20 mW excitation at 1.88 eV (plotted on the same scale). (Taken from Rivera et al. 2015)

Problems

20.1 This problem considers the density of Frenkel defects. By considering N' as the density of possible interstitial sites, show that the density of occupied interstitial sites is

$$n_i \simeq (NN')^{\frac{1}{2}} \exp\left[-\frac{E_i}{2k_B T}\right] \tag{20.26}$$

where E_i is the energy needed for removing an atom from a lattice site to form an interstitial defect site.

20.2 The band gaps of WSe_2 and $MoSe_2$ are 1.64 and 1.61 eV, respectively, and their conduction band offset is 228 meV with WSe_2 higher in energy than for $MoSe_2$.

(a) Estimate the maximum energy that an interlayer exciton could have (assuming zero binding energy)?
(b) The observed value of the interlayer exciton emission is 1.350 eV. Based on this, what is the exciton binding energy of this interlayer exciton?

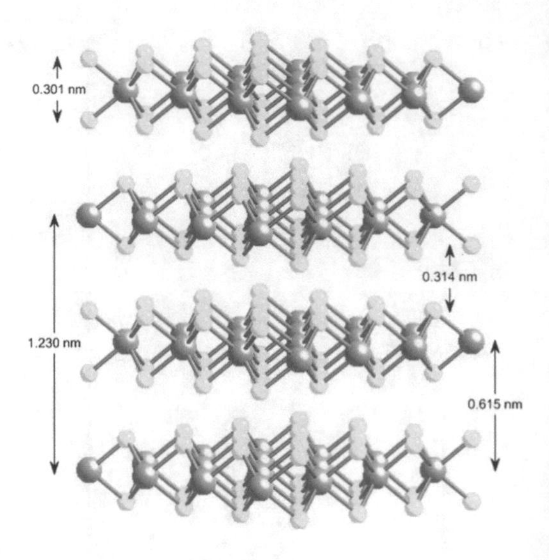

Fig. 20.21 Ball and stick model of MoS_2. (Taken from Liang et al. 2008)

(c) Based on this binding energy, do you expect these interlayer excitons to be stable at room temperature?

(d) Explain why the interlayer exciton binding energy is different from the binding energy of intralayer excitons 400 meV.

20.3 The exciton binding energy of an exciton in monolayer MoS_2 is 400 meV. Assuming that this binding energy is dominated by the ground particle-in-a-box energy corresponding to the 3.0 Å layer thickness of this material, with an effective mass of $m^* = 0.6m_o$, estimate the binding energy of bilayer MoS_2 (thickness = 9.16 Å). (See Fig. 20.21)

20.4 Calculate the bound state energies of excitons in GaAs (with both heavy and light holes) for

(a) bulk GaAs, assuming a hydrogenic model.

(b) a 50 Å quantum well, assuming the binding energy is dominated by the particle-in-a-box energy corresponding to the finite quantum well thickness.

(c) a 5 Å quantum well, assuming the binding energy is dominated by the particle-in-a-box energy corresponding to the finite quantum well thickness.

20.5 Repeat Problem 20.4 for germanium.

20.6 (a) By what factor (approximately) does a quantum well with a width of 50 Å increase the binding energy of an exciton for a quantum well of GaAs? The lattice constant for GaAs is 5.65 Å and the exciton radius of bulk GaAs is about 150 Å.

(b) The spin-orbit interaction for GaAs results in a splitting of the valence band, with the split-off band extremum at an energy 0.34 eV below the valence band extremum. Using 0.34 eV as a measure of the size of the spin-orbit interaction in GaAs, estimate the effect of the spin-orbit interaction on the binding energy of the exciton. [Hint: Consider the effect of the large Bohr radius on the magnitude of the spin-orbit interaction.] The donor impurity radius for GaAs is 136 Å. Use values of $m_{hh} = 0.62$ and $m_{\ell h} = 0.074$ for the hole masses and a dielectric constant of 15.

(c) If an exciton has no net charge, why is an exciton attracted to a charged impurity center to form a bound exciton?

20.7 Is the binding energy for an exciton in a semiconducting wire of 100 Å diameter increased or decreased relative to bulk values if the Bohr radius of the exciton for this material in the bulk is 30 Å? Under which conditions would the effect of confining the exciton within a small diameter wire be large? What is the reason for your answer?

Suggested Readings

O. Gunnarsson, Alkali-doped fullerides. Narrow-band solids with unusual properties, in *Alkali-doped Fullerides*, vol. XVII (World Scientific, Singapore, 2004), p. 282

Yu and Cardona, *Fundamentals of Semiconductors* (Springer, Berlin, 1996). Sects. 6.3 and 6.6

Bassani and Pastori–Parravicini, *Electronic States and Optical Transitions in Solids*: Chaps. 6 and 7 (1975)

References

F. Plentz, H.B. Ribeiro, A. Jorio, M.S. Strano, M.A. Pimenta, Direct experimental evidence of exciton-phonon bound states in carbon nanotubes. Phys. Rev. Lett. **95** (2005)

G. Dukovic, F. Wang, D.H. Song, M.Y. Sfeir, T.F. Heinz, L.E. Brus, Structural dependence of excitonic optical transitions and band-gap energies in carbon nanotubes. Nano Lett. **5**, 2314–2318 (2005)

K.F. Mak, C. Lee, J. Hone, J. Shan, T.F. Heinz, Atomically thin MoS_2: a new direct-gap semiconductor. Phys. Rev. Lett. **105**, 136805 (2010)

J.S. Ross, S.F. Wu, H.Y. Yu, N.J. Ghimire, A.M. Jones, G. Aivazian, J.Q. Yan, D.G. Mandrus, D. Xiao, W. Yao, X.D. Xu, Electrical control of neutral and charged excitons in a monolayer semiconductor. Nat. Commun. **4**, 1474 (2013)

P. Rivera, J.R. Schaibley, A.M. Jones, J.S. Ross, S.F. Wu, G. Aivazian, P. Klement, K. Seyler, G. Clark, N.J. Ghimire, J.Q. Yan, D.G. Mandrus, W. Yao, X.D. Xu, Observation of long-lived interlayer excitons in monolayer $MoSe_2$-WSe_2 heterostructures. Nat. Commun. **6**, 6242 (2015)

T. Liang, W.G. Sawyer, S.S. Perry, S.B. Sinnott, S.R. Phillpot, First-principles determination of static potential energy surfaces for atomic friction in MoS_2 and MoO_3. Phys. Rev. B **77**, 104105 (2008)

Chapter 21
Luminescence and Photoconductivity

21.1 Classification of Luminescence Processes

Luminescence denotes the emission of radiation by a solid in excess of the amount emitted in thermal equilibrium and can be considered as a process inverse to the absorption of radiation. Since luminescence is basically a non-equilibrium phenomena, it requires excitation by light, electron beams, current injection, etc., which generally act to create excess electrons, holes, or both. For example, the effects of electron–hole recombination give rise to **recombination radiation** or luminescence.

One classification of luminescent processes is based on the source of the excitation energy. The most important excitation sources are

1. **photoluminescence** by optical radiation,
2. **electroluminescence** by electric fields or currents,
3. **cathodoluminescence** by electron beams (or cathode rays),
4. **radioluminescence** by other energetic particles or high energy radiation sources.

A second classification of luminescent processes pertains to the time that the light is emitted relative to the initial excitation. If the emission is fast ($\leq 10^{-8}$ s is a typical lifetime for an atomic excited state), then the process is **fluorescent**. The emission from most photoconductors is of the fluorescent variety. For some materials, the emission process is slow and can last for minutes or hours. These materials are **phosphorescent** and are called **phosphors**, and is governed by the time it takes for the carriers in the excited state to reach equilibrium. In some cases there is an excited state that is well defined, so that the occupation of the excited state has a lifetime. While the electron remains in the excited state we call the excited state an excitonic state subject to the Pauli exclusion principle. If two excitons are then excited for the same atom, a biexciton is formed where the two electrons have spin up and spin down, filling that state.

Let us now consider luminescent processes of the fluorescent type with fast emission times. The electronic transitions which follow the excitation and which result in luminescent emission are generally the same for the four types of excitations

M. Dresselhaus et al., *Solid State Properties*, Graduate Texts in Physics,
https://doi.org/10.1007/978-3-662-55922-2_21

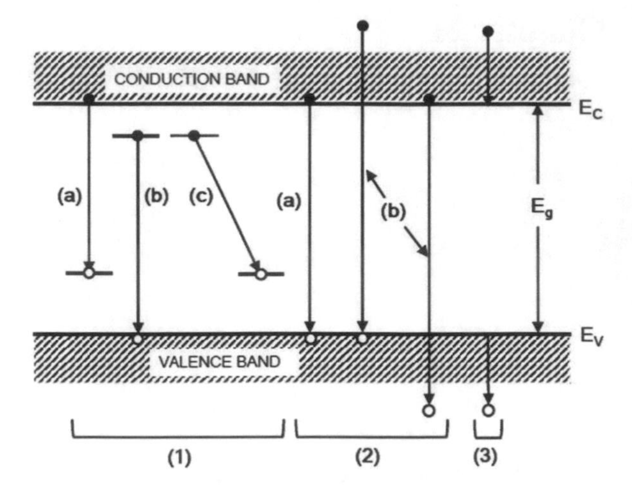

Fig. 21.1 Basic transitions in a semiconductor for stimulating the luminescent process. After H.F. Ivey, *IEEE J.Q.E.* **2**, 713 (1966). • = electrons: ○ = holes

mentioned above. Figure 21.1 shows a schematic diagram of the basic transitions in a semiconductor that are effective in exciting a luminescent process. These excitations may be classified as follows:

1. Transitions involving chemical impurities or physical defects (such as lattice vacancies):

   a. conduction band to acceptor state recombination.
   b. available electron in a donor level site makes resonant transition to a vacant conduction band state.
   c. donor to acceptor transition resulting in pair emission.

2. Interband transitions:

   a. intrinsic or edge emission, corresponding very closely in energy to the band gap, though phonons and/or excitons may also be involved.
   b. higher energy emission involving energetic or "hot" carriers, sometimes related to **avalanche** emission, where "hot" carriers refers to highly energetic carriers well above the thermal equilibrium levels.

3. Intraband transitions involving "hot" carriers, sometimes called deceleration emission.

It should be pointed out that the various transitions mentioned above do not all occur in the same material or under the same conditions. Nor are all electronic transitions radiative. Phonon emission provides a non-radiative mechanism for the relaxation of an excited state in a solid to the lowest equilibrium ground state. An efficient luminescent material is one in which radiative transitions predominate over non-radiative transitions.

When electron-hole pairs are generated by external excitations, radiative transitions resulting from the hole-electron recombination may occur. The radiative transitions in which the sum of electron and photon wavevectors is conserved are called direct transitions as opposed to indirect transitions which involve additional available scattering agents, such as phonons.

21.2 Emission and Absorption

For a given material the emission probability will depend on the photon energy and on the temperature. The emission rate $R_{vc}(\omega)$ for the transition from the conduction band (c) to the valence band (v) is related to the absorption rate $P_{vc}(\omega)$ by the relation

$$R_{cv}(\omega) = P_{vc}(\omega)\rho(\omega) \tag{21.1}$$

where $\rho(\omega)$ is the Planck distribution function at temperature T

$$\rho(\omega) = \frac{2}{\pi}\frac{\omega^2 n_r^3}{c^3[\exp(\hbar\omega/k_B T) - 1]}, \tag{21.2}$$

and the absorption rate is given by

$$P_{vc}(\omega) = \frac{\alpha(\omega)c}{n_r}, \tag{21.3}$$

where $\alpha(\omega)$ is the frequency-dependent absorption coefficient and n_r is the index of refraction. The frequency and temperature dependence of the emission rate is then given by

$$R_{cv}(\omega) = \frac{2}{\pi}\frac{\omega^2 n_r^2 \alpha(\omega)}{c^2[\exp(\hbar\omega/k_B T) - 1]}. \tag{21.4}$$

Basically, $R_{cv}(\omega)$ shows high emission at frequencies where the absorption is large, so that emission spectroscopy can be used as a technique to study various aspects of the electronic band structure.

The luminescence process involves 3 separate steps:

1. Excitation: the electron-hole (e–h) pairs are excited by an external energy source to an excited state.
2. Thermalization: the excited electron-hole pairs relax to their quasi-thermal equilibrium distributions.
3. Recombination: the thermalized electron-hole pairs recombine radiatively to produce light emission.

We now give some examples of luminescence spectra. The big picture is shown in Fig. 21.2, where luminescence spectra for InSb are presented showing several typical features. The highest energy feature in this figure is connected with luminescence from the conduction band to the valence band (the band-to-band process) at 0.234 eV,

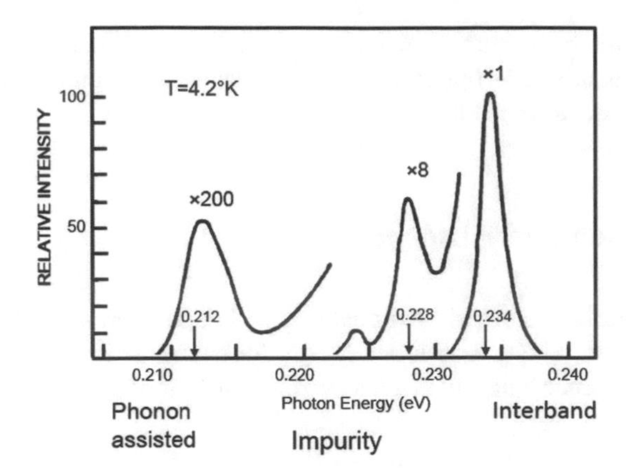

Fig. 21.2 Luminescence emission spectrum in an n-type InSb crystal with an electron concentration of $5 \times 10^{13}\,\text{cm}^{-3}$. The peak at 0.234 eV is due to interband recombinative emission, and the peak at 0.228 eV is due to an impurity state below the band edge. The peak at 0.212 eV (multiplied by 200) is due to phonon-assisted band-to-band transitions. (A. Mooradian and H.Y. Fan, *Phys. Rev.*, **148**, 873 (1966).)

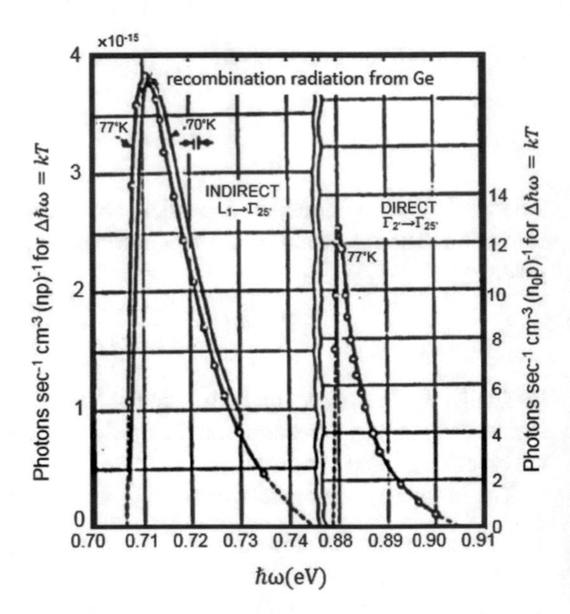

Fig. 21.3 Direct and indirect intrinsic radiation recombination in Ge. The 70 K spectrum is experimental and is in the energy range appropriate for indirect transitions assisted by longitudinal acoustic (LA) phonons. The open circles are calculated from experimental absorption data for both types of transitions. The free carrier densities at the direct and indirect conduction band minima and at the valence band maximum are denoted by n_0, n, and p, respectively

from the conduction band to an acceptor impurity level at 0.228 eV, and luminescence that is phonon assisted at 0.212 eV involving phonon absorption, so that the emitted photon has a lower energy than the excitation energy.

For intrinsic or band-to-band transitions, the peak intensity occurs near the energy gap and the width of the spectral line (at the half value of the peak intensity) is proportional to the thermal broadening energy $k_B T$. For extrinsic transitions, the peak

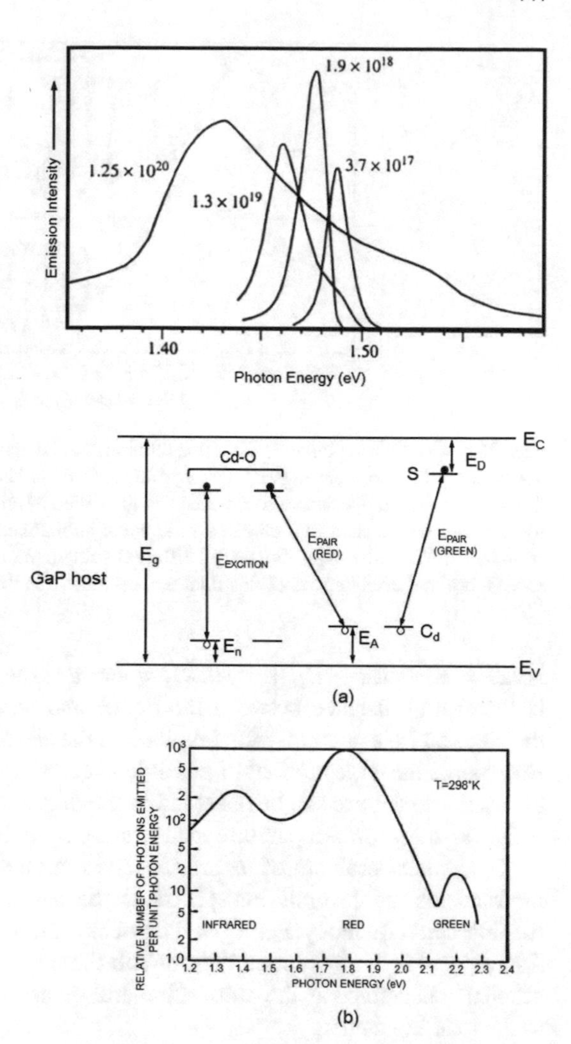

Fig. 21.4 Electroluminescence intensity of p-type (Zn-doped) GaAs at 4.2 K is shown for four different dopant concentrations, varying over 3 orders of magnitude in units of cm^{-3}. Note the increasing broadening and downshift of the emission peak with increasing dopant concentration

Fig. 21.5 Energy level diagram for a Cd-doped GaP $p-n$ junction where Cd-O denotes a cadmium-oxygen complex. Transitions between the exciton level of the Cd-O complex to the acceptor level of Cd give rise to red light emission. Transitions between the donor level (S) and acceptor level (Cd) give rise to the green light emission. **b** Measured emission spectrum from a GaP diode in which the color associated with the various luminescent peaks are shown ranging from infrared to red and green in the visible range. (After M. Gershenzon, Bell, *Sys. Tech. J.*, **45**, 1599, (1966).)

emission intensity occurs near the transition energy, but the broadening is greater than for the intrinsic band-to-band emission shown in Fig. 21.3 for both indirect and direct bandgap emission.

An example of a luminescence spectrum from a free to a bound state is presented in Fig. 21.4 where the electroluminescence is shown for p-type GaAs for various Zn dopant concentrations, given in units of cm^{-3}. As the impurity concentration increases, the luminescence emission becomes increasingly broad because of the perturbation to the crystal lattice introduced by the site-to-site potential variation in the basic periodicity of the lattice. Notice both the line broadening and increasing lineshape asymmetry at high dopant levels.

An example of donor-acceptor pair transitions is shown in Fig. 21.5 for GaP showing the exciton emission peak and structure associated with donor-acceptor pair emission. For donor-acceptor pair emission the energy of the emitted photon

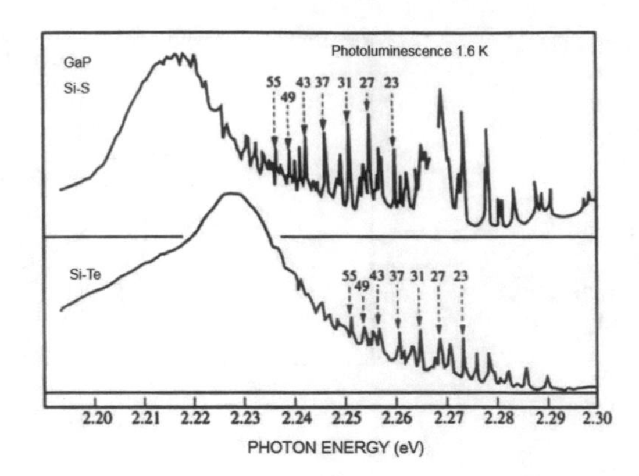

Fig. 21.6 Donor-acceptor pair (DAP) recombination spectra in GaP containing S-Si and Te-Si donor-acceptor pairs measured at low temperature (1.6 K). The integers above the discrete peaks are the shell numbers of the pairs which have been identified by comparison with theoretical predictions, for the case where each impurity is on the same sublattice (i.e., both are on the Ga or on the P sublattices). When the impurities are on different sublattices, the donor-acceptor pair recombination spectra become even more complex than what is shown in this figure

is $\hbar\omega = E_g - E_A - E_D + e^2/(\varepsilon R)$, where ε is the static dielectric constant and R is the spatial distance between the donor and acceptor impurities that constitute the electron-hole pair emission involved in the electron-hole recombination process. Because of the large number of possible sites for the donor and acceptor impurities, a very rich spectrum can be observed in the donor-acceptor pair emission, as shown in Fig. 21.6 for low temperature measurements on the wide gap semiconductor GaP.

The general problem of luminescence is not only to determine the luminescent mechanisms and the emission spectra, as discussed above, but also to determine the **luminescent efficiency**. For a given input excitation energy, the radiative recombination process is in direct competition with the non-radiative processes. Luminescent efficiency is defined as the ratio of the energy associated with the radiative process to the total input energy.

Among the fastest emission luminescent processes, electroluminescence, or excitation by an electric field or electric current, has been one of the most widely utilized for device applications. Electroluminescence is excited in a variety of ways including intrinsic, injection, avalanche, and tunneling processes, as summarized below.

1. **Intrinsic process**. When a powder of a semiconductor (e.g., ZnS) is embedded in a dielectric (plastic or glass) material, and exposed to an alternating electric field (usually chosen to be at audio frequencies), electroluminescence may occur. Generally the efficiency is low ($\sim$1%) and such materials are used primarily in display devices. The mechanism is mainly due to impact ionization by accelerated electrons and/or field emission of electrons from trapping centers.
2. **Injection process**. Under forward-bias conditions, the injection of minority carriers in a $p - n$ junction can give rise to radiative recombination. The energy level band diagram for a Cd-doped GaP $p - n$ junction is shown in Fig. 21.5.

Several different transitions for electron-hole recombination are indicated. The relative intensity of the red and the green bands can be varied by varying the impurity concentrations in the sample preparation. The red-light emission from GaP $p - n$ junctions (i.e. GaP light-emitting-diodes (LEDs)) was one of the first systems with sufficiently high efficiency to be utilized in practical applications. At a later time, high brightness, high efficiency LEDs became available throughout the IR, visible, and UV wavelength ranges, and LED technology has become a very common electronic commercial product.

3. **Avalanche process**. When a $p - n$ junction or a metal semiconductor contact is reverse-biased into an avalanche breakdown for an LED process, the electron-hole pairs generated by impact ionization may result in emission of either interband (avalanche emission) or intraband (deceleration emission) transitions, respectively.
4. **Tunneling process**. Electroluminescence can also result from tunneling into forward-biased and reverse-biased junctions. In addition, light emission can occur in reverse-biased metal-semiconductor contacts. (see M.H.P. Pfeiffer, et al. *Optics Express*, **19**, A1184–A1189, (2011).)

Fast emission luminescence also is of importance to semiconducting lasers. Luminescence is generally observed as an **incoherence** emission process in contrast with laser action, which involves the **coherent** emission of radiation in executing a radiative transition. The coherence is usually enhanced by polishing the sample faces to form an optical cavity. Examples of solid state lasers are the ruby laser and the common direct gap semiconductor lasers. Optical and electrical pumping are the most common methods of exciting laser action in solid state lasers. (see X.F. Huang et al. 2003)

Finally, we conclude the discussion of electroluminescence in semiconductors with a short discussion of slow emission luminescence, i.e. phosphorescence. Phosphorescent materials exhibit afterglow effects and are consequently important in various optical display devices. These phosphors often do not exhibit large photoconductivities. That is to say, although the electrons that were produced survive for a long time, they are bound to particular defect centers and do not readily carry charge through the crystal.

In Fig. 21.7 we show an example of how a phosphor works in an alkali halide, such as KCl with a small amount of Tl (thallium) impurities. The thallium defects act as recombination centers. If these recombination centers are very efficient at producing recombination radiation they are called **activators**; Tl doping of KCl acts as an activator. In this system, the excitation occurs at a higher energy than the emission, and therefore the excitation process is considered as an up-conversion process.

The **Franck–Condon principle** states that the atoms in the solid do not change their internuclear separations during an actual electronic transition. We now explain how emitted light is downshifted in frequency relative to the frequency of the exciting light. The Tl^+ ion in the ground or unexcited state occupies some configuration close to the symmetric center of a K^+ ion which the Tl^+ ion might be replacing. When excited, the Tl^+ ion finds a lower energy state in a lower symmetry position near one

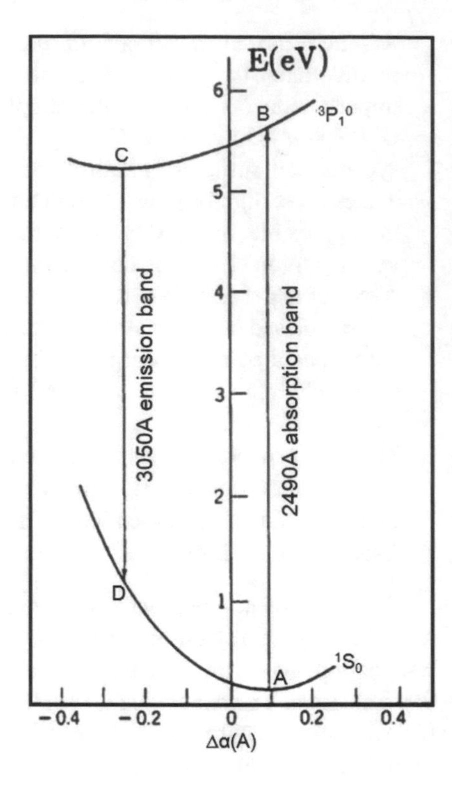

Fig. 21.7 Schematic diagram of the phosphorescence process of the thallium$^+$ activation process in KCl. The emission is downshifted from the absorption. This is an illustration of the Franck–Condon principle

of the Cl^- ions as shown schematically on the top of Fig. 21.7. The absorption is made from the ground state energy (point A in Fig. 21.7) to an excited state (point B) with the same configuration in a direct transition. Phonon interactions then will bring the electron to the equilibrium position C. Achievement of equilibrium ($B \rightarrow C$) will take a longer time than the electronic transitions ($A \rightarrow B$). Emission from $C \rightarrow D$ again occurs in accordance with the Franck–Condon principle and the readjustment to the equilibrium configuration A proceeds by phonon processes.

Photon emission is one of the main techniques for studying impurity and defect levels in semiconductors. It is an important technique also for studying new materials such as organic systems. (see Adachi et al. 2001)

One luminescent technique that has become very popular is luminescence excitation spectroscopy because of the wide variety of information that can be obtained. According to this technique the emission at a particular energy is monitored as the excitation energy is varied. This technique has become very popular for low dimensional materials systems or for very thin epitaxial layers on an opaque substrate, thereby providing much more sensitivity than absorption spectra or photoluminescence spectra offer.

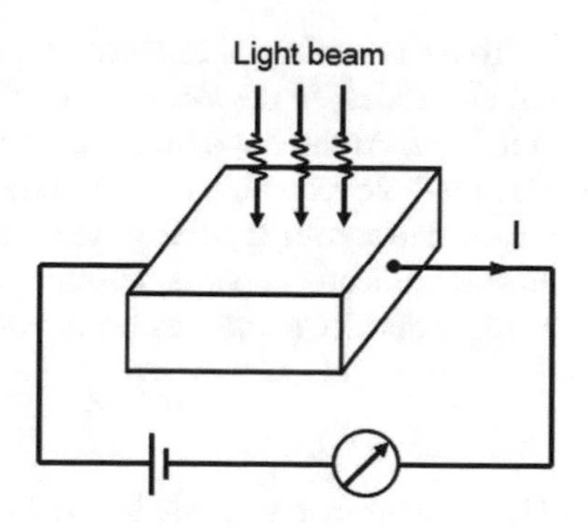

Fig. 21.8 Schematic diagram of the experimental arrangement for measuring the photoconductivity, showing a light beam incident on a sample. (see Li et al. 2014)

21.3 Photoconductivity

Photoconductivity is observed when light is incident on a poorly conducting material, (e.g., an insulator or semiconductor), and the photon energy is sufficiently high to excite an electron from an occupied valence band state to an unoccupied conduction band state. In such interband transitions, both the electron and hole will contribute to the electrical conductivity if a voltage is applied across the sample, as shown in the schematic experimental arrangement in Fig. 21.8. Since the threshold for photoconduction occurs at $\hbar\omega = E_g$, measurement of the photoconductivity can be used to determine the band gap for non-conducting materials. Photoconductivity is often the concept used for the design of practical optical detectors.

$$\Psi_{2D}(x, y) = e^{ik_x x}\phi(y) \tag{21.5}$$

The photoconduction process increases the electrical conductivity $\Delta\sigma$ due to the increase in the density of electrons (Δn) and holes (Δp) resulting from photoexcitation:

$$\frac{\Delta\sigma}{\sigma} = \frac{\Delta n\mu_n + \Delta p\mu_p}{n\mu_n + p\mu_p} \tag{21.6}$$

in which $\mu_n + \mu_p$ are respectively, the electron and hole mobilities. Since the carriers are generated in pairs in the photo-excitation process $\Delta n = \Delta p$. In preparing materials for applications as photoconductors, it is desirable to have a high mobility material with a low intrinsic carrier concentration, and long electron–hole recombination times to maximize the photo-excited carrier density concentration. Cadmium sulfide is an example of a good photoconductive material. In CdS, it is possible to change the conductivity by $\sim$**10 orders of magnitude** through carrier generation by light. These large changes in electrical conductivity can be utilized in a variety of device applications, such as light meters, photo-detectors, "electric eye" control applications, optically activated switches, and for information storage.

To measure the photo-currents, photo-excited carriers are collected at the external electrodes. In the steady state, free carriers are continually created by the incident light. At the same time, the excited free carriers annihilate each other through electron-hole recombination. To produce a large photocurrent, it is desirable to have a *long* free carrier *lifetime* τ' or a slow recombination time. If G is the rate of generation of electrons per unit volume due to photo-excitation, then the photo-excited electron density in the steady state will be given by

$$\Delta n = G\tau'. \tag{21.7}$$

The generation rate G will in turn be proportional to the photon flux incident on the photoconductor. Whereas slow recombination rates are essential to the operation of photoconductors, rapid recombination rates are necessary for luminescent materials.

In the recombination process, an electron and hole annihilate each other, emitting a photon in a radiative process. In real materials, the recombination process tends to be accelerated by certain defect sites. When such defects tend to be present in relatively greater concentrations at the surface, the process is called **surface recombination**. In bulk crystals, the density of recombination centers can be made low for a very pure and "good" crystal. A typical recombination center concentration in a high quality Si crystal would be $\sim 10^{12}\,\text{cm}^{-3}$.

Photo-excited carriers can also be eliminated from the conduction process by **electron** and **hole traps**. These traps differ from recombination centers insofar as traps preferentially eliminate a single type of carrier. In practice, hole traps seem to be more common than electron traps. For example, in the silver halides which are important in the photographic process, the hole is trapped almost as soon as it is produced and photoconduction occurs through the electrons.

Electron and neutron irradiation of materials produce both recombination centers and traps in photoconducting materials. Thus, special precautions must be exercised in using photo-detectors in high radiation environments, such as on satellites in outer space.

Trapped electrons can be released by thermal or optical excitation. For example, consider a p-type sample of Ge which has been doped with Mn, Ni, Co, or Fe. At low temperatures E_F will be near the top of the valence band and the acceptor impurity states will have very few electrons in them. Photons that are energetic enough to take an electron from the valence band to these impurity levels will result in hole carriers in the valence bands. The deep acceptor levels for these impurities are above the top of the valence band by 0.16 eV for Mn, 0.22 eV for Ni, 0.25 eV for Co and 0.35 eV for Fe. The threshold values observed for photoconductivity in these p-type Ge samples are shown in Fig. 21.9 and the experimental results are in good agreement with this interpretation. The large increase in photoconductivity at 0.7eV corresponds to the onset of an interband transition and the threshold for this process is independent of the impurity species.

The excess carrier lifetime can be measured by using light pulses and by observing the decay in the photocurrent through measurement of the voltage across a calibrated load resistor R in the external circuit, as shown in Fig. 21.10.

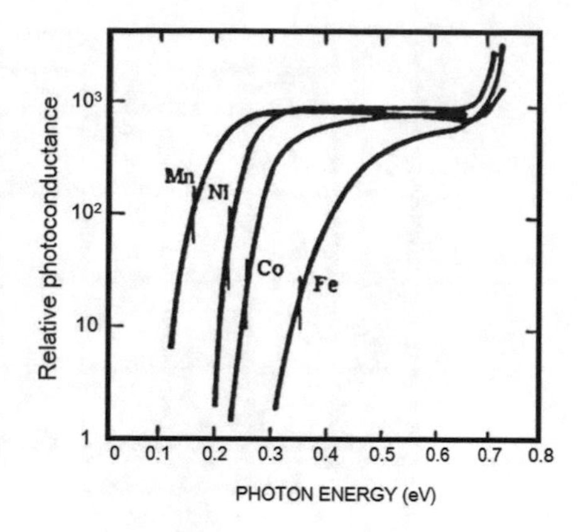

Fig. 21.9 Photoconductance spectrum occurring in the infra-red range in bulk Ge with various dopants as indicated

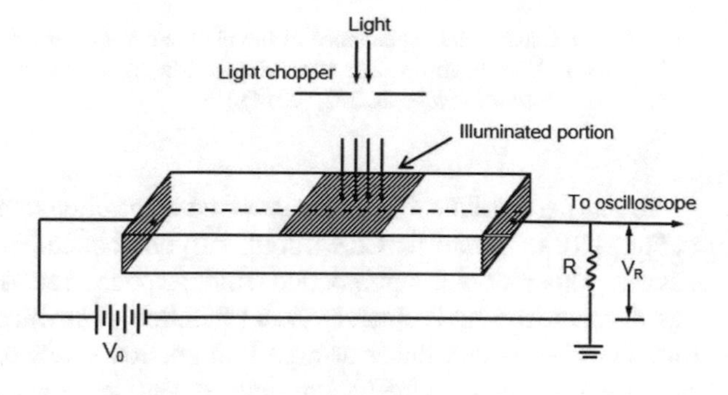

Fig. 21.10 Schematic of a circuit used to measure the excess carrier lifetime through decay in the photocurrent

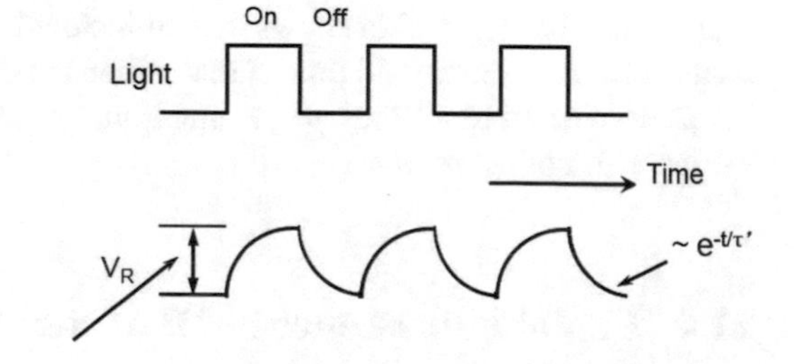

Fig. 21.11 Schematic experimental time dependence of incident light pulses and of the corresponding photoconductivity signal

For each light pulse, the carrier density will build up and then decay exponentially with a characteristic time equal to the lifetime τ' of the excess carriers generated by the light intensity coming from the photodetector signal.Using a light chopper, light pulses can be generated as indicated in Fig. 21.11.

In the interpretation of these experiments corrections must be made for surface recombination. To study a given material, the pulse repetition rate is adjusted to match approximately the excess carriers decay lifetime. For long lifetimes ($\sim 10^{-3}$ s), a mechanical chopper arrangement is appropriate. On the other hand, for short lifetimes a spark source can be used to give a light pulse of $\sim 10^{-8}$ s duration. For extremely short lifetimes, lasers with pulses well below $\sim 10^{-12}$ s are available.

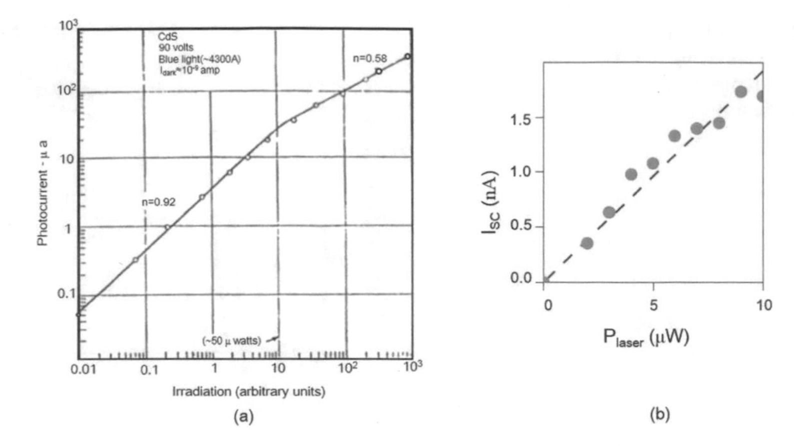

Fig. 21.12 Experimental dependence of the photocurrent on incident light irradiation for **a** CdS and **b** WSe_2. A linear response is observed for low light intensity levels. (see B.W.H. Baugher, et al. *Nature Nanotechnology*, **9**, 292, (2014).)

To get an idea of the magnitude involved in the photoconduction process, we show in Fig. 21.12 some data for CdS, a common photoconducting material used for optical devices. This plot of the photoconductive response versus illumination level shows that the photocurrent is almost a linear function of the illumination intensity for low intensities but is non-linear at high illumination levels on this log - log plot. The **dark current** refers to the background current that flows in the absence of incident light. Thus, the Fig. 21.12a shows that an incident power as small as 5×10^{-8} watts results in a photocurrent 50 times greater than the dark current. Figure 21.12b shows the photo-current of a WSe_2 $p-n$ junction, which exhibits a linear dependence on the incident optical power.

21.4 Photoluminescence in 2D Materials

Photoluminescence spectroscopy has become a very important tool for studying 2D layered materials including the monolayer transition metal dichalcogenides MoS_2 and WSe_2. Unlike the bulk material, monolayer MoS_2 emits light strongly, exhibiting an increase in luminescence quantum efficiency by more than a factor of 10^4 compared with the bulk material (see Mak et al. 2010). This is due to the effect of quantum confinement on the material's electronic structure, which induces an indirect-to-direct bandgap phase transition. This new family of atomically thin direct bandgap semiconductors have attracted much research interest, since direct bandgap semiconductors are more applicable to optoelectronic devices, such as LEDs, solar cells, and photodiodes. Because of the strong dependence on thickness, photoluminescence spectroscopy, together with Raman spectroscopy, is routinely used to determine the layer

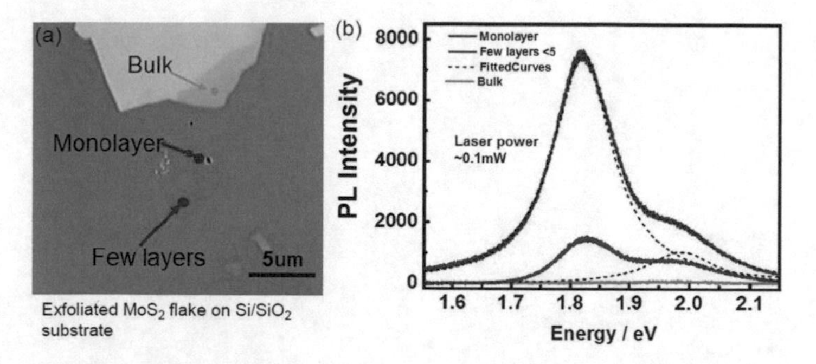

Fig. 21.13 **a** Optical microscope image and **b** photoluminescence spectra of monolayer, few-layer, and bulk MoS_2. (see Li et al. 2014)

thickness of mono- and few-layer transition metal dichalcogenides. Figure 21.13a shows an optical microscope image of MoS_2 exfoliated onto a Si/SiO_2 substrate using the conventional scotch tape method (Novoselov et al. 2004). The PL spectra of monolayer, few-layer, and bulk regions, indicated in Fig. 21.13a, are shown in Fig. 21.13b. Here, the monolayer region exhibits bright photoluminescence indicating its direct bandgap nature. The PL spectra of few layer MoS_2 shows significantly suppressed PL, and bulk MoS_2 shows almost no detectible photoluminescence.

Suggested Reading

P.Y. Yu, M. Cardona, *Fundamentals of Semiconductors* (Springer, Berlin, 1996)

References

C. Adachi, M.A. Baldo, M.E. Thompson, S.R. Forres, J. Appl. Phys. **90**, 5048–5051 (2001)
X.F. Huang, Y. Huang, R. Agarwal, C.M. Lieber, Nat. **421**, 241–245 (2003)
Z. Li, S.W. Chang, C.C. Chen, S.B. Cronin, Nano Res. **7**, 973–980 (2014)
K.F. Mak, C. Lee, J. Hone, J. Shan, T.F. Heinz, Atomically thin MoS_2: A new direct-gap semiconductor. Phys. Rev. Lett. **105**, 136805 (2010)
K.S. Novoselov et al., Electric field effect in atomically thin carbon films. Sci. **306**, 666 (2004)

Chapter 22
Optical Study of Lattice Vibrations

22.1 Lattice Vibrations in Semiconductors

22.1.1 General Considerations

The lattice vibrations in semiconductors are described in terms of $3N$ branches for the phonon dispersion relations where N is the number of atoms per primitive unit cell. Three of these branches are the acoustic branches, and the remaining $3N - 3$ are the optical branches. The optical lattice modes at $\mathbf{q} = 0$ are sensitively studied by infrared spectroscopy (optical reflectivity or transmission) for odd parity modes, including those for which the normal mode vibrations involve a dipole moment. Raman spectroscopy provides a complementary tool to infrared spectroscopy, insofar as Raman spectroscopy is sensitive to even parity modes. Since the common group IV semiconductors like silicon and germanium each have inversion symmetry, the optical phonon branch is Raman-active but is not seen in infrared spectroscopy. The III–V compound semiconductors, however, do not have inversion symmetry, so that the optical modes for semiconductors such as GaAs are both infrared-active and Raman-active. A schematic optical absorption curve of a general semiconductor is shown in Fig. 22.1.

Since the wavevector for light is very much smaller than the Brillouin zone dimensions, conservation of momentum requires the wave vector for the phonon $\mathbf{q}_{\text{phonon}}$ that is created or absorbed to be much smaller than Brillouin zone dimensions, so that the wave vectors for phonons that are observed in first order infrared or Raman processes are close to $\mathbf{q} = 0$. Since thermal neutrons can have a wide range of momentum values, neutron spectroscopy using thermal neutrons as a probe allows exploration of the phonon branches over a wide range of $\mathbf{q}_{\text{phonon}}$. Since heat in a semiconductor is dominantly carried by the acoustic phonons, information about the acoustic phonons is also provided by thermal conductivity studies.

We now review the interaction of the electromagnetic field with an oscillating dipole due to a lattice vibration. Crystals composed of two different atomic species (like NaCl) can have vibrating ions at finite temperatures. When these ions are

M. Dresselhaus et al., *Solid State Properties*, Graduate Texts in Physics,
https://doi.org/10.1007/978-3-662-55922-2_22

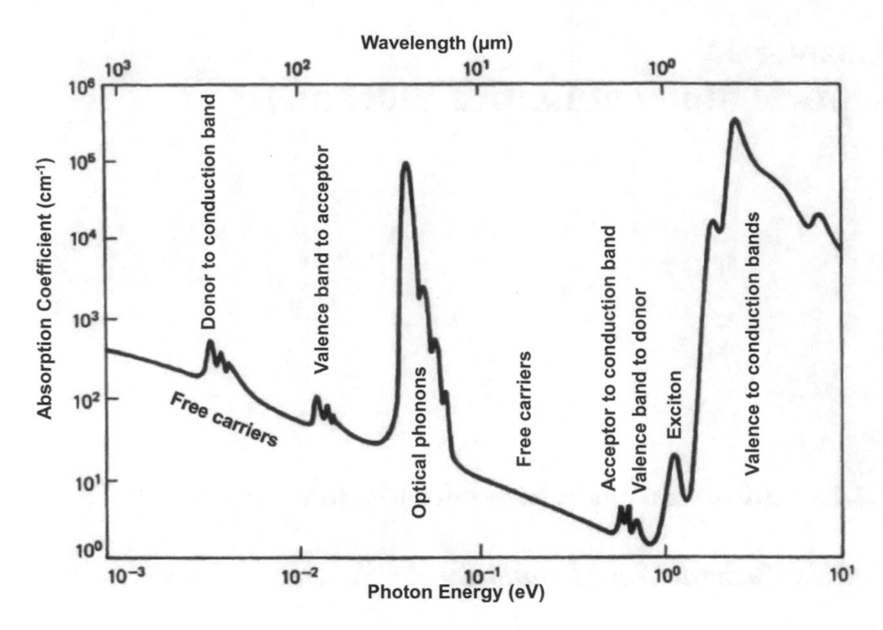

Fig. 22.1 Hypothetical absorption spectrum for a typical III–V semiconductor as a function of photon energy, over a wide range. The phenomena of dominant sensitivity are indicated for the various optical energy (or wavelength) ranges

vibrating in an optic mode $\leftarrow \oplus \rightarrow \leftarrow \ominus \rightarrow$, a vibrating dipole is created and this dipole can interact with the electromagnetic field. In discussing this interaction, we wish to focus attention on the following points which are discussed more fully in the text below:

1. The existence of two characteristic frequencies for the lattice vibrations in a solid in the presence of light:

   - ω_t = transverse optical phonon frequency (TO)
   - ω_l = longitudinal optical phonon frequency (LO)

   The description of the LO and TO phonons, distinguished from one another by the subscripts l and t, repectively, is provided by the polariton model which accounts for the interaction between light and phonon excitations. Because of the very small wavevector of the incident photons, the phonons which are optically excited will also have very small wavevectors. Therefore, ω_t and ω_l are taken as the phonon frequencies at $\mathbf{q} = 0$ for the TO and LO phonon dispersion curves.
2. The two frequencies ω_t and ω_ℓ are observable experimentally either through an infrared absorption, transmission, or reflection experiment (infrared activity) or through a light scattering experiment (Raman activity). A transparent dielectric becomes lossy as ω increases above ω_t. The transverse optical phonon frequency ω_t corresponds to a resonance in the dielectric function

$$\varepsilon(\omega) = \varepsilon_\infty + \frac{\text{const}}{\omega_t^2 - \omega^2} \tag{22.1}$$

where ε_∞ is the high frequency dielectric function (appropriate to electronic excitation processes) and a resonance in $\varepsilon(\omega)$ occurs at the TO phonon frequency $\omega = \omega_t$. The strong frequency dependence of the dielectric function (large dispersion) near ω_t is exploited in designing prisms for monochromators. The frequency ω_t is also called the *reststrahl* frequency.

3. The frequency ω_ℓ is the frequency at which the real part of the dielectric function vanishes $\varepsilon_1(\omega_\ell) = 0$. It will be shown below that ω_ℓ is the longitudinal optical phonon frequency corresponding to $\mathbf{q} = 0$ (zero wave vector). By group theory, which is used to exploit the crystalline symmetry of crystaline solid materials, it can be shown that the lattice modes at $\mathbf{q} = 0$ for a cubic crystal are threefold degenerate. This degeneracy is lifted by the electromagnetic interaction in polar materials to give a splitting between the LO and TO modes. An example of the reflectivity of a normally transparent material in the region where phonon excitation processes dominate is shown in Fig. 22.2 for three different temperatures. From the diagram, we see that for $\omega_t < \omega < \omega_\ell$, the dielectric is both highly reflective and lossy. This range between ω_t and ω_ℓ is also observed as an absorption band in infrared absorption studies.
4. The dielectric function $\varepsilon(\omega)$ approaches the static dielectric constant ε_0 as $\omega \to 0$ (large photon wavelength indicates low frequency phonons). Also, $\varepsilon(\omega)$

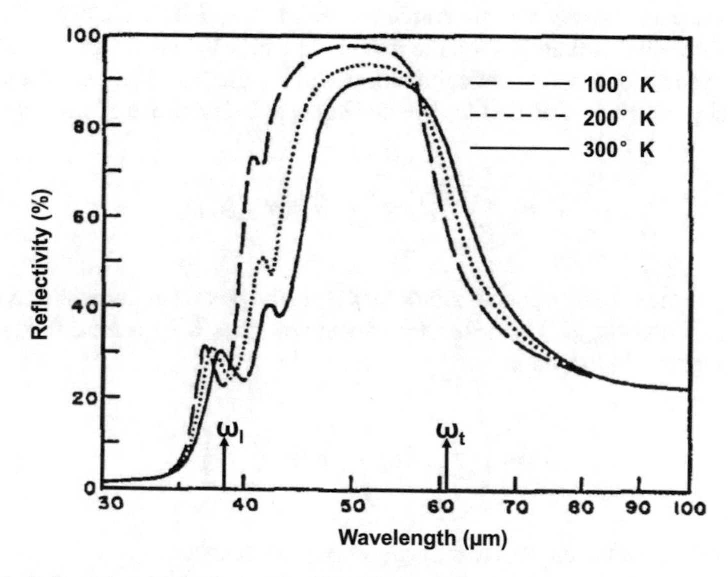

Fig. 22.2 Reflectivity of a thick crystal of NaCl versus wave length at several temperatures. The nominal values of ω_ℓ and ω_t for NaCl at room temperature correspond to wavelengths of 38 and 61 microns, respectively. The additional structure seen in the reflectivity spectrum near ω_ℓ is associated with defects

approaches the high frequency dielectric function ε_∞ as ω approaches frequencies that are large compared with ω_t and ω_ℓ. Even when we consider ω to be large, we are still thinking of ω as being very much smaller than typical interband electronic frequencies. Lattice modes typically are important in the wavelength range $10 \leq \lambda \leq 100\,\mu$m or the energy range $0.01 \leq \hbar\omega \leq 0.1\,$eV or the angular frequency range $50 \leq \omega \leq 1000\,\text{cm}^{-1}$.

5. The quantities ε_0, ε_∞, ω_t and ω_ℓ are not independent, but are related by a very general relation called the Lyddane–Sachs–Teller relation:

$$\frac{\omega_\ell^2}{\omega_t^2} = \frac{\varepsilon_0}{\varepsilon_\infty} \tag{22.2}$$

which is written here for a crystal with two atoms/unit cell, and the phonon modes have 3 accoustic and 3 optical branches.

22.2 Dielectric Constant and Polarizability

The polarizability p of an atom is defined in terms of the local electric field E_{local} at the atom site,

$$p = \alpha E_{\text{local}}. \tag{22.3}$$

The atomic polarizability α is an atomic property of the atom in the crystal, and the dielectric constant will depend on the manner in which the atoms are assembled to form a crystal. For a non-spherical atom, α will be a tensor. The polarization of a crystal may be approximated as the product of the polarizabilities of the atoms times the local electric fields,

$$P = \sum_j N_j p_j = \sum_j N_j \alpha_j E_{\text{local}}(j), \tag{22.4}$$

where N_j is the concentration of atoms, and α_j is the polarizability of atoms or ions labeled by j, and $E_{\text{local}}(j)$ is the local field at atomic sites j. If the local field is given by the Lorentz relation, then

$$P = \left(\sum_j N_j \alpha_j\right)\left(E + \frac{4\pi}{3}P\right). \tag{22.5}$$

Solving (22.5) for the electronic susceptibility χ we obtain

$$\chi = \frac{P}{E} = \frac{\sum_j N_j \alpha_j}{1 - \frac{4\pi}{3}\sum_j N_j \alpha_j}. \tag{22.6}$$

Using the definition $\varepsilon = 1 + 4\pi\chi$, one obtains the Clausius–Mossotti relation

$$\frac{\varepsilon - 1}{\varepsilon + 2} = \frac{4\pi}{3}\sum_j N_j\alpha_j. \tag{22.7}$$

This relation connects the dielectric constant ε to the electronic polarizability α, but only for crystal structures for which the Lorentz local field relation applies.

22.3 Polariton Dispersion Relations

The statements 1–5 in Sect. 22.1 provide an overview on optical studies of lattice modes. In this section, we discuss the polariton dispersion relations which describe the interaction of light with the electric dipole moment associated with infrared absorption, and the LO–TO splitting of the normal mode vibration of the atoms in the solid arising from these dispersion relations, where the atoms are perturbed by the strong electric fields generated by the neighboring species, as would occur in an ionic crystalline material. In Sect. 22.2, we discussed the limit of small local electric field whereas, here, we discuss the case of strong local electric field. We start by considering the polarizability of the ions.

Consider the equation of motion of an ion in a solid using the normal mode coordinate $\mathbf{r}$, so that harmonic motion yields

$$m\ddot{\mathbf{r}} = -\kappa\mathbf{r} + e\mathbf{E} = -m\omega^2\mathbf{r} \tag{22.8}$$

where

$$\mathbf{E} = \mathbf{E}_0 e^{-i\omega t} \tag{22.9}$$

and $-\kappa\mathbf{r}$ represents a lattice restoring force while $e\mathbf{E}$ is the force due to the actual electric field $\mathbf{E}$ at an ion site. Maxwell's equations give us

$$\nabla\times\ \mathbf{H} = \frac{1}{c}\dot{\mathbf{D}} = \frac{1}{c}(\dot{\mathbf{E}} + 4\pi\dot{\mathbf{P}}) = -\frac{i\omega}{c}(\mathbf{E} + 4\pi\mathbf{P}) \tag{22.10}$$

$$\nabla\times\ \mathbf{E} = -\frac{1}{c}\dot{\mathbf{H}} = \frac{i\omega}{c}\mathbf{H}. \tag{22.11}$$

We also have a constitutive equation which tells us that the total polarization arises from an ionic contribution $N'e\mathbf{r}$ where N' is the number of optical modes per unit volume and from an electronic contribution $n\alpha\mathbf{E}$, where n is the electron concentration and α is the electronic polarizability:

$$\mathbf{P} = N'e\mathbf{r} + n\alpha\mathbf{E}. \tag{22.12}$$

Equations (22.8), (22.10), (22.11) and (22.12) represent 4 equations in the 4 variables **r**, **H**, **E**, and **P**.

In writing (22.12) for the polarization vector **P**, we have considered two degrees of freedom: namely that of the ion system and of the electron system. We further assume that these polarizations are accomplished independently. In formulating this calculation, the electric field in all equations is the applied electric field, since it is assumed that the lattice polarization effects are weak. In more sophisticated treatments, we must consider the effect of local field corrections when the dielectric function is large, as occurs for example in the case of ferroelectrics materials.

We now seek plane wave solutions for transverse wave propagation: (**E**, **H**) in the xy plane and perpendicular to the Poynting vector, $\mathscr{S} = [c/(8\pi)]\mathrm{Re}(\mathbf{E}^* \times \mathbf{H})$, and the Poynting vector is taken along the z direction

$$E_x = E_x^0 e^{-i(\omega t - Kz)} \tag{22.13}$$

$$H_y = H_y^0 e^{-i(\omega t - Kz)} \tag{22.14}$$

$$P_x = P_x^0 e^{-i(\omega t - Kz)} \tag{22.15}$$

$$r_x = r_x^0 e^{-i(\omega t - Kz)}. \tag{22.16}$$

Here, K is the wave vector for the light, $K = 2\pi/\lambda$, and l is the wavelength. Using values for λ typical for the wavelength for lattice modes in NaCl, we have $\lambda \sim 60\,\mu$m and $K \sim 10^3\mathrm{cm}^{-1}$. Substitution of the harmonic solutions in (22.13)–(22.16) into the 4 equations given above (22.8), (22.10), (22.11) and (22.12) for the four variables **r**, **H**, **E**, and **P** yields:

$$iKH_y - \frac{i\omega}{c}E_x - \frac{4\pi i\omega}{c}P_x = 0 \tag{22.17}$$

$$-iKE_x + \frac{i\omega}{c}H_y = 0 \tag{22.18}$$

$$-\omega^2 r_x + \frac{\kappa}{m}r_x - \frac{e}{m}E_x = 0 \tag{22.19}$$

$$P_x - N'er_x - n\alpha E_x = 0. \tag{22.20}$$

Equations (22.17)–(22.20) form 4 equations in 4 unknowns. To have a non-trivial solution to (22.17)–(22.20), the coefficient determinant given in (22.21) must vanish. We arrange the four coefficient determinant following the order of the four variables in (22.13)–(22.16): ($E_x \quad H_y \quad P_x \quad r_x$):

$$\begin{vmatrix} \omega/c & -K & 4\pi\omega/c & 0 \\ K & -\omega/c & 0 & 0 \\ e/m & 0 & 0 & \omega^2 - \kappa/m \\ -n\alpha & 0 & 1 & -N'e \end{vmatrix} = 0. \tag{22.21}$$

Multiplying out the determinant in (22.21), we get a quadratic equation in ω^2

$$\omega^4[1 + 4\pi n\alpha] - \omega^2\left[c^2K^2 + \frac{\kappa}{m} + \frac{4\pi N'e^2}{m} + \frac{4\pi n\alpha\kappa}{m}\right] + K^2c^2\frac{\kappa}{m} = 0. \tag{22.22}$$

Equation (22.22) is more conveniently written in terms of the parameters ε_∞, ε_0, and ω_T where these parameters are defined in (22.23), (22.25) and (22.27) given below:

1. The high frequency dielectric constant ε_∞ is written as $\varepsilon_\infty = 1 + 4\pi P_\infty/E$, and is the parameter normally used to express the optical core dielectric constant when discussing electronic processes studied by optical techniques. From the equation of motion (22.8), we conclude that at high frequencies ($\omega \gg \omega_T$ and we show below that ω_T is identified with the transverse optical frequency), and the ionic displacement is small, for otherwise the acceleration would tend to ∞. Thus as the frequency increases, the ions contribute less and less to the polarization vector. We thus have the result $P_\infty = n\alpha E$, so that the electronic contribution dominates and

$$\varepsilon_\infty = 1 + 4\pi n\alpha. \tag{22.23}$$

2. The low frequency ($\omega \ll \omega_T$) dielectric constant is denoted by ε_0. At $\omega = 0$ the equation of motion (22.8) yields $\mathbf{r} = e\mathbf{E}/\kappa$ so that the polarization vector at zero frequency is

$$\mathbf{P}_0 = \left[\frac{N'e^2}{\kappa} + n\alpha\right]\mathbf{E}; \tag{22.24}$$

and

$$\varepsilon_0 = 1 + 4\pi\left[\frac{N'e^2}{\kappa} + n\alpha\right]. \tag{22.25}$$

At a general frequency ω, we must from (22.8) and (22.12) write $\varepsilon(\omega)$ as

$$\varepsilon(\omega) = 1 + 4\pi\left[\frac{N'e^2}{\kappa - m\omega^2} + n\alpha\right]. \tag{22.26}$$

3. Finally, we introduce a frequency ω_T defined as

$$\omega_T^2 \equiv \frac{\kappa}{m} \tag{22.27}$$

which depends only on the restoring forces and not on the externally applied field. Of course, these restoring forces will depend on internal fields, since electromagnetic interactions are responsible for producing these lattice vibrations in the first place. We will later identify ω_T with ω_t, the transverse optical phonon frequency. Substitution of ε_∞, ε_0, and ω_T into (22.22) yields the polariton dispersion relation, which is a quadratic equation in the variable ω^2 that can be written in closed form as

$$\omega^4\varepsilon_\infty - \omega^2[c^2K^2 + \omega_T^2\varepsilon_0] + \omega_T^2 c^2 K^2 = 0. \tag{22.28}$$

Equation (22.28) has two solutions

$$\omega^2 = \frac{1}{2\varepsilon_\infty}(\omega_T^2\varepsilon_0 + c^2K^2) \pm \left(\frac{1}{4\varepsilon_\infty^2}(\omega_T^2\varepsilon_0 + c^2K^2)^2 - \omega_T^2 K^2 \frac{c^2}{\varepsilon_\infty}\right)^{1/2} \tag{22.29}$$

which are shown graphically in Fig. 22.3. Each solution in (22.29) is twofold degenerate, since **E** can be chosen in any arbitrary direction perpendicular to the propagation vector. The coupled excitation of the transverse optical phonon to the electromagnetic radiation is called the **polariton** and the picture in Fig. 22.3 is identified with the polariton dispersion relation. There is also a longitudinal direction for both the light and the lattice vibrations; for this case there is no coupling between the light and the phonons and the frequency is the same as in the absence of light. We therefore obtain a total of 6 modes for the 3 coupled optical lattice modes and the three electromagnetic modes (two transverse modes representing photons and one longitudinal mode). It is of interest to examine the solutions of (22.29) for small and large K vectors, where we must remember that the scale of the K-vectors for light is a scale of $10^3 - 10^4\,\text{cm}^{-1}$ rather than 10^8cm^{-1} which corresponds to the Brillouin zone dimensions. Thus the whole picture shown in Fig. 22.3 occurs essentially at $\mathbf{q} = 0$ when plotting phonon dispersion relations $\omega_q(\mathbf{q})$ for wave vectors $\mathbf{q}$ in the Brillouin zone.

At small K vectors ($|K| \ll 10^4\,\text{cm}^{-1}$), we have two solutions to (22.29) for the polariton. The positive solution is given by

$$\omega^2 = \frac{1}{\varepsilon_\infty}(\omega_T^2\varepsilon_0 + c^2K^2) \tag{22.30}$$

defining the frequency dispersion for the polariton which is denoted by ω_T in (22.30). From Fig. 22.3 and $K = 0$, we also have the second solution

$$\omega_T^2\varepsilon_0/\varepsilon_\infty \equiv \omega_L^2, \tag{22.31}$$

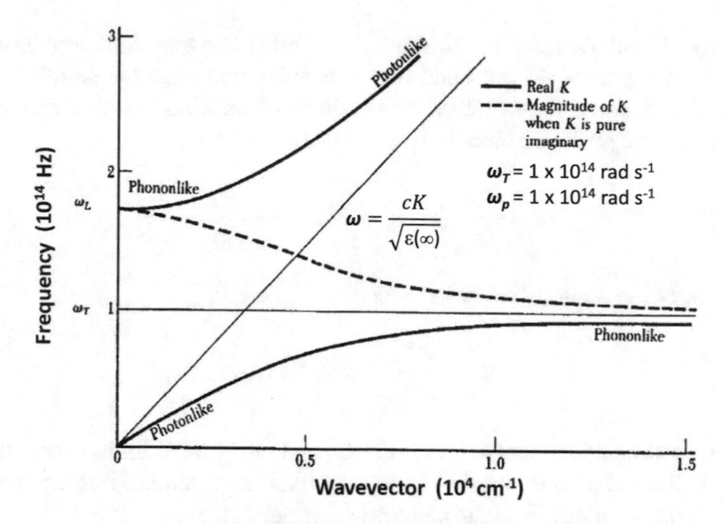

Fig. 22.3 Polariton dispersion relations showing the coupling between the transverse lattice vibrations and the electromagnetic radiation. In this figure, we clearly see the splitting of the LO and TO modes ($\omega_L - \omega_T$) induced by the ionicity of the solid which couples the two branch modes

thus defining the frequency ω_L which is identified with the phonon-like mode at large wavevector K. In writing this solution we neglected the term c^2K^2 as $K \to 0$. This solution corresponds to the phonon branch with finite frequency at $K = 0$ and hence is an optical phonon mode. We will call this frequency ω_L and later we will identify ω_L with the longitudinal optical phonon mode frequency, ω_ℓ. We shall see that the above definition is equivalent to taking the frequency ω_ℓ as the frequency where the real part of the dielectric function vanishes $\varepsilon_1(\omega_\ell) \equiv 0$. We also remember, that the longitudinal optical (LO) phonon does not interact with the electromagnetic field. For a phonon-electromagnetic interaction, we require that the electric field be transverse to the direction of propagation.

With regard to the *negative solution* of (22.29), we expand the square root term in (22.29) to obtain:

$$\omega^2 = \frac{\omega_T^2 K^2 c^2}{\omega_T^2 \varepsilon_0 + c^2 K^2} \tag{22.32}$$

or

$$\omega^2 \simeq \frac{c^2 K^2}{\varepsilon_0} \tag{22.33}$$

yielding the photon-like mode with a linear K dependence

$$\omega = \frac{cK}{\sqrt{\varepsilon_0}} \text{ for } \omega \ll \omega_T. \tag{22.34}$$

At large K values ($|K| \sim 10^5\text{cm}^{-1}$), we solve the quadratic equation given by (22.29) in the large K limit and obtain positive and negative solutions. Using a binomial expansion for (22.29), we obtain the following positive and negative solutions. For the positive solution, i.e., K is large, we obtain

$$\omega^2 \simeq \frac{1}{\varepsilon_\infty}(\omega_T^2\varepsilon_0 + c^2K^2) = \frac{c^2K^2}{\varepsilon_\infty}. \tag{22.35}$$

This is clearly the photon-like mode, since

$$\omega = \frac{cK}{\sqrt{\varepsilon_\infty}} \quad \text{for } \omega \gg \omega_T. \tag{22.36}$$

This result is almost identical to (22.34) obtained in the low K limit, except that now we have ε_∞ instead of ε_0 for the dielectric constant. Correspondingly, the phonon-like mode for large K arises from the negative solution:

$$\omega^2 \simeq \frac{\omega_T^2K^2c^2}{\omega_T^2\varepsilon_0 + c^2K^2} \simeq \omega_T^2. \tag{22.37}$$

We have thus introduced two frequencies: ω_T and ω_L and from the definition of ω_L we obtain the Lyddane–Sachs–Teller relation:

$$\frac{\omega_\ell^2}{\omega_T^2} = \frac{\varepsilon_0}{\varepsilon_\infty}. \tag{22.38}$$

Now, ω_T and ω_L have well-defined meanings with regard to the dielectric function as can be seen in Fig. 22.3. From (22.12), we have for the polarization due to ions and electrons:

$$\mathbf{P} = N'e\mathbf{r} + n\alpha\mathbf{E} \tag{22.39}$$

while the equation of motion, (22.8), ($F = ma$) gives

$$-m\omega^2\mathbf{r} = -\kappa\mathbf{r} + e\mathbf{E} \tag{22.40}$$

yielding

$$\mathbf{r} = \frac{e\mathbf{E}}{\kappa - m\omega^2} = \frac{e\mathbf{E}/m}{\omega_T^2 - \omega^2} \tag{22.41}$$

so that

$$\frac{P}{E} = \frac{\varepsilon(\omega) - 1}{4\pi} = \frac{N'e^2/m}{\omega_T^2 - \omega^2} + \frac{\varepsilon_\infty - 1}{4\pi}, \tag{22.42}$$

since the electronic polarizability term is $n\alpha = (\varepsilon_\infty - 1)/4\pi$. We therefore obtain:

$$\varepsilon(\omega) = \varepsilon_\infty + \frac{4\pi N' e^2/m}{\omega_T^2 - \omega^2} \quad (22.43)$$

where ε_∞ represents the contribution from the electronic polarizability and the resonant term represents the lattice contribution. Neglecting damping, we have the result $|\varepsilon(\omega)| \to \infty$ as $\omega \to \omega_T$, where the transverse optical phonon frequency $\omega = \omega_T$ is interpreted as the frequency at which the dielectric function $\varepsilon(\omega)$ is resonant. The name *reststrahl frequency* denotes that frequency ω_T where light is maximally absorbed by the medium.

We would now like to get a more physical idea about ω_ℓ. So far ω_ℓ has been introduced as the phonon mode of the polariton curve in Fig. 22.3 near $k = 0$. From (22.43) we have the relation

$$\varepsilon_0 = \varepsilon_\infty + \frac{4\pi N' e^2}{m\omega_T^2} \quad (22.44)$$

where ε_0 is defined by $\varepsilon_0 \equiv \varepsilon(\omega = 0)$, so that

$$\frac{4\pi N' e^2}{m} = \omega_T^2(\varepsilon_0 - \varepsilon_\infty) \quad (22.45)$$

and $\omega_t = \omega_T$ is the frequency where $\varepsilon(\omega)$ is resonant. Thus from (22.1) and (22.43), we can write

$$\varepsilon(\omega) = \varepsilon_\infty + \frac{(\varepsilon_0 - \varepsilon_\infty)}{(1 - \omega^2/\omega_t^2)} = \varepsilon_\infty + \frac{(\varepsilon_0 - \varepsilon_\infty)}{(1 - \omega^2/\omega_T^2)} \quad (22.46)$$

so that $\omega_T = \omega_t$. We define ω_ℓ as the frequency at which the dielectric function vanishes $\varepsilon(\omega_\ell) \equiv 0$ so that setting $\varepsilon(\omega) = 0$ in (22.46) yields

$$\varepsilon_\infty = \frac{(\varepsilon_\infty - \varepsilon_0)}{(1 - \omega_\ell^2/\omega_t^2)} \quad (22.47)$$

or

$$\frac{\omega_\ell^2}{\omega_t^2} = \frac{\varepsilon_0}{\varepsilon_\infty}. \quad (22.48)$$

Thus, the frequency ω_ℓ, which yields a zero in the dielectric function, also satisfies the Lyddane–Sachs–Teller relation (22.48).

We illustrate the properties of ω_ℓ and ω_t in Fig. 22.4 where we see that the frequency dependence of the dielectric function $\varepsilon(\omega)$ has two special features:

- a zero of $\varepsilon(\omega)$ occurring at ω_ℓ
- an infinity or pole of $\varepsilon(\omega)$ occurring at ω_t.

For $\omega_t < \omega < \omega_\ell$, the dielectric function $\varepsilon(\omega)$ is negative, so that losses must occur and transmission is consequently poor. The frequency difference between the

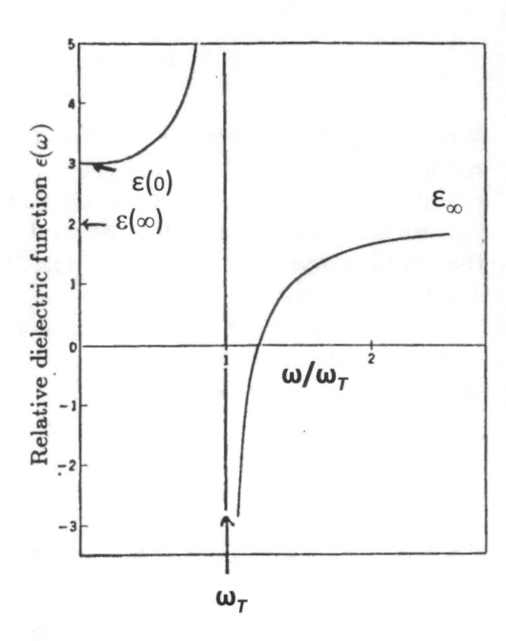

Fig. 22.4 The dielectric function $\varepsilon(\omega)$ plotted as a function of normalized frequency ω/ω_T. When damping is included, the real part of the dielectric function remains finite at ω_T

two characteristic frequencies ω_ℓ and ω_t depends on the ionicity of the crystal. Thus, predominantly covalent materials like InSb which have weak ionicity have a smaller $\omega_\ell - \omega_t$ splitting than alkali halide crystals which are highly ionic. For weakly polar materials like InSb, the treatment of the electric field given here is adequate. For highly polar materials, one must also consider the local fields, as distinct from the applied field. These local fields tend to increase the separation between ω_t and ω_ℓ, pulling ω_t to low frequencies. Since mechanically hard materials tend to have high Debye temperatures and high phonon frequencies, the passage of ω_t toward zero for ferroelectric materials (extremely high dielectric function and capable of spontaneous polarization) is referred to as the appearance of a "soft mode".

The Lyddane–Sachs–Teller relation is more general than the derivation given here would imply. This relation can be extended to cover anisotropic materials with any number of optical modes. In this context we can write the frequency dependence of the symmetrized dielectric tensor function associated with symmetry μ as

$$\varepsilon_\mu(\omega) = \varepsilon_\mu(\infty) + \sum_{j=1}^{p} \frac{f_{\mu,j}\omega_{T,j}^2}{\omega_{T,j}^2 - \omega^2 - i\gamma_j\omega} \tag{22.49}$$

where $f_{\mu,j}$ is the oscillator strength, γ_j is the damping of mode j, and p is the number of modes with symmetry μ. An example where this would apply is the case of tetragonal symmetry where μ could refer to the in-plane modes (E_u symmetry) or to the out-of-plane modes (A_{2u} symmetry). Figure 22.5 shows the measured reflectivity for the lattice modes of TeO_2 which has 4 formula units per unit cell (12 atoms/unit cell)

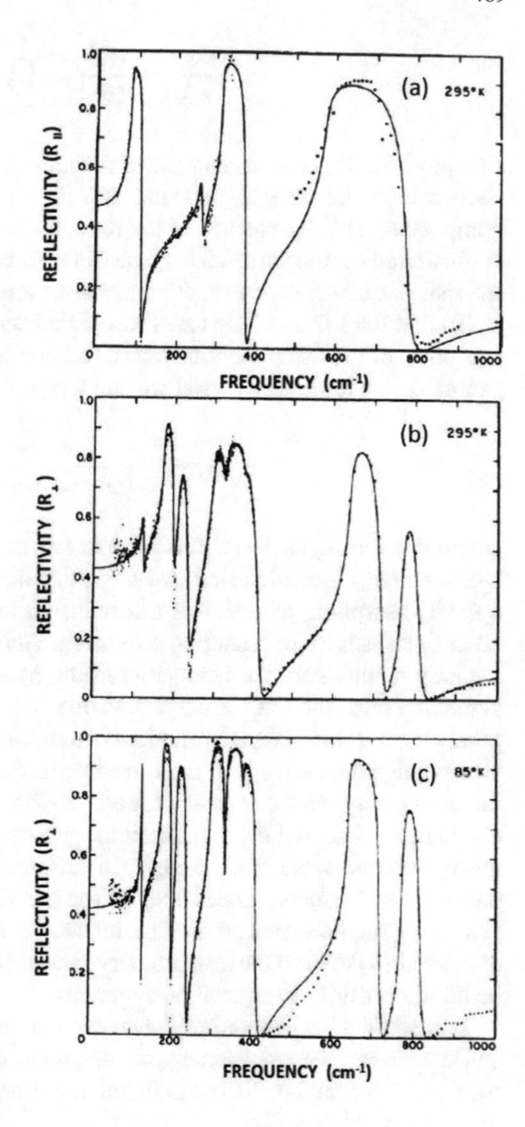

Fig. 22.5 Reflectivity in the material paratelluride, TeO_2, for (**a**) **E** parallel and (**b, c**) perpendicular to the tetragonal axis at 295 K (**b**). The corresponding data in **c** at approximately liquid nitrogen temperature ($T = 77$ K) show sharper singular behavior at a special frequency ω. The polarization $\mathbf{E} \parallel$ the tetragonal axis has only the A_{2u} optically allowed modes whereas for $\mathbf{E} \perp$ the tetragonal axis has only the E_{2u} optically allowed modes. The points are experimental and the solid line is a model based on (22.49) (After Korn, et al., Phys. Rev. B, **8**, 768 (1973).)

can be described by a model based on (22.49) for polarization of the electromagnetic field parallel and perpendicular to the tetragonal axis.

Setting the damping terms in (22.49) to zero, $\gamma_j = 0$, we obtain the result

$$\frac{\varepsilon(\omega)}{\varepsilon(\infty)} = \prod_{j=1}^{p} \left(\frac{\omega_{l,j}^2 - \omega^2}{\omega_{t,j}^2 - \omega^2} \right) \tag{22.50}$$

which leads to the generalized Lyddane–Sachs–Teller relation

$$\frac{\varepsilon_0}{\varepsilon_\infty} = \frac{\varepsilon(0)}{\varepsilon(\infty)} = \prod_{j=1}^{p}\left(\frac{\omega_{l,j}^2}{\omega_{t,j}^2}\right). \tag{22.51}$$

Equation (22.51) can be generalized for anisotropic crystals by writing (22.50) for each component, keeping in mind that the optical selection rules differ for each component. The dependence of the reflectivity on polarization and on temperature is illustrated for the tetragonal crystal TeO_2 in Fig. 22.5. For this material system we see that crystal orientation is dominant over temperature dependent effects.

To find the LO and TO modes associated with (22.49), we would look for zeros and poles of the dielectric function for a general direction of light propagation. For example, in a tetragonal crystal we can write

$$\varepsilon(\theta, \omega) = \frac{\varepsilon_\parallel(\omega)\varepsilon_\perp(\omega)}{\varepsilon_\parallel(\omega)\cos^2\theta + \varepsilon_\perp(\omega)\sin^2\theta} \tag{22.52}$$

where θ is the angle. The observation of LO and TO phonon frequencies by optical measurements is made using two basically different techniques. In one approach, optical absorption, reflection or transmission measurements are made, while in the other approach, light scattering measurements are made. These are often complementary methods for the following reason. Many important crystals have inversion symmetry (e.g., the NaCl structure). In this case, the phonon modes are purely odd or purely even. If the odd parity modes have dipole moments and couple directly to the electromagnetic fields, then these materials are infrared-active. On the other hand, the even parity modes are not infrared-active but instead may be Raman-active and can then be observed in a light scattering experiment. Thus, by doing both infrared absorption and Raman scattering measurements, we can study both even and odd parity optical phonon modes, except for the silent modes which because of other symmetry requirements are neither infrared nor Raman active. These concepts are discussed in detail in the group theory course. Spectroscopy measurements tend to be highly sensitive to crystalline symmetry

In modeling the phonon and free carrier contributions to the dielectric function, it can happen that these phenomena occur over a common frequency range. In this case, we write the complex dielectric function for an isotropic semiconductor as follows in analyzing optical data

$$\frac{\varepsilon_\mu(\omega)}{\varepsilon_\mu(\infty)} = \left(1 - \frac{\omega_p^2}{\omega(\omega + i\gamma_p)}\right) + \sum_{j=1}^{p}\frac{\omega_{L,j}^2 - \omega_{T,j}^2}{\omega_{T,j}^2 - \omega^2 - i\omega\gamma_j} \tag{22.53}$$

where the first and second terms are, respectively, the free carrier and the infrared-active phonon contributions to the dielectric function. In (22.53), ω_p is the screened electronic plasma frequency ($\omega_p^2 = 4\pi ne^2/m^*\varepsilon(\infty)$, and $\varepsilon(\infty)$ is the core dielectric constant used to approximate the higher frequency electronic polarizability). The phonon contribution to (22.53) depends on $\omega_{L,j}$ and $\omega_{T,j}$ which are the j-th longitudinal and transverse optic mode frequencies, while γ_j and γ_p are the phonon and plasma damping factors, respectively.

The model given by (22.53) can, for example, be used to model the optical properties of an anisotropic compound, such as La_2CuO_4 which becomes a high T_c superconductor upon addition of a small concentration of Sr. In this case it is important to obtain polarized reflectivity measurements on oriented single crystals, and to carry out the Kramers–Kronig analysis of the reflectivity data for each of the polarization components separately.

22.4 Light Scattering

Light scattering techniques provide an exceedingly useful tool to study fundamental excitations in solids, such as phonons, because light can be scattered from solids inelastically, whereby the incident and scattered photons have different frequencies. Inelastic light scattering became an important tool for the study of excitations in solids in the mid-1960's with the advent of laser light sources, because the inelastically scattered light is typically only $\sim 10^{-7}$ of the intensity of the incident light.

In the light scattering experiments shown schematically in Fig. 22.6, conservation of energy gives:

$$\omega = \omega_0 \pm \omega_q \tag{22.54}$$

and conservation of momentum gives:

$$\mathbf{K} = \mathbf{K}_0 \pm \mathbf{q} \tag{22.55}$$

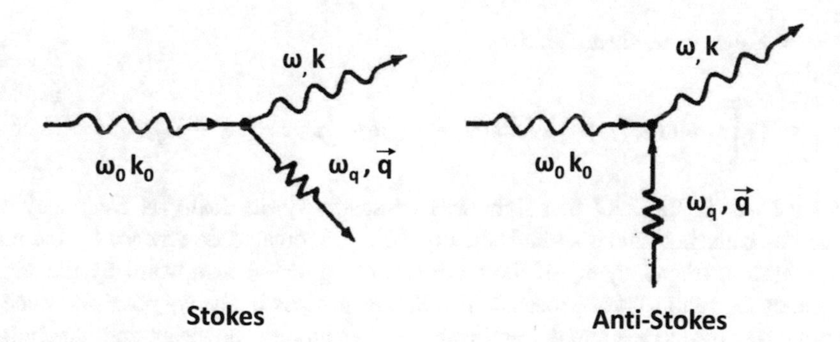

Fig. 22.6 Raman scattering of a photon showing both phonon emission (Stokes) and absorption (anti-Stokes) processes. The scattering process is called Brillouin scattering when an acoustic phonon is involved and polariton (Raman) scattering when an optical phonon is involved. Similar processes occur with magnons, plasmons or any other excitation of the solid depending on the crystal symmetry and also the dimensionality of the material under investigation

where the "0" subscript refers to the incident light, **K** refers to the wave vector of the light and "**q**" refers to the wave vector for the excitation in the solid (see Fig. 22.6). Since $K_0 = 2\pi/\lambda$ is very small compared with the Brillouin zone dimensions, measurement of the angular dependence of $\omega_q(\mathbf{q})$ can then be used to provide dispersion relations for the excitations near $\mathbf{q} = 0$. If $\omega_q \ll \omega_0$, then $|\mathbf{K}| \simeq |\mathbf{K}_0|$, and we have $|q| \simeq 2|\mathbf{K_0}| \sin(\theta/2)$ so that $|q_{max}| = 2K_0$.

If the excitation is an **acoustic** phonon, the inelastic light scattering process is called **Brillouin scattering**, while light scattering by **optical** phonons is called **Raman scattering**. Raman and Brillouin scattering also denote light scattering processes due to other elementary excitations in solids, occuring in the respective frequency ranges.

The light scattering process can be understood physically on the basis of classical electromagnetic theory. When an electric field **E** is applied to a solid, a polarization **P** results

$$\mathbf{P} = \overset{\leftrightarrow}{\alpha} \cdot \mathbf{E} \tag{22.56}$$

where $\overset{\leftrightarrow}{\alpha}$ is the polarizability tensor of the atom in the solid, indicating that the positive charge moves in one direction and the negative charge in the opposite direction under the influence of the applied field. In the light scattering experiments, the electric field is oscillating at an optical frequency ω_0

$$\mathbf{E} = \mathbf{E}_0 \sin \omega_0 t. \tag{22.57}$$

The lattice vibrations in the solid modulate the polarizability of the atoms themselves

$$\alpha = \alpha_0 + \alpha_1 \sin \omega_q t. \tag{22.58}$$

so that the polarization, which is induced by the applied electric field, is:

$$\begin{aligned} \mathbf{P} &= \mathbf{E}_0(\alpha_0 + \alpha_1 \sin \omega_q t) \sin \omega_0 t \\ &= \mathbf{E}_0 \left[\alpha_0 \sin(\omega_0 t) + \tfrac{1}{2}\alpha_1 \cos(\omega_0 - \omega_q)t - \tfrac{1}{2}\alpha_1 \cos(\omega_0 + \omega_q)t \right]. \end{aligned} \tag{22.59}$$

Thus we see in Fig. 22.7 that light will be scattered elastically at frequency ω_0 (Raleigh scattering) and also inelastically, being modulated downward by the natural vibration frequency ω_q of the atom (Stokes process) or upward by the same frequency ω_q (anti-Stokes process). The Stokes process is always possible, since a phonon can always be emitted. For the anti-Stokes process, however, enough photon energy needs to be available to provide a phonon to be absorbed.

The light scattering process can also be viewed from a quantum mechanical perspective. If the "system" is initially in a state E'', then light scattering can excite the "system" to a higher energy state E' shown in Fig. 22.8a by the absorption of an excitation energy $(E' - E'')$. Similarly, the "system" can initially be in a state E'

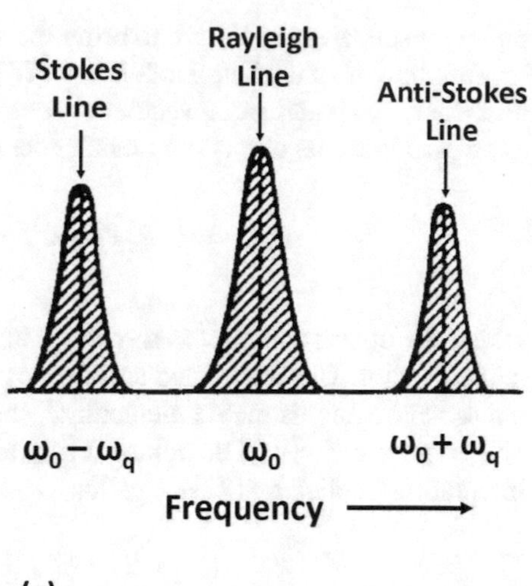

Fig. 22.7 Schematic diagram of light scattering spectrum showing the central unshifted Rayleigh line, the upshifted anti-Stokes line (emission process), and the downshifted Stokes line (absorption process). The ratio of the Stokes to anti-Stokes intensities can be used to estimate the temperature of the phonons in the material

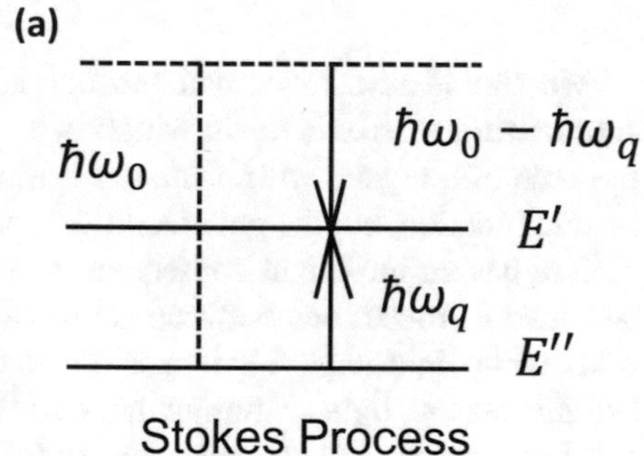

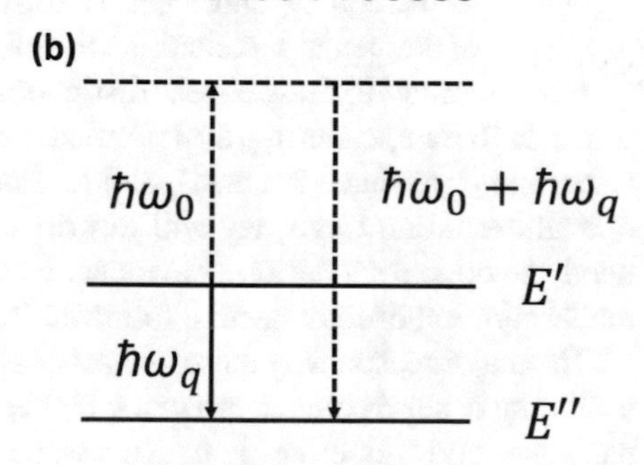

Fig. 22.8 Schematic energy level diagram for the **a** Stokes and **b** anti-Stokes processes. In this figure the solid lines denote real processes and the dashed lines denote virtual processes

and light scattering can serve to bring the system to a final state of lower energy E'' by emission of an excitation of energy $(E' - E'')$ as shown in Fig. 22.8b. The matrix element of the polarization vector between the initial and final states is written (when expressed in terms of quantum mechanics) as

$$\mathbf{P}_{nm} = \int \Psi_n^* \mathbf{P} \Psi_m d^3r = \mathbf{E} \cdot \int \Psi_n^* \overleftrightarrow{\alpha} \Psi_m d^3r \tag{22.60}$$

where the polarizability $\overleftrightarrow{\alpha}$ is a second rank symmetrical tensor and $\mathbf{P}$ is the polarization vector. The Stokes and anti-Stokes processes arise from consideration of the phase factors in this matrix element: Ψ_m has a phase factor $e^{-iE_m t/\hbar}$ while Ψ_n^* has a phase factor $e^{+iE_n t/\hbar}$. The polarizability tensor has a phase factor $e^{\pm i\omega_q t}$ so that the integration implied by (22.60) yields

$$E_m - E_n \pm \hbar\omega_q = 0. \tag{22.61}$$

We should remember that the optical absorption process is governed by the **momentum** matrix element which is a radial vector. Of particular significance is the case of a crystal with inversion symmetry whereby the momentum operator is an odd function, but the polarizability tensor is an even function. This characteristic feature has an important consequence; namely electronic absorption processes are sensitive to transitions between states of opposite parity (parity meaning even or odd), while light scattering is sensitive to transitions between states of similar parity. For this reason, light scattering and optical absorption are considered to be complementary spectroscopies, and together form basic tools for the study of the optical properties of elementary excitations in solids.

It is important to draw a clear distinction between Raman scattering and fluorescence. In Raman scattering, the intermediate states shown in Fig. 22.8a, b are "virtual" states and don't have to correspond to eigenstates of the physical "system"— any optical excitation frequency will in principle suffice. In fluorescence, on the other hand, the optical excitation state must be a real state of the system, and in this case a real absorption of light occurs, followed by a real emission at a different frequency.

The major reason why these two processes are sometimes confused is that Raman scattering in solids often has a much higher intensity when $\hbar\omega_0$ is equal to an energy band gap and this effect is called resonant Raman scattering. In such cases, the fluorescent emission differs from the Raman process because fluorescent phenomena take a finite time to occur.

Typical Raman traces are shown in Fig. 22.9 for several III-V compound semiconductors. The laser wavelength is 1.06 μm (Nd:YAG laser) which is a photon energy below the band gap for each material. The scattered light is collected at 90° with respect to the incident light and both the LO and TO phonon modes at $\mathbf{q} = 0$ are observed. For the case of the group IV semiconductors there is no LO–TO splitting and only a single optical Raman-allowed mode is observed (at 519 cm^{-1} for Si). What is measured in Fig. 22.9 is the frequency shift between the incident and scattered light

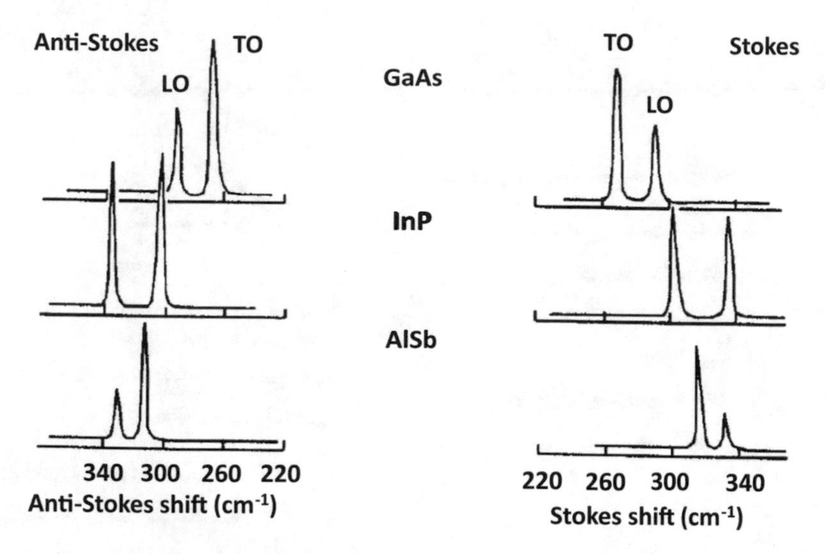

Fig. 22.9 Raman spectra of three zinc-blende-type semiconductors showing the TO and LO phonons in both Stokes and anti-Stokes scattering

beams. For the range of phonon wave vectors where Raman scattering can be carried out, this technique is the most accurate method available for the measurement of the dispersion relations near the Brillouin zone center.

By doing the Raman scattering experiment with polarized light, it is possible to get information on the symmetry of the lattice vibrations by monitoring the polarization of both the incident and scattered radiation. This approach is important in the identification of phonon frequencies with specific lattice normal modes of the material.

The inelastic neutron scattering technique, though less accurate than Raman scattering, has the advantage of providing information about phonons throughout the Brillouin zone. By using neutrons of low energy (thermal neutrons), it is possible to make the neutron wavelengths comparable to the lattice dimensions, in which case the inelastic scattering by a lattice vibration can cause a large momentum transfer to the neutron.

22.5 Feynman Diagrams for Light Scattering

Feynman diagrams are useful for keeping track of various processes that may occur in an inelastic scattering process that absorbs or creates an excitation. The basic notation used in drawing Feynman diagrams consists of propagators, such as electrons, phonons, or photons and vertices where interactions occur, as shown in Fig. 22.10g.

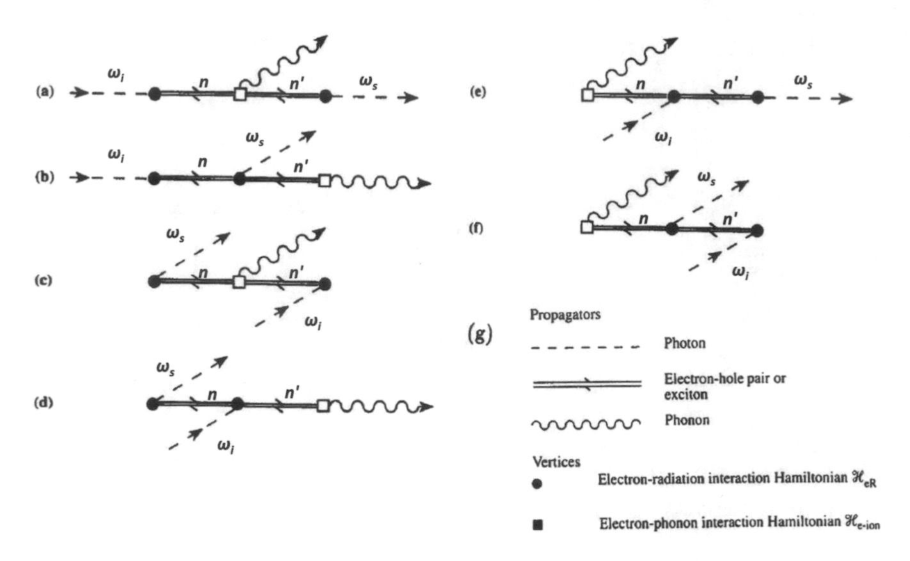

Fig. 22.10 Feynman diagrams for the six scattering processes that contribute to one-phonon (Stokes) Raman scattering. (Taken from Peter and Cardona 2010. Fundamentals of semiconductors: physics and materials properties. Springer Science & Business Media.) **g** Symbols used in drawing Feynman diagrams to represent Raman scattering

The rules in drawing Feynman diagrams are:

- Excitations such as photon, phonons and electron-hole pairs in Raman scattering are represented by lines (or propagators) as shown in Fig. 22.10g. These propagators can be labeled with properties of the excitations, such as their wavevectors, frequencies and polarizations.
- The interaction between two excitations is represented by an intersection of their propagators. This intersection is known as a *vertex* and is sometimes highlighted by a symbol such as a filled circle or empty rectangle.
- Propagators are drawn with an arrow to indicate whether they are created or annihilated in an interaction. Arrows pointing towards a vertex represent excitations which are annihilated. Those pointing away from the vertex are created.
- When several interactions are involved they are always assumed to proceed sequentially from the left to the right as a function of time.
- Once a diagram has been drawn for a certain process, other possible processes are derived by permuting the time order in which the vertices occur in this diagram.

The basic diagram for the Raman process is given in Fig. 22.10a taken from the Yu and Cardona book on "Fundamentals of Semiconductors." The other permutations of (a) obtained by different orders of the vertices are given in Fig. 22.10b–f. We then use the Fermi Golden rule for each diagram, multiplying the contributions from each vertex. For example, the first vertex in Fig. 22.10a contributes a term to the scattering probability per unit time of the form

$$\frac{\langle n|\mathcal{H}_{eR}(\omega_i)|i\rangle}{[\hbar\omega_i - (E_n - E_i)]}$$

where the sign (+) corresponds to absorption and (−) to emission and $\mathcal{H}_{eR}(\omega_i)$ denotes the interaction between the electron (e) and the electromagnetic radiation field (R). The interaction for the second vertex $\mathcal{H}_{\text{e-ion}}(\omega_i)$ between the electron and the lattice vibrations of the ion (or the electron-phonon interaction) and the corresponding energy denominator is

$$\hbar\omega_i - (E_n - E_i) - \hbar\omega_q - (E_{n'} - E_n) = [\hbar\omega_i - \hbar\omega_q - (E_{n'} - E_i)]$$

and for the third vertex the denominator becomes $[\hbar\omega_i - \hbar\omega_q - \hbar\omega_s - (E_{n'} - E_i)]$ but since the initial and final electron energies are the same, energy conservation requires $\delta(\hbar\omega_i - \hbar\omega_q - \hbar\omega_s)$ to yield the probability per unit time for Raman scattering for diagram (a):

$$P_{\text{ph}}(\omega_s) = \left(\frac{2\pi}{\hbar}\right)\left|\sum_{n,n'} \frac{\langle i|\mathcal{H}_{eR}(\omega_s)|n'\rangle\langle n'|\mathcal{H}_{\text{e-ion}}|n\rangle\langle n|\mathcal{H}_{eR}(\omega_i)|i\rangle}{[\hbar\omega_i - (E_n - E_i)][\hbar\omega_i - \hbar\omega_q - (E_{n'} - E_i)]}\right|^2 \times \delta(\hbar\omega_i - \hbar\omega_q - \hbar\omega_s). \tag{22.62}$$

Then summing over the other 5 diagrams yields the result

$$\begin{aligned} P_{\text{ph}}(\omega_s) = \left(\frac{2\pi}{\hbar}\right)\Bigg| \sum_{n,n'} & \frac{\langle i|\mathcal{H}_{eR}(\omega_i)|n\rangle\langle n|\mathcal{H}_{\text{e-ion}}|n'\rangle\langle n'|\mathcal{H}_{eR}(\omega_s)|i\rangle}{[\hbar\omega_i - (E_n - E_i)][\hbar\omega_i - \hbar\omega_q - (E_{n'} - E_i)]} \\ &+ \frac{\langle i|\mathcal{H}_{eR}(\omega_i)|n\rangle\langle n|\mathcal{H}_{eR}(\omega_s)|n'\rangle\langle n'|\mathcal{H}_{\text{e-ion}}|i\rangle}{[\hbar\omega_i - (E_n - E_i)][\hbar\omega_i - \hbar\omega_s - (E_{n'} - E_i)]} \\ &+ \frac{\langle i|\mathcal{H}_{eR}(\omega_s)|n\rangle\langle n|\mathcal{H}_{\text{e-ion}}|n'\rangle\langle n'|\mathcal{H}_{eR}(\omega_i)|i\rangle}{[-\hbar\omega_s - (E_n - E_i)][-\hbar\omega_s - \hbar\omega_q - (E_{n'} - E_i)]} \\ &+ \frac{\langle i|\mathcal{H}_{eR}(\omega_s)|n\rangle\langle n|\mathcal{H}_{eR}(\omega_i)|n'\rangle\langle n|\mathcal{H}_{\text{e-ion}}|n'\rangle}{[-\hbar\omega_s - (E_n - E_i)][-\hbar\omega_s + \hbar\omega_i - (E_{n'} - E_i)]} \\ &+ \frac{\langle i|\mathcal{H}_{\text{e-ion}}|n\rangle\langle n|\mathcal{H}_{eR}(\omega_i)|n'\rangle\langle n'|\mathcal{H}_{eR}(\omega_s)|i\rangle}{[-\hbar\omega_q - (E_n - E_i)][-\hbar\omega_q + \hbar\omega_i - (E_{n'} - E_i)]} \\ &+ \frac{\langle i|\mathcal{H}_{\text{e-ion}}|n\rangle\langle n|\mathcal{H}_{eR}(\omega_s)|n'\rangle\langle n'|\mathcal{H}_{eR}(\omega_i)|i\rangle}{[-\hbar\omega_q - (E_n - E_i)][-\hbar\omega_q - \hbar\omega_s - (E_{n'} - E_i)]}\Bigg|^2 \\ &\times \delta(\hbar\omega_i - \hbar\omega_s - \hbar\omega_q). \end{aligned} \tag{22.63}$$

Although (22.63) is not generally used to calculate scattering intensities directly, Feynman diagrams similar to those in Fig. 22.10 are widely used in physics.

22.6 Raman Spectra in Quantum Wells and Superlattices

Raman spectroscopy has also been used to study quantum well and superlattice phenomena. One important example is the use of Raman spectroscopy to elucidate zone folding phenomena in the phonon branches of a superlattice of quantum wells. Since the Raman effect is highly sensitive to phonon frequencies, this technique can be used to characterize quantum wells and superlattices with regard to the composition of an alloy constituent (e.g., the composition x of an alloy such as Si_xGe_{1-x}). The Raman effect can then be used to determine the amount of strain in each constituent from measurement of the phonon frequencies.

Zone folding effects in the phonon dispersion relations are demonstrated in a superlattice of [GaAs (13.6Å)/AlAs (11.4Å)] $\times 1720$ periods. The observed Raman spectra are shown in Fig. 22.11a, b, demonstrating the zone folding of the LA branch. The difference in the force constants between the GaAs and AlAs constituents causes splittings of the zone-folded phonon branch, as shown in Fig. 22.11c. The peaks in the Raman spectrum at $\sim 64^{-1}$ and $\sim 66\,\text{cm}^{-1}$ are identified and labeled with the zone folded modes of the LA branch with symmetries $A_1^{(1)}$ and $B_2^{(1)}$, consistent with the polarization of the incident and scattered photons. At higher frequencies the Raman spectrum of Fig. 22.11a shows additional structure related to the zone folded LO phonon branch. Here we note that the normally three-fold levels of T symmetry of the cubic crystal are split into E and B_2 symmetries in the superlattice because of its lower tetragonal symmetry. The two-fold level of E symmetry can be further split by the LO–TO splitting which occurs in ionic solids.

As another example, Raman spectroscopy can be used as a compositional characterization technique to confirm the chemical composition of a semiconductor alloy. This characterization is based on the identification of the Raman-active modes and the measurement of their frequency shifts and their relative intensities. The strain induced by the lattice mismatch at the interface between $Si_{0.5}Ge_{0.5}$ and a GaAs (110) surface is responsible for the dependence of the frequency shifts of the Ge-Ge, Si-Si and Si-Ge phonon lines on the thickness of the quantum wells in the spectra shown in Fig. 22.12 for $Si_{0.5}Ge_{0.5}$ layers of various thicknesses on a GaAs (110) surface. Since phonon frequencies depend on $(K/M)^{1/2}$ (where K represents the force constant and M is the ion mass) the mode frequencies of the Ge–Ge, Ge–Si and Si–Si optical mode vibrations are very different, as seen in Fig. 22.12. Therefore the amount of interface strain can be sensitively monitored by Raman scattering. Note the disappearance of the GaAs Raman lines (associated with the substrate) as the thickness of the $Si_{0.5}Ge_{0.5}$ overlayer increases.

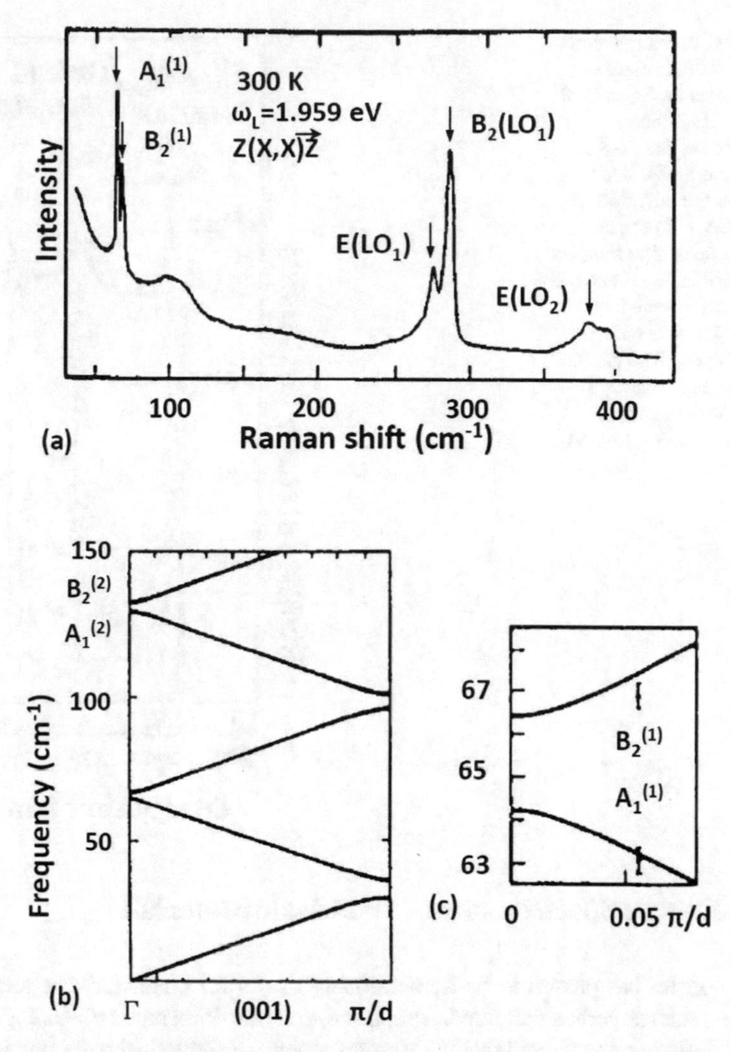

Fig. 22.11 **a** Raman spectra of a superlattice consisting of 1720 periods of a 13.6 Å GaAs quantum well and a 11.4 Å AlAs barrier. The polarizations for the incident and scattered light are arranged so that only longitudinal phonons are observed. **b** Dispersion of the LA phonons in the superlattice. **c** An expanded view of the 65 cm^{-1} region of the zone folded LA branch near $\mathbf{k} \approx 0$ (C. Colvard, T.A. Grant, M.V. Klein, R. Merlin, R. Fischer, H. Morkoc and A.C. Gossard, Phys. Rev. B, **31**, 2080 (1985).)

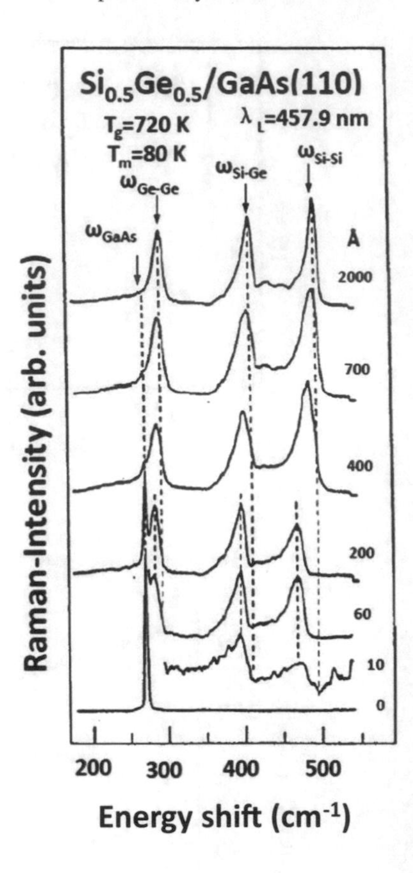

Fig. 22.12 Raman spectra for various thicknesses of $Si_{0.5}Ge_{0.5}$ on an GaAs (110) substrate. Here the dependence of the Si–Si, Ge–Ge, and Si–Ge bond lengths on the thickness of the $Si_{0.5}Ge_{0.5}$ layer can readily be seen. The samples were grown at 720 K and the measurements were made at 80 K using a laser with a wavelength of 457.9 nm (G. Abstreiter, H. Brugger, T. Wolf, H. Jorke and H.J. Herzog, *Phys. Rev. Lett.* **54**, 2441 (1985)

22.7 Raman Spectroscopy of Nanoscale Materials

Raman spectra has proven to be tremendously useful for characterizing nanoscale materials, such as carbon nanotubes, graphene, and transition metal dichalcogenides. Jorio et al. first demonstrated that the Raman spectra of individual carbon nanotubes could be measured when the incident laser energy is resonant with an optical transition in the nanotubes (Jorio et al. 2001). Figure 22.13 below shows typical Raman spectra of metallic and semiconducting carbon nanotubes. Both spectra exhibit a radial breathing mode (RBM) between 100–200 cm^{-1}, corresponding to the radial motion of atoms in the nanotube. The frequency of this mode is inversely proportional to the diameter (by the relation $\omega_{RBM} = 248\,\text{cm}^{-1}/d_t$), yielding a sensitive measure of the nanotube diameter. The G-band Raman mode corresponds to the optical phonons, which split into two peaks (G_+ and G_-) due to the curvature of the nanotube. The G-band of semiconducting nanotubes exhibits a very narrow linewidth, while the G-band of metallic nanotubes is broadened and downshifted due to the coupling of the phonon with the continuum of electronic states at the Fermi energy. The D-band corresponds to a phonon with finite momentum and is only observed in nanotubes

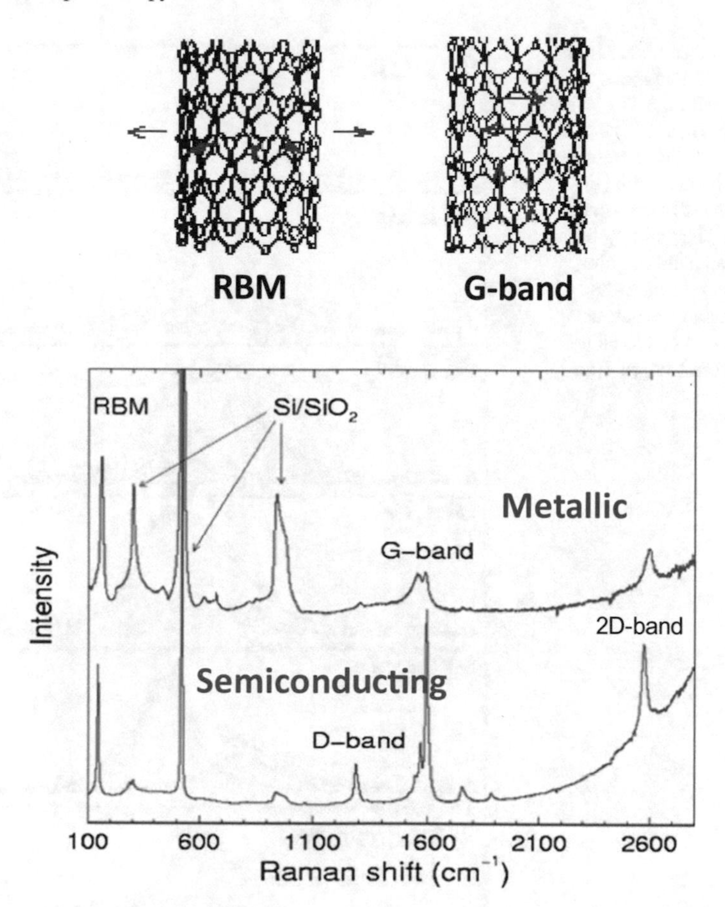

Fig. 22.13 Diagrams illustrating the radial breathing mode (RBM) and G-band optical phonon mode. Typical Raman spectra of metallic and semiconducting nanotubes (Taken from Jorio et al. (2001))

with a large amount disorder and defects, which relax the momentum conservation requirements. Lastly, the 2D-band, observed around 2600 cm^{-1}, is a two phonon process that does conserve momentum and is observed in all nanotubes.

Raman spectroscopy provides one of the easiest ways to determine the layer thickness of graphene. Figure 22.14 shows the 2D-band Raman spectra of various layer thicknesses of graphene together with highly oriented pyrolytic graphite (HOPG) and turbostratic graphite. Monolayer graphene exhibits a symmetric Lorent-zian peak, centered around 2700 cm^{-1} that can be fit with a single Lorentzian peak. Bilayer and trilayer graphene, however, require four and five Lorentzian peaks in order to fit this Raman feature. The complexity of these Raman modes arises from the electronic band structure in 2-layer and 3-layer graphene, which departs from the simple linear bands in monolayer graphene, and forms several sets of sub-bands.

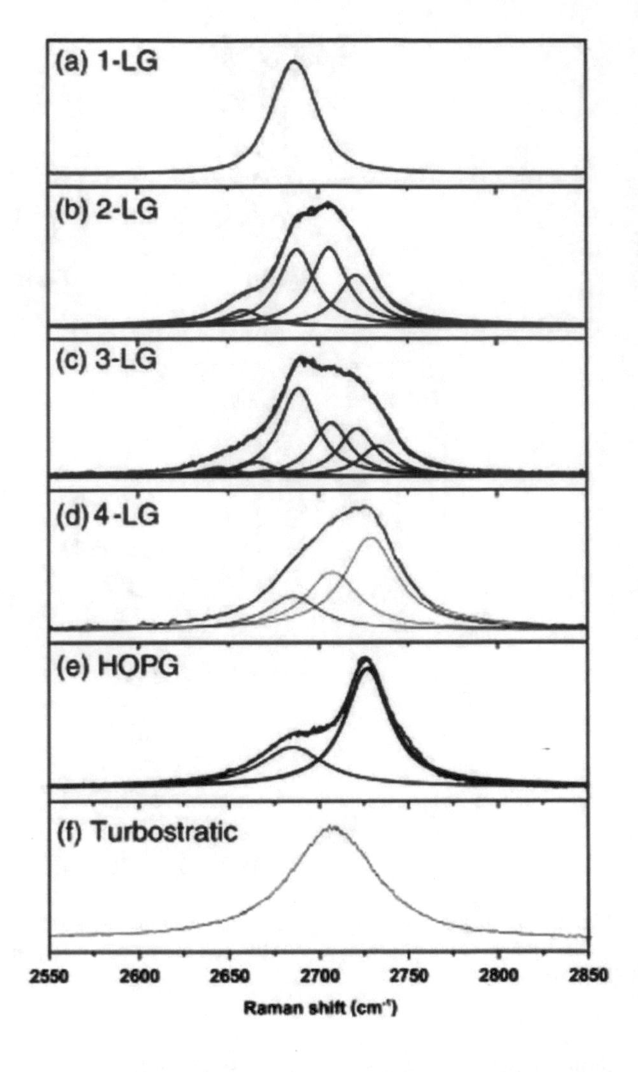

Fig. 22.14 The measured 2D-band Raman mode obtained with a 2.41 eV laser energy for (**a**) 1-, (**b**) 2-, (**c**) 3-, and (**d**) 4-layer graphene, plotted together with (**e**) HOPG and (**f**) turbostratic graphite. The splitting of the 2D Raman band in going from mono- to three-layer graphene and then closes in going from 4-LG to HOPG (Taken from Malard et al. (2009))

The Raman spectra of MoS_2 flakes exhibit two peaks corresponding to the E^1_{2g} (in-plane) and A_{1g} (out-of-plane) vibrational modes. The separation between these two peaks provides a good measure of the layer thickness of the material, varying from $19\,cm^{-1}$ for monolayer to 25 cm^{-1} for N-layer ($N > 6$) MoS_2. Figure 22.15 shows the Raman spectra of 1-, 2-, 3-, and 4-layer MoS_2 plotted together with bulk MoS_2. Here, a clear blueshift can be seen in the E^1_{2g} mode and a corresponding redshift in the A_{1g} mode for $N<4$. Other transition metal dichalcogenide materials exhibit similar features in their Raman spectra that can be related to the layer thickness.

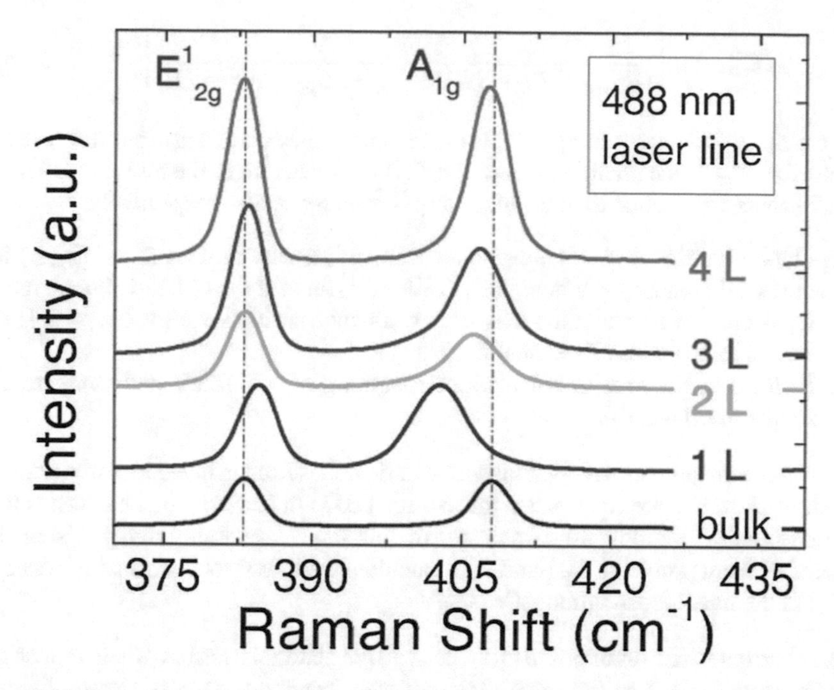

Fig. 22.15 Raman spectra of MoS_2 of various thicknesses. The left and right dashed lines indicate the positions of the E^1_{2g} and A_{1g} peaks in bulk MoS_2, respectively (Taken from Li et al. (2012))

Problems

22.1 How can the ratio of the Stokes to anti-Stokes intensities in the Raman scattering spectra of optical phonons be used to determine the lattice temperature of a 3D solid?

22.2 The ratio of the Stokes Raman intensity (phonon emission) to the anti-Stokes Raman intensity (phonon absorption) is given by the Maxwell-Boltzmann thermal factor:

$$\frac{I_{AS}}{I_S} = \exp(\frac{-E_{ph}}{k_B T}) \tag{22.64}$$

(a) Calculate the temperature of a carbon nanotube with a radial breathing mode frequency of 150 cm^{-1} exhibiting an anti-Stokes/Stokes ratio of 0.25.
(b) Calculate the temperature of the same nanotube exhibiting an anti-Stokes/Stokes ratio of 0.75.

22.3 The resonant Raman scattering process for a one-dimensional system, like a carbon nanotube, is given by the following equation:

$$I(E_{laser}) \propto \left| \frac{1}{(E_{laser} - E_{ii} - i\Gamma)(E_{laser} \pm E_{ph} - E_{ii} - i\Gamma)} \right|^2 \tag{22.65}$$

where E_{laser} is the laser energy, E_{ii} is the resonant electronic transition energy in the nanotube, E_{ph} is the phonon energy, and Γ is the linewidth of the resonance. The + and − signs correspond to anti-Stokes and Stokes processes, respectively.

(a) Plot both the Stokes and anti-Stokes Raman intensity profiles (I vs. E_{laser}) for the radial breathing phonon mode (180 cm^{-1}) and for the G-band phonon mode (1590 cm^{-1}) of a nanotube that has a resonant transition energy E_{33} = 2.41 eV. Assume a linewidth Γ = 5 meV.

(b) Perform another set of calculations assuming Γ = 17 meV and compare the results in (a) and (b).

22.4 At room temperature the Stokes and anti-Stokes Raman intensities of a phonon mode at 150 cm^{-1} are equal when taken with a 633 nm HeNe laser. This means that the nanotube is slightly off-resonance for this laser excitation energy. Using the equations from problems 24.1 and 24.2 calculate the true resonant transition energy (E_{ii}) of the nanotube assuming Γ = 8 meV.

22.5 Ferroelectrics (as opposed to dielectrics) are materials that have their atoms/molecules all polarized in the same direction even when no external electric field is applied. That is, a ferroelectric material has a built-in non-zero fixed polarization vector **P** that is independent of any external field. Some important semiconductors like gallium nitride are ferroelectric. In this problem you will explore the consequences of such a built-in polarization. Consider a circular disc of a ferroelectric material of thickness d that is much smaller than the radius R, as shown in the figure. The built-in polarization vector is given by $\mathbf{P} = P_0$Z (Fig. 22.16).

(a) Find the surface charge density due to the paired charges on the upper flat surface of the disc.

(b) Find the surface charge density due to the paired charges on the lower flat surface of the disc.

(c) Find the electric field (magnitude and direction) inside the ferroelectric disc. Hint: Use your answers from parts (a) and (b).

(d) Find the D-field (magnitude and direction) inside the disc.

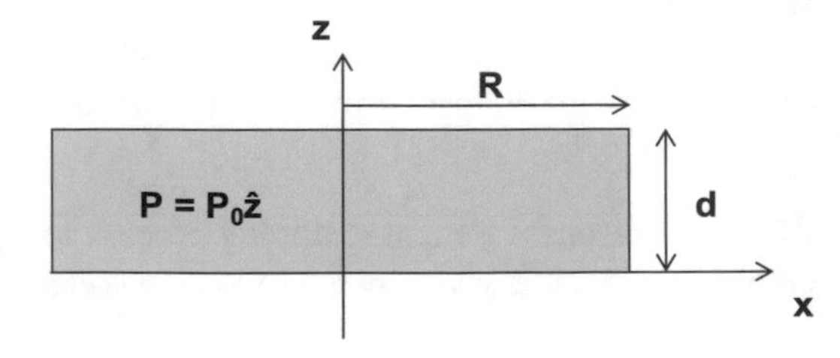

Fig. 22.16 A circular disc of a ferroelectric material of thickness d that is much smaller than the radius R

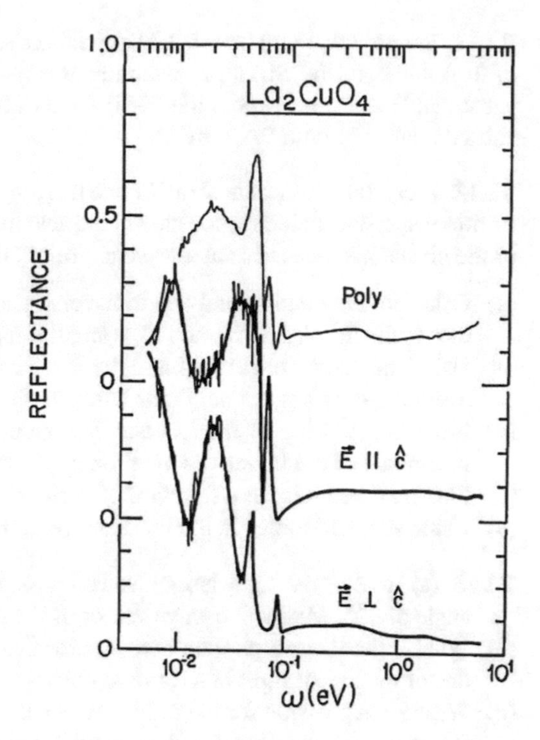

Fig. 22.17 Room temperature polarized reflectance spectra of single crystal La_2CuO_4 (Taken from Eklund et al., Journal of the Optical Society of America B, vol. 6, pp. 389 (1989).)

22.6 Derive the polariton dispersion relations for two-dimensional graphene, which has an optical phonon frequency of $1590\,\mathrm{cm}^{-1}$. Sketch these dispersion relations for graphene with carrier densities of (a) 10^{12} cm^{-2} and (b) 10^{13} cm^{-2}.

22.7 Based on the reflectance spectra of La_2CuO_4, shown in Fig. 22.17, estimate the phonon energies which dominate the dielectric function in the Lyddane–Sachs–Teller relation.

22.8 Draw the Feynman diagrams for two Raman processes in which a photon is absorbed before a phonon is emitted, and write the corresponding matrix elements for the three vertices.

22.9 The optical phonon of graphene composed of ^{12}C atoms is observed at $1590\,\mathrm{cm}^{-1}$. Estimate the phonon frequency for graphene composed of ^{13}C, based on the ratio of their atomic masses assuming that the force constants are the same in these two materials.

22.10 In addition to having a large Youngs modulus, carbon nanotubes can withstand a large amount of strain before breaking. The phonon modes of carbon nanotubes downshift with strain at a rate of $6.2\,\mathrm{cm}^{-1}$/% strain, due to the weakening of the C-C bond. Under uniaxial strain, downshifts of up to $85\,\mathrm{cm}^{-1}$ (from 1575 to 1490 cm^{-1}) have been observed. Estimate the amount of strain this corresponds to.

22.11 Suppose that you have a 2D superlattice sample of $Si_{1-x}Ge_x$/Si with a width of 10Å for both the $Si_{1-x}Ge_x$ quantum wells and the Si barriers on a $Si_{1-x}Ge_x$ substrate. Would you expect the Si–Si Raman frequency to be upshifted or down shifted relative to bulk Si? Why?

22.12 A crystal of a certain alkali halide has a static dielectric constant $\varepsilon(0) = 5.9$. Its non-dispersive dielectric constant in the near infrared is $\varepsilon = 2.25$. The reflectivity of the crystal becomes zero at a wavelength of 30.6 μm.

(a) Calculate the longitudinal and transverse phonon frequencies at $\mathbf{k}=0$. Express the results in eV, Kelvin, and s^{-1} (angular frequency).
(b) Using the results in (a) estimate the force constant κ for the TO phonon mode assuming only nearest neighbor interactions.
(c) From the splitting of the LO and TO phonon frequencies $(\omega_\ell - \omega_t)$, find the magnitude of the lattice polarization contribution to the dielectric constant.
(d) Plot the reflectivity as a function of wavelength.
(e) Using standard tables in the literature (e.g., Kittel), identify the alkali halide.

22.13 (a) If 2 ∼ eV light is incident on a semiconductor and is scattered by an angle of 60°, what is the wave vector of the phonon that is generated?
(b) What is the longest phonon wave vector that can be generated in this semiconductor by 2∼ eV light in a Raman process?
(c) Write an expression for the ratio between the Stokes (emission) and anti-Stokes (absorption) intensities for phonon (ω_q) emission and absorption by the Raman process at room temperature $T = 300$ K.
(d) Why is the Stokes process in part (c) more intense at room temperature for a 50 meV phonon?
(e) In an inelastic electron scattering process (electron energy loss spectroscopy), is the ratio of the Stokes to the anti-Stokes intensity the same or different for a 50 ∼ meV phonon with the same wave vector as in part (c)? Why?
(f) Would electron scattering or light scattering be more sensitive to probing surface oxide formation (of a few monolayers) on a semiconductor surface?

22.14 The Raman spectra of Graphite exhibits three main features: (i) the G-band feature ($\omega_G = 1580\,\text{cm}^{-1}$) that comes from the Γ point degenerate longitudinal optical (LO) and the in plane transverse optical (iTO) modes, (ii) the disorder-induced D-band ($\omega_D = 1200-1400\,\text{cm}^{-1}$) that comes from the LO phonon branch close to the K point, and (iii) its overtone, the G'-band ($\omega_{G'} \sim 2\omega_D = 2400-2800\,\text{cm}^{-1}$), which is a Raman process that involves two D-band phonons. The phonon diagram is shown in Fig. 22.18a.

(a) Explain why the G-band and G'-band appear in the Raman spectra of a perfect graphite crystal (HOPG - highly oriented pyrolytic graphite) while the D-band is observed only in defective graphitic materials.

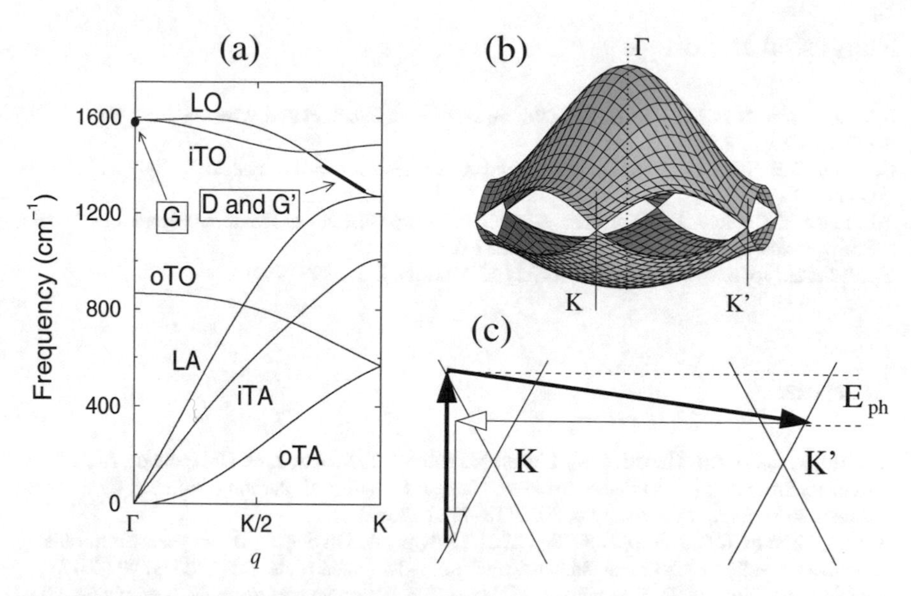

Fig. 22.18 **a** Phonon dispersion diagram of graphite. **b** Band structure of graphite. **c** Double resonance Raman effect of graphite

(b) The G-band scattering is a first-order Raman process, while the D-band and G'-band scattering are second-order Raman processes. For the D-band, one of the scattering processes is elastic due to interaction of the electron/hole with a lattice defect. Draw one Feynman diagram for the G, D and G' scattering processes (3 diagrams in all).

(c) (Optional) Graphite is a semi-metal since the valence and conduction band meet at the K point [see Fig. 22.18b]. Therefore, a resonance Raman effect is observed for electrons and phonons close to the K point. The reason why the second-order D and G' bands have enough Raman cross section to be visible in the Raman spectra with an intensity comparable to the first-order G-band is the resonance nature of their scattering processes, that involve two resonance processes, where not only the incident or scattered photons are associated with real electronic transitions, but also one of the intermediate scattering states, mediated by phonons (or by the defect in the case of the D-band), induces also an electronic transition between two real electronic states. One of these effects is illustrated in Fig. 22.18c. This process is called a double resonance process. Based on the various possible double resonance processes and on the phonon dispersion shown in Fig. 22.18a, explain why the D-band for defective graphite materials is composed of three peaks, where the intermediate frequency peak has twice the intensity of the lowest and highest frequency peaks (consider only Stokes scattering processes). Draw the Feynman diagrams for the possible Stokes processes of the D-band spectra.

Suggested Readings

N.W. Ashcroft, N.D. Mermin, Solid State Physics (Holt, Rinehart and Winston, New York, 1976) (2005). Chap. 27
C. Kittel, D.F. Holcomb, Introduction to solid state physics. Am. J. Phys. 35(6), 547–548 (1967). Chap. 10
Y.U. Peter, M. Cardona, Fundamentals of Semiconductors: Physics and Materials Properties (Media, Springer Science & Business, 2010), pp. 251–258
Eklund et al., Journal of the Optical Society of America B **6**, 389 (1989)

References

A. Jorio, R. Saito, J.H. Hafner, C.M. Lieber, M. Hunter, T. McClure, G. Dresselhaus, M.S. Dresselhaus, Structural (n, m) determination of isolated single-wall carbon nanotubes by resonant Raman scattering. Phys. Rev. Lett. **86**, 1118–1121 (2001)
H. Li, Q. Zhang, C.C.R. Yap, B.K. Tay, T.H.T. Edwin, A. Olivier, D. Baillargeat, From bulk to monolayer MoS_2: evolution of Raman scattering. Adv. Funct. Mater. **22**, 1385–1390 (2012)
L. Malard, M. Pimenta, G. Dresselhaus, M. Dresselhaus, Raman spectroscopy in graphene. Phys. Rep. **473**, 51–87 (2009)

Appendix A
Time–Independent Perturbation Theory

A.1 Introduction

Another review topic that we discuss here is time–independent perturbation theory because of its importance in experimental solid state physics in general and transport properties in particular.

There are many mathematical problems that occur in nature that cannot be solved exactly. It also happens frequently that a related problem can be solved exactly. Perturbation theory gives us a method for relating the problem that can be solved exactly to the one that cannot. This occurrence is more general than quantum mechanics –many problems in electromagnetic theory arc handled by the techniques of perturbation theory. In this book, we will however consider mostly about quantum mechanical systems.

Suppose that the Hamiltonian for our system can be written as

$$H = H_0 + H' \tag{A.1}$$

where H_0 is the part that we can solve exactly and H' is the part that we cannot solve exactly. Provided that $H' \ll H_0$ we can use perturbation theory; that is we first consider the solution of Schrödinger's equation for H_0 and then we calculate the effect of H'. For example, we can solve the hydrogen atom energy levels exactly, but when we apply an electric or a magnetic field we can no longer solve the problem exactly. For this reason, we treat the effect of the external fields as a perturbation, provided that the energy associated with the perturbing fields is small compared to the energy for the unperturbed case H_0:

$$H = \frac{p^2}{2m} - \frac{e^2}{r} - e\mathbf{r} \cdot \mathbf{E} = H_0 + H' \tag{A.2}$$

where

$$H_0 = \frac{p^2}{2m} - \frac{e^2}{r} \tag{A.3}$$

M. Dresselhaus et al., *Solid State Properties*, Graduate Texts in Physics,
https://doi.org/10.1007/978-3-662-55922-2

and

$$H' = -e\mathbf{r} \cdot \mathbf{E}. \tag{A.4}$$

As another illustration of an application of perturbation theory, consider a weak periodic potential in a solid. We can calculate the free electron energy levels (empty lattice) exactly. We would like to relate the weak potential situation to the empty lattice problem, and this can be done by considering the weak periodic potential itself as a perturbation.

A.2 Non-degenerate Perturbation Theory

In non-degenerate perturbation theory we want to solve Schrödinger's equation

$$H\psi_n = E_n\psi_n \tag{A.5}$$

where

$$H = H_0 + H' \tag{A.6}$$

and

$$H' \ll H_0. \tag{A.7}$$

It is then assumed that the solutions to the unperturbed problem

$$H_0\psi_n^0 = E_n^0\psi_n^0 \tag{A.8}$$

are known, in which we have labeled the unperturbed energy by E_n^0 and the unperturbed wave function by ψ_n^0. By non-degenerate we mean that there is only one eigenfunction ψ_n^0 associated with the eigenvalue E_n^0. We also assume that the wave functions ψ_n^0 form a complete orthonormal set

$$\int \psi_n^{*0}\psi_m^0 \mathbf{dr} = \langle\psi_n^0|\psi_m^0\rangle = \delta_{nm}. \tag{A.9}$$

in which the delta function δ_{nm} is defined to have the value 0 if $n \neq m$ and 1 if $n = m$. Since H' is small, the wave functions for the total problem ψ_n do not differ greatly from the wave functions ψ_n^0 for the unperturbed problem. So we can then expand $\psi_{n'}$ in terms of the complete set of ψ_n^0 functions

$$\psi_{n'} = \sum_n a_n\psi_n^0. \tag{A.10}$$

Such an expansion can always be made - no approximation but the set of Ψ_n^0 could be a very large number of functions. We then substitute this expansion of (A.10) into

Schrödinger's equation (A.5) to obtain

$$H\psi_{n'} = \sum_n a_n(H_0 + H')\psi_n^0 = \sum_n a_n(E_n^0 + H')\psi_n^0 = E_{n'}\sum_n a_n\psi_n^0 \quad \text{(A.11)}$$

therefore yielding

$$\sum_n a_n(E_{n'} - E_n^0)\psi_n^0 = \sum_n a_n H'\psi_n^0 \quad \text{(A.12)}$$

If we are looking for the perturbation to the level m, then we multiply on the left hand side of (A.12) by ψ_m^{0*} and integrate over all space. On the left hand side we get $\langle\psi_m^0|\psi_n^0\rangle = \delta_{mn}$ while on the right hand side we have the matrix element of the perturbation Hamiltonian taken between the n unperturbed states:

$$a_m(E_{n'} - E_m^0) = \sum_n a_n\langle\psi_m^0|H'|\psi_n^0\rangle \equiv \sum_n a_n H'_{mn} = \sum_{n\neq m} a_n H'_{mn} + a_m H'_{mm} \quad \text{(A.13)}$$

where we have written the indicated matrix element as H'_{mn}. Equation A.13 is an iterative equation on the a_n coefficients where each a_m coefficient is related to a complete set of a_n coefficients by the relation

$$a_m = \frac{1}{E_{n'} - E_m^0}\sum_n a_n\langle\psi_m^0|H'|\psi_n^0\rangle = \frac{1}{E_{n'} - E_m^0}\sum_n a_n H'_{mn} \quad \text{(A.14)}$$

in which the summation includes the $n = n'$ terms, where n includes m. We can also rewrite this expression to involve terms explicitly in the sum $n \neq m$

$$a_m(E_{n'} - E_m^0) = a_m H'_{mm} + \sum_{n\neq m} a_n H'_{mn} \quad \text{(A.15)}$$

so that the coefficient a_m is related to all the other a_n coefficients by:

$$a_m = \frac{1}{E_{n'} - E_m^0 - H'_{mm}}\sum_{n\neq m} a_n H'_{mn} \quad \text{(A.16)}$$

where n' is equal to the index denoting the energy of state we are seeking. This equation

$$a_m(E_{n'} - E_m^0 - H'_{mm}) = \sum_{n\neq m} a_n H'_{mn} \quad \text{(A.17)}$$

is an identity in the a_n coefficients. If the perturbation is small then $E_{n'} \to E_m^0$ and the first order corrections are found by setting the [coefficient of a_m] = 0 and $n' = m$. The next order of approximation is found by retaining the sum on the right hand side of the above equation and substituting for a_n the expression

$$a_n = \frac{1}{E_{n'} - E_n^0 - H'_{nn}} \sum_{n'' \neq n} a_{n''} H'_{nn''} \tag{A.18}$$

which is obtained from (A.16) the above by the transcription $m \to n$ and $n \to n''$. In the above the energy level $E_{n'} = E_m$ is the level for which we are calculating the perturbation. We now look for the a_m term explicitly in the sum $\sum_{n'' \neq n} a_{n''} H'_{nn''}$ of (A.18) and bring it to the left hand side. If we are satisfied with our solutions we end the procedure at this point. If we are not satisfied, we substitute for the $a_{n''}$ coefficients in (A.18) using the same basic equation to obtain a triple sum. We select out the a_m term, bring it to the left hand side of (A.17) the equation, etc. This procedure gives us an easy recipe to find the energy in (A.11) to any order of perturbation theory.

To illustrate the procedures, we write these iterations down more explicitly for 1st and 2nd order perturbation theory.

A.2.1 1st Order Perturbation Theory

In this case no iterations of (A.17) are needed and the sum $\sum_{n \neq m} a_n H'_{mn}$ on the right hand side of (A.17) is neglected, for the reason that if the perturbation is small $\psi_{n'} \sim \psi_{n'}^0$. Hence only a_m in (A.10) contributes significantly. We merely write $E_{n'}$ = E_m to obtain:

$$a_m(E_m - E_m^0 - H'_{mm}) = 0. \tag{A.19}$$

Since the a_m coefficients are arbitrary coefficients, this relation must hold for all a_m so that

$$(E_m - E_m^0 - H'_{mm}) = 0 \tag{A.20}$$

or

$$E_m = E_m^0 + H'_{mm}. \tag{A.21}$$

We write (A.21) this out even more explicitly so that the energy for state m for the perturbed problem E_m is related to the unperturbed energy E_m^0 by

$$E_m = E_m^0 + \langle \psi_m^0 | H' | \psi_m^0 \rangle \tag{A.22}$$

where the indicated diagonal matrix element of H' can be interpreted as the average of the perturbation in the state ψ_m^0. At this level of approximation, the wave functions to lowest order are not changed

$$\psi_m = \psi_m^0. \tag{A.23}$$

A.2.2 2nd Order Perturbation Theory

If we carry out the perturbation theory to the next order of approximation one iteration of (A.17) is required:

$$a_m(E_m - E_m^0 - H'_{mm}) = \sum_{n \neq m} \frac{1}{E_m - E_n^0 - H'_{nn}} \sum_{n'' \neq n} a_{n''} H'_{nn''} H'_{mn} \tag{A.24}$$

in which we have substituted for the a_n coefficient in (A.17) using the iteration relation given by (A.18). We now pick out the one term on the right hand side of (A.24) for which $n' = m$ and bring that term to the left hand side of (A.24). If no further iteration is to be done, we throw away what is left on the right hand side of (A.24) and get an expression for the arbitrary a_m coefficients

$$a_m \left[(E_m - E_m^0 - H'_{mm}) - \sum_{n \neq m} \frac{H'_{nm} H'_{mn}}{E_m - E_n^0 - H'_{nn}} \right] = 0 \tag{A.25}$$

Since a_m is arbitrary, the term in square brackets in (A.25) vanishes and the second order correction to the energy results:

$$E_m = E_m^0 + H'_{mm} + \sum_{n \neq m} \frac{|H'_{mn}|^2}{E_m - E_n^0 - H_{nn'}} \tag{A.26}$$

in which the sum on states $n \neq m$ represents the 2nd order correction. To this order in perturbation theory we must also consider corrections to the wave function

$$\psi_m = \sum_n a_n \psi_n^0 = \psi_m^0 + \sum_{n \neq m} a_n \psi_n^0 \tag{A.27}$$

in which ψ_m^0 is the large term and the correction terms appear as a sum over all the other states $n \neq m$. In handling the correction term, we look for the a_n coefficients, which from (A.18) are given by

$$a_n = \frac{1}{E'_n - E_n^0 - H'_{nn''}} \sum_{n'' \neq n} a_{n''} H'_{nn''}. \tag{A.28}$$

If we only wish to go to the lowest order correction terms, we will take only the most important term, i.e., $n'' = m$. We will also use the relation $a_m = 1$ in this approximation. Again using the identification n' = m, we obtain

$$a_n = \frac{H'_{nm}}{E_m - E_n^0 - H'_{nn}} \tag{A.29}$$

and

$$\psi_m = \psi_m^0 + \sum_{n \neq m} \frac{H'_{nm}\psi_n^0}{E_m - E_n^0 - H'_{nn}} \tag{A.30}$$

For better understanding, you should do the next iteration to include 3rd order perturbation theory, in order to see if you really have mastered the technique. This exercise will be useful to readers seeing this derivation for the first time.

Now look at the results for the energy E_m (A.26) and the wave function ψ_m (A.30) for the 2nd order perturbation theory and observe that these solutions are implicit solutions. That is, the correction terms are themselves dependent on E_m. To obtain an explicit solution, we can do one of two things here: (1) We can ignore the fact that the energies differ from their unperturbed values in calculating the correction terms. This approximation is known as Raleigh-Schrödinger perturbation theory. This is the usual perturbation theory given in Quantum Mechanics texts and for homework you may review the proof as given in such textbooks. (2) We can take account of the fact that E_m differs from E_m^0 by calculating the correction terms by an iteration procedure; the first time around, you put for E_m the value that comes out of 1st order perturbation theory. We then calculate the second order correction to get E_m. We next take this E_m value to compute the new second order correction term etc. until a convergent value for E_m is reached. This iterative procedure is what is used in Brillouin-Wigner perturbation theory and is a better approximation than Raleigh–Schrödinger perturbation theory to the wave function and energy eigenvalue for the same order in perturbation theory. This method is often used for practical problems in condensed matter physics. For example if you have a 2-level system, the Brillouin-Wigner perturbation theory must be carried to infinite order.

Let us summarize these ideas. If you have to compute only a small correction by perturbation theory, then Rayleigh–Schrödinger perturbation theory is usually needed because it is easier to use. If you want to do a more convergent perturbation theory (i.e., a better answer to the same order in perturbation theory), then you should use Brillouin-Wigner perturbation theory. There are other types of perturbation theory that are even more convergent and harder to use than Brillouin-Wigner perturbation theory (see Morse and Feshbach vol. 2). But these 2 types are the approaches that the reader is likely to encounter at this time.

For your convenience we summarize here the results of the Rayleigh–Schrödinger perturbation theory:

$$E_m = E_m^0 + H'_{mm} + \sum_n{}' \frac{|H'_{nm}|^2}{E_m^0 - E_n^0} + \cdots \tag{A.31}$$

$$\psi_m = \psi_m^0 + \sum_n{}' \frac{H'_{nm}\psi_n^0}{E_m^0 - E_n^0} + \tag{A.32}$$

where the sums in (A.31) and (A.32) donated by primes exclude the $m = n$ term. Thus, Brillouin-Wigner perturbation theory (A.26) and (A.30) contains terms in second order which occur in higher order in the Rayleigh–Schrödinger form. In practice, Brillouin-Wigner perturbation theory is useful when the perturbation term is too large to handle conveniently by Rayleigh–Schrödinger perturbation theory but still small enough for perturbation theory to work insofar as the perturbation expansion forms a reasonably convergent series.

A.3 Degenerate Perturbation Theory

It often happens that a number of quantum mechanical energy levels have the same or nearly the same energy. If they have exactly the same energy, we know that we can make any linear combination of these states that we like and get a new eigenstate also with the same energy. In the case of degenerate states, we have to do perturbation theory a little differently, as assumed below.

Suppose that we have an f–fold degeneracy (or near-degeneracy) such that the energy levels

$$\underbrace{\psi_1^0, \psi_2^0, \ldots \psi_f^0}_{\text{states with same or nearly the same energy}} \qquad \underbrace{\psi_{f+1}^0, \psi_{f+2}^0, \ldots}_{\text{states with quite different energies}}$$

We will here call the set of states with the same (or approximately the same) energy a "nearly degenerate set" (NDS) and this would include the f states here described. In the case of degenerate sets, the iterative equation (A.17) still holds. The only difference is that we solve for the perturbed energies by a different technique.

Starting with (A.17), we now bring to the left-hand side of the iterative equation all terms involving the f energy levels that are within the NDS. If we wish to calculate an energy within the NDS in the presence of a perturbation, we consider the a_n's within the NDS as large, and those outside the set as small. To first order in perturbation theory, we ignore the coupling to terms outside the NDS and we get f linear homogeneous equations in the $a_n' s$ where $n = 1,2,\ldots f$. We thus obtain the following equations from (A.17).

$$\begin{array}{llllll}
a_1(E_1^0 + H_{11}' - E) & + a_2 H_{12}' & + a_3 H_{13}' + \cdots & & + a_f H_{1f}' & = 0 \\
a_1 H_{21}' & + a_2(E_2^0 + H_{22}' - E) & + a_3 H_{23}' + \cdots & & + a_f H_{2f}' & = 0 \\
\vdots & \vdots & \vdots & \ddots & \vdots & \\
a_1 H_{f1}' & + a_2 H_{f2}' & + \cdots & \cdots & + a_f(E_f^0 + H_{ff}' - E) & = 0
\end{array} \tag{A.33}$$

In order to have a solution of these f linear equations, we demand that the coefficient determinant vanish:

$$\begin{vmatrix} (E_1^0 + H'_{11} - E) & H'_{12} & H'_{13} \dots & H'_{1f} \\ H'_{21} & (E_2^0 + H'_{22} - E) & H'_{23} \dots & H'_{2f} \\ \vdots & \vdots & \vdots \ \ddots & \vdots \\ H'_{f1} & \dots & \dots \dots & (E_f^0 + H'_{ff} - E) \end{vmatrix} = 0 \quad \text{(A.34)}$$

The f eigenvalues that we are looking for are the eigenvalues of the matrix in (A.34) and the set of orthogonal states are the corresponding eigenvectors. Remember that the matrix elements H'_{ij} that occur in the above determinant are taken between the unperturbed states in the NDS.

The generalization to second order degenerate perturbation theory is immediate. In this case, (A.33) and (A.34) have additional terms. For example the 1st equation in (A.33) would then become

$$a_1(E_1^0 + H'_{11} - E) + a_2 H'_{12} + a_3 H'_{13} + \cdots + a_f H'_{1f} = -\sum_{n \neq NDS} a_n H'_{1n} \quad \text{(A.35)}$$

and for the a_n in (A.35), which are now small (because they are outside the NDS) we would use our iterative form

$$a_n = \frac{1}{E - E_n^0 - H'_{nn}} \sum_{m \neq n} a_m H'_{nm}. \quad \text{(A.36)}$$

As a correction term we must only consider the terms in the above sum which are large; these terms are all in the NDS. This argument shows that every term on the left side of (A.35) above will have a correction term. For example the correction term to a_i will look as follows:

$$a_i H'_{1i} + a_i \sum_{n \neq NDS} \frac{H'_{1n} H'_{ni}}{E - E_n^0 - H'_{nn}} \quad \text{(A.37)}$$

where the first term is the original term from 1st order degenerate perturbation theory and the terms from states outside the NDS gives the 2nd order correction terms. So, if we are doing higher order degenerate perturbation theory we write for each entry in the secular equation the appropriate correction terms (A.37) that are obtained from these iterations. For example, in 2nd order degenerate perturbation theory, the (1, 1) entry to the matrix in (A.34) would be

$$E_1^0 + H'_{11} + \sum_{n \neq NDS} \frac{|H'_{1n}|^2}{E - E_n^0 - H'_{nn}} - E. \quad \text{(A.38)}$$

As a further illustration let us write down the (1, 2) entry into (A.35)

$$H'_{12} + \sum_{n \neq NDS} \frac{H'_{1n} H'_{n2}}{E - E_n^0 - H'_{nn}}. \tag{A.39}$$

Again we have an implicit dependence of the 2nd order term in (A.38) and (A.39) on the energy eigenvalue that we are looking for. To do 2nd order degenerate perturbation theory, we again have two options. If we take the energy E in (A.38) and (A.39) as the unperturbed energy in computing the correction terms, we have 2nd order degenerate Rayleigh–Schrödinger perturbation theory. On the other hand, if we iterate to get the best correction term, then we call it Brillouin-Wigner perturbation theory.

How do we know in an actual problem when to use degenerate 1st or degenerate 2nd order perturbation theory? If the matrix elements H'_{ij} coupling members of the NDS vanish then we must go to 2nd order degenerate perturbation theory. Generally speaking the first order terms will be much larger than the 2nd order terms, provided that there is no symmetry reason for the first order terms to vanish.

Let us explain this a bit further. By the matrix element H'_{12} we mean $(\psi_1^0|H'|\psi_2^0)$. Suppose the perturbation we are talking about is due to an electric field E

$$H' = -e\mathbf{r} \cdot \mathbf{E} \tag{A.40}$$

where $e\mathbf{r}$ is the dipole moment of our system. Suppose that we now we consider the effect of inversion on H'. For the $n = 2$ levels, we would treat them in degenerate perturbation theory because the $2s$ and $2p$ states are degenerate as would occur in the simple treatment of the hydrogen atom. Here first order terms only appear in entries coupling s and p states. To get corrections which split the p levels among themselves we must go to 2nd order degenerate perturbation theory. In other words, our choice of which type of perturbation theory to use depends on what it takes to life the degeneracy found remaining after applying first order perturbation theory.

Suggested Readings

Davidov - Quantum Mechanics, Chap. 7.

Morse and Feshbach, Methods of Theoretical Physics, Chap. 9.

Appendix B
Time–Dependent Perturbation Theory

B.1 General Formulation

To proceed further with the formal development of the optical properties of solids, we need to consider how to handle the effect of time-dependent electromagnetic fields quantum mechanically. The most important case of interest is the one where the external field is a sinusoidal function of time. For most practical applications, the external fields are sufficiently weak, so that their effect can be handled within the framework of perturbation theory. If the perturbation has an explicit time dependence, it must be handled by time- dependent perturbation theory. Practical problems which are handled by time-dependent perturbation theory include such subjects as magnetic resonance (nuclear and electronic spin), cyclotron resonance and optical properties of solids. We give here a brief review of the subject.

In doing time-dependent perturbation theory we write the total Hamiltonian H as:

$$H = H_0 + H'(t) \tag{B.1}$$

where H_0 is the unperturbed steady state Hamiltonian and $H'(t)$ is the time dependent external perturbation. We assume here that we know how to solve the unperturbed time-independent problem for its eigenvalues E_n and the corresponding eigenfunctions u_n

$$H_0 u_n = E_n u_n \tag{B.2}$$

Since $H'(t)$ has an explicit time dependence, then the "energy" is no longer a "constant of the motion". Since we no longer have a stationary energy time-independent solutions, we must use time-dependent form of Schrödinger's equation, which is:

$$i\hbar \frac{\partial \psi}{\partial t} = H\psi = (H_0 + H')\psi. \tag{B.3}$$

Now, if we didn't have the perturbation term $H'(t)$ to contend with, we would set

M. Dresselhaus et al., *Solid State Properties*, Graduate Texts in Physics,
https://doi.org/10.1007/978-3-662-55922-2

$$\psi(\mathbf{r}, t) = u_n(\mathbf{r})e^{-iE_nt/\hbar} \tag{B.4}$$

where $u_n(\mathbf{r})$ is independent of time and satisfies (B.2). Thus all the time dependence of $\psi(\mathbf{r}, t)$ is contained in the phase factor $e^{-iE_nt/\hbar}$. For $H'(t) = 0$, it immediately follows that

$$i\hbar\frac{\partial\psi}{\partial t} = E_n\psi \tag{B.5}$$

which yields the time-independent Schrödinger's equation. With the perturbation present, we expand the time dependent functions $\psi(\mathbf{r}, t)$ in terms of the complete set $u_n(\mathbf{r})e^{-iE_nt/\hbar}$

$$\psi(\mathbf{r}, t) = \sum_n a_n(t)u_n(\mathbf{r})e^{-iE_nt/\hbar} \tag{B.6}$$

where the $a_n(t)$ are the time-dependent expansion coefficients. Substituting (B.6) in the time-dependent Schrödinger's equation (B.3) we obtain:

$$\begin{aligned} i\hbar\sum_n \dot{a}_n(t)u_ne^{-iE_nt/\hbar} + \sum_n a_n(t)u_nE_ne^{-iE_nt/\hbar} &= \sum_n a_n(t)[H_0 + H'(t)]u_ne^{-iE_nt/\hbar} \\ &= \sum_n a_n(t)[E_n + H'(t)]u_ne^{-iE_nt/\hbar} \end{aligned} \tag{B.7}$$

as assumed below where $\dot{a_n}(t)$ denotes the time derivative $da_n(t)/dt$. We note that because of (B.2) the second term on the left hand side of (B.7) is canceled by the first term on the right hand side.

We now multiply on the left hand side of (B.7) by $u_k^*(\mathbf{r})$ and integrate over all space. If we make use of the orthogonality of the eigenfunctions

$$\int u_k^*(\mathbf{r})u_n(\mathbf{r})d^3r = \delta_{n,k} \tag{B.8}$$

we obtain from (B.7)

$$i\hbar\sum_n \dot{a}_n(t)u_ne^{-iE_nt/\hbar} = \sum_n a_n(t)H'(t)u_ne^{-iE_nt/\hbar} \tag{B.9}$$

the result:

$$i\hbar\dot{a}_ke^{-iE_kt/\hbar} = \sum_n a_n\langle k|H'(t)|n\rangle e^{-iE_nt/\hbar} \tag{B.10}$$

where we have written the matrix element

$$\langle k|H'(t)|n\rangle = \int u_k^*(\mathbf{r})H'(t)u_n(\mathbf{r})d^3r \tag{B.11}$$

Since $H'(t)$ is time-dependent, so is the matrix element time-dependent, even though the matrix element is taken between stationary states. We thus obtain the result

$$i\hbar\dot{a}_k(t) = \sum_n a_n(t)\langle k|H'(t)|n\rangle e^{i(E_k-E_n)t/\hbar}. \tag{B.12}$$

If we set

$$\hbar\omega_{kn} = E_k - E_n \tag{B.13}$$

where ω_{kn} is the Bohr frequency between states k and n, we have

$$\dot{a}_k(t) = \frac{1}{i\hbar}\sum_n a_n(t)e^{i\omega_{kn}t}\langle k|H'(t)|n\rangle \tag{B.14}$$

in which the indicated matrix element is taken between eigenstates of the unperturbed Hamiltonian H_0. So far, no perturbation theory has been used and the result given in (B.14) is exact. We notice that the unperturbed Hamiltonian is completely absent from (B.14). Nevertheless, its energy eigenvalues appear in ω_{kn} and its eigenfunctions in the matrix element $\langle k|H'(t)|n\rangle$.

In applying perturbation theory, we consider the matrix element $\langle k|H(t)|n\rangle$ to be small, and we write each time-dependent amplitude as an expansion in perturbation theory

$$a_n = a_n^{(0)} + a_n^{(1)} + a_n^{(2)} + \cdots = \sum_{i=0}^{\infty} a_n^{(i)} \tag{B.15}$$

where the superscript gives the order of the term. Thus $a_n^{(0)}$ is the *zero*th order term and $a_n^{(i)}$ is the ith order correction to a_n. From (B.14), we see that $a_k(t)$ changes its value with time only because of the time dependent perturbation. Thus, the unperturbed situation (0th order perturbation theory) must give no time dependence in *zero*th order

$$\dot{a}_m^{(0)} = 0 \tag{B.16}$$

and the first order correction yields:

$$\dot{a}_m^{(1)} = \frac{1}{i\hbar}\sum_n a_n^{(0)}\langle m|H'(t)|n\rangle e^{i\omega_{mn}t}. \tag{B.17}$$

In the application of perturbation theory we assume, for example, that if we start in an eigenstate $n = l$, only the coefficient $a_l^{(0)}$ will be appreciably large. Then all other terms in the sum can be neglected. This gives us in 1st order perturbation theory:

$$\dot{a}_m^{(1)} = \frac{1}{i\hbar}a_l^{(0)}\langle m|H'(t)|l\rangle e^{i\omega_{ml}t} \tag{B.18}$$

where $a_l^{(0)}$ is approximately unity.

For many cases of interest, this integration over the time variable can be performed and $a_m^{(1)}$ rather than its time derivative is obtained. The two simple cases that can be integrated easily are:

1. The perturbation H' is constant but is turned on at some time ($t = 0$) and we look at the amplitudes of the wave function in the various states after the perturbation has been acting for some time $t > 0$.
2. The perturbation H' has a sinusoidal time dependence with frequency ω. This is the situation for all resonant phenomena.

Let us first consider case (1). Then

$$a_m^{(1)}(t) = \frac{1}{i\hbar}\int_0^t \langle m|H'|l\rangle e^{i\omega_{ml}t'}dt' = \frac{\langle m|H'|l\rangle}{i\hbar}\frac{\left[e^{i\omega_{ml}t} - 1\right]}{i\omega_{ml}} \tag{B.19}$$

Similarly for case (2), we can write

$$H'(t) = H'(0)e^{\pm i\omega t} \tag{B.20}$$

to show the explicit time dependence, so that upon integration we obtain for the amplitudes $a_m^{(1)}(t)$

$$a_m^{(1)}(t) = \frac{1}{i\hbar}\langle m|H'(0)|l\rangle \int_0^t e^{i(\omega_{ml}\pm\omega)t'}dt' = \frac{1}{i\hbar}\langle m|H'(0)|l\rangle \frac{e^{i(\omega_{ml}\pm\omega)t} - 1}{i(\omega_{ml} \pm \omega)}. \tag{B.21}$$

We interpret the time dependent amplitudes $|a_m^{(1)}(t)|$ as the probability of finding the system in a state m after a time t has elapsed since the perturbation was applied; the system was initially in a state $l \neq m$.

We thus obtain for case (1) given by (B.19)

$$|a_m^{(1)}(t)|^2 = \left(\frac{|\langle m|H'|l\rangle|^2}{\hbar^2}\right)\left(\frac{|e^{i\omega_{ml}t} - 1|^2}{\omega_{ml}^2}\right) \tag{B.22}$$

$$|a_m^{(1)}(t)|^2 = \left(\frac{|\langle m|H'|l\rangle|^2}{\hbar^2}\right)\left(\frac{4sin^2(\omega_{ml}t/2)}{\omega_{ml}^2}\right) \tag{B.23}$$

Clearly for case (2), the same result follows except that ω_{ml} is replaced by $(\omega_{ml} \pm \omega)$ where ω is the applied frequency and a resonant denominator results for the transition probability amplitude. It is clear from the above arguments that for both cases (1) and (2), the explicit time dependence is contained in an oscillatory term of the form $[sin^2(\omega'/2)/\omega'^2]$ where $\omega' = \omega_{ml}$ for the case (1) and $\omega' = \omega_{ml} \pm \omega$ for case (2). This function was previously encountered in diffraction theory and looks like that shown in Fig. B.1. Of special interest here is the fact that the main contribution to this function comes for $\omega' \cong 0$, with the height of the main peak proportional to

$t^2/4$ and the width proportional to $1/t$. This means that the area under the central peak goes as t. If we think of $|a_m(t)|^2$ as the probability of finding the system in a state m, then for case (2), where we have a perturbation with frequency ω, the system attempts to make a transition from a state l to a state m with a transition probability proportional to the time the perturbation acts. If we then wait long enough, a system in an energy state l will make a transition to a state m, if photons of the resonant frequency ω_{lm} are present.

B.2 Fermi Golden Rule

Since the transition probability is proportional to the time the perturbation acts, it is therefore useful to deal with a quantity called the transition probability per unit time and the relation giving this quantity is called the Golden Rule (named by Fermi as the golden rule and is in fact often called Fermi's Golden Rule).

In deriving the Golden Rule from (B.19), we must consider the system exposed to the perturbation for a time sufficiently long so that we can make a meaningful measurement within the framework of the Heisenberg uncertainty principle:

$$\Delta E \Delta t \sim h \tag{B.24}$$

$$\Delta E \sim h/t \tag{B.25}$$

so that the uncertainty in energy (or frequency) during the time that the perturbation acts is or

$$\Delta\omega_{lm} \sim 2\pi/t \tag{B.26}$$

But this is precisely the period of the oscillatory function shown in Fig. B.1. In this context, we must think of the concept of transition probability/unit time as encompassing a range of energies and times consistent with the uncertainty principle. In the case of solids, it is quite natural to do this anyhow, because the wave vector **k** is a quasi-continuous variable. That is, there are a large number of k states which have energies close to a given energy. The quantum states labeled by wave vector k are close together in a solid having about 10^{22} atoms/cm^3. Since the photon source itself has a bandwidth, we would automatically want to consider a range of energies difference $\delta\hbar\omega'$. From this point of view, we introduce the transition probability/unit time W_m for making a transition to a state m

$$W_m = \frac{1}{t}\sum_{m'\approx m} |a_{m'}^{(1)}(t)|^2 \tag{B.27}$$

where the summation is carried out over a range of energy states consistent with the uncertainty principle; $\Delta\omega_{mm'} \sim 2\pi/t$.

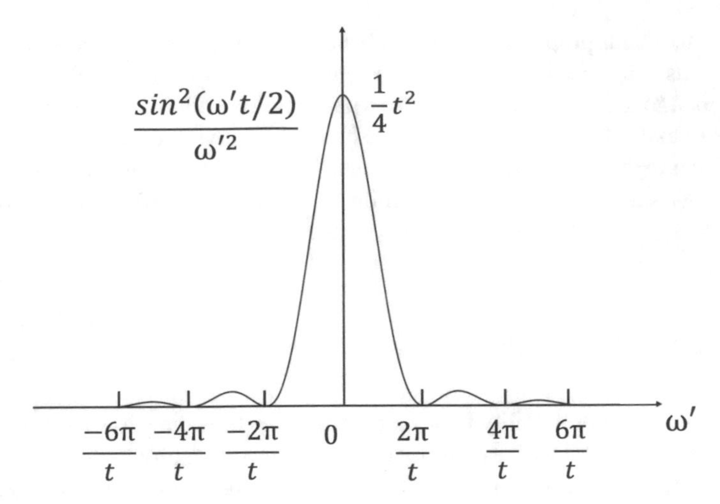

Fig. B.1 Plot of $sin^2(\omega' t/2)/\omega'^2$ versus ω', a function which enters the calculation of time dependent perturbation problems

Substituting for $|a)_{m'}^{(1)}(t)|^2$ from (B.23) we have

$$|a_m^{(1)}(t)|^2 = \left(\frac{4|\langle m|H'|l\rangle|^2}{\hbar^2}\right)\left(\frac{sin^2(\omega' t/2)}{\omega'^2}\right) \tag{B.28}$$

and the summation is replaced by an integration over a narrow energy range weighted by the density of states $\rho(E_m)$ which gives the number of states per unit energy range. We thus obtain

$$W_m = \frac{1}{\hbar^2 t}\int |4H'_{m'l}|^2 \left(\frac{sin^2(\omega_{m'l}t/2)}{\omega_{m'l}^2}\right)\rho(E_{m'})dE_{m'} \tag{B.29}$$

where we have written $H'_{m'l}$ for the matrix element $\langle m'|H|l\rangle$. But, by hypothesis, we are only considering energies within a small energy range E'_m around E_m and over this range the matrix elements and density of final states will not be varying. However, the function $[sin^2(\omega' t/2)/\omega'^2]$ will be varying rapidly, as can be seen from Fig. A.1. Therefore, it is adequate to integrate (B.29) only over the rapidly varying function $[sin^2(\omega t/2)/\omega^2]$ Writing $dE = \hbar d\omega'$, we obtain;

$$W_m \simeq \left(\frac{4|H'_{m'l}|^2\rho(E_m)}{t\hbar^2}\right)\int\left(\frac{sin^2(\omega' t/2)}{\omega'^2}\right)d\omega' \tag{B.30}$$

The most important contribution to the integral in (B.30) comes from values of ω close to ω'. On the other hand, we know how to do this integral between $-\infty$ and ∞, since

$$\int_{-\infty}^{\infty} (sin^2 x/x^2)dx = \pi \tag{B.31}$$

Therefore we can write an approximate relation from (B.30) by setting $x = \omega' t/2$

$$W_m \cong (2\pi/\hbar)|H'_{ml}|^2 \rho(E_m) \tag{B.32}$$

which is often called Fermi's Golden Rule. In the subsequent sections, we will apply the Fermi Golden Rule to calculate the optical properties of solids.

If the initial state is a discrete level (such as donor impurity level) and the final state is a continuum (such as conduction band), then the Fermi Golden Rule (B.32) as written yields the transition probability per unit time and $\rho(E_m)$ is interpreted as the density of final states. Likewise if the final state is discrete and the initial state is a continuum, W_m also gives the transition probability per unit time, only in this case $\rho(E_m)$ is interpreted as the density of initial states.

For many important applications in solid state physics, the transitions of interest are between a continuum of initial states and a continuum of final states. In this case the Fermi Golden Rule must be interpreted in terms of a joint density of states, whereby the initial and final states are separated by the photon energy $\hbar\omega$ inducing the transition. These issues are discussed in Chap. 17.

B.3 Time Dependent 2nd Order Perturbation Theory

This second order treatment is needed for indirect optical transitions, where

$$H = H_0 + \lambda H' \tag{B.33}$$

and $\lambda \ll 1$. Here $H_0\psi_o = i\hbar\frac{\partial\psi_0}{\partial t}$. Expand ψ, the solution to (B.33) in terms of the complete set of functions denoted by $\psi_0 \equiv |n, veck\rangle$

$$\psi = \sum_{n,\mathbf{k}} a_n(\mathbf{k}, t) e^{-\frac{i}{\hbar}E_n(\mathbf{k})t} |n, \mathbf{k}\rangle \tag{B.34}$$

where $|n, \mathbf{k}\rangle$ is a Bloch function describing the eigenstates of the unperturbed problem

$$H\psi = i\hbar\dot{\psi} \tag{B.35}$$

$$\begin{aligned} &\sum_{n,\mathbf{k}} a_n(\mathbf{k}, t) E_n(\mathbf{k}) e^{-\frac{i}{\hbar}E_n(\mathbf{k})t} |n, \mathbf{k}\rangle + \sum_{n,\mathbf{k}} a_n(\mathbf{k}, t) e^{-\frac{i}{\hbar}E_n(\mathbf{k})t} \lambda H' |n, \mathbf{k}\rangle \\ &= i\hbar \sum_{n,\mathbf{k}} \dot{a}_n(\mathbf{k}, t) e^{-\frac{i}{\hbar}E_n(\mathbf{k})t} |n, \mathbf{k}\rangle + \sum_{n,\mathbf{k}} a_n(\mathbf{k}, t) e^{-\frac{i}{\hbar}E_n(\mathbf{k})t} |n, \mathbf{k}\rangle \end{aligned} \tag{B.36}$$

which gives

$$\dot{a_m}(\mathbf{k}, t) = \frac{1}{i\hbar}\sum_{n,\mathbf{k}} a_n(\mathbf{k}, t) e^{\frac{i}{\hbar}(E_m(\mathbf{k}')t - E_n(\mathbf{k})t)} \langle m, \mathbf{k}'|\lambda H'|n, \mathbf{k}\rangle \tag{B.37}$$

We expand

$$a_m(\mathbf{k}', t) = a_m^{(0)} + \lambda a_m^{(1)} + \lambda^2 a_m^{(2)} + \cdots \tag{B.38}$$

and let $a_j(\mathbf{k}, 0) = 1$ and all others $a_n(\mathbf{k}, 0) = 0$ where $n \neq j$ To first order, as before, we can write,

$$\lambda^2 \dot{a}_m^{(1)}(\mathbf{k}', t) = \frac{1}{i\hbar}\lambda a_n^{(0)}(\mathbf{k}, t) exp\left[\frac{i}{\hbar}[E_m(\mathbf{k}') - E_n(\mathbf{k})]t\right] \langle m, \mathbf{k}'|\lambda H'|n, \mathbf{k}\rangle \tag{B.39}$$

or

$$a_m^{(1)}(\mathbf{k}', t) = \frac{1}{i\hbar}\int_0^t dt' exp\left[\frac{i}{\hbar}[E_m(\mathbf{k}') - E_n(\mathbf{k})]t\right] \langle m, \mathbf{k}'|\lambda H'|n, \mathbf{k}\rangle \tag{B.40}$$

To second order

$$\lambda^2 \dot{a}_m^{(2)}(\mathbf{k}', t) = \frac{1}{i\hbar}\sum_{n,\mathbf{k}} \lambda a_n^{(1)}(\mathbf{k}, t) exp\left[\frac{i}{\hbar}[E_m(\mathbf{k}') - E_n(\mathbf{k})]t\right] \langle m, \mathbf{k}'|\lambda H'|n, \mathbf{k}\rangle \tag{B.41}$$

or,

$$\begin{aligned}\dot{a}_m^{(2)}(\mathbf{k}', t) = -\frac{1}{\hbar^2}\sum_{n,\mathbf{k}} a_n^{(1)}(\mathbf{k}, t) exp\left\{\frac{i}{\hbar}[E_m(\mathbf{k}') - E_n(\mathbf{k})]t\right\} \langle m, \mathbf{k}'|\lambda H'|n, \mathbf{k}\rangle \\ \times \int_0^t dt' exp\left\{\frac{i}{\hbar}[E_n(\mathbf{k}') - E_i(\mathbf{k})]t'\right\} \langle n, \mathbf{k}'|\lambda H'|i, \mathbf{k}\rangle\end{aligned} \tag{B.42}$$

We write the time dependence of the perturbation Hamiltonian explicitly

$$H' = \sum_{\alpha} H e^{-i\omega_\alpha t} \tag{B.43}$$

and then (B.42) can be written after integrating twice.

$$|a_f^{(2)}(\mathbf{k_f}, t)|^2 = 2\pi\hbar t \sum_{m,\mathbf{k},\alpha,\alpha'} \frac{|\langle f|H'_{\alpha'}|m, \mathbf{k}\rangle|^2 \langle m, \mathbf{k}|H'_{\alpha}|i\rangle|^2}{(E_m(\mathbf{k}) - E_i - \hbar\omega_\alpha)^2} \delta(E_f - E_i - \hbar\omega_\alpha - \hbar\omega_{\alpha'}) \tag{B.44}$$

This second-order time-dependent perturbation theory expression is used to derive the probability of an indirect interband transition and we note that this term is proportional to time.

Index

M. Dresselhaus et al., *Solid State Properties*, Graduate Texts in Physics,
https://doi.org/10.1007/978-3-662-55922-2

E

P

Q

R

S

Printed in the United States
By Bookmasters